FOREWORD

These Proceedings consist of the contributed papers, keynote addresses and conclusions of the Experts' Seminar on Energy Technologies for Reducing Emissions of Greenhouse Gases, which was held in Paris on 12th-14th April 1989. The Seminar was conceived and organised as a part of the joint effort by the International Energy Agency (IEA) and the Organisation for Economic Co-operation and Development (OECD) to understand and analyse energy technologies and response strategies that would mitigate anthropogenic production of greenhouse gases. In particular, the aims were to gather and exchange updated technical knowledge, to assess and rank feasible courses of action for the different national and regional contexts and time horizons, to define issues deserving further investigation, and to propose possible research and development initiatives to governments, while promoting co-ordination of efforts and international collaboration.

The underlying assumption is that carbon dioxide and other greenhouse gases released by human activity might produce a climate change which should be avoided, though one can debate the timing, geographical distribution and magnitude of the effects. The focus of the Seminar was on all aspects of energy supply, conversion and use. Attention was given to the full range of potential or conceivable technologies which may become available within different energy scenarios and time horizons. Energy technology systems and strategies were considered that are either in use or demonstrated and which could be adopted in the event that early action is deemed necessary. Energy technologies including controversial options that can be proposed for the long-term, were also defined and some of their economic and policy implications evaluated.

The experts presented a variety of ideas and suggestions for action which do not necessarily reflect the views of the IEA, the OECD or their Member Governments. Proposals made have occasionally exceeded the traditional bounds which separate several energy technology areas in order to create those concepts and tools which appear necessary for stabilising the production of greenhouse gases, especially carbon dioxide. Some experts offered significant examples and case studies which may be of use to energy planners and decision makers. Others were more concerned with generating new insights than in rigorous analytical or technical assessments.

A broad consensus, however, seemed to emerge on a number of fundamental questions, directions and priorities for the OECD countries. The technical information brought to the Seminar and discussed appeared unique and of high quality, based on the most recent evaluation and research. Thus, it is the hope of the IEA and OECD that the availability of these Proceedings will promote more in-depth analysis, and contribute to the identification of acceptable responses to the emerging global environmental challenge.

Bill L. Long
Director
Environment Directorate
OECD

Sergio F. Garribba
Director
Energy Research, Development
and Technology Applications
IEA

ENERGY TECHNOLOGIES FOR REDUCING EMISSIONS OF GREENHOUSE GASES

PROCEEDINGS
OF AN EXPERTS' SEMINAR

PARIS, 12th-14th APRIL 1989

VOLUME 1

ORGANISATION FOR ECONOMIC CO-OPERATION AND DEVELOPMENT

INTERNATIONAL ENERGY AGENCY

INTERNATIONAL ENERGY AGENCY

2, RUE ANDRÉ-PASCAL, 75775 PARIS CEDEX 16, FRANCE

The International Energy Agency (IEA) is an autonomous body which was established in November 1974 within the framework of the Organisation for Economic Co-operation and Development (OECD) to implement an international energy programme.

It carries out a comprehensive programme of energy co-operation among twenty-one* of the OECD's twenty-four Member countries. The basic aims of IEA are:

i) co-operation among IEA participating countries to reduce excessive dependence on oil through energy conservation, development of alternative energy sources and energy research and development;

ii) an information system on the international oil market as well as consultation with oil companies;

iii) co-operation with oil producing and other oil consuming countries with a view to developing a stable international energy trade as well as the rational management and use of world energy resources in the interest of all countries;

iv) a plan to prepare Participating Countries against the risk of a major disruption of oil supplies and to share available oil in the event of an emergency.

*IEA Participating Countries are: Australia, Austria, Belgium, Canada, Denmark, Germany, Greece, Ireland, Italy, Japan, Luxembourg, the Netherlands, New Zealand, Norway, Portugal, Spain, Sweden, Switzerland, Turkey, United Kingdom, United States.

Pursuant to article 1 of the Convention signed in Paris on 14th December 1960, and which came into force on 30th September, 1961, the Organisation for Economic Co-operation and Development (OECD) shall promote policies designed:

– to achieve the highest sustainable economic growth and employment and a rising standard of living in Member countries, while maintaining financial stability, and thus to contribute to the development of the world economy;

– to contribute to sound economic expansion in Member as well as non-member countries in the process of economic development; and

– to contribute to the expansion of world trade on a multilateral, non-discriminatory basis in accordance with international obligations.

The original Member countries of the OECD are Austria, Belgium, Canada, Denmark, France, the Federal Republic of Germany, Greece, Iceland, Ireland, Italy, Luxembourg, the Netherlands, Norway, Portugal, Spain, Sweden, Switzerland, Turkey, the United Kingdom and the United States. The following countries became Members subsequently through accession at the dates indicated hereafter: Japan (28th April 1964), Finland (28th January 1969), Australia (7th June 1971) and New Zealand (29th May 1973).

The Socialist Federal Republic of Yugoslavia takes part in some of the work of the OECD (agreement of 28th October 1961).

TABLE OF CONTENTS

VOLUME 1

CONTRIBUTED PAPERS

APPENDICES

OPENING ADDRESS
by
Helga Steeg
Executive Director
International Energy Agency

Good morning Ladies and Gentlemen. It is a great pleasure for me to welcome this distinguished assembly of experts, scientists and government officers. Over the next three days, you will analyse and discuss possible energy technology approaches to climate change issues. In doing so, you will address the technical feasibility of these approaches, their projected costs and time horizons, and means and opportunities for international collaboration. This Seminar has been organised under the auspices of both the IEA and the OECD Environment Directorate, as part of our joint study to evaluate energy options, policies and technologies that could help reduce the production of man-made greenhouse gases, especially carbon dioxide.

Because of the increasing importance of concerns about the environment, governments are reappraising policy options so that environmental issues are incorporated into a wide range of economic decisions, including those which relate to the energy sector. Energy policy makers have long been aware that energy production, transformation and use can be significant causes of environmental degradation. A marriage of energy and environmental concerns thus exists, which can and must be further solidified in identifying policies to deal with climate change. It will not be an easy task.

The environmental effects of energy systems may entail different types of impact, and influence a variety of subjects. While some energy policies and initiatives (such as greater energy efficiency) can achieve both energy and environmental objectives, others may not (such as significant constraints on the use of fossil fuels). As a result, appropriate mechanisms and solutions have to be found which equitably balance energy and environmental requirements within the context of the general economy.

As policy-makers, we cannot afford to look at energy and the environment in a vacuum. We are obliged to evaluate the macroeconomic and microeconomic impact of policy options, particularly those which are proposed in the face of climate change. It seems to me that policies would have to evaluate carefully the impact on societies' needs to develop, as well as on existing living standards. Yet, it is clear we must find answers, although there will be no "quick-fixes" to the problem, nor one single answer, nor a solution by western democracies alone.

It now seems reasonably clear that growing accumulation in the atmosphere of carbon dioxide and other greenhouse gases could result in climate change at some time in the future. It is equally clear that effective action to avoid or reduce the risk will have strong implications for the energy sector and the economy in general. Relative to the earlier energy challenges, this emerging set of problems seems characterized by lack of scientific knowledge, difficult decisions and high costs. Time scales easily go beyond most feasible and useful projections and the global dimension involved certainly transcends the capacity of national institutions.

In an area where scientific knowledge is still far from complete and often highly contested, it is critical that policy-makers identify least-cost solutions. This is why technological responses are so important, because they can lead to emission reductions and/or efficiency gains. A well co-ordinated and strengthened international collaborative effort on energy technology is possibly the most significant response we have at present.

During the last two days an Expert Group Meeting on Methodologies and Analytical Tools was held here in Paris, under the auspices of the Intergovernmental Panel on Climate Change, Response Strategies Working Group (Sub-Group on Energy and Industry). One purpose of this Expert Group Meeting was to try to identify some international criteria and instruments which could be adopted when evaluating policy options that may reduce man-made energy-related greenhouse gas emissions.

Energy technology options and technology strategies to reduce the emissions of greenhouse gases are the subject of this Seminar. The Seminar represents a highly qualified forum in which we expect an exchange of updated knowledge on available and potential energy technologies. This assembly faces many challenges, not least those of how to co-ordinate, manage and accelerate, if needed, progress in energy technology, and to arrange for its subsequent transfer.

My hope is that this Seminar will both help to put more clarity into the climate change debate and to identify in more concrete form some of the policy options. The number of international conferences on this subject, while heightening awareness of the issue, have so far largely failed to do this. The presentations and discussions here will provide a useful imput into the work of the OECD and IEA, both of whom have the environmental issue high on their Ministerial agendas later next month. The process of finding solutions will extend far beyond this date. But we do hope that this Seminar will produce the building blocks in the technology area out of which we can construct further useful work.

Finally, I would like to thank all those who contributed to the organisation of this event, the Member countries, the OECD Delegations, including the IEA's Committee for Energy Research and Development, the Standing Group on Long-Term Co-operation, the OECD's Environment Committee and its Energy and Environment Group, as well as that of the invited experts and the staff in the IEA and OECD Environment Directorate.

Bill Long, OECD Director of Environment, will tell you more about the environmental side and the ensuing need to identify technologies and practices that could limit greenhouse gas emissions, while Sergio Garribba, IEA Director of Energy R&D and Technology Applications will shortly describe how the Seminar was arranged and how it will develop. The effort has been a notable one. I am confident that the results can match our expectations.

OPENING COMMENTS
by
Bill L. Long
Director for Environment
OECD

On behalf of the OECD, I would like to join Mrs. Helga Steeg, the Executive Director of the IEA, in welcoming you here today. I can be very brief since I believe that she has done an excellent job already in presenting the "environmental" perspective. I fully share her views on the importance of balancing energy growth and environmetal management policies and the difficulties of achieving this and also on the importance and objectives of this Seminar.

For three decades the OECD has pursued its basic mandate to "promote economic growth" within Member countries. Today, in the face of the upsurge of concern about environmental threats that has reached the very highest levels of government, the OECD is addressing not only the magnitude of economic growth but also its quality and its sustainability. By the quality of growth we are really talking about the quality of life aspects of economic growth over the near term, with all of its environmental implications. By sustainability we are concerned with the longer-term problem of maintaining economic growth, with all of its energy implications.

In carrying out the OECD's environmental programme, the trend is to emphasize the integration of environmental policy with economic policy-making. And, as we engage other parts of the OECD in this work - on transportation policy, on agriculture, on urban affairs, on international development assistance - energy policy and the choice of energy technologies is a recurrent theme. Thus, while we in the Environment Directorate view environmental issues through our own prism, we are convinced that the relationship between energy and the environment will be the dominant concern of the next decade.

Another perception is that there is considerable confusion and uncertainty over the proper energy pathways to the future. Within the quite small OECD family of nations, twenty-four in all, there are widely divergent views on subjects such as the future of nuclear energy after Chernobyl, prospects for coal in the face of global warming, and the potential of the "soft" energy path given its uncertain economics.

At the same time there is increasing public pressure on governments to do something to cope with the array of environmental problems and threats which have burst upon the world scene. This, in turn, has stimulated a competition for political leadership, enveloping legislatures, ministers and heads of state.

This uncertainty, coupled with pressure for action, I submit, provides the ingredients for bad public policy. Bad in the sense that neither environmental nor economic and social development goals will likely be served well over the long term. Put another way, the area of energy policy and technological choices to cope with climate change and also to mitigate a range of other environmental problems cries out today for good information and solid analysis.

It is the desire of the OECD to contribute to sound policy analysis and planning by working with, in this case, the International Energy Agency on forward-looking, effective and efficient policy options and strategies. This accounts for the special importance we attach to this Seminar.

For our part, there are four aspects of energy technologies which we believe deserve attention:

- the state-of-the-art, and prospects for the future;
- environmental costs, and how they might be ameliorated;
- the economic aspects, including the internalization of environmental costs; and
- developing country needs and technology transfer considerations.

The importance of the last concern is reflected in the debate at the recent meeting convened by the UK on stratospheric ozone depletion. Speaking on behalf of 1.1 billion citizens of the world, the Chinese delegate said "We want to help you solve this global problem, but ..." The global demographic considerations for the climate change threat are equally clear.

We clearly cannot expect to deal fully with each of these subjects at this Seminar. This, however, is certainly not the last opportunity, nor are we starting afresh. A seminar on climate change was co-sponsored by the IEA and OECD in 1981. As Mrs. Steeg said, "this Seminar will produce the building blocks in the technology area out of which we can construct further useful work".

So, while we see immediate benefits from the Seminar in terms of the insights and information you will provide, we also see the opportunity to create an expanded network of contacts who might help us take the necessary next steps. Here I would echo Mrs. Steeg's view that we are not embarked on an easy task, not will it be resolved with "quick fixes" and facile solutions. Again, on behalf of the OECD, I would like to thank you for coming.

INTRODUCTORY REMARKS
by
Sergio F. Garribba
Director
Energy R&D and Technology Applications
International Energy Agency

Good morning to all of you. I would like to explain how this IEA/OECD Seminar on Energy Technologies for Reducing Emissions of Greenhouse Gases has been organised and how it will develop. The Seminar has seventy-two speakers, plus observers, delegates and scientists coming from twenty-four different countries and international organisations.

The oral presentations have been broadly divided into five Sessions, as you may see from the programme. I am saying "broadly divided" because the allocation of several papers was a matter of judgment since they were covering simultaneously a number of issues or problems. Other speakers have time constraints and their papers had to be accommodated within a different grouping of subjects. Some of the Sessions begin or terminate with a Specialist Panel Session: these are Sessions I, III, IV. Specalist Panel Sessions are held in parallel.

Session I will consider technology options and systems for immediate or short-term action. These are energy technologies for the reduction of carbon dioxide and other greenhouse gases which are available on the market (for instance, technologies for increased energy end-use efficiency, nuclear power, shift to natural gas, use of biomass and afforestation schemes), as well as technologies which are demonstrated but have not reached full market penetration, both demand and supply side. Means would also be outlined to improve emission characteristics of existing or deployed energy technologies and systems. A Specialist Panel will be dealing with greenhouse gas concerns in coal use and conversion technologies.

Session II will refer to energy technologies that might be developed for the next century, including enhanced energy end-use efficiency, storage in biomass, other carbon-free renewable energy resources, advances in power generation and nuclear energy, and so forth. Clearly, future scenarios might depend upon and interact with policy measures, particularly R&D "rush" programmes that are addressed towards special sub-sets of energy technologies and systems.

Session III has been conceived to deal with technology options and systems that might be proposed for the principal energy use and conversion sectors, such as electricity, transport, industry, residential consumers and commerce. There will be some analysis of technology strategies and sectoral energy policies, and debate on how to rank technologies on the basis of multiple relevant attributes (environment, energy security, economics, market barriers). A Specialist Panel is dedicated to the presentation and evaluation of a number of sectoral cases with the solutions that may be envisioned.

Session IV will generally refer to the application of the energy technologies to significant contexts and regions. The question of the transfer and adoption of technological knowledge would receive some attention. Recent and important national technology studies will be presented in the ad-hoc Specialist Panel.

Session V should offer the opportunity to discuss national R&D initiatives and to present, as we hope, ways and means to foster co-ordination of activities, re-direction of ongoing R&D collaboration and new mechanisms to support actions as required. In the light of possible threats posed by climate change, it is imperative that new thinking be brought into public and private decision-making.

Amongst the participants in this Seminar I can also see Mr. Keiichi Yokobori, Co-chairman of the Energy and Industry Subgroup, Response Strategies Working Group of the Intergovernmental Panel on Climate Change. Mr. Yokobori is former Director, Energy Information and Emergency Systems Operation of the IEA Secretariat. Although there is at present no formal link between the IEA or OECD and the Intergovernmental Panel, he may wish to transfer the substance of our debate and the conclusions into that context.

Each Session and Specialist Panel will have a Chairman, a Rapporteur and a Co-Rapporteur. The Chairman will briefly describe the general subject of the Session or Specialist Panel, will introduce the speakers, make comments if appropriate and keep the events on schedule. After each paper is presented, (twenty minutes is the time allowed for it) there will be five minutes for discussion.

The Rapporteur will begin the discussion by addressing a couple of questions to the expert, in order to extract more information and to underline relevant or controversial facts and opinions. Should time be left, additional questions from the audience will be called for by the Chairman. The Rapporteur and Co-Rapporteur will make an oral summary of the main issues which emerged during the Session (or the Specialist Panel).

Finally, for each one of the three days of the Seminar, we have nominated Programme Co-ordinators. These are the three OECD/IEA Directors who were responsible for the conception, design, structure and organisation of the event. Also on their behalf I would like to express warmest gratitude and appreciation to all those from the staff, secretaries included, who gave help and extraordinary support. A special mention is due to Mr. Denis Kearney of the IEA's Combined Energy Staff, Technical Secretary of the Seminar, for his efforts and work, particularly in the preparation of the details and documentation of the Seminar programme.

Thanks for the attention and my best wishes for good and effective work.

CLOSING COMMENTS

by

Sergio G. Garribba
Director
Energy R&D and Technology Applications
IEA

Bill L. Long
Director
Environment Directorate
OECD

In convening this Expert Seminar, the IEA and OECD wished to create a forum for the exchange of the best information currently available on energy technology options, systems and strategies, as they might apply to national and international efforts to stabilise and mitigate the production and emission of greenhouse gases. The Seminar was probably the largest international meeting ever held entirely dedicated to the subject. More than two hundred experts, heads of research departments, government officers and scientists, coming from twenty-four different countries and international organisations attended. The seventy-two papers and reports presented gave a unique opportunity to gather and discuss, at first-hand, highly qualified knowledge on energy technologies. The analysis of technology options and how to pursue them was extensive and some of the economic and R&D implications underwent close scrutiny. There was continuing interaction between participants coming from the energy technology side, and those who are more involved with environmental and energy policy. This interaction proceeded smoothly and led to a well-balanced and constructive debate.

In concluding the Seminar, it seems appropriate to recall that it was conceived with four specific objectives:

a) identification of energy technologies and systems to reduce the emissions of greenhouse gases;

b) consideration of possible energy technology approaches for the different energy use and conversion sectors, and adaptation of technologies to particular national or regional contexts, including the developing countries;

c) generation of practical recommendations for R&D and demonstration initiatives in the main energy areas;

d) development of suggestions for the coordination of energy technology programmes and international collaboration.

The Seminar has generally achieved all its objectives. While the Seminar was not designed to reach consensus on the various issues involved with energy technologies as they may relate to climate change, the Seminar Organisers detected an apparent agreement on a number of points, and noted a number of suggestions for potentially useful follow-up work by the IEA, OECD and other international or national organisations. In particular, the material collected in this Seminar and the advice given by the experts, represent valuable contributions to the joint activity that IEA and OECD Environment Directorate now have and may plan on the climate change problem.

Information and advice received will also be reported to the Response Strategies Working Group of the Intergovernmental Panel on Climate Change and its Sub-Groups.

It now seems appropriate to briefly review the specific objectives against the outcomes of the Seminar, and summarise the most significant issues presented and discussed. First was the identification of energy technologies and systems. The Seminar provided comprehensive and quite detailed coverage of the state-of-the-art regarding a very wide spectrum of energy technologies for both short-term and long-term action. Many experts seemed to agree that:

- the greenhouse problem is complex and goes well beyond solely carbon dioxide emissions; other greenhouse gases contribute, and not all their sources are man-made and energy related;

- greenhouse gas emissions involve almost all the aspects of the different energy cycles; thus an integrated total fuel cycle approach is necessary in appraising the effectiveness of energy technologies and confronting uncertainties;

- there is no quick technology fix or single technical solution to reduce greenhouse gas emissions; rather, technological advance and progress should be viewed as an opportunity and key element in shaping integrated policies which promote energy security, environmental protection and economic growth; therefore, technology options should converge with other energy and environmental policy measures into well-balanced, scaled and sustainable courses of action:

- long-term progress and breakthroughs in energy technology are difficult to anticipate and forecast; as a consequence, the very long-term modelling that some energy and environmental policy analysts are proposing to represent the distant future of the planet, cannot reliably anticipate and incorporate technological advances and innovation; given this note of caution, it was recommended that careful attention be devoted to the linkage between energy technology assessments on the one hand, and energy and environmental modelling on the other;

- important gaps exist in the knowledge of the environmental effluents and impacts resulting from practical energy technology systems in current use; some experts suggested the idea of setting up an integrated data network and a consistent reference base concerning the various environmental flows and emission factors resulting from energy technologies and systems, as these technologies and systems are applied and available in the industrialised and possibly in developing nations.

Second there was a considerable debate on the possible technological courses of action and strategies. On the whole, experts observed that:

- no one set of energy technologies should be regarded as the single and most effective way of reducing emissions of greenhouse gases; several technologies should combine into composite courses of action and the optimum combination would vary from country to country according to each one's economic and physical circumstances, and its specific energy, environmental and economic policies;

- there is an array of energy technologies and courses of actions which are already available and demonstrated, although their large-scale and accelerated adoption might involve additional costs for the energy users and the governments; these costs, including all the environmental components, need to be more carefully assessed along with other barriers which are constraining the introduction of the energy technologies into the market;

- advances in energy technology require continuing R&D effort; several energy technologies designed to increase energy security have advantages from the viewpoint of greenhouse gas emissions; while it was underlined that sectoral efforts (or "rush" actions and programmes conceived in isolation) might have negative consequences on the economic and social context;

- case studies of technologies for individual energy conversion and use sectors, as well as country studies, would be a useful instrument to understand how energy technologies could contribute to the reduction of greenhouse gas emissions and be adopted in practice; analysis of cases and country monitoring would also help in establishing coherent technical and economic criteria for technology evaluation, and lead to the appraisal of effective lines of action to achieve significant progress.

Many experts suggested selecting energy technology options and planning R&D initiatives by assuming that the solutions and technologies are first adopted which are easist to implement (or to deploy into the different national and regional markets) and show the best ratio between benefit and cost. Solutions and technologies in this class would comprehend substitute chemicals for chlorofluorocarbons, preserving forests and biomass, conserving energy, fuel switching and so forth. A strategy of progressive and phased actions might be envisioned for the removal of increasing fractions of greenhouse gases while buying time and flexibility in the responses. R&D efforts tend to augment the number of available options and decrease some of their costs. Within a longer time horizon it would then become possible to fully exploit potential and conceivable energy technologies which now look perhaps more attractive and effective, but are slow to emerge. Two points were emphasised:

- at present (and in the mid-term), technologies for improving energy efficiency seem to be the best option for reducing emissions, particularly of carbon dioxide, the single biggest source of greenhouse gases; development and diffusion of technologies for more efficient uses of energy in manufacturing processes, buildings and especially in the transportation sector would offer the important additional advantage that they often increase productivity and economic growth; furthermore, more efficient uses of energy also reduce the waste heat which is generated directly and indirectly by all energy production and use processes;

- other essential technology options are now available and demonstrated for reducing production of carbon dioxide; experts recurrently referred to technologies which set conditions for protecting forests and introducing reforestation schemes; those which facilitate switching to less carbon-intensive fuels such as natural gas, though security of supply implications need appropriate analysis; further key technology options would be those which improve the greenhouse emission characteristics of energy systems and fuel cycles; those which ameliorate the opportunities for increased use of carbon-free renewable sources, such as hydro, geothermal, solar and wind energy; and, where national and international conditions so allow and contemplate, those technologies which lead to continuing and expanding use of nuclear power.

With regard to the third specific objective of the Seminar, generation of practical recommendations for R&D and demonstration initiatives in the main energy areas, there was widespread recognition of the need to re-examine R&D and demonstration programmes and their priorities. Along this view, the Seminar discussed potential energy technologies that may become adopted and largely used during the next decades or the next century, but for which further research, development and demonstration efforts, national and international, government and private, are still required and urgent. Experts seemed to share the view that promising and priority directions include:

- techniques for carbon dioxide recirculation and its storage into biomass; industrial large-scale uses of CO_2 were found to be worth exploring as well as possibilities offered by advances in species selection, genetic engineering, artificial photosynthesis and photochemistry;

- next-generation technologies for increased energy end-use efficiency and fuel substitution; the observation was made that new waves of technologies are in sight and progress in energy end-uses requires continuing demonstrative effort and dissemination of experiences through appropriate mechanisms;

- technological advances in critical stages of the natural gas chain, particularly technologies that may increase the transportability of natural gas and its marketable supply, and technologies that would permit the exploration and exploitation of geopressurised and very deep resources;

- technologies for low-greenhouse gas (and low-CO_2) production in the conversion of and use of coal and other fossil fuels;

- next-generation technologies and systems for electric power supply, transmission and distribution;

- development of advanced nuclear systems and the associated fuel cycle so that nuclear reactors become easier to construct, to maintain and to adopt for a variety of applications also beyond electricity supply; nuclear fusion power may add further to this direction;

- better integration of carbon-free renewable energy sources, energy storage and hydrogen fuels into the existing systems, and the changing energy supply and demand context.

The fourth (and last) objective of the Seminar was the development of suggestions for the co-ordination of national efforts and international collaboration. The problem of greenhouse gas emissions has consequences which clearly surpass the physical boundaries of OECD Member countries. New forms of international co-ordination may become necessary, a course that has been often advocated. The existing co-operation on energy technologies amongst OECD and IEA Member countries appears to offer firm ground towards further initiatives. Ad hoc and intensified international R&D collaboration would help shorten long lead times, share the risk entailed by capital-intensive energy technology projects, define common requirements which may enlarge market perspectives. Again, the effort made by single countries, although important (and giving notable examples of possible action) might not lead to significant results on a global scale and/or might not display all its potential spin-offs in the event that it lacks international links and connections.

Several experts suggested the desirability for OECD Member countries of preparing and establishing suitable means (the concept of a clearinghouse might be worth investigating) to collect and exchange information and knowledge concerning available, demonstrated and innovative technologies for the stabilisation and mitigation of greenhouse gas production. In this perspective, mention was made of problems of proprietary know-how, resources, market responses and timing. The possible adaptation of energy technology to developing countries was also considered, as well as the technical advice which might be required if technologies are to be conceived and designed to meet the demands of those nations.

The two Directors finally want to thank all the experts and those who participated for their continuing attention and support. It must be acknowledged that experts received a preliminary notice of the Seminar four months ago, they were formally invited towards the end of February and they were all able to attend and present written contributed papers and reports, most of them of outstanding scientific content. Numerous experts also asked to maintain contacts after the Seminar. Participation was offered in informal and advisory working groups that the IEA and OECD Environment Directorate might wish to establish for dealing with more specialised subjects, as new actions will be planned and develop.

Yet a particular expression of gratitude goes to the ones who helped in the successful organisational effort and in providing all the support and service which was needed. We think that all together we accomplished an extraordinary and very much appreciated job, so that those who took part might say in the future, "I was there". Thanks for your attention. The Seminar is closed.

PART A

TECHNOLOGY OPTIONS AND SYSTEMS FOR
IMMEDIATE OR SHORT-TERM ACTION

ENERGY TECHNOLOGY OPTIONS FOR CLIMATE
WARMING: PRELIMINARY RESULTS

by Paul Schwengels
U.S. Environmental Protection Agency

INTRODUCTION

The U.S. Environmental Protection Agency has recently
completed a draft report titled <u>Policy Options for Stabilizing
Global Climate.</u> The report is based on a two year study
initiated at the request of the U.S. Congress. It focuses on
technological and policy options for reducing greenhouse gas
emissions. This paper presents an overview of the results
contained in the draft report and outlines some of the key
implications for technology options and policies for promoting
them. The next section of the paper provides a very brief
summary of the approaches used and key results of the EPA draft
report. The following section presents more detail on results
related to new energy technologies and their impact on the
greenhouse problem. The final section discusses examples of
policy approaches which could be effective in promoting or the
rapid deployment of key energy technologies for reducing
greenhouse gas emissions.

The analytic results presented should be considered
preliminary. The draft report is currently undergoing extensive
review by the EPA's Science Advisory Board as well as by experts
in other Federal Agencies. Numerous other experts in the U.S.
and other countries have been asked to review and comment on the
report. Thus, these results are clearly subject to change in the
final version. It is nonetheless useful to present these
preliminary results in some detail for two reasons. First, the
explanation of relevant components of this analysis to groups of
experts in particular technical areas encompassed by the study
may stimulate discussion of key assumptions and results which
will help improve the final product. Secondly, even though
preliminary, the results provide some insights which may be
helpful to the group in thinking broadly about the roles of
various energy technologies in reducing greenhouse emissions.

THE STABILIZATION REPORT

The study includes a comprehensive summary of estimated
greenhouse gas sources and trends in concentrations. The major
sectors contributing to emissions and the major greenhouse gases
are considered on a global basis. The study identifies a broad
range of technological options for each of the major source

3

categories and examines various U.S. and international technology and policy options that, if implemented, could slow the buildup of greenhouse gases which contribute to global climate.

In developing the "Stabilization Report", EPA established several goals: (1) assemble data on global trends in emissions and concentrations of greenhouse gases; (2) develop an integrated analytic framework to study how economic, policy and scientific assumptions influence warming estimates; identify promising technologies and practices that have the potential to limit greenhouse gas emissions; and (4) identify policy options that could slow the buildup of greenhouse gases which may contribute to climate change. One major approach to achieving these goals was to carry out an extensive literature review and data gathering process. The Agency held several informal panel meeting and enlisted the help of leading experts in the governmental, non-governmental and academic research communities. EPA also sponsored five workshops to gather information and ideas regarding factors affecting atmospheric composition, and options for reducing greenhouse gas emissions from agriculture and land use change, electric supply technologies, end-uses of energy, and developing countries.

Based on results of this process, and on prior analytic efforts by many organizations, EPA developed an integrated analytic framework for analysis of climate warming issues and options. As indicated in figure 1, the framework links components which:

(1) represent current estimates of emissions from major source categories and projected changes in those emissions over time based on alternative input assumptions;

(2) estimate the concentrations of greenhouse gas emissions based on a parameterized model of atmospheric composition which represents relationships in much larger atmospheric research models, and several alternative models of atmosphere-ocean interactions: and

(3) translates changing greenhouse gas concentration into estimated ranges of changes in global average temperatures reflecting the current scientific uncertainty about this relationship.

The framework, though highly simplified, permits examination of the effects of alternative assumptions about a range of factors which drive future emission rates, emission reduction strategies and uncertainties in scientific relationships. Limitations of this current analysis include simplified representations of complex and uncertain scientific and economic relationships. Detailed economic and technology cost or penetration analyses have not been conducted. We did not attempt

4

to calculate the macroeconomic impacts of policy options or total costs associated with various strategies. Limitations on available capital for investment were not explicitly considered. All of these ares need to be evaluated in more detail before policy decisions can be intelligently evaluated.

In developing the report, EPA recognized that many factors which are difficult to project will influence future greenhouse gas missions. Important driving factors include rate of population growth, economic growth, technological change, fuel mix, and energy, industrial and agricultural policies. To broadly account for these factors, EPA developed two "no response" scenarios of possible future worlds which could occur if no major policies are adopted to reduce the emissions of greenhouse gases. One of these scenarios assumes that future economic growth and technological changes are rapid and that population growth slows in conjunction with rising per capita incomes in developing countries. This scenario is referred to as the Rapidly Changing World (RCW). A second no response scenario, the Slowly Changing World (SCW), assumes that economic growth and technological change are slow and population growth remains higher.

For each of these broad alternative futures, a corresponding policy scenario was constructed which assumes that a number of stabilizing policy measures are adopted resulting in substantial changes in technologies and fuels. These scenarios are referred to as the Rapidly Changing World with Stabilizing Policies (RCWP) and Slowly Changing World with Stabilizing Policies (SCWP). Key assumptions for all four of the above scenarios are shown in table 1.

In addition to these four major scenarios, a range of additional cases were evaluted in two basic categories. First, sensitivity analyses were conducted to evaluate the effects of changes in input assumptions and ranges of uncertainty around many of the key scientific relationships. Second, alternative policy cases were examined incorporating more stringent measures to reduce emissions as well as the effects of some alternative policy assumptions which could cause more rapid increases in emissions the shown in the no response scenarios. Other policy variations examined included delay in implementation of stabilizing policies and effects of non-participation by developing countries.

GENERAL RESULTS

The review of current science and the modeling results presented in the draft report indicate that uncertainties regarding the magnitude and timing are quite large as illustrated in figure. There is a growing consensus in the scientific community that concentrations of greenhouse gases are increasing and that, as a result, global warming and climate change, perhaps of a significant degree are likely to occur. In the EPA scenarios with no policies to limit emissions of greenhouse gases, the equivalent doubling of CO2 is estimated to occur between 2030 and 2040, and warming of $2-6^{\circ}C$ is projected by the end of the 21st century.

Very large reductions in emissions would be necessary to stabilize atmospheric concentrations. EPA policy scenarios do not achieve reductions large enough to ensure stabilization. However, the mix of policy and technological change measures assumed in these scenarios does appear effective in significantly reducing the rate and magnitude of warming. Figure 2 illustrates the range of results for the no response and policy scenarios. Scientific uncertainties are reflected in the ranges of projected impacts of alternative scenarios.

Policy measures and technological improvements to limit greenhouse gas emissions are estimated to result in reductions in all categories of emissions, in many cases reductions of much more than 50 percent. Accounting for the scientific uncertainties about the response of the climate system, these emission reductions are estimated to decrease the rate and magnitude of projected warming, perhaps by one half or more, relative to the projected results without policy measures.

Global cooperation would be required in any effective strategy to reduce greenhouse gas emissions and warming potential. No single country or source category contributes more than a fraction of the greenhouse gases. Obviously, any program to address the climate change problem will require active involvement of many countries and measures affecting a wide range of source activities. One illustration of the importance of international cooperation is provided in the draft report. As shown in figure 3, failure to involve the major emitting developing countries in effective emission reduction measures would severely limit the ability of developed countries to reduce the potential effects of warming.

IMPLICATIONS FOR ENERGY TECHNOLOGY

The results of the analysis prepared for EPA's draft stabilization report indicate that energy technology choices in both the next decade or so and over the next century will be extremely important in influencing the potential for global climate warming. Figure 4 illustrates that energy production and consumption will be likely to continue as the largest contributing sector to greenhouse gas emissions, mainly CO_2 and CH_4. In fact, some other source categories, such as CFC production, are already being controlled and others such as agriculture and deforestation are expected to grow at a slower rate than energy use. Thus, energy use is expected to represent an increasing share of the greenhouse problem over time unless significant policy and technology measures are implemented.

To bring about a significant reductions in the contribution of energy use to greenhouse gas emissions will require complementary technology based strategies for both the near term (i.e. the next decade) and the longer term. Pressures to reduce the greenhouse gas emissions related to the global climate problem over the next several decades may significantly increase the cost of energy production and become a constraint on economic development improvements in quality of life. Alternatively, these pressures could be viewed as an opportunity to improve international cooperation and to carry out an orderly transition away from fossil fuels to cleaner more environmentally acceptable energy sources. The success in developing and implementing energy technology will be a major factor in determining which of these outcomes is closer to reality.

It does not appear that the currently available technologies offer options for massive reductions in fossil fuel use and attendant greenhouse gas emissions, without equally massive increases in energy costs and widespread disruption of energy markets and economic activities. Nonetheless, many technologies are available which could be implemented beginning in the next decade to slow the rate of growth in greenhouse gas emissions.

Implementation of measures to deploy currently available technologies may have significant benefits in terms of "buying time" by slowing growth in greenhouse gas emissions while more advanced technologies to allow greater,cost-effective reductions are being developed. This is illustrated by one of the sensitivity scenarios presented in the draft report. As shown in figure 4, if the same measures incorporated in the policy scenarios to reduce emissions are implemented but delayed until 2010 for industrialized countries and 2025 for developing countries, cumulative greenhouse gas emissions and warming potential are substantially increased.

Because of the scientific uncertainty about the seriousness of the global warming problem and the long term nature of potential effects, it would be difficult to justify measures which impose significant costs on societies in the near term, particularly in the developing countries. However, it should be much more acceptable to implement measures which are relatively cost-effective in reducing greenhouse gas emissions, and are compatible with other near term policy goals such as promoting economic development, mitigating other environmental problems such as acid rain and urban air pollution, and promoting stability in international energy markets.

7

A number of technologies are identified in EPA's draft
report which may fit the desired criteria. In particular, there
is considerable evidence that major technological improvements in
the efficiency of energy end uses could be achieved at life cycle
costs equivalent to or less than the value of energy savings.
(see, for example, IEA, 1987 and Goldemberg, et al., 1988) Three
specific components of current and projected energy end use may
be particularly attractive targets for efficiency improvements.

Automobile fuel use is one key component of current energy
use which is expected to rise significantly over time as incomes
rise in developing countries and Centrally Planned Europe. Light
duty vehicles represent a single basic type of technology which
accounts for a significant fraction of global greenhouse gas
emissions. The technologies are also relatively standardized
globally and likely to remain so in the near future. Thus there
is an apparent opportunity to implement technology improvements
which could be rapidly disseminated around the globe and have s
significant near term impact on emissions.

There is also considerable evidence that technological
improvements have been identified and in many cases tested in
prototype which could substantially reduce the energy intensity
of light duty vehicle use. (IEA, 1987, Bleviss, 1988) There is
also considerable debate about how much efficiency improvement is
practically possible and over what time frame. Real world
constraints of mass production, concerns about cost and potential
sacrifices relative to other goals such as safety, performance,
and other air pollution could limit the fuel economy improvements
which would be achievable in the next 10-15 years. All of these
issues need to be evaluated in some detail and should be as a
matter of some priority, given the potential importance of these
technologies.

Major improvements in efficiency may also be possible in end
uses of electricity. Particularly in countries, like the U.S.,
where the dominant fuel for electricity is coal, reductions in
electricity consumption can be extremely effective in reducing
greenhouse gas emissions. Major technological opportunities for
reductions in energy intensity have been identified in buildings
for space conditioning, lighting and appliances (IEA, 1987,
Rosenfeld and Hafemeister, 1985, Geller, 1988) Questions remain
over how much of the technical potential is actually achievable
in practice, actual cost of some measures and how to encourage
greater deployment. It is likely, however, that substantial
cost-effective improvements in efficiency of electrical energy
use can be implemented in most countries over the next decade.

Technologies are also already available for industrial
production of energy intensive basic materials such as steel and
cement. (Ross, 1985) These technologies are not being rapidly
deployed in developed countries because basic materials

industries are not generally growing rapidly. In developing countries, on the other hand, basic infrastructure development is occurring or projected as a component of rapid economic development. This infrastructure development will require production of increasing amounts of basic industrial materials. It may be extremely important from the perspective of global warming (as well as from the perspective of sustainable economic growth) to ensure that near term investments in expanding basic industries in developing countries incorporate advanced energy efficient technology rather than outmoded and much more energy intensive technology still in common use in many industrialized nations. (Goldemberg, et al., 1988)

Over the next century, the policy scenarios developed by EPA assume implementation of alternative energy technologies on a massive scale to reduce greenhouse gas emissions. Technologies are not currently commercially available which would allow the reductions to be obtained without also incurring major increases in the cost of energy and possibly sacrifices in other policy goals such as economic development and environmental quality. These scenarios clearly assume significant advancements in the available menu of energy technologies through research, development and commercial demonstration programs. Perhaps the most critical component of a long-term strategy to reduce greenhouse gas emissions in the effective identification and prioritization of research, development and demonstration opportunities and needs, and the continued or increasing investment by currently industrialized countries in these programs.

Some key areas for potential R,D, and D attention are suggested in the EPA draft report. These include advanced technologies for improving efficiency of energy end uses, renewable energy technologies -- including solar photovoltaic conversion, enhanced biomass production, and biomass conversion, advanced nuclear fission technologies, and over the very long term nuclear fusion and hydrogen as an alternative energy carrier. These specific technology areas are intended to be suggestive of possible opportunities rather than comprehensive. Systematic evaluation of technology R, D and D alternatives was not done in the study, but this need to be done by individual countries as well as in international discussions.

POLICY OPTIONS FOR PROMOTING TECHNOLOGICAL CHANGE

On of the most potentially effective ways of promoting energy efficiency and less carbon intensive supply technologies is to increase the cost of activities responsible for emissions. In the energy sector this could theoretically be accomplished most directly through measures to increase the prices of fossil fuels, which are responsible for energy related emissions of CO_2

and CH_4. Ideally, fuel taxes would be applied equivalent to the costs and risks associated with contributions to climate warming and to other environmental externalities. This quantity cannot be quantified in practice, of course, and there are many other impacts of significant fuel price increases which make such measures difficult policy options for many countries to adopt.

Nonetheless, the evidence in the U.S. and other OECD countries during the last 15 years demonstrates the significant impact that prices can have in encouraging energy conservation and affecting fuel choices. In the last few years, of course, lower world oil prices have resulted in a levelling off of energy efficiency improvements in developed countries. (EEC, 1988, Geller, 1989) If low fossil fuel prices continue in the future, it will be much more difficult to encourage deployment of more efficient or low emitting technology.

In addition to the difficulty in implementing policies which fully "internalize" the costs of environmental damage into market energy prices, there are a number of other limits to the ability of prices alone to achieve energy policy objectives. Demand may be inelastic due to lack of information, the absence of short term alternatives, and capital constraints. In addition, energy investment decisions are often made be actors, such as landlords and homebuilders, who are not affected by operating costs. Significant segments of energy markets, such as electric utility production and distribution, are regulated industries in the U.S. and most other countries. These regulated industries do not generally respond to price changes in the ways economic markets would dictate. For these and other reasons, other policy approaches for promoting technological change in specific applications should be considered as well as broad fuel price options.

Several alternative policy approaches are available for encouraging reductions in the energy intensity of light duty vehicle travel. In the U.S. the principal policy approach in recent years has been efficiency standards structured as corporate average fuel economy (CAFE) requirements. Such standards provide a straightforward way of establishing and enforcing energy goals for new vehicles. However, the U.S. experience has raised a number of questions about the efficacy of this approach. Critics have argued that standards have been inefficient and have imposed substantial costs on manufacturers and consumers and that higher fuel taxes would be the most efficient way to promote efficiency improvements. CAFE standards can leave manufacturers with considerable uncertainty about future markets, particularly in low fuel prices are encouraging consumers to demand larger, less-efficient cars. Regulation combined with fuel taxes could provide a clearer target to industry, reducing the need for hedging strategies that dilute efforts and increase costs. (Bleviss, 1988)

Other problems with the current structure of CAFE standards have been identified (McNutt and Patterson, 1986) The fleet average concept may unduly penalize large cars when technologies could allow improvements in all classes. Requirements for annual improvements may be inconsistent with the need to incorporate improvements into long term design changes. The attempt to reintroduce flexibility through credits encouraged a search for administrative exemption at the expense of long term improvement. A lower fuel economy standard for light duty trucks also undercuts the effectiveness of the program by encouraging consumers to switch from large cars to these less efficient vehicles. Economic incentives might be more effective than the cumbersome enforcement mechanisms in the program.(Bleviss, 1988)

Gas guzzler excise taxes on inefficient cars are also in effect in the U.S. This approach could also be effective in encouraging consumer choices of more efficient vehicles as it is directly relevant to the "first-cost sensitivity" (or high implicit consumer discount rate). Properly structured such a tax system could substitute for a gasoline tax increase in a non-regressive way. One problem with the excise tax approach, however, is that it may discourage fleet turnover, keeping less efficient vehicles in service longer. Government funding of demonstrations of automotive technology and use of government procurement programs to provide a market for very efficient models could also be helpful in promoting use of more energy efficient technology. It appears from a preliminary review, that none of the current policy options for promoting automobile fuel economy by itself is effective. More detailed evaluation is needed to identify alternative mixes of policies which could be more effective.

Policies for encouraging more cost-effective planning by electric utilities is receiving a great deal of attention under the heading of least cost utility planning (LCUP). As regulated public utilities, electric companies often have not been provided with incentives to make least cost decisions. Many states are now requiring utilities to define energy options much more broadly and to include decentralized generation options as well as demand side management measures in the evaluation of proposals to build new a generating capacity. This is also being encouraged in some states through the use of bidding systems for new capacity and for conservation programs. Bidding systems should have the effect of introducing more competition into the electricity systems and therefore encouraging efficiency.

Another related policy concern is that existing financial rate-of-return systems may penalize utilities which invest in conservation measures even when these measures are much less costly than corresponding supply options. (Moskovitz, 1988) Several states are considering major changes in their regulatory systems to remove this disincentive. In addition, several states including Wisconsin and New York are considering the inclusion of environmental weighting factors in competitive bidding processes or other approaches to defining least cost options. These concepts may ultimately provide a mechanism for incorporating significant incentives to reduce greenhouse gas emissions directly and efficiently.into utility planning processes.

Over the long term major emphasis is needed on priorities and resource allocations for energy research, development and demonstration. Long term goals of reducing greenhouse gas emissions should be factored into R,D and D planning and management in conjunction with other national and international energy policy goals. Long term stable commitments of funds to key research areas may be necessary to achieve the technological goals incorporated into the EPA long term scenarios. At times in the past, the demonstration phase has been underemphasized with the result that attractive technologies identified in research may take an inordinately long time to reach commercial application. Greater emphasis on demonstration is likely to be required to achieve the technology goals implied by EPA analysis.

In the international context, perhaps the most dramatic issue is the transfer of technology to developing countries. These countries have the potential to produce significant increases in fossil energy consumption and greenhouse gas emissions over the next century as they strive for rapid economic growth. On the other hand, because they are in the process of building up infrastructure, opportunities may exist for developing countries to reach higher income levels without the massive increases in fossil energy use which accompanied this transition in the currently industrialized countries. This may require changes in development planning processes to more explicitly consider energy implications and options, strong support from the international lending community to finance the front end capital requirements of efficiency improvement and renewable energy supply, and transfer of technology from the OECD countries which currently control most of the most advanced energy efficiency and alternative supply technologies.

In summary, an effective approach to reducing greenhouse gas emissions may need to incorporate policy options to accelerate deployment of currently available technology in key areas in the near term, while simultaneously aggressively pursuing development of advanced technologies for the long term. A major focus of such a strategy must be to encourage transfer of energy efficient and low carbon supply technologies to the developing countries consistent with their own development objectives. Finally, a search for innovative policy approaches is also important to more effectively overcome the barriers which have retarded deployment of alternative energy technologies over that last decade.

REFERENCES

Bleviss, D. 1988. The New Oil Crisis and Fuel Economy Technologies: Preparing the Light Transportation Industry for the 1990's. Quorum Press, New York.

EEC (Commission of European Communities). 1988. The Main Findings of the Commissions's Review of Member States Energy Policies, the 1995 Community Objectives, COM(88) 174, Brussels.

Geller, H.S. 1988. Residential Equipment Efficiency: 1988 Update. American Council for an Energy-Efficient Economy, Washington, D.C.

Geller, H.S. 1989. U.S. Energy Demand: Back to Robust Growth? Energy Efficiency Issues Paper No. 1. American Council for an Energy-Efficient Economy, Washington, D.C.

Goldemberg, J., T.B. Johansson, A.K.N. Reddy, and R.H. Williams. 1988. Energy for a Sustainable World. Wiley Eastern Limited, New Dehli.

IEA (International Energy Agency). 1987. Energy Conservation in IEA Countries. Organization for Economic Cooperation and Development, Paris.

McNutt, B. and P. Patterson. 1986 CAFE Standards - Is a Change in Form Needed? Paper presented to the Society for Automotive Engineers, September 22-25, Dearborn, Michigan.

Moskovitz, D. 1988. Will Least-Cost Planning Work Without Significant Utility Reform? Paper presented to the Least-Cost Planning Seminar of the National Association of Regulatory Utility Commissioners, Aspen, Colorado, April 12 (Revised June 10).

Rosenfeld, A.H., and D. Hafemeister. 1985. Energy Conservation in Large Buildings. In D. Hafemeister et al., (eds.). Energy Sources: Conservation and Renewables. American Institute of Physics, New York.

Ross, M. 1985. Industrial Energy Conservation. In Hafemeister, D. et al., (eds.). Energy Sources: Conservation and Renewables. American Institute of Physics, New York.

Table 1

Overview of Scenario Assumptions

Slowly Changing World

Slow GNP Growth

Continued Rapid Population Growth

Minimal Energy Price Increases

Slow Technological Change

Carbon-Intensive Fuel Mix

Increasing Deforestation

Montreal Protocol/Low Participation

Rapidly Changing World

Rapid GNP Growth

Moderated Population Growth

Modest Energy Price Increases

Rapid Technological Improvements

Very Carbon-Intensive Fuel Mix

Moderate Deforestation

Montreal Protocol/High Participation

Slowly Changing World with Stabilizing Policies

Slow GNP Growth

Continued Rapid Population Growth

Minimal Energy Price Increases/Taxes

Rapid Efficiency Improvements

Moderate Solar/Biomass Penetration

Rapid Reforestation

CFC Phase-Out

Rapidly Changing World with Stabilizing Policies

Rapid GNP Growth

Moderated Population Growth

Modest Energy Price Increases/Taxes

Very Rapid Efficiency Improvements

Rapid Solar/Biomass Penetration

Rapid Reforestation

CFC Phase-Out

14

Figure 1

STRUCTURE OF THE ATMOSPHERIC STABILIZATION FRAMEWORK

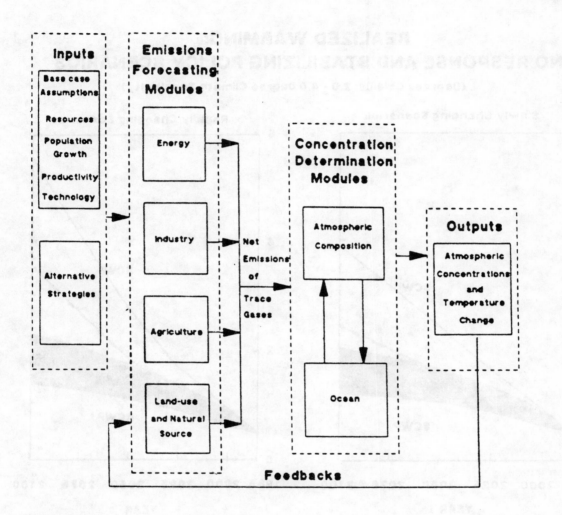

Figure 2

REALIZED WARMING:
NO RESPONSE AND STABILIZING POLICY SCENARIOS
(Degrees Celsius; 2.0 - 4.0 Degree Climate Sensitivity)

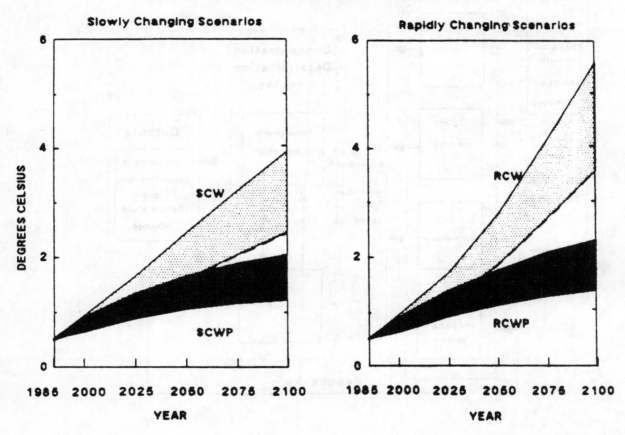

Shaded areas represent the range based on an equilibrium climate sensitivity to doubling CO_2 of 2-4°C.

Figure 3

INCREASE IN REALIZED WARMING
WHEN DEVELOPING COUNTRIES DO NOT PARTICIPATE
(Degrees Celsius; Based on 3.0 Degree Sensitivity)

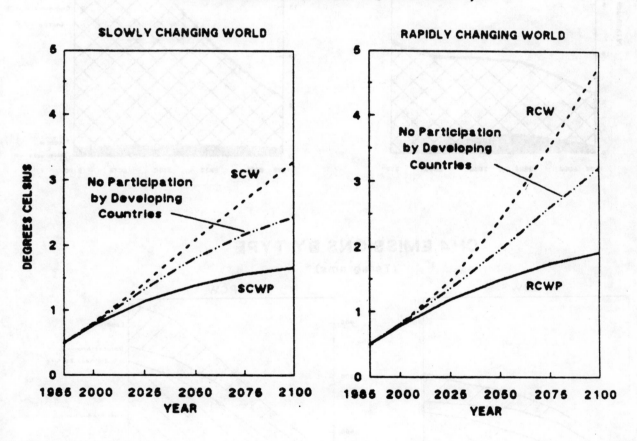

FIGURE 4

CO2 EMISSIONS BY TYPE
(Petagrams Carbon)

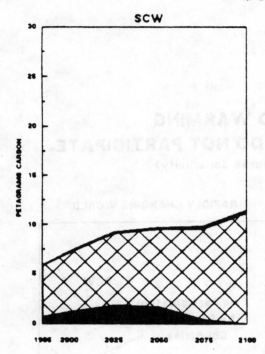

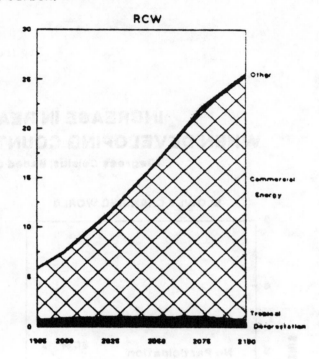

CH4 EMISSIONS BY TYPE
(Teragrams)

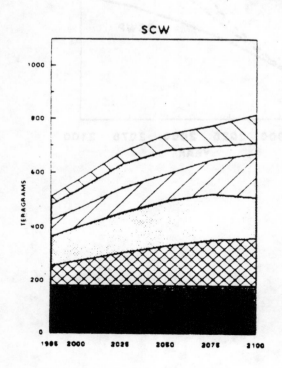

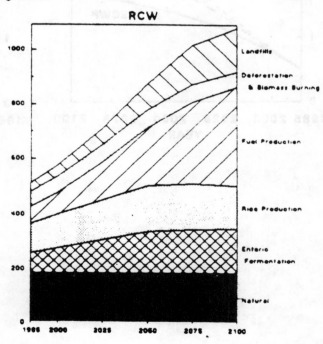

18

Figure 5

INCREASE IN REALIZED WARMING
DUE TO GLOBAL DELAY IN POLICY ADOPTION
(Degrees Celsius; Based on 3.0 Degree Sensitivity)

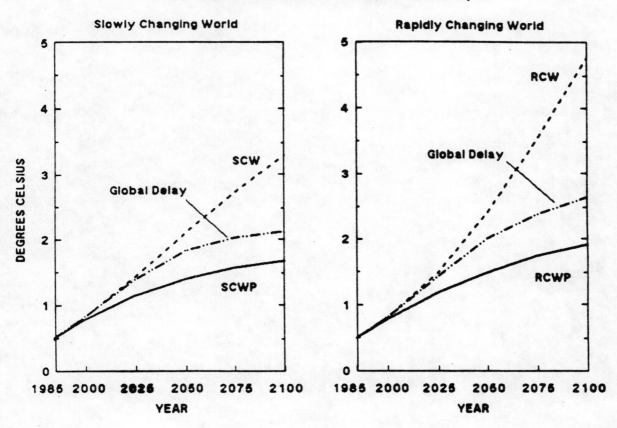

Assumes that industrialized countries delay action until 2010 and that developing countries delay action until 2025. Once action is initiated, policies are assumed to be implemented at roughly the same rate as in the Stabilizing Policy cases.

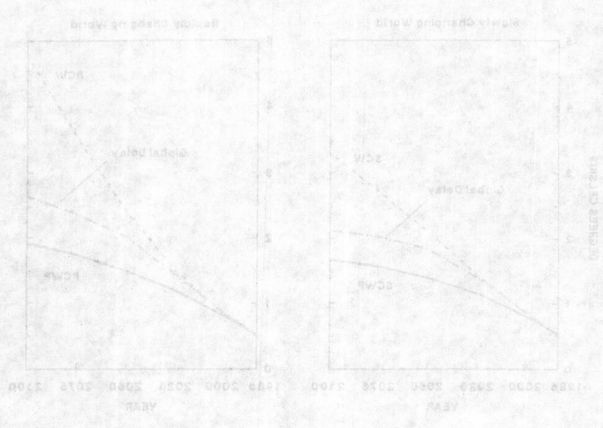

FIGURE 2

INCREASE IN REALIZED WARMING
DUE TO GLOBAL DELAY IN POLICY ADOPTION
(Degrees Celsius Based on 3 C Degree Sensitivity)

Slowly Changing World   Rapidly Changing World

Assumes that industrialized countries delay action until 2010 and that developing countries delay action until 2025. Once action is initiated, policies are assumed to be implemented at roughly the same rate as in the Stabilizing Policy cases.

PRESENT STATUS AND FUTURE PROSPECT OF ENERGY UTILIZATION TECHNOLOGY IN JAPAN FOR GREENHOUSE GAS MITIGATION

Takao KASHIWAGI, Dr.-Eng.
Professor
Department of Mechanical Systems Engineering
Tokyo University of Agriculture & Technology
KOGANEI, TOKYO, 184 JAPAN

SUMMARY

This paper summarizes the recent development of energy utilization technologies in Japan and also refers to the future prospect for the greenhouse gas mitigation. From this viewpoint, it is represented that the LNG should have received more attention and the creation of an ideal exergetic flow by the industrial processes rationalizaion will be essential.

In another level, it is pointed out that the development of elementary regeneration techniques is highly important for taking the electric power from the low grade waste heat sources.

1. INTRODUCTION

Recently, a new concept for the energy saving technology has received a great attention especially in a field of industrial application from the viewpoint of global greenhouse effect by CO_2 emission. In Japan various techniques of industrial waste heats recovery for energy saving have progressed a lot in the past ten years under the influence of national big projects, and one may now consider that the high temperature waste heats are fully recovered.

And now, although new stage for the rationalization of the industrial processes has come in order to use the effective thermal energy completely, quite a lot of problems remain and have to be solved from now on in the field of the relatively low grade waste heats.

With the development of the so-called "Super Heat Pump Energy Accumulation System" which is a part of the Moonlight project initiated by the MITI of Japanese Government, every kind of energy is aimed to be stored at a high density, high efficiency and used as a heating or a cooling sources at the time it is needed, and then used for big buildings air-conditioning or large scale domestic heating or industrial processes. And also, all the techniques developed in this project and applied to the industrial existing low and medium temperature waste heats are liable to enlarge drastically the domain of accessible waste heats.

In the present paper, author wants to put the accent mainly on industrial use, and will for this purpose examine the ideal flow of energy and present low grade waste heat recovery elementary techniques and their problems, including the future prospect of energy utilization technologies in Japan.

2.PRESENT STATUS OF THE INDUSTRIAL ENERGY FLOW IN JAPAN

Let us begin this study by classifying the various processes used for industrial waste heat recovery and upgrade. Fig. 1 shows a schematic representation of the energy flow in case of an industrial plant. The greater part of waste heats above the environmental temperature level lies in refrigeration waste or combustion exhaust gases. For instance, there is a large quantity of gases evacuated over 250°C in industries such as steal, oil or coke industries. Higher temperature gases are usually recuperated in such systems as waste heat boilers, supplying other parts of the factory with buildings heating, providing hot water for various facilities.

Globally, it can be said that the recovery techniques for what is commonly called high level waste heats have been thought and investigated with success, but the domain of low-medium temperatures from 30 to 200°C has still just slightly been recovered and one may say that high temperature boosting with high efficiency or easily adaptable techniques are greatly expected. As shown in Fig. 1, heat pumps appear as one of the recovery techniques for industry which has gathered a lot of attention.

The industrial heat pumps have been developed from the earlier progresses of air-conditioning heat pumps. As for them, a heat source is required, as you know, in order to upgrade the heating temperature; the low grade waste heat which has been unused is one of the most desirable heat sources for the industrial heat pump.

Besides this heat pump, waste heat boilers using heat pipes can be mentioned, and also high performance heat exchangers working with small temperature differences, and various kinds of heat storage techniques can be effectively applied to low grade waste heat, so, a lot of basic techniques that may be combined with each other will enlarge the heat recovery application field without any doubt.

The development tendencies of these numerous techniques will be seen in 5th chapter. Then in the following chapter, author would like to focus on the energetic ideal flow for the best use of heat in industries.

3.IDEAL ENERGY FLOW IN INDUSTRIAL PROCESSES FOCUSING ON LNG

In Japan the impulse for a positive action towards energy saving was given after the second oil crisis and this time is still present in our memories. At the initial stage, we tried to develop the energy saving technology in terms of thermodynamic first principle, but let us now think in terms of second principle, focusing on the effective use of high quality heat sources. As thermodynamics indicates it, it is quite obvious that one is able to get a higher work quantity or power when using a given heat quantity from a high temperature level to the environmental temperature than the same heat quantity taken at a lower temperature. In other terms, focusing on the operation that would consist in using a high enough temperature heat source such as fuel, gas, etc. to warm up a low temperature heat sink near the environmental temperature, even if the heat transfer is total, this operation can appear to be good in terms of first principle but not in terms of second principle; according to this second principle, this heat should be used at its highest efficiency limit. Based on this point, the effort should be made to form the ideal energy flow among industrial processes classified by their temperature level including the lower temperature below the environmental condition. In fact, we believe that it is necessary for industry to get closer to this global point of view of an ideal form. From the point of CO_2 gas mitigation, author would like to recommend to use LNG more widely. LNG stored at the temperature of -160°C is usually evaporated by the sea water, but this exergy should be recovered by some industrial processes with its highest efficiency limit. With this point of view, author suggests the representative example of the ideal exergetic utilization of LNG in the industrial processes with a cascade form as shown in Fig. 2 including the suggestion by French Prof. Le Goff. Until now, a lot of efforts were spent to reduce the energy consumption in Japan. However most of the efforts were only focused on one industrial process with a relatively short-sighted aspect as shown in Fig. 1, and then analyzing the best recovery possibilities inside of this process. But from now on author wants to emphasize here from the point of CO_2 mitigation that the ideal exergetic flow and cascaded dispositions of industrial processes are needed in which their temperature levels are taken into consideration in a global way when a new facility will be created.

After these considerations about the second principle of thermodynamics, let us return to industrial rationalization and to the complete use of waste heats as presented in the beginning of this paper. In France, it is said that some utility processes have already appeared, which take in account the characteristics of many processes and aim to approach the most effective use of heat, as presented in its ideal form previously. It can be said for other countries also that a serious analysis of industrial processes for their combination is an important challenge to engineers, in order to progress in the way of rationalization.

4.MIDDLE-LOW TEMPERATURE LEVEL HEATS AND THEIR PROBLEMS

In the last chapter author could draw some general conclusions, but in the present one, let us return to the real state of factories and analyze the recovery techniques of unused waste heats and the means to solve the inherent problems.

In table 1, various industrial processes liable to use directly are listed, classified by their temperature level, and in each case, a representative example of recuperation is given. A common definition of the temperature ranges is 20 to 80°C for the low temperature level and 80-250°C for the medium temperature level; processes which can recuperate this heat directly like air conditioning, domestic heating, drying processes, food industry, etc. are fairly limited. From this reason the recovery techniques should not only limit to the direct thermally applicable processes but also include power generation techniques, even if a lot of problems are still to be solved in this field.

In the present time, it is not exaggerated to say that all the relatively easily recoverable waste heats have been recuperated and it means that the ones which are still not recovered have been left for some essential reasons. Among these reasons,one can mention in a general way:

(1) the low energy level of these heat sources which implies a high recovery cost and a low profit.

(2) in case of exhaust gases, the problem lies in the temperature limitation due to the acid condensation. Especially with exhaust gases containing a high SO_x concentration, the recuperation by a waste heat boiler should be done above the condensing point due to the corrosion problem.

(3) problems of recuperation instability due to large variation in time of the waste heat flow.

(4) problem of no available space for the recovery equipment.

(5) because of the bad effect on the products or the main equipment, a modification of this equipment may be needed if the recuperation technique is inserted.

(6) no use of the upgraded heat in the proximity of the heat source.

But away these existing problems, the ones that can be solved by the recent remarkable progresses of recuperation techniques are not rare. For instance, with a high efficiency flon turbine or an oil free screw expander, recuperation of power from low grade waste heats becomes possible. And furthermore, the progresses in the field of acid corrosion resistant materials for heat exchangers, high performance heat storage system, low cost energy transportation system, etc. make the thermal recovery from the low grade waste heats possible. In the coming chapter, the recent trend of these techniques in Japan will be examined focusing on a few representative recuperation techniques.

5. TENDENCIES OF PROGRESSES IN RECUPERATION TECHNIQUES APPLIED TO LOW GRADE WASTE HEATS

Waste heat boilers are considered to be the most widely spread technique for waste heat recuperation in Japan. In case of high temperature, for instance gas turbine or diesel engine exhaust gases are easy to recuperate with high efficiency, but in case of low-medium temperature heat recovery, various problems appear: one of the most important defects is the accumulation of dust on the heat transfer surface, its adherence and the corrosion that follows. Particularly in case of waste heats below 200°C, corrosion problems can not remain unsolved. The phenomenon is due to the sulfur content of the combustible, which turns in sulfuric acid at low temperature; the lower the fuel quality is, the higher the sulfur content and also the acid condensation temperature become. For instance, in case of type C oil, oxidation appears at 160°C and it is largely discussed now, whether this corrosion can be overcome or not by some means. Here lies the reason of the numerous studies of pre-treatment techniques, or heat transfer surface temperature control or studies to develop new corrosion resistant materials. For what concerns the machine, the recent tendencies of research have been directed towards the use of heat pipes which enables the wall temperature control on the side of the hot gases, making it possible to maintain it above the condensation point, and also enables an easy dust removal. The same advantage is obtained if using a heat exchanger made of heat resistant glass, which was realized recently and the main purpose of which is to recuperate heat at a lower temperature than the condensing temperature of the acid fraction; this system has received a lot of attention.

There is no doubt that these techniques are going to get a great power in the future.

But low grade recovery techniques are not limited to these; heat pumps also exist with a large variety and enable to convert the low grade waste heat into a high temperature - high quality heat. Heat pumps are of various types, compression, absorption

25

and chemical types, but it may be said that chemical heat pumps are still at a development stage.

One of the well known characteristics of heat pumps is that their efficiency becomes very high if the temperature boosting remains in a low domain, in other words it means that the objective temperature is close enough to the waste heat temperature or a waste heat source is available at a temperature near the objective one.

Let us begin with recent progress in the compression heat pumps, a multistage compression type was developed in order to keep the temperature boosting small for each stage, and also non-azeotropic refrigerant mixtures was adopted as a working fluid. Their development tendency towards higher efficiencies is the base of a lot of efforts. An example of a recent development using the multistage compression concept is the coupling of 2 screw compressors with direct contact heat exchange system between lower and higher stages in order to reach high temperatures from a low grade heat source; the realization stage of this technique is awaited with a lot of interest.

A vapor recompression type heat pump (VRC) shown in Fig. 3 is one of the most successful ones, which is fully established and gets a big market in chemical and food industries in Japan.

Until now it was very difficult to recover the waste heat of vapor generated in the distillation-concentration process, because the heat was transferred to a final stage of relatively low temperature stream. But, with the introduction of open cycle type heat pumps, this steam can be directly compressed, reach a high temperature and return to the process if one uses the oil free screw compressor just described or a turbo-compressor, which realizes a highly performant energy saving system. Theoretically, the temperature difference required for heat exchange can be provided directly by the compressor work, without any additional external heat, which is why this can be described as a very effective technique of heat recuperation.

Besides these compression type heat pumps, absorption heat pumps have also been paid a lot of attention from the recent background of the worldwide ozone problem. The ozone layer depletion causes apparently in the presence of flon gases, which also belong to the greenhouse effect gases. In order to reduce drastically the use of these flon gases, refrigerants such as ammonia and water (in Japan mainly water) to be used in absorption heat pumps should receive the greatest attention. For waste heat recuperation, it is the temperature boosting type (type 2) which is concerned.

A major difference between absorption and compression heat pumps is the existence for the absorption type of 2 extra heat exchangers besides the usual condenser and evaporator, which are called absorber and generator and realize the same function as the compressor. In the temperature boosting type, waste heat is used at the evaporator and at the generator, where as high quality heat is supplied at the absorber; a disadvantage is that the system can not provide heat at a higher temperature than the one determined by the solution boiling point elevation related to the waste heat used. In case of the lithium-bromide-water combination, this temperature elevation reaches 60°C. An advantage is that it dose not need a working power, so the machine size becomes larger, absorption type has a great advantage comparing with the compression one. For what concerns the temperature limit, it can be taken equal to 170°C due to the corrosion problem.

A recent technique developed in the U.S. and using a water-nitrate salts ($LiNO_3$, $NaNO_3$ etc.) combination is reported to reach the temperature level of 260°C for the heat output. Also, the U.S. DOE and Japan have proposed a heat pump for low grade waste heat recuperation, which uses the combination of TFE (trifluoro ethanol, alcohol of the flon category) and E181, and it knows an energetic development.

6. CO-GENERATION AND THE POWER RECUPERATION BY BOTTOMING CYCLE

Techniques enabling to recuperate low level waste heat in a convenient form such as electric power are very important and mean possible great progresses in energy problems for the future. Cycles converting primary energy into power can be separated into 2 main groups, the usual topping cycle and the bottoming cycle. Focusing on the principle of co-generation recently in the spotlights, the concept of these cycles is shown in Fig. 4. One may distinguish them by the primary operation they do with the primary combustible electric power generation or heating process. The former of these 2 cycles has been paid a lot of attention in Japan for co-generation. The latter, bottoming cycle, can be used for industrial waste heat recuperation. It is very famous that the U.S. regulation PURPA for the tax credit classifies co-generation as shown in Fig. 4.

As shown in the Table 1, in case of relatively high temperature waste heats, one can rather easily take a power with high efficiency by using a Rankine cycle or the bottoming cycle of a gas-turbine, but in case of low grade waste heats, it is not exaggerated to say that no-adapted technique, neither using an expander nore a bottoming cycle exists in the present time. Flon turbines have received for a time a great attention and have known a strong development, and the fact they have been rejected is economically regrettable. Furthermore with the world wide problem of flon gases, one has to turn his efforts towards another direction for working fluids.

27

During many years, the author has paid attention to ammonia as a working fluid. In Japan to use ammonia is restricted by a lot of regulations especially in the metropolitan area. But the author believes that numerous scientists and engineers insist upon the fact we should change our mentality concerning this fluid. And it should be noticed that ammonia is not only interesting for freezing applications, but also in combination with other substances and used in chemical heat pumps, for reaching temperature over 200°C. Compared with flons, the transportation characteristics and the heat capacity of ammonia can not be overlooked. Also, besides the heat recuperation and transfer qualities, power recuperation is also possible with ammonia used in bottoming cycle with high efficiency. For instance a Rankine cycle using ammonia and working between a waste heat source at 45°C only and cooling water at 20°C would have pressure levels of 19 and 9 atm at the inlet and outlet of the expander, and if one uses an oil free expander which is being developed now, one would get power with rather high efficiency.

From now on, one should turn his eyes towards the high efficiency low grade waste heat recuperating bottoming cycle, and combined efforts of government, industry and university are definitely needed for the development of these high efficiency expanders using high performance working fluids from the global view point of greenhouse gases mitigation.

7. CONCLUDING REMARKS AND FUTURE PROSPECTS

This paper summarized the recent development of energy utilization technologies in Japan and also referred to the future technologies taking into consideration the effect of global environmental problem. From this view point, the author represented that the LNG should have received more attention for a primary energy sources. The ultimate strategy for the energy utilization is to make the efforts for the industrial rationalization for waste heat recuperation, trying to approach the ideal flow of energy including the low temperature processes from the exergetic view point, which is no matter of efforts inside a given industry but rather of gathering effectively many industries and processes in order to progress for the benefit of each of them for multipurpose.

And, for what concerns existing conditions with low grade waste heats, the author would like to conclude with the conviction that great efforts should be paid to thermal and power recuperation with high efficiency by using a bottoming cycle with appropriate working fluids, and the regeneration techniques.

Table 1. Representative applications for waste heats

Temperature level of waste heats	Application processes for heat recuperation	Power recuperation
20∿30°C	Culture, Heat pumps sources, Cultivation	Difficult
30∿50°C		Rankin cycle using screw expander with ammonia as a working fluid
50∿70°C	Heating, Drying, Sterilizatinon, Hot water washing, Heat pumps sources	Rankin cycle with organic working fluid if a sufficient of heat is obtainable
100°C	Washing, Dyeing, Pulping, Heating, Drying, Process hot water, Sterilizatinon, Condensation	
200°C	Heating, Drying, Distillation, Process vapor	Rankin cycle with organic working fluids
300°C	Industrial heating processes	
500°C		Rankin cycle with water working fluids
700°C		
1000°C		Gas turbine
1200°C		

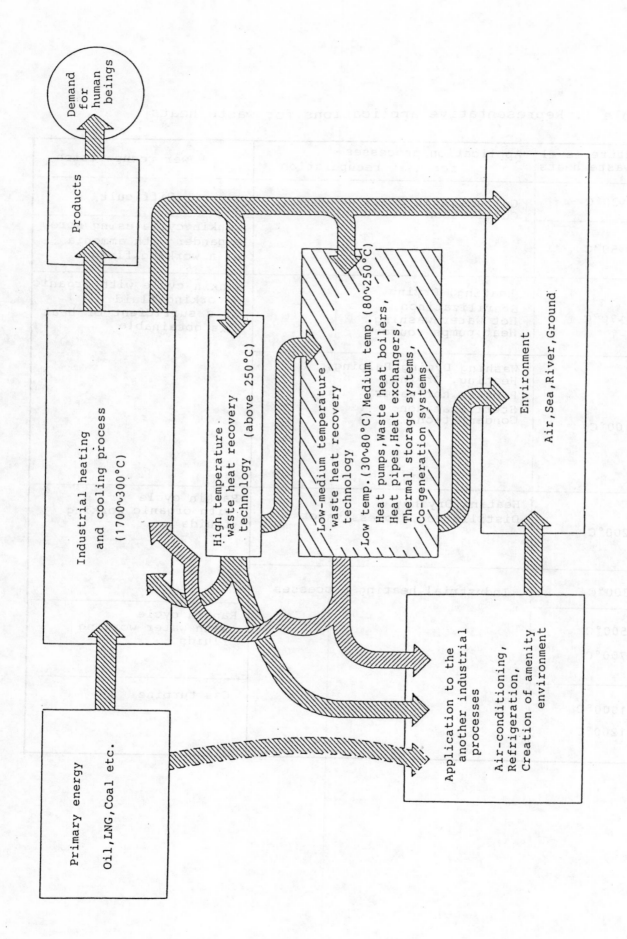

FIg.1 The industrial energy flow

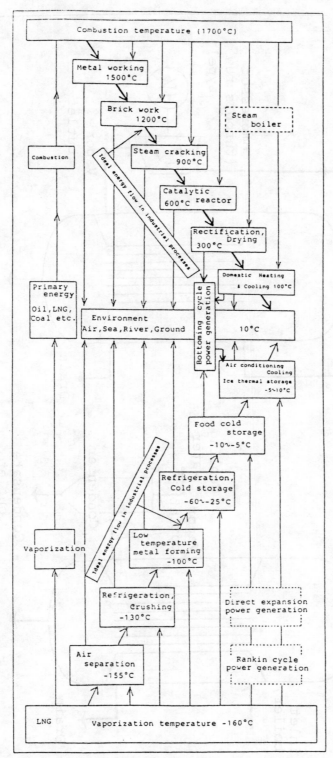

Fig.2 Ideal energy flow by the rationalization
of industrial processes

31

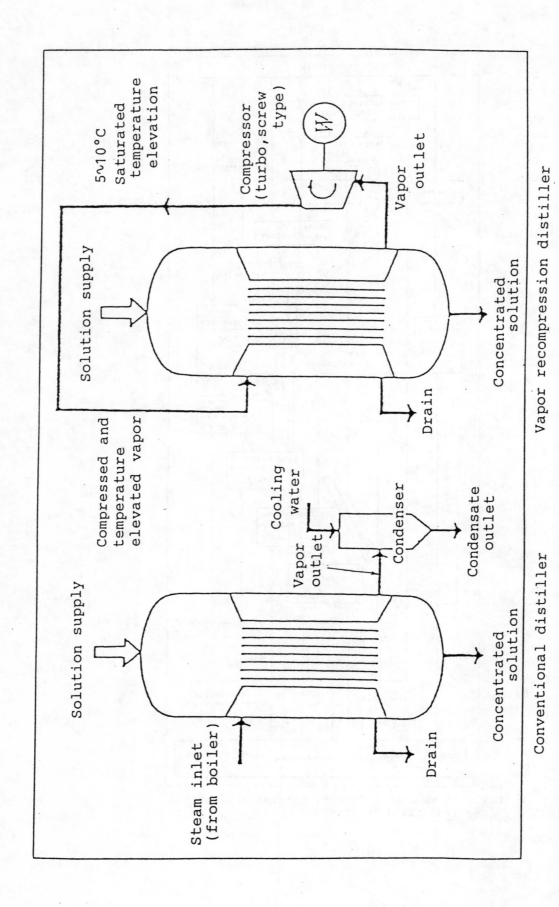

Fig.3 Vapor recompression type heat pump

Conventional distiller

Vapor recompression distiller

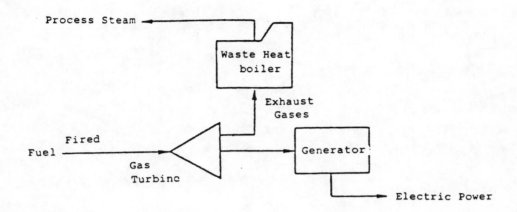

TOPPING CYCLE

Gas Turbine/Waste Heat Recovery

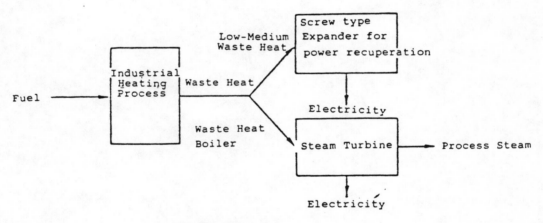

BOTTOMING CYCLE

Waste Heat Recovery/Electric Generation

Fig.4 Topping cycle and Bottoming cycle for power recuperation

POTENTIAL FOSSIL ENERGY—RELATED TECHNOLOGY OPTIONS TO REDUCE GREENHOUSE GAS EMISSIONS

R. Kane
Office of Fossil Energy
U.S. Department of Energy

D. W. South
Energy and Environmental Systems Division
Argonne National Laboratory

presented at:
IEA/OECD Expert Seminar on Energy Technologies
for Reducing Emissions of Greenhouse Gases

Paris, France
April 12—14, 1989.

FOSSIL FUEL TECHNOLOGIES CAN PLAY A ROLE IN REDUCING GREENHOUSE GAS EMISSIONS

- One factor contributing to the growth of atmospheric greenhouse gases is fossil fuel consumption

- However, worldwide fossil energy resource base, economic development policies and energy/economic growth relationships indicate continued reliance on these fuels

- If it is determined that greenhouse gas emissions should be reduced, more efficient fossil energy technologies can play a role

- Deployment of advanced fossil fuel technologies would reduce global fossil fuel consumption and greenhouse gas emissions relative to conventional technologies

WHILE OTHER GREENHOUSE GASES ARE OF GROWING IMPORTANCE, ONLY CO_2 IS EXAMINED HERE DUE TO ITS LARGE (70%) FOSSIL FUEL CONTRIBUTION

- Fossil fuel contributes to CO_2, CH_4, and N_2O emissions

- Limited data suggest N_2O emissions from fossil fuel power plants have been overstated, 5–10 parts per million versus 200 parts per million

- NO_x controls are commercially available for most applications

- CH_4 is principally an area source problem without technological solutions

ANNUAL FOSSIL FUEL–RELATED CO$_2$ EMISSIONS ILLUSTRATE SHIFTING PATTERN OF GLOBAL SOURCES

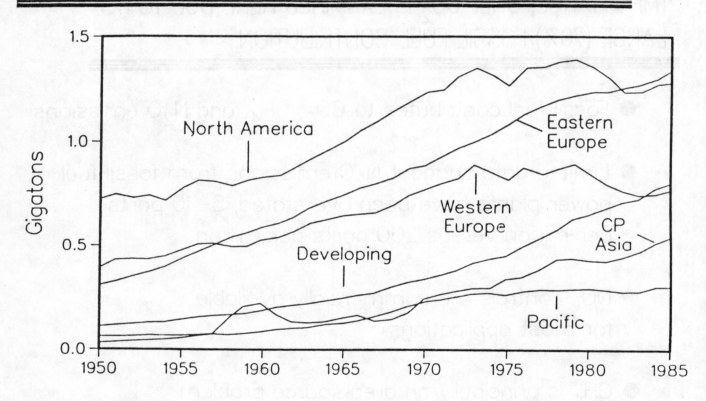

Source: USEPA, Policy Options for Stabilizing Global Climate (1989)

FOSSIL FUELS PRODUCE MORE THAN 60% OF WORLD-WIDE ELECTRICITY

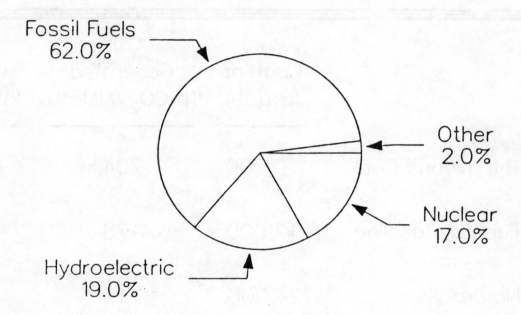

Fossil Fuels
62.0%

Other
2.0%

Nuclear
17.0%

Hydroelectric
19.0%

Source: USDOE/EIA, International Energy Annual 1987;
IEA Energy Statistics Annual

NATURAL GAS GENERATES ABOUT TWICE AS MUCH ELECTRICITY PER UNIT OF CO_2 AS COAL

	Heat Content (Btu/lb)	CO_2 Generated lb CO_2/MMBtu	Energy Generated kWh$_e$/lb CO_2
Bituminous Coal	12,700	204	0.56
Fuel Oil/Gasoline	17,600	178	0.70
Natural Gas	24,000	115	1.01
SNG from Coal Gas	24,000	329	0.34

Source: Steinberg, et al., Air Pollution Control Association Meeting (June 1988)

RECOVERABLE WORLD-WIDE FOSSIL ENERGY RESERVES ARE EXTENSIVE; AT CURRENT PRODUCTION RATES EXISTING OIL/GAS RESERVES WOULD LAST <50 YEARS

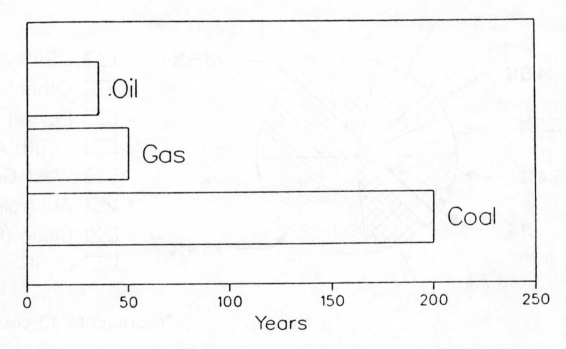

Source: USDOE/EIA, International Energy Annual 1987

SEVEN COUNTRIES ACCOUNT FOR OVER 90% OF ESTIMATED WORLDWIDE RECOVERABLE COAL RESERVES

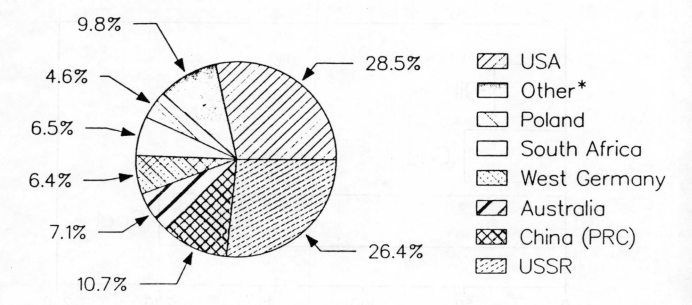

9.8%
4.6%
6.5%
6.4%
7.1%
10.7%
28.5%
26.4%

▨	USA
▭	Other*
◩	Poland
▭	South Africa
▨	West Germany
▨	Australia
▨	China (PRC)
▨	USSR

*represents 33 countries

Source: USDOE/EIA, International Energy Annual, 1987

ADVANCED FOSSIL ENERGY (FE) TECHNOLOGIES WILL REDUCE GREENHOUSE GAS EMISSIONS RELATIVE TO CONVENTIONAL FE TECHNOLOGIES

- U.S. Department of Energy, Office of Fossil Energy (USDOE/FE) is developing clean coal technologies (CCT) for greenfield and repowering applications
  - pressurized fluidized bed combustion (PFBC)
  - atmospheric fluidized bed combustion (AFBC)
  - integrated gasification combined cycle (IGCC)

- While these technologies are still under development, through an accelerated commercial demonstration program they are likely to be available by 2005

ADVANCED FOSSIL ENERGY
(FE) TECHNOLOGIES (cont'd)

- CCT conversion efficiencies will be in 40–45% range, conventional coal–fired power plants with scrubbers have 30–35% efficiencies

- For each 5% efficiency improvement, CO_2 emissions are reduced by approximately 15%

- In the more distant future, fossil energy based fuel cells and magnetohydrodynamics will be available with thermodynamic efficiencies of 45–60%

ADVANCED FOSSIL ENERGY TECHNOLOGIES WILL REDUCE CO₂ EMISSIONS THROUGH IMPROVED CONVERSION EFFICIENCIES

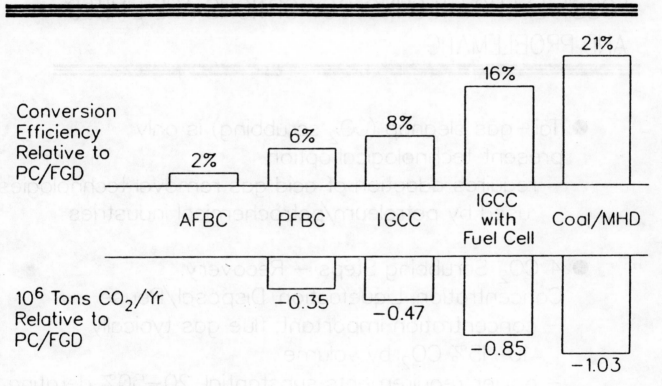

Conversion Efficiency Relative to PC/FGD

10^6 Tons CO_2/Yr Relative to PC/FGD

Source: USDOE/FE

IF *IMMEDIATE* CO_2 REDUCTIONS ARE REQUIRED, FOSSIL ENERGY TECHNOLOGY OPTIONS ARE LIMITED AND PROBLEMATIC

- Tail—gas cleanup (CO_2 scrubbing) is only present technological option
  - requires adaption of acid gas removal technologies used by petroleum/petrochemical industries

- 4 CO_2 Scrubbing Steps — Recovery, Concentration, Liquefaction, Disposal/Reuse
  - concentration important, flue gas typically 10–15% CO_2 by volume
  - power requirements substantial, 20–50% derating
  - CO_2 disposal/reuse constrained by access and storage/market size

FOUR CO$_2$ SCRUBBING TECHNIQUES ARE POSSIBLE, UNDER VARIOUS STAGES OF DEVELOPMENT

- Adsorption
  - uses materials such as clay, stored in clay pits after saturated
  - for 90% CO$_2$ concentration, power derating approaches 50% of plant capacity

- Absorption
  - used extensively by petroleum industry for acid gas removal
  - scrubs flue gas with liquid solvents (amines, seawater, etc.)
  - possible to recover/concentrate 90% of CO$_2$, power derating only 30%
  - more concentrated amine solutions could produce 30% efficiency gain and reduce power derating to 20%

47

SCRUBBING TECHNIQUES (cont'd)

- Condensation
  - capable of removing 90% CO_2, 20–30% power derating
  - least mature of CO_2 scrubbing options

- Chemical/Biochemical Reactivity
  - not available until post–2000 under current development plans
  - relies on plankton/algae for photosynthesis
  - captures CO_2 and converts to useable form (e.g., cellulose)

IMPORTANT OPERATIONAL CONSIDERATIONS EXIST WITH CO_2 SCRUBBING

- With CO_2 removal limited to 50%, generation efficiency at a conventional coal fired power plant would be reduced from 34% to 25% (26% derating)
  — integration of CO_2 removal/recovery process into power plant would reduce potential efficiency loss by 50%

- System power requirements would increase, unit generation costs could double

- If conventional coal—fired power plants supply derated power, additional SO_x, NO_x and CO_2 would be produced

- CO_2 disposal problematic, associated costs substantial

CO$_2$ DISPOSAL/REUSE IS LIMITED

- Storage in depleted gas wells is constrained by location, access and capacity
    - current U.S. capacity estimate, 48 gigatons (10^9 tons)
    - only 25 years of capacity exist at current CO$_2$ emission rates from U.S. fossil fuel power plants (assumes no location constraint)

- Ocean disposal at 500 and 3000 meter depth is possible
    - limited to power plants along coastal zones
    - uncertainties exist regarding absorptive capacity of ocean and CO$_2$ release rate

CO_2 DISPOSAL/REUSE (cont'd)

- Enhanced oil recovery (EOR) is a potential market for 1–3 trillion cubic feet/year of CO_2 recovered from power plants
    - equivalent to CO_2 recovered from 50 gigawatts of fossil fuel capacity, approximately 10% of equivalent U.S. capacity
    - limited to plants in oil/gas producing regions or with pipeline access
    - CO_2 use for EOR could expand from current level, 600–900 pounds CO_2/barrel
    - CO_2 recycle power plant has been linked with EOR market, technical problems exist

EPRI ESTIMATES CAPITAL COST OF CO_2 CONTROL ON U.S. POWER PLANTS TO BE $584 BILLION, $1230/kW

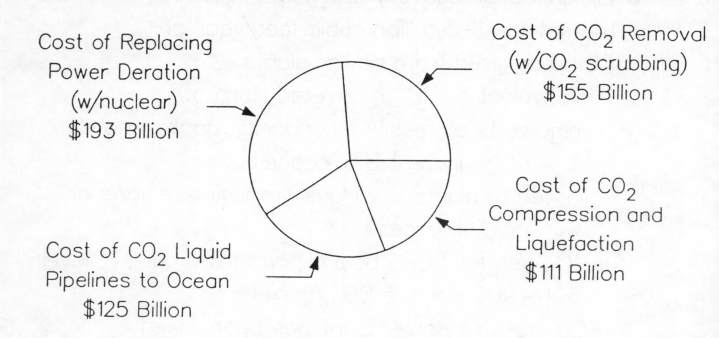

Cost of Replacing
Power Deration
(w/nuclear)
$193 Billion

Cost of CO_2 Removal
(w/CO_2 scrubbing)
$155 Billion

Cost of CO_2
Compression and
Liquefaction
$111 Billion

Cost of CO_2 Liquid
Pipelines to Ocean
$125 Billion

Source: Torrens, Conf. on Global Climate Change Linkages (1988), Derived from Steinberg and Cheng (1984)

SEVERAL USDOE/FE PROGRAMMATIC INITIATIVES WILL RESULT IN REDUCED GREENHOUSE GAS EMISSIONS

- Research into new unconventional gas recovery techniques will ensure long—term, economic supplies of natural gas

- Cooperative investigation with U.S. natural gas industry on the potential for co—firing natural gas at coal—fired power plants

INITITATIVES (cont'd)

- $5 billion industry—government program to promote deployment/commercialization of technologies capable of reducing acid rain emissions
  - first 2 rounds of program completed
  - approximately 30 projects selected totaling $2.6 billion (40% government funds)

- Innovative Control Technology Advisory Panel (ICTAP) explored opportunities to remove disincentives and provide effective incentives for adoption of high risk, low emission technologies

USDOE/FE INTERNATIONAL PROGRAM CAN ALSO PLAY A ROLE IN REDUCING GREENHOUSE GAS EMISSIONS

- Program consists of bilateral and multi-lateral research, information exchange projects. Projects directed toward:
  - collaborative R&D to increase CCT utilization in industrialized countries
  - CCT deployment in developing countries (e.g., cooperative technical agreement with Costa Rica under negotiation)

- On-going projects include:
  - initiatives to determine potential competitiveness of U.S. coal and CCTs in Pacific Basin
  - initiative to promote/coordinate intra- and inter-agency activities related to coal and CCT exports

IN CONCLUSION, FOSSIL ENERGY TECHNOLOGIES CAN PLAY A ROLE

- Deployment of advanced fossil fuel technologies would reduce global fossil fuel consumption and greenhouse gas emissions relative to conventional technologies

- Combustion efficiency improvements in countries currently using fossil energy technologies, together with reliance on advanced fossil technologies in developing countries, could achieve long−term reductions in greenhouse gas emissions

CONCLUSIONS (cont'd)

- Evidence of technological improvements exist; several advanced fossil energy technologies are reaching pre-commercial stage of development.

- In the short term, tail-gas cleanup opportunities are possible, but several problems exist
  - not presently cost-effective
  - power derating approximately offsets CO_2 reduction acheived
  - disposal/reuse options are regionally and demand constrained

- USDOE/FE is involved in several domestic and international initiatives that could play a role in reducing greenhouse gas emissions

GREENHOUSE GASES OTHER THAN CO_2 ARE NOW RESPONSIBLE FOR APPROXIMATELY 50% OF THE INCREASES IN GREENHOUSE EFFECT

1980's

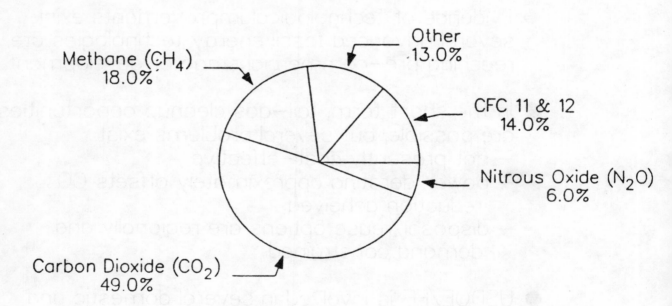

Methane (CH_4)
18.0%

Other
13.0%

CFC 11 & 12
14.0%

Nitrous Oxide (N_2O)
6.0%

Carbon Dioxide (CO_2)
49.0%

Source: USEPA, Policy Options for Stabilizing Global Climate (1989)

ENERGY TECHNOLOGIES FOR REDUCING EMISSIONS OF GREENHOUSE GASES

THE NEAR TERM CONTRIBUTION OF NUCLEAR ENERGY IN
REDUCING CO_2 EMISSIONS IN OECD COUNTRIES

by

K. Todani, Y.M. Park, G.H. Stevens
Nuclear Development Division
OECD Nuclear Energy Agency

THE NEAR TERM CONTRIBUTION OF NUCLEAR ENERGY IN REDUCING CO_2 EMISSIONS IN OECD COUNTRIES

INTRODUCTION

1.   There is much uncertainty at the present time about various aspects of the Greenhouse Effect, but there is general agreement that CO_2 has a major role among greenhouse gases, and that it is technically difficult to reduce CO_2 emissions as they are inseparably related to fossil fuel burning. The Toronto Conference (27 - 30 June 1988) recommended that CO_2 emissions be reduced by approximately 20 per cent of the 1988 level by the year 2005 as an initial overall goal for industrialised nations. This will be technically and politically a formidable objective.

2.   Among several approaches, the greater use of nuclear generation is certainly one which could be expected on technical and economic grounds to reduce these emissions. This brief paper sets out the contribution that nuclear generation makes towards the reduction of CO_2 emissions at present and in the near future, up to the year 2005.

3.   The calculations are based on (a) the 1988 Brown Book (Electricity, Nuclear Power and Fuel Cycle in OECD Countries) issued by the NEA (Nuclear Energy Agency), which presents forecasts of nuclear generation and total electricity generation in OECD countries, including that from the use of fossil fuels, up to 2005, which is based on statistics of OECD Member governments, with gaps filled by the Secretariat after discussion with the Member countries, and (b) published CO_2 emission factors.[1]

FORECASTS OF NUCLEAR GENERATION

4.   In 1987, nuclear power plants were operated for electricity generation in 13 of the 24 OECD countries (Belgium, Canada, Finland, France, F.R. Germany, Italy, Japan, Netherlands, Spain, Sweden, Switzerland, the United Kingdom and the United States). The total electricity generation in OECD countries was 5 838.2 TWH, including 1 312.6 TWH from nuclear generation. The proportion of nuclear generation relative to the total electricity generation has reached 22.5 per cent.

5.   According to the NEA's latest published statistics (1988 Brown Book), the electricity generation of all OECD countries will be 8 690 TWH by 2005, including 1 990 TWH from nuclear generation in 2005, or 22.9% of the total generation, as shown in Table 1. The forecast shows that nuclear generation will not expand very much in intervening years, and will remain at about the same proportion during the 1990s.

6.   In OECD America, OECD Europe and also OECD Pacific, the total
electricity generation will increase by 1 010 TWH to 3 030 TWH and by 400 TWH
to 1 150 TWH, and by 1 440 TWH to 4 500 TWH respectively, as shown in Table 2,
Table 3 and Table 4. The proportion of nuclear generation to total
electricity generation in OECD Europe was the highest among the three regions
in 1987. However, it is predicted to decrease by 1.8 per cent to
28.3 per cent in 2005. Also in OECD America, the share of nuclear generation
is predicted to decrease by 1.1 per cent. On the other hand, in OECD Pacific
nuclear generation is expected to increase with a rather high pace. The
forecast shows that nuclear generation in OECD Pacific will increase 230 TWH
to around 400 TWH in 2005, meaning 128.8 per cent increase compared with the
year 1987. The proportion of nuclear generation to the total electricity
generation will be expected to reach 34.7 per cent in 2005 in OECD Pacific.

7.   The generating capacity is also proportional to the electricity
generation. The nuclear generating capacity in OECD countries is predicted to
increase by 80.9 GWe. Its proportion to the total electricity capacity in
2005 is predicted to keep almost the same level of 16.8 per cent compared with
16.0 in 1987 as shown in Table 5.

8.   This forecast did not, however, take into consideration the current
environmental problems. Rather, in many countries, since there is a lack of
public confidence in nuclear safety it has become difficult to expand the
nuclear programme. In addition, the recent fall in the price of oil and coal
has reduced the economic attractiveness of nuclear generation. Several
countries have moratoria, either in practice or by political choice. Sweden
has decided to phase out nuclear generation by 2010. Taking into account
these factors, Member Countries estimate their growth of nuclear generation,
as shown in Table 1. We believe that the forecast is highly realistic,
particularly to 2000, although there is still some risk the plants included in
the forecast will not be completed. The as yet incomplete figure from the
1989 survey show that there has been a further increase in the nuclear share
of growing electricity demand.

9.   If the circumstances around nuclear generation are changed, and every
country makes great efforts to expand nuclear generation, its growth would be
certainly greater than that predicted by the 1988 Brown Book. However, little
could be done by 2000 because the lead time required for construction of new
nuclear power plants from the announcement of plans to commercial operation is
likely to be from 8 to 10 years in most countries. By 2005 more change could
be brought about but only by a few per cent.

CONTRIBUTION OF NUCLEAR GENERATION TO THE REDUCTION OF CO_2 EMISSIONS

10. CO_2 emissions factors depend on the characteristics of oil, gas and
coal. There are individual factors for various kinds of oil, gas or coal from
different places of origin. This calculation was based on the following
emission factors: 0.43 Gtons of carbon per TWyr for gas, 0.62 Gtons of carbon

per TWyr for oil, and 0.75 Gtons of carbon per TWyr for coal.[1] According to our survey, the emission factor ranges as shown below for different types of fuel: [1] [2] [3]

Gas   0.43 to 0.49 Gtons/TWyr
Oil   0.62 to 0.66 Gtons/TWyr
Coal 0.75 to 0.88 Gtons/TWyr

We adopted the set of conservative and rather wider used figures to estimate CO_2 emissions as above.

11.   In 1987, nuclear generation in OECD countries was 1 312.6 TWH. If nuclear power had not been used and been replaced by a country's non-nuclear fuel mix, including oil, coal and gas, the CO_2 emissions from electricity generation of the OECD countries would be increased by 0.3 Gtons of carbon from 0.74 Gtons of carbon to around 1.04 Gtons of carbon, as shown in Table 6. Then it can be said that the CO_2 emissions from electricity generation was decreased by around 30 per cent by using nuclear generation. It is said that CO_2 emissions due to fossil fuel burning in the whole world reach around 5 Gtons of carbon per year. It is not too much to say that nuclear generation in OECD countries has already played a noticeable role in reducing the emission of CO_2, because the CO_2 emissions decreased by around 6 per cent of the whole world CO_2 emissions from fossil fuel burning.

12.   Our forecast shows that nuclear generation in OECD countries will expand from 1 312.6 TWH in 1987 to around 1 990 TWH in 2005. Consequently, the CO_2 emissions avoided by using nuclear generation will be expected to increase to 0.44 Gtons of carbon per year. However, at the same time, the total electricity generation of OECD countries will increase by around 2 850 TWH, therefore CO_2 emissions from electricity generation of OECD countries will increase by 0.39 Gtons of carbon to 1.13 Gtons of carbon per year in 2005, meaning the emissions will increase by around 50% as compared to 1987, as shown in Fig. 1 and Table 6.

13.   Although many kinds of efforts to reduce CO_2 emissions will be made, assuming that only nuclear generation would be expected to play a role in these reductions, in order not to increase current emissions, nuclear generating capacity in OECD countries would have to be increased by around 410 GWe from 240 GWe in 1987 to 650 GWe in 2005. The 1988 NEA Brown Book predicted that nuclear generation will increase by only 80 GWe to 321 GWe in 2005. The difference is too great: 410 GWe may be impractical.

14.   However, as described before, this forecast did not take into consideration the current environmental problem. If a decision to expand the nuclear programme is made, more growth of nuclear generating capacity will be expected by 2005. The forecast predicted that nuclear generating capacity in 1995, 2000 and 2005 would be 280 GWe, 300 GWe and 321 GWe respectively, meaning the growth rate of electricity capacity would be only 4 GWe per year from 1995 to 2005. Among OECD countries, only Japan is predicting a larger expansion of nuclear generating capacity to 38.5 GWe in 1995, 49.8 GWe in 2000, and 61.0 GWe in 2005. The growth rate reaches around 2 GWe per year, but if additional orders were to be placed immediately, the manufacturing capability of nuclear power plants could be expected to increase. France has already reached a rather high share of nuclear generating capacity to total

electricity capacity and its nuclear generating capacity is not expected to increase at such a high rate as previously. Therefore, we can expect reserve power in its plant manufacturing capability. Plant construction capacity is also under-used in the USA, Federal Republic of Germany and Sweden. Taking into consideration those factors, an additional nuclear capacity growth rate of around 8 GWe per year could well be achieved by the late 1990s if it were required. The nuclear generating capacity in OECD countries would then be expected to be around 340 GWe in 2000 and around 400 GWe in 2005. That means CO_2 emissions from electricity generation in 2005 would decrease by around 0.1 Gtons of carbon per year from that predicted by the forecast.

15. However, without these or other remedies, CO_2 emissions from electricity generation in OECD countries will still increase at rather a high pace. Nuclear generation has already played an important role and has the potential to play an even greater one, but it will not be enough to stop the increase of CO_2 emissions by 2005 under current predictions.

TABLE 1
ESTIMATES OF TOTAL, FOSSIL AND NUCLEAR ELECTRICITY GENERATION OF OECD COUNTRIES

	1987		1990		1995		2000		2005	
	TWH	%	TWH	%	TWH	%	TWH	%	TWH	%
Nuclear	1312.6	22.5	1529.5	24.6	1687.9	23.9	1829.6	23.3	1990.8	22.9
Fossil	3343.4	57.3	3417.3	55	4008.7	56.7	4528.2	57.8	5101.1	58.8
Others	1182.2	20.2	1268.1	20.4	1368.3	19.4	1482.4	18.9	1593.4	18.3
TOTAL	5838.2	100	6214.9	100	7064.9	100	7840.2	100	8685.3	100

TABLE 2
ESTIMATES OF TOTAL, FOSSIL AND NUCLEAR ELECTRICITY GENERATION OF OECD AMERICA

	1987		1990		1995		2000		2005	
	TWH	%	TWH	%	TWH	%	TWH	%	TWH	%
Nuclear	528.2	17.3	626.9	19.4	685	18.4	711	17.4	731	16.2
Fossil	1949.7	63.8	1992.4	61.6	2395.5	64.2	2693.9	66.1	3075.5	68.4
Others	576.2	18.9	613.4	19	648.7	17.4	671.3	16.5	692.2	15.4
TOTAL	3054.1	100	3232.7	100	3729.2	100	4076.2	100	4498.7	100

TABLE 3
ESTIMATES OF TOTAL, FOSSIL AND NUCLEAR ELECTRICITY GENERATION OF OECD EUROPE

	1987		1990		1995		2000		2005	
	TWH	%	TWH	%	TWH	%	TWH	%	TWH	%
Nuclear	609.6	30.1	697.7	32.2	749.1	30.9	790.5	29	859.8	28.3
Fossil	921	45.4	934.7	43.2	1098	45.2	1285.2	47.2	1445.3	47.7
Others	496.2	24.5	533.5	24.6	579.2	23.9	648.3	23.8	729	24
TOTAL	2026.8	100	2165.9	100	2426.3	100	2724	100	3034.1	100

TABLE 4
ESTIMATES OF TOTAL, FOSSIL AND NUCLEAR ELECTRICITY GENERATION OF OECD PACIFIC

	1987		1990		1995		2000		2005	
	TWH	%	TWH	%	TWH	%	TWH	%	TWH	%
Nuclear	174.8	23.1	204.9	25.1	253.8	27.9	328.1	31.5	400	34.7
Fossil	472.7	62.4	490.2	60.1	515.2	56.7	549.2	52.8	580.3	50.4
Others	109.8	14.5	121.2	14.8	140.4	15.4	162.8	15.7	172.2	14.9
TOTAL	757.3	100	816.3	100	909.4	100	1040.1	100	1152.5	100

TABLE 5
ESTIMATES OF TOTAL AND NUCLEAR CAPACITY

(Net GWE)

	1987			1990			1995		
	TOTAL	NUCLEAR	%	TOTAL	NUCLEAR	%	TOTAL	NUCLEAR	%
OECD America	774.6	106.2	13.7	784.4	113.6	14.5	806.2	117.3	14.5
OECD Europe	532	107.6	20.2	557.4	119	21.3	593.4	125.1	21.1
OECD Pacific	195.4	26.3	13.5	209.7	31	14.8	227.8	38.5	16.9
OECD TOTAL	1501.9	240.1	16	1551.7	263.6	17	1627.4	280.9	17.3

	2000			2005		
	TOTAL	NUCLEAR	%	TOTAL	NUCLEAR	%
OECD America	862.5	119.1	13.8	943.5	120.9	12.8
OECD Europe	642.4	131	20.4	697.7	139.1	19.9
OECD Pacific	244.2	49.8	20	271.3	61	22.5
OECD TOTAL	1749.1	299.9	17.1	1912.5	321	16.8

TABLE 6
CO_2 EMISSIONS AVOIDED BY NUCLEAR GENERATION IN OECD COUNTRIES

Gtons of Carbon per year

	1987			2005		
	CO2 Emissions	CO2 Emissions Avoided (1)	CO2 Emissions Avoided (2)	CO2 Emissions	CO2 Emissions Avoided (1)	CO2 Emissions Avoided (2)
Nuclear Countries	0.68	0.3	0.34	1	0.44	0.5
Non-Nuclear Countries	0.06	—	—	0.13	—	—
OECD Total	0.74	0.3	0.34	1.13	0.44	0.5

(Note)

1. CO2 Emissions Avoided (1): The increase of CO2 emissions assuming that nuclear
generation is replaced by a country's fossil fuel mix

2. CO2 Emissions Avoided (2): The increase of CO2 emissions assuming that nuclear
generation is replaced by coal-fired power plants

FIGURE 1
PREDICTED CO₂ EMISSIONS FROM ELECTRICITY GENERATION IN OECD COUNTRIES

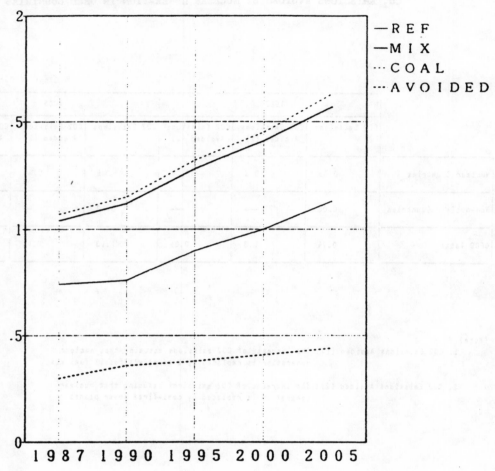

[Gtons of Carbon/year]

[year]

Ref;   CO₂ emissions predicted by the current forecast

Mix;   CO₂ emissions predicted assuming that nuclear generation
          were replaced by a country's fossil fuel mix

Coal;   CO₂ emissions predicted assuming that nuclear generation
          were replaced by coal-fird generation

Avoided; CO₂ emissions avoided compared by the case MIX

68

NOTES

(1) The Greenhouse Effect, Climate Change, and Ecosystems (Edited by
     Bert Bolin, B.R. Döös, Jill Jüger and Richard A. Warrick. Published on
     behalf of the SCOPE of the ISCO.).

(2) Environmental Effects of Electricity Generation (OECD, 1985).

(3) OECD/IEA/ETSAP Energy Environment Systems Analyses Towards Fossil
     Nuclear Symbioses (Shigeru Yasukawa, etc., Japan Atomic Energy Research
     Institute, 1988).

RESPONDING TO THE CHALLENGE OF GLOBAL WARMING:
THE ROLE OF ENERGY EFFICIENT TECHNOLOGIES

by: G R Davis
      Group Planning Co-ordination
      Shell International Petroleum Company
      London, United Kingdom

SUMMARY

Energy is a basic requirement of a modern industrial society and, for the foreseeable future, the combustion of fossil fuels will be the main source of this energy. However, combustion of carbon produces greenhouse gases (GHG) and global warming, and the increased concentration of GHGs in the atmosphere has emerged as a major public issue. A first line of defence to containing greenhouse gas emissions will be to improve the efficiency of energy use. The potential for energy saving is shown to be large. How to tap this technological potential for energy saving will be a major challenge to society and the way it frames policy. The policy context and the mobilising of possible policy options is briefly outlined.

Contents

Energy and Economic Growth

Economic growth has been powered by increased use of energy. Since 1860, there has been a four-fold increase in population, a forty-fold increase in gross world economic output and a twenty-fold increase in energy consumption; eighty-fold excluding wood (Figure 1).

Figure 1

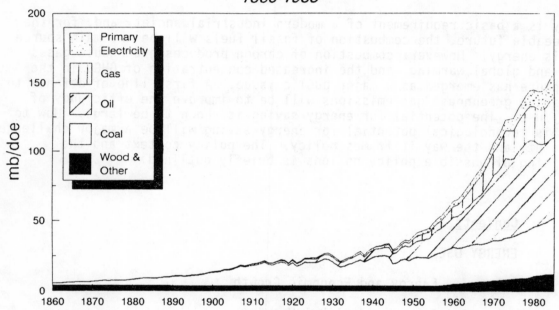

World Primary Energy Consumption
1860-1985

Energy growth exhibits patterns. As economic development takes place, there is first, in any country, a gradual switch from traditional to commercial fuels in households and, then, progressive dependence on commercial fuels for industry and transportation. Central to the development process is the building of infrastructure (roads, railways, ports, etc) and urbanisation. Often, economic growth has depended on exports of processed raw materials and manufactured goods. Such energy-intensive activities imply rapid growth in energy use through the early stages of industrialisation[1]. As a country

---

[1] 'Energy-Intensive Materials and the Developing Countries', A M Strout, ERG Review Paper No 53, MIT, March 1985.

increases its wealth and per capita income, it increases its energy use for basic materials in a regular fashion (Figure 2).

Figure 2   Figure 3

ENERGY USE IN MATERIALS CONSUMPTION

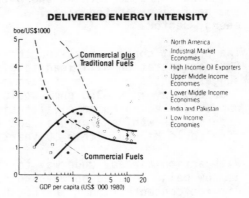

Energy intensity rises as a country enters industrialisation and is progressively lower for middle and high income economies (Figure 3). As an economy matures, it becomes less energy intensive. Much of the future growth in energy demand will be for countries now in the early stages of development.

Energy and Global Warming

Meeting energy needs requires the combustion of fossil fuels and transformation to useful energy forms such as thermal, mechanical and electrical energy. Current emissions of carbon dioxide due to fossil fuel combustion amount to some 5½ billion tonnes of carbon per annum (Figure 4).

Figure 4

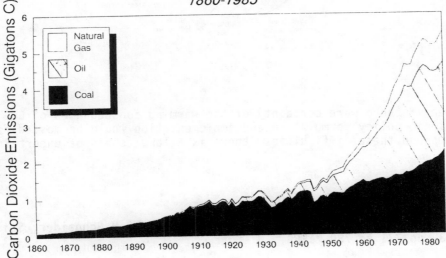

Of the GHGs, carbon dioxide, methane, chloroflorocarbons (CFCs), nitrous oxide and ozone, carbon dioxide accounts for about 55 per cent of projected global warming; of this some 15 to 30 per cent is due to deforestation. Allowing for some methane and nitrous oxide emissions in the order of one-half of projected global warming can be related to GHGs emitted from the energy sector.

The present concern rests on the strengthening consensus that increases in GHG concentration will lead to global warming. Global climate models suggest that a doubling of carbon dioxide concentration from 300 to 600 ppm (current level is ~350 ppm) would warm the planet $3 \pm 1.5°C$; a level unprecedented in human civilisation[2]. Given present trends in demography, economic growth, energy use, and technology many believe such a temperature rise could happen within 50 years.

With action already underway to reduce the level of CFC emissions attention is now being focused on ways to control the increase in atmospheric carbon dioxide concentration. In the short to medium term, suggestions emphasise 'energy saving' technologies but include the substitution of hydrogen-rich for carbon-rich fossil fuels and reforestation. To move away from fossil fuels will be a massive task since some 85 per cent of primary energy consumption is fossil fuel combustion (Figure 5); the leading short-term option would be nuclear fission.

Figure 5

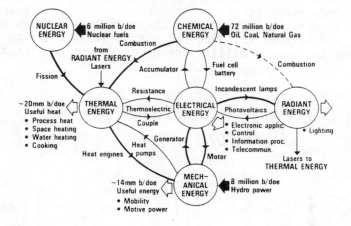

ENERGY USE AND CONVERSION TECHNOLOGIES

So even if there were certainty of the warming consequences of GHG emissions, policy formulation and implementation would be most difficult. To compound these difficulties, there is a long 'chain of uncertainty', from

[2] 'The Greenhouse Effect: Science and Policy', S H Schneider, Science, 243, pp 771 to 780, 1989.

the guesstimates of future emission levels to the magnitude of their likely
regional impact on natural and human systems (Figure 6).

Figure 6

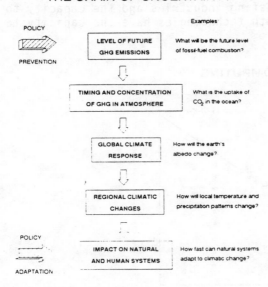

GLOBAL WARMING
THE CHAIN OF UNCERTAINTY

Note GHG is greenhouse gases

THE POTENTIAL FOR ENERGY SAVINGS

The Long-Term Technological Push

Industrial development since the 18th century can be considered to have
taken place in three phases:

		Key Technologies
I	Late 18th and 19th century	Steam engine and iron smelting. Coal provided motive power and heat for railways and factory steam; iron and steel were principal materials
II	Late 19th and 20th century	Electric motor, turbine generator, internal combustion and jet engine, metallurgy, chemicals and hydrocarbons. Oil, especially, the basis for mobile mass transportation.
III	Late 20th and 21st century	Micro electronics/information technology, materials, biotechnology, catalysis, ...

The first two phases had a system of core technologies which interrelated and reinforced one another. Energy use and interfuel competition have been strongly influenced by contemporary technology; the preferred energy form was closely related to the dominant core technologies.

In the technological era we now enter there is much speculation as to which will be the dominant technologies. But undoubtedly information technology[3] and new materials[4] will strongly influence the energy sector. Both offer potential to transform existing industries and the capacity to create new ones (Figure 7 and 8). Both technologies have the capacity to bring about

Figure 7

A CENTURY OF COMPUTING

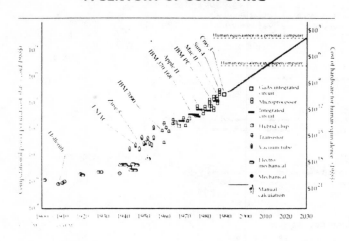

Figure 8

A CENTURY OF ENGINEERING MATERIALS

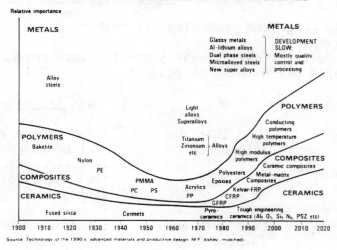

[3] 'Mind Children', H Moravec, Harvard Univ. Press, Cambridge, Mass. 1988.

[4] 'Technology of the 1990s: Advanced Materials and Predictive Design'.
M F Ashby, Phil Trans Royal Society, London, A322, pp 393 to 407, 1987.

significant reduction in energy use. Their impact on the design and configuration of energy-using artifacts, such as cars, aircraft and buildings could be very substantial[5].

Improvements in Energy Efficiency by 2010

Of the 156 million b/doe (mb/doe) of primary energy (including traditional fuels) consumed by the world in 1985, only 101 mb/doe was delivered to consumers. The major loss on the way to consumers is due to conversion in electricity generation but significant amounts are consumed in gas and electricity transmission and non-energy uses (Figure 9). Delivered energy is converted to useful energy, necessary for the enjoyment of some particular energy service (Figure 10).

Figure 9

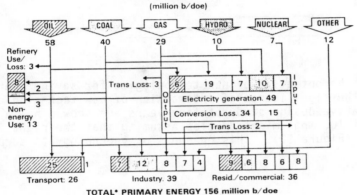

Figure 10

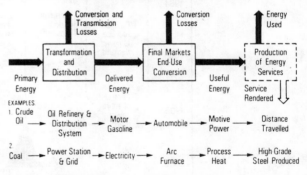

The potential for improving efficiency is along the whole chain from primary energy consumption to delivered energy services. Our studies of end-use technologies indicate that we could improve average efficiencies by 10 to

---

[5] 'Technological Development, Effect on Demand and Use of Oil and Gas'. G R Davis, Sanderstølen Energy Policy Seminar, 1988.

77

25 per cent by employing today's best available practice and save this much again within 20 years (Figure 11) as 'best-practice' improves.

Figure 11

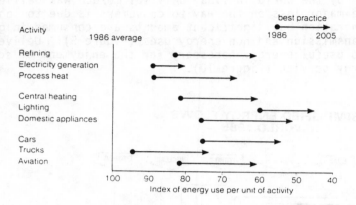

POTENTIAL ENERGY EFFICIENCY IMPROVEMENTS – OECD COUNTRIES

Many estimates of the potential of new technologies for saving energy tend to underestimate the significance of interaction of new technologies. For example, a speculative design study of five years ago showed how a possible 'average US car of 100 mpg' might develop assuming that the right incentives were put in place (Figure 12). This openness of technology to surprises

Figure 12

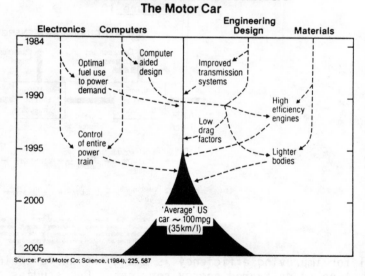

TECHNOLOGICAL SYNERGY
The Motor Car

Source: Ford Motor Co: Science. (1984), 225, 587

tends to make prediction of potential gains very uncertain. Developments in materials, combustion technologies, information technology and in other areas could interact to greatly reduce energy needs in residences, public

78

and commercial building and factories (Figure 13).

Figure 13

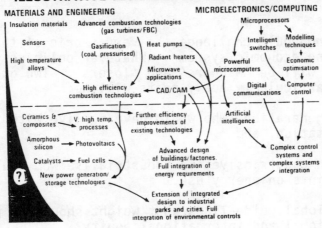

**ENERGY MARKETS
ILLUSTRATIVE TECHNOLOGY DEVELOPMENTS**

In the recent study[5] (earlier referred to), we examined the potential for energy saving in the World outside Centrally Planned Economies. We used as a reference case a scenario with strong economic growth (3 per cent per annum GDP growth in OECD and 5 per cent in the developing countries) and increasing energy prices. By widespread application of '1990 Technology' fuel efficiencies would rise 0.7 per cent per annum through turnover of plant in industry and power generation and energy-using artifacts in end-use. However, if 'High Technology' equipment were introduced in the 1990s average fuel efficiencies would improve at 1.7 per cent per annum (Table 1). In the 'High Technology' world, developments would include average efficiencies of new cars reaching 60 mpg by 2010 (Figure 14) and prop-fan jets becoming commercial in the early 1990s. This may be faster than the present consensus, but could be achieved if adequate incentives were put in place.

Table 1   Figure 14

TECHNOLOGICAL CHANGE AND WORLD ENERGY MARKETS

	Efficiency Improvements (% per annum)		
	Direct Markets		Primary Energy
	Physical Basis	Economic Basis	Economic Basis
1970 to 1985	2.0	1.9	1.1
1985 to 2010:			
- "1990 Technology"	0.7	1.7	1.2
- "Reference Case"	1.2	2.2	1.7
- "High Technology"	1.7	2.7	2.0

Note   1 World excludes Centrally Planned Economies
       2 Physical basis is weighted sectoral fuel efficiency indices
       3 Economic basis is GDP per unit of energy consumption

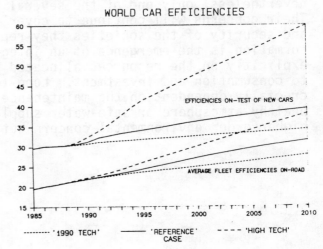

WORLD CAR EFFICIENCIES

EFFICIENCIES ON-TEST OF NEW CARS

AVERAGE FLEET EFFICIENCIES ON-ROAD

-------- '1990 TECH'   ——— 'REFERENCE' CASE   - - - - 'HIGH TECH'

79

HARNESSING ENERGY SAVING TECHNOLOGIES

The Policy Context

Global warming issues are complex and pose a number of basic dilemmas for policy makers:

- Global warming could challenge the very fabric of the world's ecological and economic systems. Whatever policies are chosen there will be 'winners' and 'losers'. Two groups who could bear particularly heavy costs will be:

   Future generations who would have to live with the costs of adaptation, and

   Those in countries yet to industrialise who would face constraints on energy use.

   In framing global guide-lines what weight should be placed on intergenerational and international equity?

- Action or inaction on implementing global warming policies in one country will affect the many.

   How in a world of nation-states can we ensure that policies will be implemented to resolve this global problem?

- We are relatively ignorant of the cause-effect mechanisms that relate GHG emissions to regional consequences; although for some the issue is only one of timing.

   How should we allocate resources between prevention and adaptation?

Even with global warming as a major public issue it will remain, nevertheless, only one of the several challenges faced by policy-makers. Their concerns will continue to cover the many issues of economic well-being and security of the societies they represent. Of importance to policy formation is the emergence of an integrated approach which focuses explicitly on the resources allocated to environmental maintenance compared to consumption and investment. Long-term economic prospects are viewed as crucially dependent on the maintenance of environmental resources, such as a healthy atmosphere or safe-water supplies. Global warming may be a higher-order environmental concern but might reasonably fit into this policy framework shown in Figure 15. This could assist in addressing the dilemmas raised above.

Figure 15

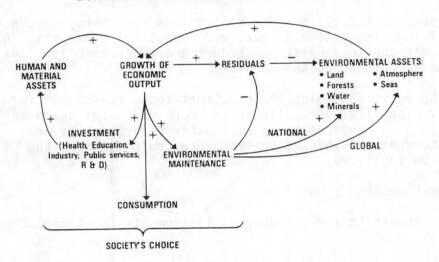

ENVIRONMENT AND ECONOMIC GROWTH: LINKAGES

However, if global warming is identified as the vital challenge to policy-makers the dilemmas will need to be addressed. A minimum response should include:

- **More knowledge** to reduce the uncertainties where possible and to identify those areas of irreducible uncertainty.

- **Some institution building** both formally and informally. This would include not only the move towards global agreements between nations, including the need to establish world-wide monitoring of emissions, but the establishment of wider networks between interested groups, such as energy and environmental professionals.

- **Taking a position on the equity issue.** At the root of this issue lies the question of personal and societal values. Finding the 'golden mean' in this area will not be easy. However, with energy/capita use in the developing countries one-tenth that of the OECD countries we can expect that any major programme to constrain fossil fuel combustion will require substantial assistance from the developed countries.

- **Some agreement on appropriate prevention levels.** Given the likelihood of a variety of views on the urgency needed to tackle the control of GHG emissions, some nations may well take unilateral action and embark on 'accelerated' prevention programmes while others lag behind. The most useful contribution of the 'leaders' may well be to develop relevant technological options which can both reduce GHG emissions and meet the cost-benefit criteria of the 'laggards'.

Even if public support for global warming policies is limited, short-term concerns over acid rain, urban air pollution and energy security may provide the arguments for introducing policies which would release the potential of energy-efficient technologies.

Mobilising Options

Virtually every investment decision has implications for energy use, e.g. the design of buildings, selection of cars and industrial processes. As indicated above the technology exists to reduce energy consumption levels if we are prepared to accept the costs.

Even at present the micro-economics of investment seems reasonable for fluorescent lamps, high-miles-per-gallon cars, more efficient space heaters and appliances, although many consumers have other priorities or are influenced by factors others than economics. How could consumer and producer response be mobilised?

■ Policy and Instruments

Policy makers have access to a wide range of instruments for influencing markets (Figure 16).

Figure 16

ENVIRONMENTAL POLICY INSTRUMENTS

INSTRUMENT	APPLICATIONS (examples)
Economic Instruments	
Taxes (polluter pays principle)	☐ Greenhouse Gases ☐ Acid Deposition
Subsidy/Tax relief	☐ Energy conserving investments ☐ Energy research (private) ☐ Unleaded fuels; catalytic converters
Direct public investment	☐ Afforestation programmes ☐ Energy research (private)
Grant aid	☐ LDC afforestation, soil conservation, agrarian reform and land settlement programmes ☐ Regional programmes in industrial countries
Law	
Public regulation; Law; legal liability	☐ CFCs ☐ Toxic wastes ☐ Community property rights ☐ Unleaded fuels; catalytic converters
Negotiation	☐ Localised issues: Emissions; wastes; compliance with local land use and building codes
Values	
Self-regulation	☐ 'Green' consumers and producers
Positive Side-effects of Policy	
Economic policies with environmental benefits	☐ Urban congestion charges ☐ Forestry Policy (LDCs) ☐ Agrarian reform (LDCs)

In modern, open societies all market systems operate within a regulatory framework established by law and government and the values of their citizens. This framework includes the setting of legal standards, regulations and safeguards and is enforced by legal liability. In environmental matters, voluntary negotiations by producers, communities and individuals on safeguards, standards and matters of compensation are a significant aspect of this regulation. In addition, the government's capacity to tax subsidise and grant aid offer a direct way of modifying incentives in the market.

We can anticipate that these instruments will continue to be further developed to tackle environmental problems. A key tenet in formulating such policy is likely to be the 'polluter pays principle'. Ideally, in implementation, this would require consistency of policy and uniformity of compliance across countries. Clearly the enforcement of such a principle at a global level will be no easy matter.

■ National Energy Policies

Energy policies tend to focus on the production and transformation of energy; much less on useful energy and energy services. In an energy-efficient world one might expect the establishment of incentive systems that promote the effective supply of energy services. Provision of energy services (whether comfort in the home or personal mobility) is influenced by a host of factors from building and city design to the structure of the transport system. Matters such as infrastructure development and city planning should be considered as important components in a 'National Energy Services Plan'. A more detailed understanding of the factors which determine a society's demand for energy services would be valuable in identifying new options for saving energy.

■ Response Times

Energy systems are large and complex with considerable momentum. Much of the energy-using equipment has lives of 10 to 25 years, although retrofitting is always possible.

When there is a strong public commitment to action on environmental issues and technology is available response times can be fast (Figure 17). A key element is the existence of organisations capable of mobilising know-how

Figure 17

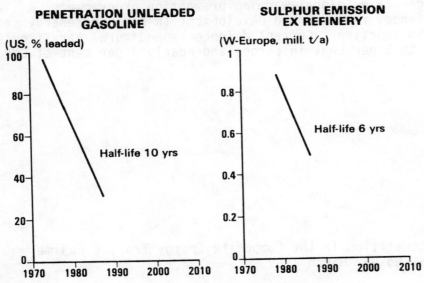

ENVIRONMENTAL CHALLENGES
RESPONSE TIMES

and capital to tackle the relevant problems. This channelling of resources in cost-effective ways is a major contribution of industrial corporations.

In this regard in some countries, for example Japan[6] (Reference 6), energy industry strategies, strong interfuel competition and a services orientation could lead to many end-use innovations relevant for an energy-efficient economy (Figure 18).

Figure 18

JAPANESE ENERGY INDUSTRY STRATEGIES TO COMPETE IN THE "COMPOSITE ENERGY ERA"

Oil industry

o Improve quality of petroleum products
o Target sales to "total energy systems", district heating/cooling and building heating/cooling
o Develop oil-burning home heating equipment and "clean" kerosene supply systems, and (longer term) oil-using fuel cells, stratified engines and ceramic gas turbines

Gas industry

o Implement new pricing systems to meet more diverse customer needs
o Develop/improve cogeneration systems, gas cooling/heating equipment, gas engines and heat pumps, and multi-functional and smaller household equipment

Electricity industry

o Stabilize electricity prices and implement new pricing systems to meet more diverse customer needs
o Offer new services including equipment supply, after-sale service, and total energy management/maintenance packages
o Develop/improve multi-functional heat pumps, storage systems, and new types of electrical heating/cooling systems

Many of these innovations will be important for developing countries. Organisations which can diffuse their technologies from OECD countries towards the regions of high energy growth will be of significance.

■   Costs

There are no reliable estimates of the additional costs that might be needed to implement a large-scale 'global warming prevention programme'. A comparison with defence, research and development (R & D) and energy sector investment may be instructive. Current defence expenditures vary from 1 per cent in Japan to 3 to 5 per cent in Europe and nearly 7 per cent of GNP in

---

[6] 'Inter-energy Competition in the Composite Energy Era', K Fujime, Energy in Japan, No 96, March 1989.

USA (Table 2). This can be compared with R & D expenditures of $2\frac{1}{2}$ to 3 per cent and total investment in the energy sector of some $1\frac{1}{2}$ per cent GNP in OECD. One might envisage a major prevention programme costing this order of magnitude. If so with investment levels at 20 per cent of GNP (~ US$ 3 trillion in OECD) such resource allocation could be handled without crippling economic growth.

Table 2

DEFENCE EXPENDITURES IN 1986

| | Total Defence Expenditures | | R & D Expenditures | |
	US$ bn	% GNP	All purposes % GNP	per cent for military purposes
US	288	6.7	2.7	27.8
Japan	20	1.0	2.8	0.6
Germany	28	3.1	2.8	
France	28	3.9	2.5	13.5 *
UK	27	4.7	2.8	
LDCs	75	3.3	n.a.	n.a.

* An average for Europe

Sources: Institute of Strategic Studies (1988) **The Military Balance**
World Bank **World Development Report, 1987**
Dosi (1988)

CONCLUSION

Because of the uncertainties surrounding the issue of global warming, a period of debate about policy actions is likely. However, if there is any further evidence of the adverse effect of rising GHG concentrations in the atmosphere, then harnessing technologies to save energy would be a priority target for policy makers.

The costs for prevention programmes could appear high but the benefits of effective joint action may well be higher. If nations can forge a common vision and purpose to tackle this problem, then a basis for progress on a wider agenda may have been laid.

ENERGY EFFICIENCY AND GLOBAL WARMING

Alan J. Streb
Deputy Assistant Secretary
for Conservation
U.S. Department of Energy

A. INTRODUCTION

The prospect of global warming resulting from increasing atmospheric concentration of the greenhouse gases has emerged as one of the most serious environmental issues of our time -- one with significant international implications. It is therefore appropriate that we look toward cooperative international efforts to identify technology options to help mitigate this problem. While considerable scientific uncertainty remains on the timing and potential impacts of global warming, it is clear that greenhouse gas concentrations continue to increase. Government policymakers are confronted with the task of using the limited available information to determine appropriate responses to a problem of uncertain timing, magnitude and impact.

There is a growing scientific consensus that the largest contributors to the global warming threat are the greenhouse gases that result from energy production and use. Accordingly, attention has focused on trends in global energy production and use and the factors that can affect these trends. The search for technology solutions, particularly more efficient energy use and fuel substitution, is recognized as a key element in any worldwide greenhouse mitigation strategy.

The so called greenhouse gases, CO_2, CH_4, O_3, N_2O and chlorofluorocarbons (CFC's) are emitted from both natural and man-made (anthropogenic) sources. Among these gases, carbon dioxide is the largest contributor to the greenhouse phenomenon. While mitigation of CO_2 is the focus of this paper, we recognize that energy efficiency and fuel switching measures that reduce fossil fuel demand may also reduce emissions of other key greenhouse gases (CH_4, and N_2O). While actual emissions of these secondary gases are relatively small, their ability to absorb and reradiate infrared radiation is much higher than for carbon dioxide. When considered in conjunction with emissions from other anthropogenic sources, reducing emissions of these secondary gases may become increasingly important.

Over the past one and a half centuries, atmospheric concentration of CO_2 has risen about 25 percent with half of this occurring in the last 50 years. Future concentration levels will be affected by a variety of factors, including trends in global energy intensities - the amount of energy consumed per unit output. To understand what contribution energy-efficient technology can make to reduce future CO_2 levels, it is important to first consider the current and likely future sources of greenhouse gases and then establish a baseline scenario of energy and economic growth, including an outlook for carbon dioxide emissions. From here, alternative energy efficiency scenarios can be evaluated and their effect on carbon dioxide emissions can be estimated.

The following sections of this paper describe the sources of greenhouse gases, the potential effect of these gases on global warming, and the growth rate of emissions and atmospheric concentration of CO_2 under three alternative energy efficiency scenarios. Technologies whose development would improve energy efficiency and reduce CO_2 emissions are then discussed. Finally, options for international cooperation to develop and promote energy efficiency measures are explored.

B.   SOURCES OF GREENHOUSE GASES

The potential for global warming occurs as a result of increased atmospheric concentrations of several gases, CO_2, CH_4, O_3, N_2O and CFC's, emitted from anthropogenic sources. These gases are termed "greenhouse gases" because of their ability to trap heat due to their high infrared reradiative potential. Among these, carbon dioxide is the largest contributor to current warming potential, accounting for roughly half of all contributions (Exhibit 1). Furthermore, studies estimate that this contribution could grow to 70 percent or more over the next century. Anthropogenic carbon dioxide results from energy production and use, deforestation, agriculture, and non-energy industrial activities, mostly cement production. By far, the dominant source of anthropogenic carbon dioxide is fossil energy production and use, accounting for about 70 percent of these CO_2 emissions (Exhibit 2). Accordingly, the reduction of fossil-based energy demand is critical to slowing current warming trends.

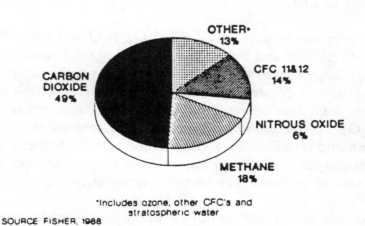

*Includes ozone, other CFC's and stratospheric water

SOURCE: FISHER, 1988.

Exhibit 1. Current Greenhouse Gas Contributions to Global Warming

88

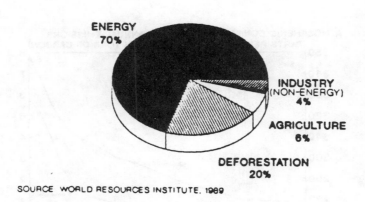

SOURCE WORLD RESOURCES INSTITUTE, 1989

Exhibit 2. Current CO_2 Emissions by Anthropogenic Sources

Before the beginning of the industrial revolution it appears that atmospheric carbon dioxide concentration had been nearly constant for millennia. Since the 1800's, global atmospheric concentration of carbon dioxide has risen. Exhibit 3 shows estimates of atmospheric concentrations over the past 250 years based on analysis of ice core samples (Neftel et al. 1985). Carbon dioxide emissions[*] from fossil fuels over the past 150 years are also shown in Exhibit 3, which illustrate the impact such emissions have had on increasing atmospheric concentrations (Rotty and Masters 1985).

Estimates of temperature change associated with past increases in greenhouse gas concentrations are harder to determine due to complexities in the relationship of the many factors that determine climate. Exhibit 4 provides one estimate of realized[**] temperature increases associated with selected time periods between 1850 and the 1980's, before consideration of climate feedback effects (Hansen et al. 1988). Cumulative temperature change due to all greenhouse gases since 1850 is approximately 0.4°C, with slightly over half

---

[*] Emission of CO_2 is expressed in units of metric tons of carbon emitted. Atmospheric concentration of CO_2 is expressed in parts per million.

[**] Temperature increases caused by global warming can be expressed in two ways. Realized warming is the actual temperature increase that occurs from greenhouse warming in a given year. Equilibrium warming is the temperature increase that would occur in equilibrium if atmospheric composition were to remain fixed in that year.

of this due to CO_2 alone. The temperature increase that is estimated to have occurred over the past three decades is the basis for concern among many scientists.

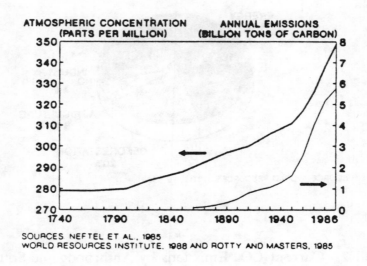

ATMOSPHERIC CONCENTRATION (PARTS PER MILLION) ANNUAL EMISSIONS (BILLION TONS OF CARBON)

SOURCES NEFTEL ET AL., 1985
WORLD RESOURCES INSTITUTE, 1988 AND ROTTY AND MASTERS, 1985

Exhibit 3. Global Atmospheric Concentrations and Annual Fossil Fuel Emissions of CO_2: 1740 - 1986

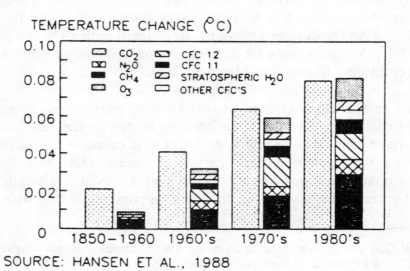

TEMPERATURE CHANGE ($^{\circ}$C)

SOURCE: HANSEN ET AL., 1988

Exhibit 4. Estimated Realized Global Surface Temperature Increases Attributable to Greenhouse Gas Emissions

90

The uncertainty in the ability of existing CO_2 sinks to absorb additional emissions, as well as uncertainties in other scientific parameters, make it difficult to evaluate the impact of energy use on global warming. Natural sinks of CO_2, which include ocean and plant uptake, are thought to have the capacity to absorb somewhat more CO_2 than is produced by natural sources, although the degree of this additional absorption capacity is unknown (Wuebbles et al. 1988). Factors that further complicate predictions of CO_2 concentrations are physical and biogeochemical climate feedback mechanisms, which are not well understood. Physical climate feedback mechanisms - cloud cover, atmospheric water vapor content, a reduction in the snow and ice cover - are expected to have the greatest warming impact; other warming effects are expected from biogeochemical mechanisms such as changes in global vegetation, ocean chemistry and biology.

Carbon dioxide is a normal byproduct of fossil fuel combustion, but not all fossil fuels release carbon at the same rate per unit of energy produced. The release rates are well known for most fuel types and must be considered in the development of any mitigation strategy. The carbon release coefficients for various fuels are shown in Exhibit 5. These values represent average release rates for each type of fuel. Also shown are the total 1986 global consumption levels for each fuel type.

Exhibit 5. Carbon Release Coefficients From Fossil Fuel Combustion and Oil Shale Mining

Fuel	Million Tons of Carbon per Exajoule*	1986 Total Global Consumption (Exajoules)
Gases	13.8	68.2
Liquids	19.7	128.0
Coal	23.9	93.1
Oil Shale Mining	27.9	Negligible

Sources: Edmonds and Reilly 1986, EIA 1988.

---

* One ton = one metric ton; one billion tons = 10^{15} grams = one petagram. One exajoule = 10^{18} joules = 9.48×10^{14} Btu = .948 quads.

C. THE BASELINE SCENARIO

To best illustrate what impact energy efficiency can have on reducing carbon dioxide emissions, it is useful to establish a baseline or reference scenario of future global activity. Several studies have attempted to characterize future global activity and determine the corresponding CO_2 emission levels. For the purposes of this paper, the Edmonds/Reilly model, developed at the Institute for Energy Analysis, is used to create a baseline reference scenario (Edmonds and Reilly 1985, Edmonds et al. 1986, Edmonds et al. 1984). This model was developed in the late 1970's and has evolved to better reflect global economic and energy systems. It has been widely used over the past ten years in a number of energy studies and has been modified by other modelers to study the impact of a variety of factors. When run using the default assumptions, the model provides a future global outlook that is intended to represent a likely "business-as-usual" scenario.

1. Energy Demand Assumptions

The major factors that determine energy demand include population, economic activity (GNP), energy prices, the relationship between energy consumption and economic growth, energy efficiency levels, and government policy. Due to the large uncertainties in predicting these factors, the model should be considered as a tool for evaluating consistent scenarios and measuring the relative impacts of changes in input factors. Exhibit 6 lists assumptions about several important factors that affect energy demand in the baseline scenario.

Exhibit 6. Key Global Energy Demand Assumptions - Baseline Scenario

Factor	Assumption
Population	8.2 billion globally in 2050
GNP Growth Rate	2.4% aggregate annual increase
End-Use Energy Efficiency Improvement Rate	1.0% annual improvement

Source: Edmonds and Reilly 1986.

Energy efficiency is an important factor that can affect future greenhouse gas emissions. The baseline scenario assumes a one percent annual improvement in end-use efficiency that is intended to reflect the historical decline in the energy intensity of activities in the end-use sectors. This improvement rate captures changes in technology, the mix of goods and services in the economy that use energy, and stock turnover (i.e., replacement of equipment,

92

appliances, vehicles, and housing). The energy efficiency improvement rate is speculative because of uncertainties in the state of technology, in consumer preferences, and in energy policies throughout the world. One study that examined uncertainty factors in future global energy use found the rate of energy efficiency improvement to have the second highest uncertainty of 79 factors that affect future CO_2 emissions (Edmonds et al. 1986).

2. Energy Supply Assumptions

The energy supply assumptions in the baseline scenario determine the potential availability of each fuel source. The technologies and costs associated with exploiting each resource are included in the analysis; other supply factors, such as pollution and transportation, are considered if they affect the future viability of the resource. Exhibit 7 lists key supply assumptions about global energy resources by fuel type.

Exhibit 7. Key Energy Supply Assumptions - Baseline Scenario

Fuel Type	Estimated Resources
Oil	2.3 trillion barrels
Natural Gas	10,000 trillion cubic feet
Coal	Unconstrained in the foreseeable future
Oil Shale	Capable of yielding at least as much oil as from conventional resources
Tar Sands; Heavy Oils	Highly uncertain; base rate of production expansion assumed to be zero through 2050

Source: Edmonds and Reilly 1986.

3. Baseline Scenario Results

The baseline scenario generated using the Edmonds/Reilly model provides projections of energy demand, energy supply, and CO_2 emissions by fuel type for 1975, 2000, 2025 and 2050. Exhibit 8 shows annual fossil-based CO_2 emissions to 2050 for the baseline scenario. Emissions in 2050 are projected to be 15.0 billion tons of carbon per year, which is over three times the 1975 emissions rate.

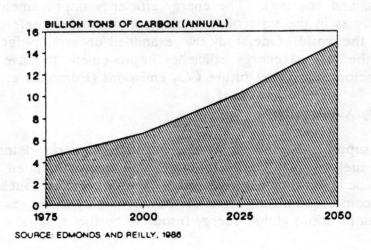

BILLION TONS OF CARBON (ANNUAL)

SOURCE: EDMONDS AND REILLY, 1986

Exhibit 8. Annual Global Fossil-Based CO_2 Emissions - Baseline Scenario

There is a wide diversity of views concerning future CO_2 emissions due to different scientific, economic and policy outlooks. While the baseline is used for illustrative purposes, it is instructive to see how it compares with other studies. Edmonds and Reilly developed two alternative scenarios to represent suitable upper and lower bounds for CO_2 emissions with confidence levels of at least 50 percent (Edmonds et. al. 1984). These scenarios, designated as Case A and C in Exhibit 9, are shown along with selected other studies.

The range of emissions estimates among forecasters underscores the uncertainty and controversy regarding future energy use and emission trends. This uncertainty stems from a combination of factors. Patterns of global economic development involve complex interactions among key factors such as political stability, labor productivity, technological innovation, capital formation, and the economic policies of nations, regions and the world. The complexity and unpredictability of these interactions contribute to the substantial uncertainty about future levels of energy production, use, and carbon dioxide emissions.

94

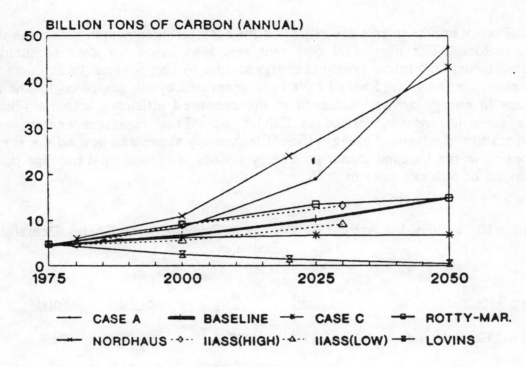

BILLION TONS OF CARBON (ANNUAL)

| | CASE A | | BASELINE | | CASE C | | ROTTY-MAR. |
| NORDHAUS | | IIASS(HIGH) | | IIASS(LOW) | | LOVINS |

SOURCES: EDMONDS AND REILLY, 1986
EDMONDS ET AL., 1984

Exhibit 9. Estimates of Future Fossil-Based CO_2 Emissions:
Baseline Scenario and Other Studies

D. ALTERNATIVE ENERGY EFFICIENCY SCENARIOS: TWO CASES

The baseline scenario provides the basis for comparing potential impacts of different energy efficiency levels on global energy demand and annual CO_2 emissions. Two cases are presented that alternately assume no end-use efficiency improvement and enhanced efficiency improvement using the best available, cost-effective technology.

The "no efficiency improvement" scenario assumes no net improvements in end-use efficiency will occur; any efficiency gains would be offset by efficiency losses elsewhere in the global economy. The impacts of this scenario are determined by running the Edmonds/Reilly model using an end-use efficiency improvement rate of zero percent.

The "enhanced efficiency" scenario reflects a global economy that makes full use of the best available, energy-efficient technology. It reflects the technically achievable energy savings potential assuming full implementation of all known and potentially cost-effective energy-efficient technologies, including those still under development. The estimate of potential efficiency improvement is derived from analysis of future economic and technical opportunities made by the staff of the U.S. DOE's Office of Conservation. This analysis is based on estimates of energy savings potential of established and emerging technologies in each of the end-use sectors.

Efficiency improvements are estimated for each technology area within the three major end-use sectors. The likely fuel type that would be saved for each technology area is considered in deriving future potential energy savings by fuel source. In addition, we assume that all electric energy saved would have been generated by oil, gas, or coal. The anticipated reductions in energy demand achieved in the enhanced efficiency scenario relative to the baseline scenario are summarized in Exhibit 10. They represent energy savings from implementation of advanced energy-efficient technology above and beyond the energy savings incorporated in the baseline scenario. Energy savings are based on a baseline primary fossil fuel demand of 803 exajoules in 2050.

Exhibit 10. Achievable Annual Reductions in Primary Fossil Energy Demand (2050) (Exajoules)

End-Use Sector	Liquid	Gas	Solid	Total	Baseline Demand
Residential/Commercial	33.5	37.0	56.2	126.7	297.1
Industrial	15.1	19.6	56.0	90.7	370.9
Transportation	59.0	0.0	0.0	59.0	134.9
Total Reductions	107.6	56.6	112.2	276.4	
Baseline Demand	263.6	170.1	369.2	802.9	

Source: Energetics, Inc. 1989.

The corresponding change in CO_2 emissions is calculated using the energy savings potentials in Exhibit 10 and the appropriate emission coefficients[*]. This results in a reduction of 6.4 billion tons of carbon per year by 2050 under the enhanced efficiency scenario when compared to the baseline scenario (Exhibit 11). Under the no efficiency improvement scenario, an additional 14.5 billion tons per year would be emitted by 2050 compared to the baseline scenario.

The second graph in Exhibit 11 shows atmospheric concentrations of CO_2 that would result from emissions under the baseline and the two alternative scenarios. Atmospheric concentrations would be 60 parts per million (ppm) lower than the baseline in 2050 under the enhanced efficiency scenario and 129 ppm higher than the baseline in the no efficiency improvement scenario.

---

[*] A portion of the liquid and gas fuels saved are assumed to come from coal liquefaction and coal gasification, which release additional CO_2 in the fuel production process.

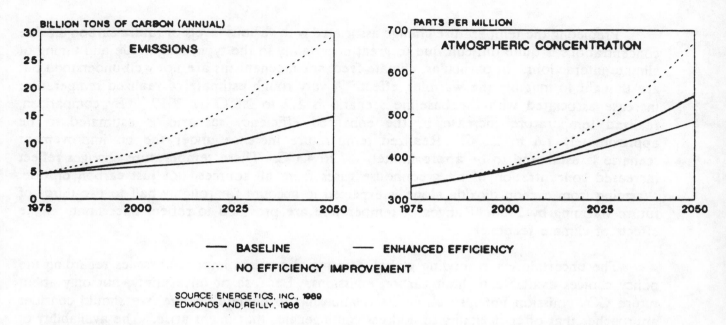

BASELINE ———— ENHANCED EFFICIENCY

···· NO EFFICIENCY IMPROVEMENT

SOURCE: ENERGETICS, INC., 1989
EDMONDS AND REILLY, 1986

Exhibit 11. Reductions in Annual CO_2 Emissions and Atmospheric Concentrations Achievable Through Enhanced Efficiency

The achievable reductions in CO_2 emissions by end-use sector and fuel type are provided in Exhibit 12. The fuel type that contributes most to the reductions is coal. This is partly due to the electricity used in the industrial and residential and commercial sectors that is generated using coal as a fuel. Reductions in oil use in transportation and residential and commercial buildings are also large contributors.

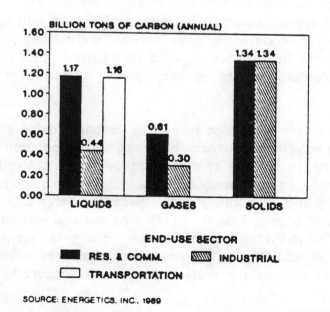

SOURCE: ENERGETICS, INC., 1989

Exhibit 12. Achievable Annual Reductions in CO_2 Emissions by End-use Sector and Fuel Type (2050)

97

Predicting the temperature change associated with various levels of future carbon dioxide concentrations is quite difficult due to great uncertainty in the type, magnitude and timing of climate interactions. In particular, climate feedback mechanisms are not well understood but are thought to magnify the warming effect. A very rough estimate of realized temperature increase associated with the baseline scenario is 1.9 to 2.8°C by 2050.[*] By comparison, realized temperature increase in the enhanced efficiency scenario is estimated to be approximately 1.6 to 2.5°C. Realized temperature increase under the no improvement scenario is estimated to be approximately 2.7 to 4.1°C. These temperature changes reflect increased concentration of all greenhouse gases from all sources, not just carbon dioxide. Warming from carbon dioxide alone is expected to account for roughly half to two-thirds of future warming by 2050. Ranges of temperature are provided to reflect uncertainty in the effects of climate feedback.

The uncertainty surrounding global warming also raises important issues regarding the policy choices available to limit carbon emissions. Because of uncertainty, not only about future CO_2 emission but also about its relationship to climate change, we should consider approaches that offer flexibility to address contingencies that might arise. The availability of advanced energy technology, which proved so important during the energy crises of the 1970's, may again become essential regardless of the warming mitigation strategies that are pursued. Continued R&D activities in advanced energy technologies on both the supply and demand sides need to be pursued as a key element of the response to global warming.

E. CONSERVATION OPPORTUNITIES FOR REDUCING FOSSIL-FUEL-DERIVED EMISSIONS OF GREENHOUSE GASES

The task of reducing global carbon dioxide emissions is complex and multi-faceted. A panacea is not likely to appear because the release of carbon into the atmosphere results from almost every aspect of modern economic activity: the manufacturing processes used to create goods and services; the vehicles required to transport goods and services; and the lighting, heating, and cooling needed to keep people comfortable in their homes and productive at work.

Of the options available to reduce man-made carbon dioxide emissions, the adoption of more energy-efficient vehicles, appliances, buildings, and manufacturing technologies offers several important advantages. Energy efficiency improvements can reduce emissions without reducing economic growth. In fact, many energy saving technologies have the added benefit of increasing productivity and economic growth. Because energy efficiency makes economic sense in a large number of applications, it can often be pursued without policy mandates that require forced choices by individuals or businesses. Energy efficiency improvements have environmental benefits in addition to the potential for reduced carbon dioxide emissions. Sulfur dioxide and other acid rain precursors can also be reduced. Ultimately, however,

---

[*] These temperature increases are relative to pre-industrial levels. It is estimated that approximately 0.5°C of greenhouse warming has already occurred since 1850.

human behavior will play an important role in how and to what extent efficiency improvements are made.

The estimated technically achievable energy savings of 276 exajoules per year by 2050, and the concomitant reduction in fossil-based carbon dioxide emissions of 6.4 billion tons, can be achieved through the use of advanced energy-efficient technologies that are briefly described in this section. These technologies are discussed in relation to the specific end-use sector areas to which they apply.

1. Industrial Sector

Energy savings in the industrial sector will be achieved through the use of advanced technologies that improve process energy efficiency and productivity. These technologies may also expand energy options through use of multiple fuels that allow switching, preferably to fuels that emit less carbon dioxide per unit of energy released. These new technologies will reduce waste energy in the form of heat and materials, allow simultaneous production of electric power and process heat, reduce energy requirements of industrial processes, and in some cases eliminate process steps. The estimates of global efficiency potential of these technologies are shown in Exhibit 13 by functional area. Examples of advanced technologies that will contribute to savings in energy are: ceramic composite reactor vessels and heat exchangers; advanced industrial heat pumps; innovative combustion systems; biotechnology and other conversion processes for upgrading industrial wastes to higher value products; stable electrodes for electrolytic processes; new separation technologies such as continuous chromatography, membranes, freeze crystallization, and chemical air separation; new metals, paper and textile processing technologies; advanced sensors and controls; and new high-pressure/high-temperature cogeneration systems.

Exhibit 13. Estimated Annual Energy Savings Potential
Industrial Sector (2050)

	Savings in Primary Fossil Energy (Exajoules)*
Improved Combustion Technologies	16.9
Waste Heat Recovery	
High Temperature (Enhanced Heat Exchangers)	15.5
Low Temperature (Heat Pumps)	14.9
Energy Cascading	
Topping Cycles	16.0
Bottoming Cycles	4.4
Industrial Waste Utilization	8.1
Improved Industrial Processes	
Process Electrolysis	1.4
Carbothermic Reduction	2.0
Catalysis	1.5
Separation and Concentration	10.0
TOTALS	90.7

Primary fossil energy demand includes fossil fuels used directly and fossil fuels used in the generation of electricity.

Source: Energetics, Inc. 1989.

Energy efficiency opportunities in the industrial sector can be associated with five functional areas: improved combustion, energy cascading, waste heat recovery, industrial waste utilization, and improved industrial processing. Opportunities in each of these areas are discussed.

Improved Combustion. Over half of the energy currently consumed in the industrial sector involves the burning of fuels to obtain thermal energy. Burning fossil fuels directly results in the production of greenhouse gases and other environmentally significant gases such as CO, SO_x, and hydrocarbons. Improved combustion technologies that affect CO_2 should also affect the other gases that exist in equilibrium in the atmosphere and influence global warming. Advanced combustors will eliminate or reduce emissions of particulates, carbon monoxide, sulfur oxides, other hydrocarbons, and nitrogen oxides while maintaining high efficiency.

Energy Cascading. Cogeneration, the simultaneous production of electric power and heat captures more useful energy from fuels than can be achieved when they are produced separately, thus increasing energy productivity and efficiency. Many industries do not take full advantage of cogeneration technology because the electric-to-thermal output ratios of current systems are not high enough to meet their needs. In addition to advanced cogeneration technology, direct conversion of thermal energy to electrical energy will improve future industrial energy efficiency.

Advanced cogeneration -- The higher electrical-to-thermal ratios of advanced cogeneration systems could achieve large energy savings for industry and provide excess electrical power for sale. These advanced systems will employ efficient prime movers that can be sized to satisfy the varying plant thermal loads and be interconnected with the grid.

Direct energy conversion technologies -- Direct thermal-to-electric conversion will improve the overall efficiency of conversion systems by allowing electricity to be generated from thermal streams at various temperatures as well as from waste heat. The power provided by these types of devices may be used in electrolysis and other industrial processes.

Waste Heat Recovery. Much of the vast amount of waste heat generated by industry at a wide range of temperatures is recoverable. Improved heat exchangers will continue to be developed to recover heat at high temperatures. Waste heat released at temperatures too low for direct use in processes will be recovered using advanced heat pump technology.

Advanced heat exchangers -- Advanced recuperators and other equipment such as process reactors made of ceramic materials will be able to withstand the high temperatures and corrosive environments typical of many industrial processes. Improved heat transfer will result from the use of surface modification techniques that increase the overall heat transfer area and heat transfer coefficient. Enhanced effectiveness will be possible with liquid-to-liquid, liquid-to-gas, and gas-to-gas heat exchangers.

Advanced heat pumps -- Innovative chemical and magnetic heat pumps offer greater temperature lift and a wider operating range to meet industrial needs. These advanced technologies can replace heat pump and refrigeration equipment that currently use CFC's as working fluids.

Industrial Waste Utilization. Essentially every stage of industrial production, from raw material input through product distribution and servicing, produces wastes. Combustion of these wastes is the least desirable use of their embodied energy. New processes using biotechnology, membrane systems, and innovative pyrolysis techniques will allow conversion of industrial wastes to higher value products or feedstocks. Plants will review their relevant processes to arrive at near-zero discharge of wastes. An example of industrial waste utilization with particular relevance to global warming is a new carbon dioxide recovery process.

Carbon dioxide recovery process -- A novel process that will recover CO_2 from industrial furnaces has been developed that is more energy-efficient than conventional CO_2 recovery or generation technologies. The process replaces normal combustion air in large industry and utility furnaces with a mixture of CO_2 and O_2 to combust the fuel. The resulting flue gas, which is about 90 percent carbon dioxide, can be used directly or employed in higher-valued applications following purification. Seventy-five percent of the flue gas is returned to the furnace and mixed with 25 percent oxygen to repeat the cycle. The remaining 25 percent CO_2 (undiluted by the nitrogen that would exist in normal air-fuel combustion) is available for economical pipeline transport to oil fields for enhanced oil recovery, the largest volume use. Carbon dioxide can also be used for industrial refrigeration, as a fire extinguishing agent, and for carbonation of beverages. This process is competitive with the best current technology for recovery of CO_2 from flue gases, but limited markets make it unlikely that this could have a very significant impact on a global scale.

Improved Industrial Processing. There are numerous potential improvements in unit operations that could contribute to overall reduction in the energy-intensiveness of production and manufacturing if adopted industry-wide. The areas of opportunity include: process electrolysis; carbothermic processing; chemical catalysis; separation and concentration systems; and advanced sensors and process control systems.

Process electrolysis -- Industrial materials such as aluminum, magnesium, sodium, chromium, manganese, copper, and chlorine are produced by energy-intensive electrolytic reduction processes. Nonferrous metals reduction and chlorine production processes typically operate at energy efficiencies of less than 50 percent. Improved electrolytic reduction processes using advanced anode and cathode materials, as well as improved electrolytic cell designs, will significantly increase process efficiency.

Carbothermic processing -- Energy-intensive carbothermic processes are common in iron and steel production. Energy consumption will be reduced through better understanding of the chemical reactions that occur in new steel reduction processes, which replace the blast furnace and eliminate cokemaking. The carbothermic reduction of aluminum oxide to produce aluminum metal may replace the use of the energy-intensive electrolytic cell.

Chemical catalysis -- Catalysis represents a significant opportunity for energy efficiency improvement in industrial applications employing chemical reactions, including chemical production, petroleum refining, and pharmaceutical manufacturing. Advanced catalysts will allow chemical processes to proceed at lower operating temperatures and pressures and at increased reaction rates, thus improving the energy efficiency and productivity of many processes.

102

Separation and concentration systems -- The separation and concentration of one or more chemicals or components in a solution or mixture are critical steps in many industrial processes. Most industrial separation processes, including distillation, evaporation, drying, absorption, adsorption, and extraction are highly energy-intensive and use much more energy than is thermodynamically required to perform the separation. For example, an estimated one-third of the energy consumed in petrochemical processing is used for separations, principally by distillation. Alternative advanced technologies that will be less energy intensive include membrane separation, freeze concentration, chemical separation, and advanced dewatering and drying concepts.

Advanced sensors and control systems -- Advanced process control technologies will be a significant factor in improving industrial energy efficiency. More precise and efficient process control systems will be possible with sensors that can operate in the environmental extremes of many industrial processes. Innovative new sensors, coupled with existing data processing capabilities and artificial intelligence techniques, will replace subjective judgments. Improved process monitoring and enhanced real-time control will increase energy productivity and decrease the amount of waste and pollutants associated with the production process. Energy conservation opportunities for sensors in high-temperature processes include measuring melting temperature and heat flux in glassmaking and measuring alumina saturation in aluminum production. Examples of low-temperature applications include measuring chemical compositions and agricultural growth conditions.

2.   Residential and Commercial Sector

Improvements in the energy efficiency of building systems and building energy conversion equipment will lead to significant energy savings, especially in the form of electricity, in the residential and commercial sector. The global conservation potentials in the year 2050 for the major uses of energy in this sector are shown in Exhibit 14. Achieving these potentials could result in reduced overall energy demand in the buildings sector despite expected sectoral growth. Key new technologies that will contribute to improved energy efficiency include: innovative and thermally efficient building materials and components; efficient combustion heating systems; thermally activated heat pumps; advanced refrigeration systems; improved building thermal distribution systems; advanced lighting concepts; and district heating and cooling systems.

Exhibit 14. Estimated Annual Energy Savings Potential
Residential and Commercial Sector (2050)

	Savings in Primary Fossil Energy (Exajoules)[*]
Space Heating	57.4
Space Cooling	23.7
Water Heating	7.9
Refrigeration	7.9
Lighting	17.2
Other	12.6
TOTALS	126.7

[*] *Primary fossil energy demand includes fossil fuels used directly and fossil fuels used in the generation of electricity.*

Source: Energetics, Inc. 1989.

Energy savings opportunities in the residential and commercial sector occur in four major functional areas: efficient space heating and cooling, advanced refrigeration systems, improved water heating, and improved lighting.

Efficient Space Heating and Cooling. Reducing energy demand for space heating and cooling of residential and commercial buildings can be achieved through the use of improved building materials and components, increased understanding of the interactions of building subsystems, and advanced space conditioning equipment.

Improved building materials and components -- The thermal efficiency of the building envelope can be improved by use of innovative and thermally efficient walls, roofs, foundation systems, insulation, and windows with improved capabilities to regulate heat transfer. The use of vapor retardants, variable and switchable insulation materials, composite materials, sealants, coatings, and other advanced materials will also contribute to increased efficiency. In the area of insulation, new thermally insulating materials that do not use CFC's in their manufacture will be developed. Evacuated panels filled with fumed silica powder of varying sizes may be used for both appliance and building insulation applications.

Building subsystem integration -- Integration of the various subsystems of a building (e.g., heating, ventilating, air conditioning, lighting) can optimize building energy performance. Factors such as subsystem interactions, climate, building characteristics, and occupant behavior will be integrated to reduce space conditioning loads.

Advanced space conditioning equipment -- Further reductions in the energy demand for space heating and cooling applications will be accomplished with more efficient space conditioning equipment, including advanced furnaces, heat pumps and air conditioning systems. Current oil and gas heating systems typically have seasonal heating efficiencies that range from 63 to 75 percent. Future combustion heating systems will operate at much higher efficiencies for all types of fuels. The use of thermally activated heat pumps, in which the electrical compressor is replaced by a system that burns fuel to drive the refrigeration cycle, will allow the substitution of natural gas and liquid fuels for electricity in space conditioning applications. Waste heat from the fuel combustion process will be used for additional space or water heating. In the heating mode, such heat pumps can deliver two or more units of thermal energy for each unit of fuel energy, thus potentially achieving a three-fold increase in energy efficiency over combustion heating systems.

Advanced Refrigeration Systems. Significant technological changes in refrigeration systems will result in energy savings of one-third or more over currently available technology in many applications. Advanced electrically driven refrigeration (and heat pump HVAC systems) with capacity modulation capabilities will optimally match system performance to cooling/refrigeration demand. The thermodynamic performance of the refrigeration cycle will be improved through the use of new nonazeotropic refrigerant mixtures instead of pure CFC fluids. Such mixtures can result in less ozone depletion than conventional refrigerants.

Improved Water Heating. The use of improved oil and gas burners will make advanced water heaters more energy efficient than current heaters. Compared to conventional oil and gas burners with 70 to 80 percent efficiency, combustion efficiencies of up to 95 percent can be achieved with the use of liquid- and gas-fueled pulse combustors. Heat pumps will be used to provide supplemental energy to water heating systems for improved energy efficiency. In some applications, a heat pump water heater can reduce energy requirements by a factor of three or more while providing dehumidification and space cooling at the same time.

Improved Lighting. Lighting currently accounts for over one-fourth of commercial building energy use. Conventional incandescent and fluorescent lights operate far below the theoretical efficiency limit for a white light source (approximately 350 lumens/watt). Light sources capable of achieving operating efficiencies of close to 200 lumens/watt will significantly reduce energy consumption in buildings. Advanced lighting technologies will include magnetic phosphor and two-photon phosphor fluorescents, isotopically enriched fluorescents, and electrodeless and surface wave high intensity discharge lamps. Widespread use of advanced technologies could reduce lighting energy requirements by a factor of three or more.

3. Transportation Sector

Energy savings opportunities in the transportation sector will be realized with advanced, energy-efficient and fuel-flexible engine options, more efficient power transmission systems, adoption of alternative fuels, and electric and hybrid vehicles. Advanced engine technologies will be at least 30 percent more fuel-efficient than current automotive engines and will be capable of meeting stringent emissions, safety, and noise standards. Exhibit 17 shows the potential global savings in liquid fuels for the transportation sector in the year 2050. This represents potential petroleum saving opportunities within three technological areas of the transportation sector: advanced heat engines; electric and hybrid vehicles; and advanced vehicle design.

Exhibit 17. Estimated Annual Energy Savings Potential
Transportation Sector (2050)

	Savings in Primary Fossil Energy (Exajoules)*
Advanced Heat Engines	37.0
Electric and Hybrid Vehicles	17.0
Advanced Vehicle Design	5.0
TOTAL	59.0

* *Primary fossil energy demand includes fossil fuels used directly and fossil fuels used in the generation of electricity.*

Source: Energetics, Inc. 1989.

Advanced Heat Engines. Technologies that will reduce energy use and allow the use of alternative fuels in the transportation sector include advanced internal combustion engines and gas turbine engines for automobiles, and high-efficiency adiabatic diesel engines for heavy-duty truck and rail applications. Piston engines will most likely continue to evolve and remain a major consumer of fuels in the sector. New, lightweight ceramic and ceramic composite engine components will enable engines to operate at higher temperatures, at higher thermal efficiency and for longer lifetimes.

Ceramic components are already developed and being used in current production engines because of their reduced mass, reduced wear, lower friction, and greater resistance to corrosion. These applications of ceramic components will allow current engines to have improved performance, improved fuel economy, reduced emissions, and reduced maintenance.

Advanced internal combustion engines -- These engines will be significantly more fuel-efficient than currently available engines. One example is the Direct-Injection Stratified-Charge (DISC) engine. This engine approaches diesel engine efficiencies without the corresponding problem of particulate emissions. The advanced DISC engine will have spray characteristics tailored with load, will reduce cycle-to-cycle variation, and will control high unburned hydrocarbon emissions. Alternate fuel technology is under development to allow the use of fuels such as compressed natural gas (CNG) and methanol in modified conventional engines.

Gas turbine engine -- Gas turbine heat engines for automotive applications will be somewhat similar to gas turbines used in electric power generation and aviation. Low-maintenance, multifuel ceramic turbine engines will operate at higher temperatures than conventional heat engines with fuel efficiencies that could be 30 to 50 percent higher. In addition to energy savings through increased efficiency, the ability of these engines to use a variety of alternative fuels that emit less CO_2 will reduce petroleum consumption and CO_2 emissions in the transportation sector.

Adiabatic diesel engine -- A low-heat-rejection diesel engine for heavy duty trucks and rail locomotives will operate approximately 30 percent more efficiently than conventional engines. The adiabatic diesel engine, which requires no water cooling, will have thermally insulated combustion chambers and exhaust gas passages. Recovery of waste heat from the exhaust contributes to its improved fuel efficiency.

Electric and Hybrid Vehicles. Electric and hybrid vehicles have the potential to alter the almost total dependence of the transportation sector on petroleum-based fuels. Electric vehicles use rechargeable, battery-driven propulsion systems; hybrid vehicles combine the features of electric and internal combustion vehicles. The use of full-performance electric and hybrid vehicles, equipped with advanced batteries and propulsion systems and powered with electricity generated from non-fossil fuel resources, will reduce both petroleum consumption and CO_2 emissions in this sector.

Advanced batteries -- New rechargeable batteries will provide electric vehicles with the necessary range, cost-effectiveness and reliability to become competitive with conventional heat engine-driven vehicles. Advanced batteries under study that may meet the performance requirements include sodium-sulfur, zinc-bromine, lithium-aluminum-iron sulfide, and iron-air.

Advanced propulsion systems -- Advanced controls and electric-to-mechanical energy conversion equipment for use in electric and hybrid vehicles will reduce battery storage requirements by up to 40 percent. The integration of the battery and propulsion systems will be optimized for maximum energy efficiency. Hybrid electric propulsion systems using fuel cells, which have higher maximum theoretical efficiencies than heat engines, offer a potential opportunity to achieve long vehicle range combined with the benefits of electric propulsion.

Advanced Vehicle Design. Fuel economy will be improved by reducing engine power requirements through vehicle downsizing and materials substitution. Downsizing will be accomplished through more efficient use of the vehicular envelope, including: closer packaging of the main vehicle components; reducing interior waste space; shortening the trunk overhang; and reducing styling-induced waste space. Further weight reduction will be accomplished through the extensive use of lightweight materials such as high-strength, lightweight alloys and composite materials.

4. Infrastructure and Integrated Energy Planning

The way in which the world's infrastructure -- highways, railways, utilities, and cities -- is planned and provided could have a substantial impact on energy use and carbon dioxide emissions. This is particularly important for emerging economies whose infrastructure is just developing. Making energy efficiency a priority in land use, community, utility and transportation planning could lead to substantial reductions in carbon dioxide emissions.

Mass transit, for example, is a more efficient mode of transport for people, goods, and services. New technologies such as magnetically levitated trains can increase the desirability and speed of mass transit to and from work and between cities. Planning for the maximum use of telecommunications and computers is another way to reduce the growth of automotive and air travel.

Taking advantage of energy efficiency opportunities within cities and communities can involve available technologies such as district heating and cooling. Advanced technologies such as use of ice slurries and additives to reduce friction in fluids will greatly increase the cost-effectiveness of district cooling and facilitate its penetration into developing countries and hot climate regions.

A simpler community approach involves greater use of trees in cities. When planned properly, tree planting can reduce cooling requirements through shading, thus off-setting some of the deleterious effects of the "heat island" phenomena. Planting more trees has the added benefit of contributing to lower atmospheric concentrations of carbon dioxide by enlarging the world's inventory of carbon dioxide sinks.

Least-cost utility planning (LCUP) is another approach for achieving greater energy efficiency. LCUP involves the side-by-side evaluation of supply- and demand-side energy resources for meeting the energy service needs of a community, region, nation, or potentially, the world. The Electric Power Research Institute projects that by 2000, U.S. electric utilities can use demand-side management to save 45,000 megawatts of peak power and 25,000 megawatts of off-peak power (EPRI 1986). Such planning helps ensure that the proper mix of resources are considered in meeting our energy needs. Including environmental considerations the LCUP framework could help mitigate carbon dioxide emissions.

F.   OPPORTUNITIES FOR INTERNATIONAL COLLABORATION

Amid the uncertainties of the global warming problem one thing is certain: we need to better understand the relationship of technology improvements to reduced warming. We also must recognize that we will never have all the scientific answers but must continue to work on technology improvements and begin to make decisions on priorities.

Because global warming affects all nations, it is appropriate that international collaboration play a central role in developing mitigation strategies. Projections of future greenhouse gas emissions indicate that the less developed countries will become increasingly larger contributors to the global warming problem. As these countries develop, they will be adopting the technologies developed by the industrialized nations, including the members of the OECD. The particular economic, political, and social structures of these countries will determine what role energy will play in their economic development and which fuels will be used. Therefore, we must look beyond ourselves in considering solutions to the global warming problem and at the same time accept a more important role - that of providing technology options for the emerging nations, not just the developed ones.

There are several steps we can take in the IEA to help address global warming. First, the technology R&D being pursued in the various implementing agreements of the End-Use Technologies Working Party should be continued and strengthened. In particular, we should evaluate existing and emerging technologies currently being studied in these agreements to determine their potential for reducing global warming.

Second, when proposing additional agreements and annexes, we should evaluate the potential impact the proposed research may have on global warming. This will help us balance our research agenda to promote a technology mix that satisfies both our energy and environmental objectives.

Third, it is clear that more work needs to be done to better understand the opportunities and limits of energy efficiency improvements on greenhouse gas emissions and global warming. We know energy efficiency is part of the solution but we do not have a good basis for quantifying the contribution. The foregoing analysis presented two "what if" scenarios but these do not provide an adequate basis on which to make future decisions. Some additional analysis, drawing upon the combined expertise of the OECD countries, would go a long way toward focusing our policy choices.

Fourth, we need to address how technologies will be adopted in the markets, particularly in the less developed countries. Advanced technology will only be valuable if it is readily available and cost-effective for consumers. The IEA's CADDET Program is a good way to help disseminate and share information on new technologies. If expanded, this could prove to be an important tool for assisting developing nations in choosing energy efficient technologies that meet their needs. One role the IEA can perform is that of technology advisors to existing international groups such as the United Nations Environment Programme (UNEP). This offers the advantage of joining the combined technical expertise of the IEA with an existing international group involved in the global warming problem.

Finally, we should think more broadly about ways to combat atmospheric warming. Because of the long-term, global nature of this problem, we should also consider opportunities beyond the purely technical - ones that focus on complete systems that recognize differences and corresponding opportunities in country infrastructures. These include system approaches such as community planning, transportation systems planning and least-cost utility planning. Many of these strategies are particularly attractive for emerging economies that are not restricted by an existing infrastructure.

We have learned much from our cooperation, about both the benefits of an integrated systems approach to conserve energy, and about more efficient end-use technologies. A question that the IEA should address is whether and how this knowledge and experience should be conveyed to non-member countries, particularly those that can be expected to experience the largest increase in per capita energy consumption in the coming decades and the associated production of greenhouse gases.

110

REFERENCES

Cheng, H.C., et al. 1986. *Effects of Energy Technology on Global CO$_2$ Emissions.* TR030, DOE/NBB-0076. U.S. Department of Energy, Washington, D.C.

Edmonds, J., and J. Reilly. 1986. *The IEA/ORAU Long-Term Global Energy-CO$_2$ Model: Personal Computer Version.* A84PC, CMP-002/PC, ORNL/CDIC-16. Oak Ridge National Laboratory, Oak Ridge, TN.

Edmonds, J., and J. Reilly. 1985. *Global Energy: Assessing the Future.* Oxford University Press: New York.

Edmonds, J., et al. 1986. *Uncertainty in Future Global Energy Use and Fossil Fuel CO$_2$ Emissions: 1975 to 2075,* TR036, DOE/NBB-0081. U.S. Department of Energy, Washington, D.C.

Edmonds, J., et al. 1984. *An Analysis of Possible Future Atmospheric Retention of Fossil Fuel CO$_2$.* TR013, DOE/OR/21400-1. U.S. Department of Energy, Washington, D.C.

EIA. 1988. *International Energy Annual 1987,* DOE/EIA-0219(87). Energy Information Administration, U.S. Department of Energy, Washington, D.C.

Energetics, Inc. 1989. (*Unpublished data*). Energetics, Inc., Columbia, MD.

EPA. 1989. *Policy Options for Stabilizing Global Climate* (Draft Report to Congress). Daniel Lashoft and Dennis Tirpak, eds. Environmental Protection Agency, Office of Policy, Planning and Evaluation, Washington, D.C.

EPRI. 1986. *Impacts of Demand-Side Management on Future Customer Electricity Demand,* EPRI EM-4815-SR. Electric Power Research Institute, Palo Alto, CA.

Fisher, L.J. 1988. Testimony before the House Subcommittee on Water and Power Resources, September 27, 1988. Washington, D.C.

Hansen, J. and S. Lebedeff. 1988. "Global Surface Air Temperatures: Update through 1987." *Geophysical Research Letters.* 15:323-326.

Hansen, J. et al. 1988. "Global Climate Changes as Forecast by Goddard Institute for Space Studies Three Dimensional Model." *Journal of Geophysical Research.* 93(D8):9341-9364.

Marland, G. et al. 1983. *Carbon Dioxide Emissions from Fossil Fuels: A Procedure for Estimation and Results for 1950-1981.* TR003, DOE/NBB-0036. U.S. Department of Energy, Washington, D.C.

Neftel, A. et al. 1985. "Evidence from Polar Ice Cores for the Increase in Atmospheric CO_2 in the Past Two Centuries." *Nature*. 315:45-47.

Ramanthan et al. 1985. "Trace Gas Trends and Their Potential Role in Climate Change." *Journal of Geophysical Research*. 90:5547-5566.

Raynaud, D. and J.M. Barnola. 1985. "An Antarctic Ice Core Reveals Atmospheric CO_2 Variations Over the Past Few Centuries." *Nature*. 315:309-311.

Rotty, R.M., and C.D. Masters. 1985. "Carbon Dioxide from Fossil Fuel Combustion: Trends, Resources, and Technological Implications." In *Atmospheric Carbon Dioxide and the Global Carbon Cycle*, Trabalka, J.R., ed. U.S. Department of Energy, Washington, D.C.

Rotty, R.M., et al. 1984. *The Changing Pattern of Fossil Fuel CO_2 Emissions*. TR014, DOE/OR/21400-2. U.S. Department of Energy, Washington, D.C.

Schneider, S.H. 1989. "The Greenhouse Effect: Science and Policy." *Science*. 243:771-781.

Stevens, W.K. 1989. "With Cloudy Crystal Balls, Scientists Race to Assess Global Warming." *New York Times*. Feb. 7, 1989.

World Resources Institute. 1989. Presentation to the Biomass Energy Research Association by W.R. Moomau, January 18, 1989. Washington, D.C.

World Resources Institute. 1988. *World Resources: 1988-89*. Basic: New York.

Wuebbles, D.J., et al. 1984. *A Proposed Reference Set of Scenarios for Radiatively Active Atmospheric Constituents*. TR015, DOE/NBB-0066. U.S. Department of Energy, Washington, D.C.

Wuebbles, D.J., et al. 1988. *A Primer on Greenhouse Gases*. TR040, DOE/NBB-0083. U.S. Department of Energy, Washington, D.C.

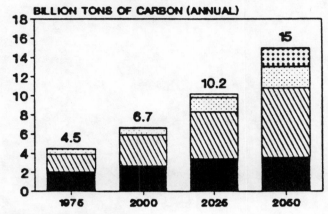

GLOBAL CO2 EMISSIONS FROM FOSSIL

BILLION TONS OF CARBON (ANNUAL)

SOURCE: EDMONDS ET AL., 1984

LE DEFI DE L'EFFET DE SERRE :

l'efficacité énergétique est-elle une réponse ?

-=-

Philippe CHARTIER,
Directeur Scientifique de l'A.F.M.E.

Maurice CLAVERIE,
Directeur du PIRSEM/C.N.R.S.

François MOISAN,
Chef du Service Programmation de la Recherche de l'A.F.M.E.

Quelles sont les options technologiques immédiates et les stratégies possibles à court terme vis-à-vis du problème de l'effet de serre ? Telle est la question-titre de cette session.

I.   L'EFFET DE SERRE ET L'ENERGIE

Avant d'avancer une contribution de réponse il convient de préciser le dilemme de cette interrogation :

- le problème de l'effet de serre reste encore incertain et les effets sont inconnus : toute stratégie pour le court terme doit donc se situer dans une logique *d'anticipation* et non *d'urgence* "à tout prix" ; ainsi il s'agit davantage de choisir des options à bas profil d'émissions et des orientations sans bouleversement radical du système technique, bouleversement dont l'impact ne saurait être davantage maîtrisé que le problème environnemental potentiel . ces options doivent permettre de gagner du temps afin de mieux comprendre les phénomènes mis en jeu.

- une stratégie à "court terme" doit bien évidemment s'inscrire dans une orientation à plus long terme et s'avérer compatible avec les échéances de temps qu'implique l'évolution progressive de la concentration en gaz à effets de serre : toute option technologique nouvelle ne pourra se concrétiser de façon tangible qu'au terme d'un certain laps de temps correspondant au taux de renouvellement des équipements (dans les pays industrialisés). Par ailleurs certaines options technologiques nécessitent encore des efforts de développements même si leur faisabilité technique est prouvée : il s'agit alors d'amélioration des performances et de baisse des coûts.

Plusieurs gaz contribuent au phénomène de l'effet de serre : CO_2, méthane, oxyde nitreux, CFC... les cycles que subissent ces gaz font l'objet de modélisation afin de mesurer l'impact d'une croissance des émissions sur la température (cf. figure 1).

Bien que son augmentation relative soit plus faible que celle d'autres gaz le CO_2 est imputable pour 80 % à la consommation d'énergie ; l'énergie participe également de façon notable à l'augmentation des émissions d'autres gaz : le méthane pour 35 % et les oxydes nitreux pour 25 %.

Il est certain que si l'effet de serre devient un défi mondial ce sera principalement en raison de l'accroissement de la consommation d'énergie et le CO_2 sera d'ici 30 à 50 ans le principal vecteur de cet effet de serre.

Afin de situer un horizon significatif pour la mise en oeuvre d'une stratégie technologique et d'assurer une cohérence avec des scénarios existants nous avons choisi l'horizon 2020. L'action à court terme (à l'horizon 2000) s'inscrirait ainsi dans une stratégie à moyen terme (2020) seule susceptible de nous procurer des scénarios réellement contrastés.

II.   2020 : L'ETAT DE LA QUESTION

Nous avons retenu comme hypothèse de travail le scénario élaboré dans le cadre de l'étude ATRE (Avenir des Tensions sur les Ressources Energétiques) de la Conférence Mondiale de l'Energie présenté à Cannes en 1986.
Parmi les trois scénarios retenus dans cette étude le scénario "médian" est souvent retenu comme référence avec une estimation de la demande d'énergie mondiale en 2020 de 15,1 Gtep (milliard de tonne équivalent pétrole). Les experts qui ont collaboré à cette étude estiment que 10 Gtep de combustibles fossiles répondraient en partie à ces besoins ainsi que 1,8 Gtep d'énergie nucléaire. Les émissions de CO_2 résultant de ce recours aux énergies fossiles seraient alors de 8,5 Gtonnes d'équivalent carbone.
Alors qu'actuellement les émissions "énergétiques" de CO_2 sont de 5,4 Gtonnes équivalentes avec une consommation mondiale d'énergie de 7,4 Gtep (base 1985), la concentration en CO_2 de l'atmosphère pourrait passer de 345 ppm à une valeur comprise entre 450 et 600 ppm.
Ce scénario suppose, il convient de le souligner, une augmentation sensible de la capacité nucléaire installée puisqu'elle serait multipliée approximativement par 6 entre 1986 et 2020.

Parallèlement à ce scénario on peut également restituer les estimations qui résultent du scénario élaboré par l'AIE pour l'année 2025 (scénario moyen élaboré par EDMONDS et REILLY) [figure 2] ; la consommation mondiale d'énergie serait en 2025 de l'ordre de 16 Gtep dont 13 Gtep de combustibles fossiles et 0,5 Gtep d'énergie nucléaire ; les émissions de CO_2 sont alors estimées à environ 11 Gtonnes équivalentes.
Il s'agit là d'un scénario impliquant une croissance de la consommation d'énergie assez voisine du scénario précédent et un recours beaucoup plus faible au nucléaire (doublement de la capacité installée entre 1985 et 2025).

Ces scénarios peuvent être considérés comme les hypothèses "hautes" de la consommation d'énergie. En effet, le groupe ATRE estime que les élasticités retenues dans leur étude (basées sur les observations antérieures à 1980), s'avèrent trop élevées et actualisent pour la prochaine conférence mondiale leurs scénarios à la baisse.
On peut également constater que malgré la forte différence en matière de croissance du parc nucléaire entre les scénarios ATRE et AIE (capacité trois fois plus importante dans le premier), les émissions de CO_2 à l'horizon 2025 ne diffèrent que de l'ordre de 20 %.

De façon contrastée on peut représenter les résultats obtenus par J. Goldemberg & al. dans le cadre d'un scénario volontariste de maîtrise des consommations d'énergie. Ce scénario suppose une généralisation des technologies les plus économes en énergie, actuellement disponibles, à l'ensemble des activités y compris dans le tiers monde, et ceci avec un niveau de vie dans les pays en développement comparable au niveau de vie actuel des pays industrialisés.

Ce scénario conduit à une estimation de la consommation mondiale d'énergie de 7,8 Gtep soit approximativement une stabilisation par rapport à la consommation actuelle ; les émissions de CO_2 seraient même en légère décroissance (moindre recours au charbon). Ce scénario ne découle pas d'une évolution tendancielle et relève davantage des scénarios normatifs ; pour les auteurs il s'agit d'un objectif potentiellement réalisable à l'horizon 2020. Nous considérons que ce scénario peut constituer l'hypothèse basse de l'évolution énergétique à l'horizon 2020 ; en effet si les solutions technologiques retenues ne sont pas révolutionnaires ce scénario suppose par contre deux conditions essentielles :

- que les technologies les plus performantes actuellement disponibles le soient en 2020 à un coût sensiblement équivalent au coût "moyen" des technologies actuelles ce qui suppose un effort de R et D conséquent.

- que les consommateurs d'énergie maîtrisent leurs comportements ce qui suppose des politiques nationales incitatives et une cohérence dans les grands choix d'infrastructure et de mode de vie (notamment vis-à-vis des transports).

Dans le cadre de ce scénario les émissions de CO_2 sont fortement réduites par rapport aux deux autres scénarios (40 à 60 %).

Cette fourchette d'évolution de la consommation d'énergie à long terme (et des émissions "énergétiques" de CO_2) nous permet d'apprécier un moyen terme en l'an 2000 ; en première approximation on peut estimer qu'en l'an 2000 les émissions de CO_2 devraient se situer entre 5 et 7 Gtonnes équivalentes. Ce point de "rendez-vous" devrait permettre de situer les politiques énergétiques à l'horizon 2000 vis-à-vis d'une évolution tendancielle à plus long terme des consommations d'énergie et des émissions de CO_2.

Le tableau 1 indique, pour deux des scénarios contrastés les bilans en consommation d'énergie et en émissions de CO_2 à l'horizon 2020 (scénario CME et scénario Goldemberg). Si l'on examine la contribution aux émissions de CO_2 (et à la demande mondiale de l'énergie) respectivement des pays industrialisés (NORD) et des pays en développement (SUD) on constate que les deux scénarios sont très contrastés pour ce qui concerne le bilan pur les pays NORD mais s'accordent relativement vis-à-vis des pays SUD.

Le scénario "Goldemberg" suppose une croissance plus importante du niveau de vie des pays en développement mais avec un profil énergétique plus économe par rapport au scénario CME mais ils convergent sur le niveau de la demande d'énergie et sur le fait que l'essentiel de cette demande serait satisfaite par les combustibles.

Finalement c'est donc bien dans les pays industrialisés qu'existe la véritable marge de liberté.

Le cas de la France, où la filière nucléaire joue un rôle particulièrement important, fera l'objet de cette analyse.

III.   1973-2000 : LES MARGES DE LIBERTE EN FRANCE

Depuis 1973, la politique énergétique de la France, a contribué à limiter les rejets de CO_2 dans l'atmosphère à la fois en raison de la progression de la part du nucléaire et de la maîtrise des consommations d'énergie ; il s'agit donc certainement d'un cas exemplaire.

Entre 1973 et 1985 on peut estimer que l'énergie nucléaire s'est substituée à environ 40 Mtep de combustibles. Approximativement il en résulte une diminution des rejets de CO_2 de l'ordre de 40 Mtonnes équivalentes (en supposant que le combustible déplacé est du charbon).

Pendant la même période la maîtrise des consommations d'énergie a permis d'économiser environ la même quantité d'énergie (36 Mtep) et les rejets ont donc été diminués de 30 Mtonnes équivalentes (en supposant que les économies ont concerné essentiellement le fuel).

En ce qui concerne l'évolution d'ici l'an 2000 on peut estimer les gains potentiels encore à venir sur la base des trois éléments suivants : achèvement de l'installation de la capacité nucléaire, politique de maîtrise de l'énergie, reboisement (cf. figure 3).

- en l'an 2000 la contribution du nucléaire sera en France de l'ordre de 80 Mtep soit une augmentation de 37 Mtep par rapport à 1985. La réduction des émissions de CO_2 est une fois de plus de 37 Mtonnes équivalent carbone (en conservant l'hypothèse d'une substitution au charbon).

- En ce qui concerne les économies d'énergie réalisables entre 1985 et 2000, elles ont été estimées à une valeur de l'ordre de 30 Mtep. Les scénarios présentés par le Ministère de l'Industrie afin de cadrer la politique énergétique de la France à l'horizon 2000 recouvrent différentes hypothèses quant à l'effort de maîtrise de l'énergie ; entre les scénarios "maîtrise de l'énergie soutenue" (prolongation des tendances des années 1983-86) et "maîtrise de l'énergie atténuée", l'évaluation de la consommation primaire globale diffère de 15 Mtep ; mais le scénario maîtrise de l'énergie atténuée implique toutefois déjà une réduction de la consommation d'énergie (sans incitation publique soutenue) de l'ordre de 15 Mtep. A l'horizon 2000 les économies d'énergie potentiellement réalisables au delà des acquis de 1985 sont donc de 30 Mtep.

En supposant une fois encore que ces économies portent essentiellement sur du fuel la diminution des rejets de CO_2 est alors de l'ordre de 25 Mtonnes équivalent carbone.

- Enfin, un programme de reboisement peut conduire à fixer une partie du carbone émis : les experts estiment qu'en France à l'horizon 2000, 5 millions d'hectares de terres agricoles seront "libérées" en raison des excédents agricoles structurels de l'Europe et du vieillissement de la population agricole ; le reboisement constitue une alternative aux friches et peut fournir des débouchés économiques (bois matière première et bois énergie). Des plantations ayant un rendement en matière sèche de 6 T/ha/an (moyenne de la forêt française) permettrait de fixer l'équivalent de 2 tonnes de carbone par hectare et par an pendant la période de plantation donc 10 millions de tonnes équivalent carbone pour les 5 millions d'hectares libérables.

Au total entre 1973 et 2000 la diminution des émissions CO_2 serait donc de 140 Mtonnes équivalent carbone dont 54 % imputables au nucléaire 39 % aux économies d'énergie et 7 % à la reforestation.

Ce bilan prospectif permet d'apprécier les contributions potentielles à la limitation des émissions de CO_2, notamment dans le cas d'un pays dont la politique énergétique s'est appuyée sur le développement très important de la filière nucléaire : les réductions d'émission obtenues grâce au nucléaire ne constituent pas bien sûr une "marge de liberté" en 1989 puisque nous sommes en aval des décisions qui ont conduit à cet équipement mais il est ainsi possible d'estimer le gain par rapport à une situation hypothétique "sans nucléaire" ; il faut souligner que le gain obtenu en l'an 2000 représente, proportionnellement, la contribution maximale de la filière nucléaire (face au suréquipement constaté).

En ce qui concerne les économies d'énergie il est à noter que les gains obtenus pendant la période 1973-85 sont plus élevés que ceux estimés entre 1985 et 2000 ; ceci en raison des perspectives de prix du pétrole plus basses que pendant la période précédente.

Enfin l'hypothèse d'un reboisement important d'ici l'an 2000 apporte une contribution importante à la réduction des émissions : il faut toutefois préciser que la réduction des émissions ne peut être comptabilisée que pendant le programme de reboisement, au delà il s'agit d'une rotation avec impact nul. Par contre la contribution éventuelle de ces plantations au bilan énergétique en tant que bois énergie n'a pas été prise en compte ; en se substituant à des combustibles fossiles elle diminuerait les émissions de CO_2 en proportion.

IV.   LES MARGES DE LIBERTE OUVERTES PAR LES ECONOMIES D'ENERGIE

4.1.   *Economie d'énergie : un gisement renouvelable*

Les estimations des économies d'énergie réalisables à l'horizon 2000 identifiées précédemment se fondent sur des estimations d'évolution des consommations spécifiques moyennes de chacune des activités consommatrices d'énergie dans chaque secteur consommateur.

Pour une activité donnée (chauffage d'une maison individuelle, production d'acier, transport d'une tonne de marchandise par camion, pour citer des exemples) il existe une diversité assez importante de solutions technologiques mises en oeuvre par les acteurs économiques (et une diversité de comportements). La diversité des solutions technologiques tient souvent au fait que les équipements ne se renouvellent pas instantanément avec l'apparition de nouvelles technologies plus performantes : ainsi dans un parc d'équipements relevant d'une activité coexiste une diversité de technologies que l'on peut schématiquement représenter sur un graphique en fonction de leur efficacité :

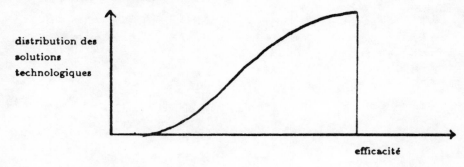

Le plus souvent les technologies moins efficaces ont été mises en place par le passé ; on trouve par exemple cette distribution dans le parc de véhicules automobiles d'une gamme donnée : en France le parc est constitué de véhicules dont l'âge varie jusqu'à 20 ans même si la majorité des véhicules a moins de 10 ans.

Si on observe l'évolution de la consommation spécifique conventionnelle des véhicules mis sur le marché depuis 15 ans on note une baisse régulière due au progrès technique :

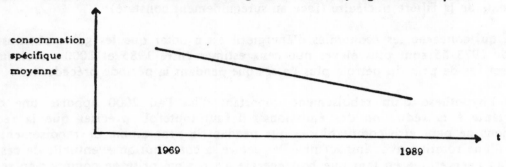

Le gisement d'économie d'énergie (et d'émission de CO_2) peut alors être représenté par l'économie qui résulterait d'un alignement de la consommation de tous les véhicules du parc sur les véhicules les plus performants ; ainsi le gisement peut se représenter en pondérant la courbe 1 par la courbe 2 : on obtient une courbe d'allure gaussienne.

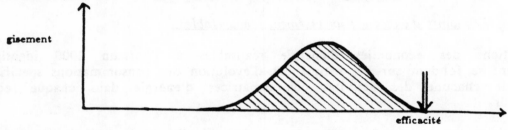

La surface située sous la courbe représente le gisement ; à droite de la courbe se trouve l'efficacité "maximale" disponible et le renouvellement progressif du parc va tendre à épuiser le gisement en réduisant l'écart entre les performances des technologies.

Une politique d'économie d'énergie conduit ainsi à promouvoir la diffusion des technologies les plus efficaces disponibles et à réduire l'inertie du parc.

Toutefois le processus ne prend pas en compte le progrès technique dont l'effet conduit à introduire continûment des nouvelles technologies plus efficaces que les précédentes : le seuil d'efficacité maximale disponible se déplace donc sur l'axe à droite de la courbe et engendre ainsi un nouveau gisement.

La recherche conduit ainsi à renouveler le gisement que la diffusion tend à épuiser.

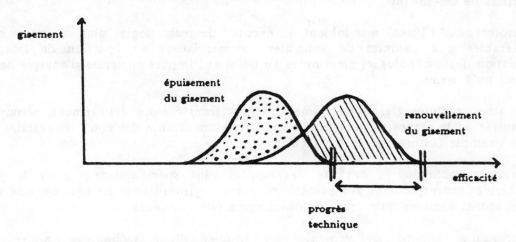

Une politique d'économie d'énergie doit donc conjuguer ces deux axes : incitation à la diffusion des techniques, incitation à la recherche et au développement.

Les perspectives d'évolution du parc d'équipement pour une activité donnée à un horizon de dix ans peuvent bien souvent être appréciées (et l'on peut ainsi mesurer les gains d'efficacité par alignement technologique). Par contre au delà d'une dizaine d'années l'introduction de nouvelles techniques issues de la R et D rend beaucoup plus difficile les estimations et une partie seulement de l'impact des programmes de recherche est prise en compte dans les modélisations.

Ainsi l'AFME a estimé quelles pouvaient être les économies d'énergie résultant de ses programmes de R et D à l'occasion de l'élaboration de son programme de recherche pluriannuel pour les années 1988-1992.

Pour la quarantaine de projets de recherche soutenus par l'AFME l'impact en terme d'économie d'énergie a été estimé à une valeur de l'ordre de 15 Mtep à un horizon de l'ordre de 15 ans.

Il nous semble donc que l'impact du progrès technique se poursuivra bien au delà de l'an 2000 en terme de diminution des consommations spécifiques et que les estimations d'économies d'énergie à l'horizon 2000 restent conservatoires à son égard.

4.2. *R et D et progrès technologique, le cas des programmes de recherche sur les véhicules économes*

Afin d'illustrer le potentiel d'économie d'énergie résultant de ses programmes de recherche, l'AFME conduit des évaluations périodiques de ses actions. Ainsi en 1987 une évaluation a été conduite sur les projets de R et D concernant les véhicules routiers.

Ces projets ont notamment débouché sur des véhicules prototypes sous le concept de "véhicules 3 litres" (ces véhicules ont des consommations conventionnelles inférieures à 3 l/100 km).

Cette réussite technologique s'est déjà concrétisée au niveau commercial puisque la Citroën AX Diesel détient le record du monde en terme de consommation et bénéficie des retombées de ces projets.

Les prototypes "3 litres" rassemblent un éventail de technologies plus ou moins rapidement transférables aux gammes de véhicules commercialisés en fonction de leur coût. La pénétration des technologies ainsi mises au point et l'impact en terme d'énergie ne pourra se mesurer qu'à terme :

- certaines technologies sont immédiatement transférables (allègement aérodynamique), d'autres ne le seront qu'au terme d'une baisse importante des coûts de certains matériaux par exemple (utilisation de Kevlar).

- certaines technologies ont été développées plus spécifiquement pour la gamme de véhicules correspondant aux prototypes (petites cylindrées) et ne peuvent être transférées aux autres gammes qu'au prix de développements ultérieurs.

- Lorsqu'une gamme de véhicules est équipée d'une technologie nouvelle l'impact énergétique ne sera sensible qu'en fonction du taux de renouvellement du parc concerné.

Une enquête conduite auprès des constructeurs automobiles qui ont mené ces programmes de R et D permet d'apprécier les échéances de pénétration des technologies dans leurs différentes gammes et d'estimer l'impact énergétique des retombées du programme ; cet impact est illustré dans la figure 4. L'effet de ces projets de recherche, en terme d'économie d'énergie ne se manifestera pleinement qu'au delà de l'an 2000 alors qu'il s'agit de projets dont la phase de R et D s'est achevée en 1987.

Il est bien certain que la vitesse de pénétration des technologies, ainsi mesurée, n'est pas une donnée intangible : le soutien public à la recherche peut permettre d'accélérer le processus (par exemple en diminuant le coût d'obtention de matériaux nouveaux), à l'inverse une moindre sensibilité générale aux problèmes de l'énergie peut conduire les constructeurs à repousser le transfert de certaines technologies.

Dans cet exemple le délai entre le résultat de la recherche (une technologie disponible) et l'impact énergétique (sa généralisation) est très lié à la nature du parc automobile : diversité des gammes et taux de renouvellement ; dans l'habitat le taux de renouvellement du parc est encore plus faible même si les techniques de réhabilitation sont conduites à se développer ; dans l'industrie le taux de renouvellement dépend bien entendu de la nature de l'investissement (élevé pour un composant mais faible pour un process).

Il est certain que le progrès technologique induit généralement une baisse des consommations spécifiques (baisse des consommations conventionnelles des véhicules, baisse des consommations des logements neufs, baisse des consommation des procédés industriels) ; mais l'incertitude sur l'évolution des consommations réellement observée se cristallise autour des problèmes d'aménagement de l'espace notamment urbain et de modes de vie ; dans le domaine des transports routiers par exemple on observe une importante augmentation de consommation en France en 1988 par rapport à 1987 (+ 5,8 %) et ceci malgré les efforts "technologiques".

En ce qui concerne le secteur des véhicules particuliers entre 1975 et 1985 la consommation a augmenté en France de 30 % ; cette augmentation est essentiellement due à une croissance du trafic et une détérioration des conditions de circulation (bien que les consommations spécifiques des véhicules aient baissé) ; si la technologie n'avait pas évolué, la consommation de pétrole aurait été supérieure de 19 % à celle observée en 1985. Si les conditions d'utilisation étaient restées les mêmes qu'en 1975, pour le volume de déplacements de 1985, la consommation d'énergie aurait été inférieure de 16 %.

La dégradation des conditions de circulation a ainsi un impact de plus en plus important sur les consommations d'énergie : une réponse "technologique" est actuellement élaborée par le biais des grands projets d'aide à la navigation routière dans le cadre des projets EUREKA (Carminat, Europolis, Prometheus) ; ces projets visent à améliorer les conditions de circulation, à diminuer les temps de trajet et par là les consommations d'énergie. Ces techniques ne peuvent pourtant à elles seules résoudre les problèmes structurels qui sous tendent ces phénomènes : il s'agit de l'aménagement de l'espace, des modes de vie et des grands choix d'infrastructures pour lesquels les pouvoirs publics jouent un rôle majeur.

V. ECONOMIES D'ENERGIE : UNE REPONSE APPROPRIEE

En conclusion il apparaît que parmi les orientations technologiques majeures qui peuvent contribuer à limiter les émissions de CO_2 les technologies économes en énergie peuvent y participer de façon significative.

La filière nucléaire constitue bien sûr une réponse du côté de la production d'énergie et sa contribution à la réduction des émissions de CO_2 en France entre 1973 et 2000 apparaît comme majeure. Toutefois la proportion d'énergie primaire nucléaire dépasse en France l'optimum économique interne (surcapacité) et le même taux d'équipement ne pourrait être retenu pour un ensemble de pays voisin.
Par ailleurs les conditions socio-politiques favorables qui ont accompagné la mise en oeuvre du programme électronucléaire français ne semblent pas se rencontrer dans la plupart des pays industrialisés (sans évoquer le problème du développement d'importants programmes nucléaires dans le tiers monde).

Il importe également de mesurer le rôle que peut jouer une politique active de reboisement dont les impacts ne se limitent bien évidemment pas seulement à l'absorption de CO_2 ; pour la France cette contribution potentielle à l'horizon 2000 n'est pas du tout négligeable.

En ce qui concerne les énergies renouvelables leur contribution au problème de l'effet de serre à court ou moyen terme nous semble plus faible : le bois énergie représente une part très importante des consommations des zones rurales dans le tiers monde ; cette "offre" pourrait se stabiliser avec des programmes de reforestation. Les autres énergies renouvelables trouvent et trouveront des créneaux de marché de plus en plus importants mais leur contribution en terme de *quantité* d'énergie (ou de diminution d'émissions) restera faible à l'horizon 2020 ; c'est bien davantage dans le service rendu aux utilisateurs notamment en tant que facteur de développement dans le tiers monde qu'elles trouveront leur justification.

Les économies d'énergie ont joué un rôle très important dans la réduction des émissions de CO_2 en France entre 1973 et 2000. Au plan *mondial* elles apparaissent comme le seul scénario cohérent susceptible de réduire de façon significative les émissions de CO_2 : elles

123

ne mettent pas en jeu les mêmes risques majeurs que la prolifération des technologies nucléaires et rencontrent dans les pays industrialisés une acceptation sociale beaucoup plus grande que le nucléaire. Par ailleurs elles participent activement à la modernisation de l'économie notamment des activités industrielles. Sur le plan économique il est généralement plus rationnel d'éviter le rejet d'une tonne de CO_2 en économisant l'énergie qu'en substituant l'énergie nucléaire aux combustibles fossiles.

Par contre les programmes publics d'économie d'énergie relèvent d'une gestion complexe (multiplicité d'acteurs, de technologies, de décideurs...) ; la complexité des technologies que l'on observe au niveau de la production dans les grandes unités (nucléaire) se retrouve dispersé dans les technologies économes en énergie. Ces programmes sont également moins spectaculaires (objectifs multiples, résultats peu médiatiques). Enfin ils doivent conjuguer les aspects technologiques (R et D) et les aspects sociopolitiques (comportements, modes de vie). Ils impliquent la persévérance, l'évaluation en continue et une souplesse d'adaptation.

124

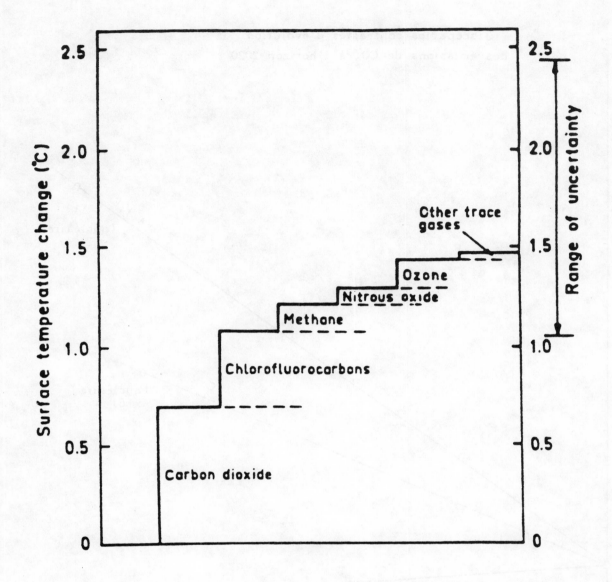

Figure 1 : Participation des différents gaz à l'effet de serre à l'horizon 2030 en terme d'élévation de la température.

(source : M. SADOURNY d'après modèle de RAMANATHAN et al.)

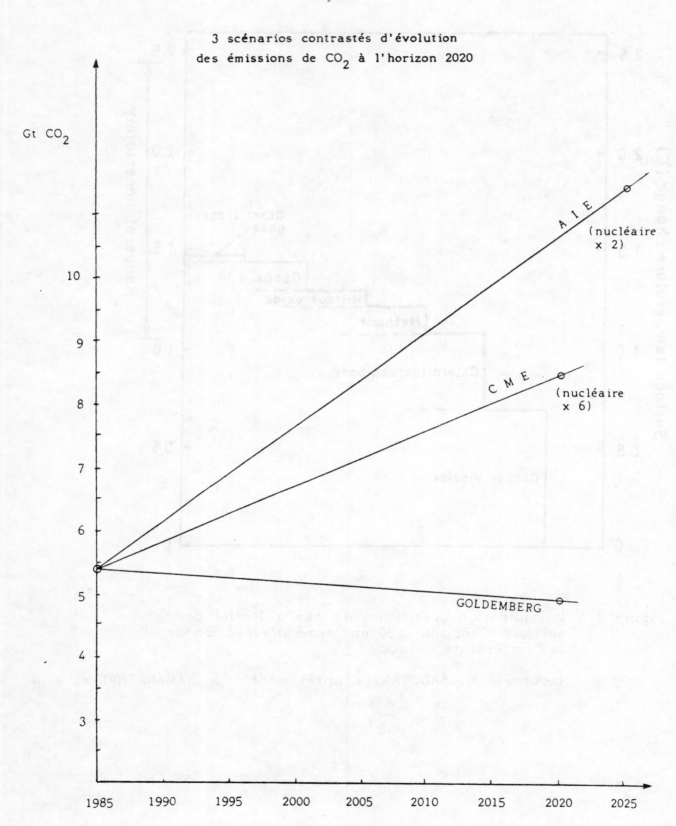

Figure 2

3 scénarios contrastés d'évolution
des émissions de CO_2 à l'horizon 2020

Gt CO_2

A I E
(nucléaire x 2)

C M E
(nucléaire x 6)

GOLDEMBERG

Tableau 1

EMISSIONS DE CO_2 DUES A L'ENERGIE A L'HORIZON 2020 :

2 scénarios contrastés

		NORD	SUD	MONDE
	ENERGIE (Gtep)	5,7	1,7	7,4
1985	dont nucléaire	0,3	–	0,3
	comb. fossiles	5,0	1,6	6,6
	CO_2 (Gt)	4,1	1,4	5,5
	concentration CO_2 p.p.m.			345 ppm
	1 – CME			
	Energie (Gtep)	9,8	5,4	15,1
	dont nucléaire	1,6	0,2	1,8
	comb. fossiles	7,0	3,0	10,0
2020	CO_2 (Gt)	6,1	2,4	8,5
*	Concentration CO_2 (estimations) p.p.m.			460 à 580 ppm
	II – GOLDEMBERG			
	Energie (Gtep)	2,7	5,1	7,8
	Dont nucléaire	0,4	0,1	0,5
	comb. fossiles	1,9	3,9	5,8
	CO_2 (Gt)	1,5	3,1	4,6

* Légende :   1 – CME : Scénario J.R. FRISH Conférence Mondiale
de l'Energie Cannes 1986 – Hypothèse centrale

II – GOLDEMBERG et al. : World Resource Institute –
Scénario à bas profil énergétique

ENERGIE

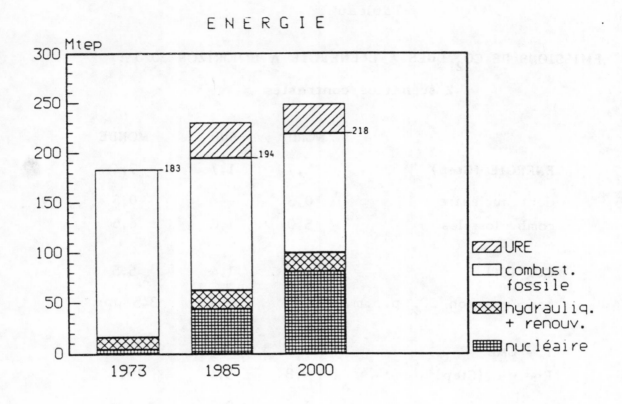

C O 2

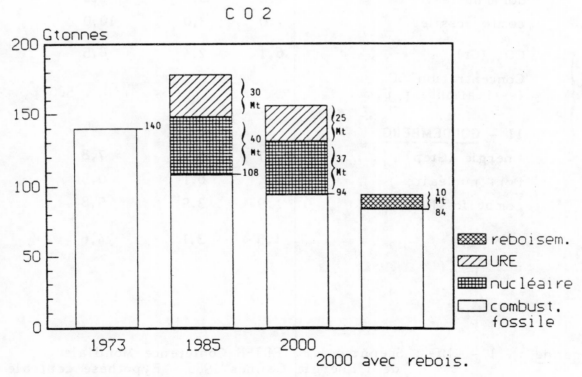

Figure 3 : Bilans énergétiques et réduction des émissions
de CO_2 en France 1973-2000.

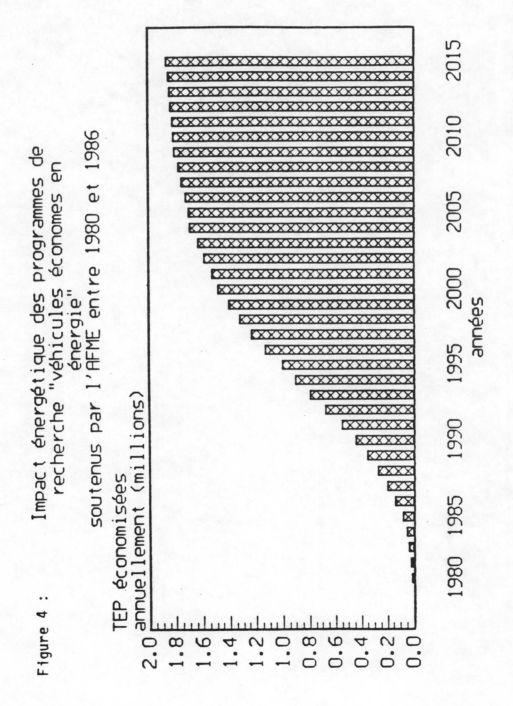

Figure 4 :   Impact énergétique des programmes de
recherche "véhicules économes en
énergie"

soutenus par l'AFME entre 1980 et 1986

THE POTENTIAL ENHANCED
GREENHOUSE EFFECT:

STATUS/PROJECTIONS/CONCERNS
AND
NEEDS FOR CONSTRUCTIVE APPROACHES

Duane G. Levine
EXXON CORPORATION

IEA/OECD Seminar
April 12-14, 1989
Paris, France

THE GREENHOUSE EFFECT EXERTS A PROFOUND INFLUENCE ON CLIMATE. WITHOUT IT TEMPERATURES WOULD BE SOME 30 ^oC COOLER AND CURRENT LIFE COULD NOT EXIST. TODAY'S CONCERNS ARE ABOUT AN <u>ENHANCEMENT</u> OF THIS EFFECT DUE TO HUMAN ACTIVITIES. SO I'LL REFER TO THESE CONCERNS COLLECTIVELY AS "POTENTIAL ENHANCED GREENHOUSE" OR PEG. IT HAS BEEN CITED AS ONE OF THE BIGGEST ENVIRONMENTAL CHALLENGES WE FACE.

ALTHOUGH SOME DECLARE THAT SCIENCE HAS DEMONSTRATED THE EXISTENCE OF PEG-INDUCED CLIMATE CHANGE TODAY...I DO NOT BELIEVE SUCH IS THE CASE. WE WILL REQUIRE SUBSTANTIAL ADDITIONAL SCIENTIFIC INVESTIGATION TO DETERMINE HOW ITS EFFECTS MIGHT BE EXPERIENCED IN THE FUTURE.

THE PURPOSE OF THIS REVIEW IS TO COVER THE STATE OF OUR <u>UNDERSTANDING</u> OF PEG, OUR <u>CONCERNS</u> ABOUT POTENTIAL CLIMATE CHANGE... AND OUR <u>NEEDS</u> IN DEVELOPING STRATEGIES TO LIMIT AND ADAPT TO POTENTIAL CLIMATE CHANGE.

WE WILL BEGIN WITH A BRIEF REVIEW OF THE STATUS OF THE SCIENCE...INCLUDING WHAT IS KNOWN/UNKNOWN, WHAT IS PROJECTION, AND WHAT IS UNCERTAIN.

NEXT WE WILL REVIEW THE INITIATIVES UNDERWAY TO ORGANIZE NATIONAL AND INTERNATIONAL STUDIES ON PEG.

FINALLY, I WILL OUTLINE BROAD RESPONSE
OPTIONS AND NEEDS IN THE CONTEXT OF THE
IMPLICIT SCIENTIFIC UNCERTAINTY AND THE LONG
LEAD TIMES REQUIRED TO ADVANCE UNDERSTAND-
ING.

GLOBAL ATMOSPHERIC CONCERNS

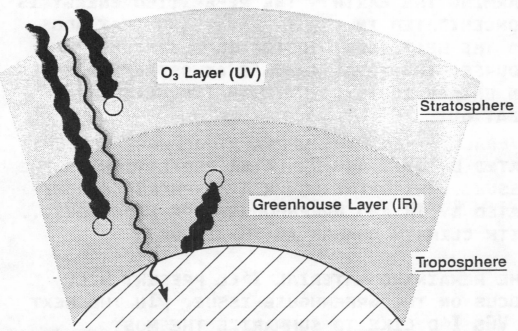

O_3 Layer (UV)

Stratosphere

Greenhouse Layer (IR)

Troposphere

TO PUT PEG IN PERSPECTIVE, LET'S START WITH
A SCHEMATIC OF THE EARTH'S ATMOSPHERE AND 2
OF THE MAJOR GLOBAL CONCERNS. ON THE LEFT,
SOLAR RADIANT ENERGY IS REPRESENTED BY A
BEAM RANGING FROM INFRARED (IR IN RED) TO
ULTRAVIOLET (UV IN BLUE)...WITH THE BULK OF
THE ENERGY IN THE VISIBLE RANGE SHOWN IN
YELLOW. MOST OF THE UV IS ABSORBED BY THE
O_3 LAYER IN THE STRATOSPHERE. MAN-MADE
CHLOROFLUOROCARBONS (CFCs) CAN ATTACK O_3
AND REDUCE ITS CONCENTRATION IN THE STRATO-
SPHERE. SUCH DETERIORATION COULD ALLOW MORE
UV TO PENETRATE THE ATMOSPHERE AND IMPACT
LIFE ON THE EARTH'S SURFACE. PROTECTION OF
THE O_3 LAYER BY REDUCING CFCs IS THE GOAL
OF THE RECENT MONTREAL PROTOCOLS.

133

MOST OF THE IR PENETRATES THE STRATOSPHERE AND IS ABSORBED BY GREENHOUSE GASES IN THE LOWER ATMOSPHERE OR TROPOSPHERE, HELPING TO WARM THE EARTH. THE BULK OF THE RADIANT ENERGY REACHING THE EARTH IS ABSORBED IN THE VISIBLE RANGE (SHOWN IN YELLOW)...DIRECTLY WARMING THE EARTH. THE RE-EMITTED ENERGY IS CONCENTRATED IN THE IR RANGE AND THAT ADDS TO THE HEAT INPUT TO THE TROPOSPHERE. OF COURSE, THE TOTAL EARTH'S ATMOSPHERE MUST BE IN ENERGY EQUILIBRIUM WITH THE HEAT IN = HEAT OUT.

OVERALL, THEN, THE UPPER ATMOSPHERE IS DOMI-NATED BY UV...AND O3 LAYER PROTECTION IS THE ISSUE...WHILE THE LOWER ATMOSPHERE IS DOMI-NATED BY IR AND GREENHOUSE-TYPE PROCESSES... WITH CLIMATE CHANGE AS THE ISSUE.

THE REMAINING MATERIAL I'LL PRESENT WILL FOCUS ON THE GREENHOUSE ISSUE. IN THE NEXT 3 VGs I'D LIKE TO SUMMARIZE THE MOST IMPORTANT AVAILABLE DATA RELATING TO PEG.

GROWTH OF ATMOSPHERIC GREENHOUSE GASES

Record of Modern CO_2 Variation

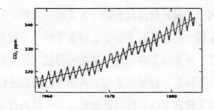

Data on CO_2 and Major Trace Gases

GAS	CONCENTRATION (PPM)		GROWTH RATE %/Year	MOLECULAR FORCING (CO_2=1)
	1850	1985		
CO_2	280	345	0.4	1
CH_4	0.7	1.7	1.0	25
N_2O	0.29	0.31	0.35	250
CFC11/12	0.0	0.0006	5.0	20,000

THIS FIGURE DISPLAYS THE FAMOUS MAUNA LOA (HAWAII) DATA SHOWING THE INCREASING CONCENTRATION OF ATMOSPHERIC CO_2. OBSERVATIONS AT THE SCRIPPS INSTITUTE BEGAN IN 1958 AS PART OF THE INTERNATIONAL GEOPHYSICAL YEAR. THE DATA SHOW A STEADY INCREASE IN CO_2 WITH AN OBVIOUS ANNUAL OSCILLATION. CO_2 IS DRAWN DOWN DURING THE GROWING SEASON AND RELEASED IN THE FALL AND WINTER. SINCE 1958, CO_2 HAS RISEN FROM 315 PPM TO OVER 345 PPM, TODAY.

CO_2 MIXES RAPIDLY IN THE ATMOSPHERE, SO RESULTS ON THE AVERAGE GROWTH OF CO_2 ARE SIMILAR ACROSS THE GLOBE. THESE RESULTS HAVE BEEN CONFIRMED AT NUMEROUS SITES TO ±1 PPM.

THE NET ACCUMULATION RATE OF CARBON CORRESPONDING TO THIS GROWTH IN THE ATMOSPHERE IS ABOUT 3 GIGATONS (BILLIONS OF METRIC TONS) PER YEAR. CO_2 EMISSION FROM COMBUSTION OF FOSSIL FUEL ADDS ABOUT 5 GIGATONS OF CARBON AND DEFORESTATION INCREASES THAT AMOUNT TO MORE THAN 6 G TONS OF CARBON PER YEAR. ONLY ABOUT HALF THE EMITTED CO_2 REMAINS AIRBORNE. A QUANTITATIVE EXPLANATION OF THE BUILDUP IS NOT YET AVAILABLE. THIS IS JUST ONE OF A NUMBER OF UNKNOWNS SURROUNDING PEG.

FROM MEASUREMENTS OF AIR BUBBLES TRAPPED IN GLACIAL ICE CORES, IT IS ESTABLISHED THAT THE RECENT BUILDUP OF CO_2 BEGAN WITH THE INDUSTRIAL ERA. IN THE MID 1800S THE CO_2 LEVEL WAS ABOUT 280 PPM (±10 PPM). CO_2 GROWTH IS ABOUT 25% SINCE THEN.

BESIDES CO_2 SEVERAL OTHER GASES ARE INCREAS-
ING IN THE ATMOSPHERE AND CONTRIBUTING TO
GREENHOUSE RADIATIVE FORCING, WHICH INCLUDES
THE SPECIFIC IR ABSORPTIVE CAPACITY OF THE
GASES AS WELL AS THEIR ATMOSPHERIC
CONCENTRATION. THOUGH PRESENT AT MUCH LOWER
CONCENTRATIONS, THESE OTHER GASES HAVE
STRONG ABSORPTION BANDS IN IR REGIONS THAT
PROMOTE GREENHOUSE. THE TABLE SUMMARIZES
DATA FOR THE MOST IMPORTANT GREENHOUSE
GASES: CO_2, METHANE, NITROUS OXIDE, AND THE
CHLOROFLUOROCARBONS (CFCs).

NOTE THAT THE GROWTH RATE OF THE OTHER TRACE
GASES TYPICALLY EQUALS OR EXCEEDS THE GROWTH
RATE FOR CO_2, AND THE INTRINSIC MOLECULAR
FORCING IS MUCH HIGHER THAN FOR CO_2. CFCs
IN PARTICULAR ARE POTENT GREENHOUSE GASES,
AS WELL AS AGENTS FOR OZONE DEPLETION.

THE RELATIVE CONTRIBUTIONS OF THESE GASES TO
GREENHOUSE RADIATIVE FORCING AS OF 1985 CAN
BE SUMMARIZED AS FOLLOWS. IN SUM, THE TOTAL
EFFECT OF ALL TRACE GASES ON INCREMENTAL
WARMING IS COMPARABLE TO CO_2 TODAY.

HOWEVER, SINCE SOME OF THESE GASES ARE
GROWING MORE RAPIDLY THAN CO_2, THEY MAY
BECOME MORE IMPORTANT IN THE NEXT CENTURY.
THE CFCs, WHICH ARE ALREADY UNDER RESTRIC-
TION FOR THEIR ROLE IN OZONE DEPLETION, ARE
VERY STRONG GREENHOUSE GASES. IF THEY ARE
NOT FURTHER RESTRICTED...THEIR BUILDUP WILL
BE A MAJOR CONTRIBUTOR TO GREENHOUSE
RADIATIVE FORCING IN THE FUTURE.

RECENT STUDIES SHOW THAT METHANE IS GROWING RAPIDLY. METHANE HAS NUMEROUS SOURCES... FROM BOGS AND AGRICULTURE TO LANDFILLS. METHANE'S SOURCES AND SINKS ARE NOT CURRENTLY UNDERSTOOD WITH MUCH PREDICTIVE CAPABILITY. THIS IS YET ANOTHER SIGNIFICANT UNKNOWN.

THE TYPES OF ACTIVITIES RESPONSIBLE FOR THE EMISSION OF GREENHOUSE GASES COVER A WIDE RANGE OF HUMAN ENDEAVORS. CO_2 DERIVES MOSTLY FROM ENERGY PRODUCTION AND USE. ABOUT 10% OF GREENHOUSE RADIATIVE FORCING IS DUE TO LAND USE, INCLUDING DEFORESTATION. AGRICULTURAL PRACTICES AND LAND USE CONTRIBUTE TO METHANE AND N_2O EMISSIONS. CFCs COME FROM SEVERAL ACTIVITIES INCLUDING: FOAM BLOWING, AEROSOLS, REFRIGERATION, AND SOLVENTS.

HISTORICAL RECORD
OF GLOBAL TEMPERATURE CHANGE

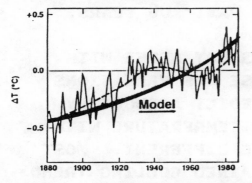

**Noteworthy Differences
With Greenhouse Predictions**

- **Recent Warming Reverses Cooling From 1940-1970s**
  - 1980s Warmest Decade on Record
  - Persistent Trend Could Signal Greenhouse Warming

- **US 1988 Summer Heat and Drought, A Critical Media Event**
  - Not a Predicted Consequence of Greenhouse Warming
  - Due to a Natural Weather Fluctuation
  - Cited as an Example of Future Trends

SINCE THE BUILDUP OF GREENHOUSE GASES HAS
PROCEEDED FOR OVER 100 YEARS, IT IS REASON-
ABLE TO ASK WHETHER THE ACCUMULATION HAS
PRODUCED A RECOGNIZABLE WARMING TREND. THIS
FINAL DATA SET COVERS PAST VARIATION IN
GLOBAL AVERAGE TEMPERATURE. OF COURSE,
GLOBAL AVERAGE TEMPERATURE IS A STATISTICAL
CONCEPT. IT CANNOT BE DIRECTLY MEASURED.
THERE ARE SERIOUS ISSUES CONCERNING COM-
PLETENESS, ACCURACY, AND INTERPRETATION OF
THIS HISTORICAL DATA. HOWEVER, MOST STUDIES
SHOW RESULTS SIMILAR TO THESE.

THE DATA SHOW A LARGE SCATTER OF A FEW
TENTHS OF A DEGREE FROM YEAR TO YEAR. THIS
"NOISE" OCCURS FROM NATURAL FLUCTUATIONS
THAT ARE NOT COMPLETELY UNDERSTOOD. IT IS
KNOWN THAT EVENTS LIKE VOLCANIC ERUPTIONS
AND CHANGES IN OCEANIC UPWELLING (SUCH AS EL
NINO), CAUSE PART OF THE VARIABILITY. THE
DASHED TREND LINE ILLUSTRATES THE GENERAL
BEHAVIOR. THE RECORD SHOWS AN APPARENT RISE
OF ABOUT 1/2 ºC OVER THE PAST 100 YEARS.

HOWEVER, THE WARMING DOES NOT AGREE WITH
MODELS BASED ON GREENHOUSE GAS VARIATIONS.
IN PARTICULAR, MODELS PREDICT A SMOOTHLY
ACCELERATING INCREASE OF TEMPERATURE WITH
TIME. THE DATA ARE QUITE DIFFERENT. MOST
NOTICEABLE IS THE UNEXPLAINED COOLING TREND
BETWEEN THE 1930S AND LATE 1970S WHEN THE
MODELS PREDICT WARMING.

DATA ON TEMPERATURE VARIATION IN THE 1980S
SHOW A REVERSAL OF THE RECENT COOLING TREND.

IN FACT, THE 3 WARMEST YEARS ON RECORD OCCURRED IN THE 1980s...AND 1988 APPEARS TO BE EVEN WARMER. IF THIS TREND PERSISTS GREENHOUSE WARMING MUST CERTAINLY BE CONSIDERED AS A POSSIBLE CAUSE.

THE 1988 HEAT AND DROUGHT IN THE US WERE A SIGNIFICANT EVENT IN THE GREENHOUSE ISSUE. THEY STIRRED PUBLIC ATTENTION, AND BROUGHT HOME POTENTIAL CONSEQUENCES OF CLIMATE WARMING. HOWEVER, 1988 WAS NOT A PREDICTED CONSEQUENCE OF GREENHOUSE MODELS. METEOROLOGISTS INTERPRET THE SUMMER AS AN INFREQUENT, BUT NOT UNEXPECTED FLUCTUATION IN WEATHER. GREENHOUSE SCIENTISTS BY AND LARGE HAVE NOT CLAIMED THE US HEAT AND DROUGHT WERE CAUSED BY ENHANCED GREENHOUSE, BUT THEY HAVE CITED IT AS AN EXAMPLE OF WHAT THE FUTURE MIGHT BRING.

SO FAR WE HAVE DISCUSSED THE HISTORICAL RECORD. GREENHOUSE IMPACTS OCCUR IN THE FUTURE, SO WE MUST LOOK AT PROJECTIONS.

PROJECTED SOURCES OF ENERGY AND CO_2 EMISSION

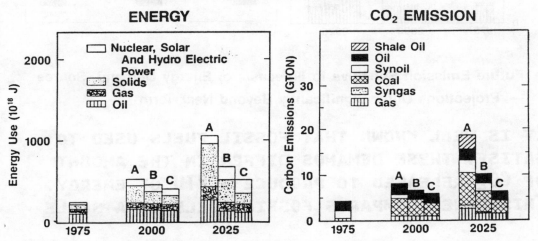

- Future Emission Sensitive to Forecasts of Energy Demand, Source
  — Projections Differ Significantly Beyond Near Term

To forecast future levels of CO_2, projections of energy needs and sources are made well into the next century. The figure shows three forecasts of energy demand taken from a DOE study...labeled Case A,B,C for high, medium, and low growth. For each model of energy demand they also compute CO_2 emissions as shown in the second panel. Cases with greater energy demand also produce much higher CO_2 emissions, since in this energy scenario, they disproportionately rely on fuels with greater emissions.

PROJECTED SOURCES OF ENERGY AND CO_2 EMISSION

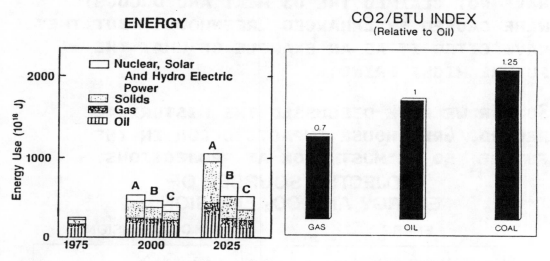

ENERGY

CO2/BTU INDEX
(Relative to Oil)

- Future Emission Sensitive to Forecasts of Energy Demand, Source
  — Projections Differ Significantly Beyond Near Term

It is well known that fossil fuels used to satisfy these demands differ in the amount of CO_2 released to produce a BTU of energy. This panel compares fossil fuels on a scale

WHERE EMISSIONS PER BTU FROM OIL ARE ONE, GAS IS 0.7, AND COAL IS 1.25. BASICALLY, THIS VARIATION FOLLOWS THE CARBON-TO-HYDROGEN RATIO OF EACH FUEL.

THE FORECASTS DIFFER WIDELY BEYOND THE NEAR TERM. ALTHOUGH NOT SHOWN HERE, EXTENDED OUT TO 2100, THEY VARY BY FACTORS OF 20 IN CO_2 EMISSION LEVELS. ENHANCED GREENHOUSE REPRESENTS AN INTERNATIONAL PROBLEM WHOSE SOLUTIONS WILL REQUIRE UNPRECEDENTED WORLD-WIDE COOPERATION. LESS DEVELOPED COUNTRIES PLAY AN INCREASINGLY IMPORTANT ROLE IN CO_2 EMISSIONS BECAUSE OF GROWING ENERGY DEMAND AND...IN SOME COUNTRIES... RELIANCE ON HIGHER CO_2 PRODUCING FUELS.

GENERAL CIRCULATION CLIMATE MODELS

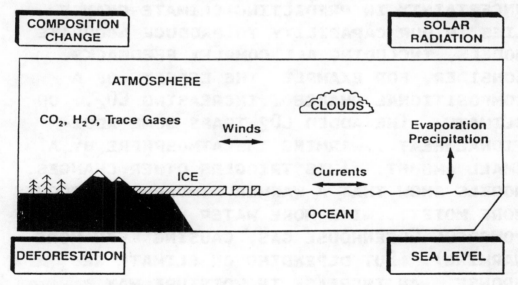

- Known Limitations
    - Coarse Resolution
    - Missing Science
    - Incomplete Calibration Data
    - Poor Agreement on Regional Scales

THESE PROJECTED AND HIGHLY VARIABLE LEVELS OF GREENHOUSE GASES PROVIDE INPUT TO COMPLEX MODELS USED TO PREDICT CLIMATE...THE AVERAGE TREND OF WEATHER, INCLUDING ITS OWN VARIABILITY. THIS SCHEMATIC ILLUSTRATES MANY OF THE PROCESSES IN ACTUAL CLIMATE. DYNAMIC EFFECTS OF WINDS AND CURRENTS CONTROL GLOBAL CLIMATE BY TRANSPORTING HEAT OVER LARGE DISTANCES. IMPORTANT FEATURES OF CLIMATE INCLUDE EVAPORATION, PRECIPITATION, AND CLOUDS...ALL OF WHICH DISPLAY ENORMOUS VARIABILITY. HERE AGAIN WE SEE SIGNIFICANT UNCERTAINTY IN THE BASIC SCIENCE.

FOR CLIMATE FORECASTS THESE PROCESSES MUST BE DESCRIBED MATHEMATICALLY AND SOLVED ON A LARGE COMPUTER. CLIMATE MODELS ARE KNOWN AS GENERAL CIRCULATION MODELS...GCMs. THE UNCERTAINTY IN PREDICTING CLIMATE CHANGE LIES IN OUR CAPABILITY TO PRODUCE ACCURATE MODELS, INCLUDING ALL COMPLEX FEEDBACKS. CONSIDER, FOR EXAMPLE, THE EFFECTS OF A COMPOSITIONAL CHANGE...INCREASING CO_2...ON CLIMATE. THE ADDED CO_2 TRAPS SOME ADDITIONAL HEAT...WARMING THE ATMOSPHERE BY A SMALL AMOUNT. THIS TRIGGERS OTHER CHANGES. MODELS SHOW THAT A WARMER ATMOSPHERE BECOMES MORE MOIST...WITH MORE WATER VAPOR (A MORE POWERFUL GREENHOUSE GAS, CAUSING EVEN MORE WARMING). BUT DEPENDING ON CLIMATE RESPONSE...AN INCREASE IN MOISTURE MAY INCREASE CLOUD FORMATION...SHIELDING PARTS OF THE EARTH'S SURFACE FROM DIRECT SOLAR RADIATION. OTHER EFFECTS NOT WELL UNDERSTOOD...LIKE OCEAN CURRENTS...CAN ALSO AMPLIFY OR REDUCE WARMING.

WE ARE ALSO CONCERNED TODAY WITH MASSIVE DEFORESTATION. THIS DESTRUCTION ADDS TO THE CO_2 LOADING TO THE ATMOSPHERE AND REMOVES SOME CO_2 SINK CAPACITY THROUGH PHOTOSYNTHESIS. OVER VERY LONG GEOLOGICAL TIME SCALES...CONTINENTAL DRIFT, SHIFTS IN THE EARTH'S ORBIT, AND OTHER SOLAR VARIATIONS...MAY CAUSE CHANGES IN SEA LEVEL AND SOLAR RADIATION WHICH CAN HAVE ENORMOUS EFFECTS ON CLIMATE. BUT THESE ARE USUALLY NOT PART OF CURRENT PEG MODELS.

GCMs HAVE BEEN CLOSELY SCRUTINIZED...SCIENTISTS RECOGNIZE THEIR STRENGTHS AND THEIR LIMITATIONS. FOR THE FORESEEABLE FUTURE THE MODELS REMAIN LIMITED FOR SEVERAL REASONS.

TO FIT IN EVEN THE LARGEST COMPUTER, RESOLUTION MUST BE COARSE. GRID BLOCKS ARE HUNDREDS OF KM ON A SIDE, SAY ONE BLOCK FOR FRANCE, OR A DOZEN FOR THE US. CONSEQUENTLY, THE MODELS CAN ONLY APPROXIMATE CRUCIAL PROCESSES, LIKE CLOUDS, THAT OCCUR ON SCALES SMALLER THAN THE GRID.

SCIENTIFIC UNDERSTANDING OF THE BEHAVIOR OF CLOUDS AND OCEANS SIGNIFICANTLY LIMIT THE MODELS. FOR INSTANCE...GCMs HAVE LIMITED ABILITY TO MAKE FORECASTS THAT FOLLOW PROGRESSIVE WARMING OVER THE NEXT 100 YEARS. INSTEAD THEY JUST COMPARE TWO TYPES OF CLIMATE: ONE AT TODAY'S CONDITIONS, AND ONE AT DOUBLED CO_2. PREDICTIONS OF TRANSIENT CLIMATE CHANGE REQUIRE ACCURATE MODELS THAT COUPLE OCEANS AND ATMOSPHERES.

DATA AND METHODS TO VALIDATE THE MODELS ARE QUITE INCOMPLETE. DATA IS ESPECIALLY LACKING OVER AND UNDER AREAS OF THE EARTH COVERED BY OCEANS. THIS IS A CRITICAL GAP IN THE SCIENCE.

FINALLY, DIRECT COMPARISON OF RESULTS FROM DIFFERENT MODELING GROUPS CLEARLY DEMONSTRATES THEIR INADEQUACY IN MAKING REGIONAL FORECASTS. THIS LIMITS OUR ABILITY TO PERFORM MEANINGFUL ASSESSMENTS OF THE IMPACT OF CLIMATE CHANGE.

POTENTIAL CLIMATE IMPACT FROM CO_2: NEXT 100 YEARS

"CHANGING CLIMATE", NATIONAL RESEARCH COUNCIL 1983

- Temperature
  — Global Mean Temperature Increase ... 1.5 – 4.5°C
  — Greater Warming in Polar Regions (2–3×)

- Sea Level/Sea Ice
  — Coverage and Thickness of Sea Ice/Glaciation Will Decrease
  — Sea Level Rise (Meltwater + Thermal Expansion) ... 70 cm

- Natural Ecosystems and Agriculture
  — Regional Climate Change: Temperature, Hydrology
  — Enhanced Productivity From Increased CO_2
  — Global Net Effect Uncertain

IN SPITE OF THE LIMITATIONS, GCMS HAVE BEEN USED TO PREDICT POTENTIAL CLIMATE CHANGES.

CONSENSUS PREDICTIONS OVER THE NEXT 100 YEARS, TAKEN FROM THE NATIONAL RESEARCH COUNCIL REPORT "CHANGING CLIMATE", CALL FOR WARMING BETWEEN 1.5-4.5 OC FOR DOUBLED CO_2. NOTE THAT THESE NUMBERS REFLECT THE RANGE PRODUCED BY AVAILABLE MODELS. NO ONE KNOWS HOW TO EVALUATE THE ABSOLUTE UNCERTAINTY IN THE NUMBERS.

WARMING AND MELTING OF GLACIERS LEADS TO SEA LEVEL RISE. THE NRC REPORT CHOSE A MOST LIKELY VALUE OF 70 CM SEA LEVEL RISE. OTHER STUDIES SUGGEST A BROADER RANGE FROM 30-200 CM. ESPECIALLY WHEN COUPLED TO A PREDICTION THAT THE FREQUENCY AND MAGNITUDE OF LARGE STORMS COULD INCREASE IN A WARMER CLIMATE, RISING SEA LEVEL AND ENHANCED STORM SURGES COULD POSE SERIOUS IMPACTS FOR COASTAL REGIONS.

FINALLY, CLIMATE CHANGE AND HIGHER LEVELS OF ATMOSPHERIC CO_2 AFFECT AGRICULTURE AND ECOSYSTEMS. THERE ARE TWO ASPECTS. FIRST, THE DIRECT EFFECT OF CHANGED CLIMATE ALTERS THE LENGTH OF THE GROWING SEASON AND THE AVAILABILITY OF WATER. SECOND, NEARLY ALL PLANTS GROW MORE RAPIDLY, AND USE LESS WATER, IN A HIGH CO_2 ENVIRONMENT. THE SECOND EFFECT CAN BE QUITE POSITIVE FOR MANAGED AGRICULTURE.

MODELS CANNOT YET PREDICT REGIONAL CLIMATE CHANGE WITH MUCH ACCURACY. ESPECIALLY FOR PRECIPITATION, DIFFERENT MODELS CANNOT EVEN AGREE ON THE DIRECTION OF REGIONAL CHANGES.

THE IMPACTS OF GREENHOUSE ARE NOT ALL NEGATIVE, AND THEY WILL AFFECT DIFFERENT PARTS OF THE GLOBE UNEQUALLY. THE MODELS PREDICT WINNERS AND LOSERS. IT IS ALSO IMPORTANT TO NOTE THAT THE RATE AS WELL AS THE MAGNITUDE OF CHANGE MATTERS IN IMPACT ASSESSMENT.

NUMEROUS NATIONAL AND INTERNATIONAL BODIES
ARE ACTIVELY ENGAGED ON PEG. INITIALLY, THE
US AND UN HAVE TAKEN LEADING ROLES. THE UN
HAS ARRANGED FOR THE FORMATION OF AN
INTERGOVERNMENTAL PANEL ON CLIMATE CHANGE
(IPCC) WHICH PROMISES TO BE PARTICULARLY
INFLUENTIAL. IN ADDITION, THE CANADIAN AND
WEST GERMAN GOVERNMENTS HAVE ESTABLISHED
HIGH PROFILES SPONSORING INTERNATIONAL
CONFERENCES IN 1988 CALLING FOR SUBSTANTIAL
REDUCTIONS IN CO_2 EMISSIONS IN THE NEAR
TERM. OTHER IMPORTANT PLAYERS INCLUDE THE
INTERNATIONAL COUNCIL OF SCIENTIFIC UNIONS
(ICSU), EEC, VARIOUS ENVIRONMENTAL GROUPS,
AND (OF COURSE) OECD.

SCHEDULED ACTIVITIES

US
- EPA REPORTS ON   1989
    - POTENTIAL CONSEQUENCES FOR US/WORLD
    - MITIGATION/STABILIZATION APPROACHES

- DOE   1989
    - ASSESSMENT OF R&D ON ALTERNATE ENERGY
    - INVENTORY OF GREENHOUSE GASES
    - ANALYSIS OF PRIVATE SECTOR OPTIONS
    - POLICY OPTIONS TO LIMIT EMISSIONS

- US/INTERNATIONAL SCIENCE ASSESSMENT   1990

UN
- COORDINATE WORLDWIDE SCIENTIFIC ASSESSMENT WMO/ICSU   1990

- REGIONAL IMPACT ASSESSMENTS   1990

- POLICY OPTIONS TO LIMIT CLIMATE CHANGE   1990

- INTERNATIONAL CONVENTION TO LIMIT CLIMATE CHANGES   MID '90'S

IPCC
- WORKING GROUP REPORTS: SCIENCE, IMPACTS,   1990
  RESPONSE STRATEGY

146

THE INTENSITY OF THESE EFFORTS IS REFLECTED
IN THE SLATE OF ACTIVITIES CURRENTLY SCHED-
ULED. FOR EXAMPLE, THE EPA WILL SOON REPORT
TO CONGRESS ON BOTH POTENTIAL CONSEQUENCES
OF PEG FOR THE US/WORLD AND MITIGATION/
STABILIZATION APPROACHES TO LIMIT CLIMATE
CHANGE. DOE WILL BE VERY BUSY. IN 1989, IT
IS DUE TO DEVELOP REPORTS ON VARIOUS ASPECTS
OF PEG INCLUDING:

    o ASSESSING ALTERNATE ENERGY R&D
    o CATALOGUING GREENHOUSE GASES
    o ANALYZING OPTIONS FOR THE PRIVATE
       SECTOR

AND

    o EVALUATING POLICY OPTIONS FOR LIMITING
       CO_2 EMISSIONS

FINALLY, NSF IS PLANNING A US/INTERNATIONAL
SCIENCE ASSESSMENT BY 1990.

THE UN HAS RECENTLY COMMISSIONED A WORLDWIDE
SCIENTIFIC ASSESSMENT DESIGNATING WMO TO
WORK CLOSELY WITH ICSU. REGIONAL IMPACT
STUDIES ARE BEING SET UP USING CLIMATE
MODELS TO PROJECT THE VARIOUS WINNERS AND
LOSERS AS RAINFALL, WIND, STORMS, AND SEA
LEVEL PATTERNS CHANGE. THIS COULD BE ESPE-
CIALLY CONTENTIOUS BECAUSE OF THE WIDE ROOM
FOR INTERPRETATION IN APPLYING THE MODELS
AND THE ENORMOUS POTENTIAL POLITICAL CONSE-
QUENCES. CONCURRENTLY, THE UN ENVIRONMENTAL
PROGRAM (UNEP) WILL DEVELOP ITS VIEW OF
POLICY OPTIONS TO LIMIT CLIMATE CHANGE.

UNEP HAS BEEN URGED TO AIM FOR AN INTER-
NATIONAL CONVENTION ON PEG BY ABOUT 1995.

PARTLY AS A BASIS FOR THIS INTERNATIONAL
CONVENTION, THE IPCC ESTABLISHED AN AMBI-
TIOUS SCHEDULE FOR ITS WORKING GROUPS TO
ISSUE REPORTS IN 1990 ON SCIENCE, IMPACTS,
AND RESPONSE STRATEGIES.

GIVEN, THE...

O   COMPLEXITY OF THE SCIENCE
O   ENORMOUS POTENTIAL GLOBAL IMPACTS
O   DIVERSITY OF THE PLAYERS
AND
O   INTENSITY OF THEIR ACTIVITIES...

WHERE IS ALL THIS HEADED? I BELIEVE THERE
IS A PATTERN...AND IT'S ROOTED IN THE EVOLU-
TION OF THE JUST-COMPLETED MONTREAL PROTO-
COLS TO PROTECT THE STRATOSPHERIC O3 LAYER
BY LIMITING MAN-MADE CFCs.

STRATOSPHERIC OZONE/PEG ANALOGY

OZONE LAYER		ENHANCED GREENHOUSE
'74 ✓	ATMOSPHERIC CHEM/PHYS	✓ '75
✓	GROWTH IN [CFC'S] / [CO2] & [TRACE GASES]	✓
✓	INDUSTRIAL SOURCES	✓
✓	MODELS: END EFFECT PROJECTIONS	✓
✓	CONCEPT OF "DELAY"	✓
✓	ENVIRONMENTAL CAUSE	✓
✓	INTERNATIONAL OWNERSHIP	✓
✓	US/UN AXIS	✓
✓	CRITICAL EVENT	?
✓	CALL FOR ACTION	?
VIENNA CONVENTION		?
'87 MONTREAL PROTOCOLS		?

148

ABOUT 20 YEARS AGO, THE ATMOSPHERIC CHEMIS-
TRY AND PHYSICS BEGAN WITH CONCERNS OVER THE
EFFECT OF SUPERSONIC TRANSPORTS ON THE O3
LAYER. AS THESE FEARS ABATED...IN 1974...
LABORATORY TESTS INDICATED THAT THE VERY
STABLE CLASS OF MAN-MADE CHEMICALS (CFCs)
BREAK DOWN UNDER THE KIND OF INTENSE UV
RADIATION THAT IS PRESENT IN THE STRATO-
SPHERE...WITH THE RESULTING CL AND BR ATOMS
ATTACKING AND DESTROYING O3.

SUBSEQUENTLY, IT WAS ESTABLISHED THAT THE
CFC CONCENTRATIONS WERE INCREASING IN THE
STRATOSPHERE. THESE CFCs WERE EASILY
TRACED TO MAN-MADE SOURCES AND INDUSTRIAL
APPLICATIONS.

EXTENSIVE MODELING EXERCISES PREDICTED...

   o THE LONG RANGE RATE OF O3 LAYER DETERI-
      ORATION
   o THE RESULTING INCREASED UV PENETRATION
      OF THE EARTH'S ATMOSPHERE
AND
   o THE SERIOUS POTENTIAL REPERCUSSIONS
      LIKE INCREASED HUMAN CANCER RATES AND
      PLANT DAMAGE

NOW THESE PREDICTED EFFECTS WERE WELL INTO
THE FUTURE. SO A CRUCIAL STEP WAS THE
INTRODUCTION OF "A DELAY CONCEPT" BASED ON
THE UNUSUAL CHEMICAL STABILITY OF CFCs IN
THE LOWER ATMOSPHERE AND THE VERY LONG
TRANSPORT TIMES FOR CFCs TO REACH THE UPPER

STRATOSPHERE. THE REASONING WAS...THERE WAS AN <u>ALREADY</u> COMMITTED O3 LAYER DETERIORATION BASED ON CFCS <u>ALREADY</u> "IN THE PIPELINE".

THIS GAVE RISE TO AN ENVIRONMENTAL CAUSE WHICH WAS QUICKLY ADOPTED AS AN INTERNATIONAL ISSUE. THE PLAYERS WERE BASICALLY SIMILAR TO THOSE ORGANIZING AROUND THE CURRENT GREENHOUSE ISSUE...PRIMARILY IN THE US AND UN. WHEN THE US BECAME ACTIVELY INVOLVED ...INITIATING COOPERATION WITH THE UN TO LIMIT WORLDWIDE CFC PRODUCTION AND SALES...ALL OF THE ELEMENTS WERE IN PLACE.

BUT WITH ALL OF THIS, PROGRESS BEGAN TO LANGUISH AND THE EFFORT MIGHT WELL HAVE FOUNDERED, EXCEPT FOR THE DISCOVERY OF THE SO-CALLED "O3 LAYER HOLE" OVER ANTARCTICA. THIS WAS A MOST CRITICAL EVENT - ALTHOUGH ITS EXACT RELEVANCE TO CFC RELATED O3 LAYER DETERIORATION REMAINS UNEXPLAINED. ITS DISCOVERY RE-ENERGIZED THE EFFORT LEADING TO A "CALL FOR ACTION" AND ADOPTION OF THE VIENNA CONVENTION AND SHORTLY THEREAFTER...IN 1987...THE MONTREAL PROTOCOLS TO LIMIT CFCS WITH A PHASED 50% REDUCTION BY THE TURN OF THE CENTURY. I SHOULD ADD THAT ONLY IN EARLY 1988...DID THE SCIENTIFIC COMMUNITY ESTABLISH CONVINCING SCIENTIFIC EVIDENCE THAT O3 LAYER DETERIORATION HAS OCCURRED SO THAT <u>AFTER THE FACT</u>...SOME ACTION SEEMS JUSTIFIED.

FROM THE MATERIAL COVERED ON GREENHOUSE SO
FAR THIS MORNING...IT IS CLEAR THAT WE HAVE
ADVANCED THROUGH SIMILAR STAGES. SOME CLAIM
THAT THE "LONG HOT SUMMER OF '88" IN THE US
IS A CRITICAL EVENT AND WE'RE STARTING TO
HEAR THE INEVITABLE CALL FOR ACTION. EXACT-
LY WHAT HAPPENS NOW IS NOT CLEAR...BUT THIS
EVENT HAS ENERGIZED THE GREENHOUSE EFFORT
AND RAISED PUBLIC CONCERN OVER PEG.

OPTIONS TO LIMIT CLIMATE CHANGE

- REDUCE PROJECTED FUTURE GROWTH OF CO_2 EMISSIONS
  - IMPROVE EFFICIENCY
  - IMPROVE AND DEVELOP TECHNOLOGY
  - SHIFT ENERGY RESOURCE MIX

- REDUCE ENERGY
  - RESTRICT
  - RATION
  - ABANDON RESOURCES

- REDUCE OTHER GREENHOUSE GASES

- PURSUE OTHER COMPENSATING OPTIONS
  - REFORESTATION
  - MITIGATING STRATEGIES

THE OPTIONS TO DEAL WITH PEG AT THIS STAGE
REFLECT THE DIFFICULT CHALLENGE...NONE ARE
EASY OR STRAIGHTFORWARD...AND SOME MAY
RESULT IN SERIOUS DISRUPTIONS TO THE WORLD'S
SOCIO/POLITICAL/ECONOMIC STRUCTURE. THEY
INCLUDE:

151

O   REDUCING THE GROWTH RATE OF FUTURE CO_2
    EMISSIONS...
    BY
    -   IMPROVING EFFICIENCIES
    -   IMPROVING AND DEVELOPING NEW ENERGY
        TECHNOLOGY...
    OR
    -   SHIFTING THE ENERGY RESOURCE MIX

O   CONTROLLING OR REDUCING ENERGY DIRECT-
    LY...
    BY
    -   RESTRICTING ITS USE...WHICH HAS A
        DIRECT EFFECT ON ECONOMIC ACTIVITY
    -   RATIONING ITS AVAILABILITY, E.G.,
        BETWEEN DEVELOPED AND DEVELOPING
        COUNTRIES
    -   AND EVEN ABANDONING RESOURCES

O   REDUCING OTHER GREENHOUSE GASES

AND FINALLY

O   IMPLEMENTING AND DEVELOPING VARIOUS
    COMPENSATING EFFECTS, INCLUDING WIDE
    RANGING OPTIONS...
    FROM
    -   REFORESTATION...WHICH IS DOABLE NOW
    TO
    -   OTHER MITIGATING APPROACHES, LIKE
        INCREASING THE EARTH'S REFLECTIVE
        CAPABILITY BY CLOUD SEEDING OR
        REFLECTORS IN SPACE

AMONG RECENT PROPOSALS TO LIMIT PEG ARE
CALLS TO REDUCE CO_2 EMISSIONS BY AT LEAST
20% FROM 1986 LEVELS BY 2000-2010. IN THE
NEXT TWO VGS WE WILL LOOK AT SOME OF THE
IMPLICATIONS OF SUCH A STEP.

WORLD CARBON EMISSIONS
FROM FOSSIL FUELS

5.0 G TONS   6.8 G TONS

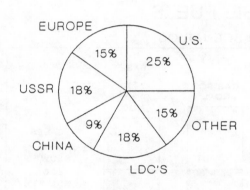

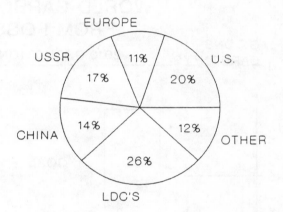

1986   2010

THIS CHART ILLUSTRATES WORLD CARBON EMIS-
SIONS FROM FOSSIL FUELS BY COUNTRY AND
REGION IN 1986 AND A FORECAST THAT SHOWS
PROJECTED EMISSIONS IN 2010.

FIRST, NOTE THE EXPECTED RISE FROM 5.0 TO
6.8 GTON PER YEAR.

SECOND, NOTE THAT THE SHARE OF EMISSIONS
FROM THE US, EUROPE, AND USSR DECREASES FROM

60% IN 1986 TO LESS THAN 50% IN 2010. CHINA
AND THE LESS DEVELOPED COUNTRIES (LDCs)
INCREASE THEIR SHARE SIGNIFICANTLY. RAPID
ENERGY GROWTH AND THE USE OF INDIGENOUS
FUELS ARE CRITICAL TO THE EXPANSION OF THEIR
ECONOMIES.

TO DECREASE CO_2 EMISSIONS IN 2010 BY 20%
FROM 1986 LEVELS REQUIRES SLIGHTLY MORE THAN
A 40% REDUCTION FROM THE EXPECTED LEVEL OF
6.8 GTON/YEAR.

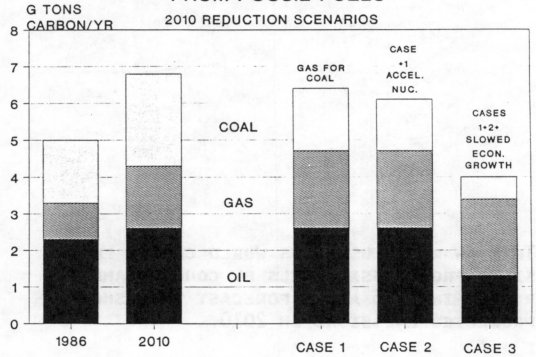

WORLD CARBON EMISSIONS
FROM FOSSIL FUELS

IN THIS CHART WE SHOW A SCENARIO AIMED AT
ACHIEVING THE TARGET CARBON LEVEL OF 4
GTON/YEAR BY 2010. THE TWO BARS ON THE LEFT
SHOW EMISSIONS IN 1986 AND A FORECASTED
LEVEL FOR 2010 BROKEN OUT BY SOURCE (OIL,
GAS, AND COAL). IMPLICIT IN THE 2010 CASE
ARE SIGNIFICANT IMPROVEMENTS IN EFFICIENCY.

ONE MEANS OF REDUCING CO_2 EMISSIONS COULD BE THROUGH FUEL SUBSTITUTION. IN CASE 1, GAS IS SUBSTITUTED FOR COAL IN RESIDENTIAL/ COMMERCIAL, INDUSTRIAL, AND UTILITY SECTORS TO THE MAXIMUM EXTENT THOUGHT FEASIBLE, CONSIDERING RESOURCE LIMITATIONS. THIS STEP REDUCES CO_2 EMISSIONS BY ONLY ABOUT 6%.

IN CASE 2, NUCLEAR IS INTRODUCED AS A SUBSTITUTE FOR COAL IN ELECTRIC POWER GENERATION WORLDWIDE. THE RESULT ASSUMES WORLD-WIDE NUCLEAR POWER WILL GROW AT 4%/YEAR, SIMILAR TO JAPAN'S CURRENTLY FORECAST NUCLEAR GROWTH (AND ABOUT 2.5 TIMES THE PACE FORECAST FOR INDUSTRIAL NATIONS AS A WHOLE). THIS PRODUCES A FURTHER 15% REDUCTION IN COAL USE BELOW CASE 1, DROPPING TOTAL EMISSIONS TO 6.1 GTON/YEAR. THERE IS NO INDICATION THAT SUCH A REJUVENATION OF NUCLEAR IS POLITICALLY FEASIBLE.

TO ACHIEVE THE GOAL OF 4 GTON/YEAR IN CASE 3, THE REMAINING COAL AND OIL UTILIZATIONS WERE FURTHER REDUCED BY 45% TO MATCH THE TARGET. ONE WAY TO ACHIEVE THIS WOULD BE TO REDUCE THE PROJECTED WORLD ECONOMIC GROWTH RATE FROM 2.7 %/YEAR TO 2.0 %/YEAR. IF THE TARGET WERE TO BE ACHIEVED SOLELY BY REDUCTIONS IN THE US, EUROPE, AND JAPAN, THEN THEIR PER CAPITA ECONOMIC GROWTH DECREASES SIGNIFICANTLY...FOR EXAMPLE...NEARLY TO 0 IN THE U.S. OF COURSE IF THE EXTRA GAS AND NUCLEAR ENERGY ARE NOT AVAILABLE, THEN THE SLOWDOWN IN ECONOMIC GROWTH WOULD BE MORE DRAMATIC.

IT IS CLEAR FROM THIS DISCUSSION, THAT THE PATH TO REDUCED CO_2 EMISSIONS IS A DIFFICULT ONE...REQUIRING BASIC CHANGES IN THE ENERGY INTENSITY, FUEL CHOICE, AND GROWTH RATE OF THE WORLD'S ECONOMY.

R&D NEEDS FOR POLICY DEVELOPMENT

- Science of Climate Change
  Improve Understanding of Climatic Forcing
  Establish Global Monitoring to Confirm and Validate

- Impact Assessments
  Develop Reliable Regional Forecasts
  Improve Scientific Understanding of Stressed Systems
  Establish Appropriate Monitoring

- Policies
  Assure Effectiveness in Stabilizing Climate Change
  Determine Costs and Feasibility
  Assess Associated Impacts on Society

IN VIEW OF ALL THE UNCERTAINTIES IN THE SCIENCE OF GLOBAL CLIMATE CHANGE, AND THE SERIOUS CONSEQUENCES TO SOCIETY OF DRASTIC STEPS TO LIMIT CLIMATE CHANGE, THERE IS AN ESSENTIAL NEED FOR AN <u>INTEGRATED</u> PROGRAM OF R&D IN POLICY DEVELOPMENT.

FIRST AND FOREMOST, DEVELOPMENT OF POLICY ON CLIMATE CHANGE DEMANDS RECOGNITION OF THE CURRENT STATE OF SCIENTIFIC UNCERTAINTY. WE DO NOT KNOW THE MAGNITUDE OR TIMING OF CLIMATE CHANGE. IGNORANCE COULD LEAD TO EXPENSIVE, MISTAKEN POLICIES. WE MUST CONTINUE TO SUPPORT AND EXPAND RESEARCH ON THE SCIENCE TO IMPROVE OUR UNDERSTANDING. PART OF THAT EFFORT REQUIRES THAT WE

ESTABLISH A NETWORK OF GLOBAL MONITORING
SYSTEMS FROM SPACE, ON LAND, AND OVER THE
OCEANS. THE TWIN PURPOSES OF THE DATA ARE
(1) TO CONFIRM WHETHER CLIMATE IS INDEED
CHANGING, AND (2) TO VALIDATE THE INTERNAL
MECHANISMS OF THE MODELS SO THAT WE CAN GAIN
CONFIDENCE IN THEIR PREDICTIVE CAPABILITIES.

SECOND, WE MUST IMPROVE THE CURRENT STATUS
OF IMPACT ASSESSMENTS. IT IS ONE THING TO
PROJECT THAT TEMPERATURE WILL INCREASE AND
SEA LEVEL RISE. IT IS ANOTHER TO ESTABLISH
A RELIABLE ASSESSMENT OF THE IMPACT ON
MANAGED AND NATURAL ECO-SYSTEMS. FUNDA-
MENTAL TO THAT CAPABILITY IS THE NEED TO
DEVELOP TECHNIQUES FOR MAKING RELIABLE RE-
GIONAL FORECASTS OF CLIMATE CHANGE. THE
SUBSEQUENT IMPACT ASSESSMENTS OFTEN REQUIRE
THEIR OWN COMPLEX MODELS OF RESULTING
STRESSED SYSTEMS. AND, THESE STRESSED
SYSTEM MODELS ALSO REQUIRE MONITORING TO
CONFIRM THEIR VALIDITY.

WITH ALL THE UNCERTAINTY IN PEG SCIENCE...IN
OUTLINING PROPOSED POLICIES WE SHOULD BE
CONFIDENT THAT THEY WILL BE EFFECTIVE,
AFFORDABLE, AND DOABLE. ALSO, WE MUST
ESTABLISH ASSOCIATED IMPACTS OF POLICY
OPTIONS ON SOCIETY SO THAT THESE ARE
RECOGNIZED AND ACCEPTED IN SELECTING A
COURSE OF ACTION.

NEAR TERM APPROACHES:
THE TIE-IN STRATEGY

Recognizing Uncertainty and the Long Lead Times to Confirm
Greenhouse Impacts, Pursue Approaches Which Make Sense
in Their Own Right, and Mitigate Climate Change

- Improve Energy Efficiency

- Protect and Refurbish Major Global Forest Resources

- Continue Initiatives to Limit CFCs

- Consider Cost Effective Limits on Other Trace Gases

FROM THE DISCUSSION SO FAR WE SEE THAT
CLIMATE CHANGE PRESENTS US WITH AN UNCOM-
FORTABLE DILEMMA: POTENTIAL SIGNIFICANT
RISK, LARGE UNCERTAINTY, AND DIFFICULT
CHOICES FOR RESPONSE. RECOGNIZING BOTH THE
UNCERTAINTY AND THE LONG LEAD TIMES TO
CONFIRM PEG IMPACTS, IN THE NEAR TERM IT
MAKES SENSE TO PURSUE APPROACHES WHICH
IMPROVE THE GENERAL ENVIRONMENT...THAT
IS...THEY STAND ON THEIR OWN MERIT WITH
RESPECT TO HELPING RESOLVE OTHER IMPORTANT
ENVIRONMENTAL CONCERNS...BUT ALSO MITIGATE
OR AMELIORATE PEG. THIS APPROACH IS OFTEN
REFERRED TO AS THE "TIE-IN STRATEGY." IT
INCLUDES:

- CONTINUING IMPROVEMENTS IN ENERGY
  EFFICIENCY.
- PROTECTING AND REFURBISHING MAJOR
  GLOBAL FOREST RESOURCES.
- CONTINUING THE ONGOING INITIATIVES TO
  LIMIT CFCs.

-   **EXTENDING THESE INITIATIVES TO OTHER TRACE GREENHOUSE GASES...WHERE COST/EFFECTIVE.**

LONGER TERM APPROACHES

• Balanced Approach to Developing New Technology for Fossil Fuel Energy Resources as Well as Renewable and Nuclear Energy Resources

- Mitigate Impacts of CO2 from Fossil Fuels
- Achieve Cost Competitive Renewable Energy
- Resolve Nuclear Safety/Waste Disposal/Cost Problems So That Nuclear Is a Viable Option

NEXT WE SHOULD CONSIDER LONGER TERM APPROACHES TO PEG INDUCED CLIMATE CHANGE AS REQUIRED BY GROWING EVIDENCE AND ANALYSES...EMPHASIZING A BALANCED APPROACH TO DEVELOPING NEW TECHNOLOGIES FOR FOSSIL FUEL ENERGY RESOURCES AS WELL AS RENEWABLE AND NUCLEAR ENERGY RESOURCES.

o   **FOR FOSSIL FUELS, THAT MEANS MITIGATING POTENTIAL CO2 IMPACTS.**

o   **FOR RENEWABLE ENERGY, IT MEANS ACHIEVING THE MAJOR STEP-OUTS NECESSARY TO MAKE THESE RESOURCES COST COMPETITIVE.**

AND

o   **FOR NUCLEAR, IT MEANS RESOLVING CURRENT CONCERNS ABOUT SAFETY, WASTE DISPOSAL, AND COSTS.**

ALL OF THIS AIMED AT A REALISTIC, LONG-RANGE, STAGED, STRATEGIC...GLOBAL APPROACH, RESISTING PREMATURE POLICIES AND DRACONIAN MEASURES.

HOW TO SOLVE THE CO$_2$ PROBLEM WITHOUT TEARS [*]

Cesare Marchetti

International Institute for Applied Systems Analysis

A-2361 Laxenburg, Austria

(*)   Published with the permission of the International Association for Hydrogen Energy, PO Box 248226,Coral Gables, FL 33124, USA

How to Solve the CO_2 Problem Without Tears

Cesare Marchetti

The preoccupation for the radiation balance of the earth atmosphere, and the effects of human activities on it has started about three decades ago, giving rise to a small set of very interesting work and a huge amount of "recycled paper" in the wake of a successful keyword.

A sample of this literature collected by the Institute for Scientific Information (ISI) of Philadelophia, analyzed using the "fashion wave equations", shows that the peak of this literature wave was reached in 1984 and the ebb is beginning (Figure 1). I did not have the possibility to monitor the real work, of which I give an interesting example in Figure 2, where air trapped in glacier ice has been measured for its CO_2 content (Neftel et al., 1985).

The methodology followed is very interesting because it permits monitoring air "put in the freezer" way back in the past to get the grand view and try to establish a correlation between CO_2 content and climate. This also because climate models are still too rough and too complicated to generate the fine print of what happens *if*. History may serve to calibrate them, as many situations did occur in the most various context (Bolin, 1985).

The fact is, the CO_2 content of the atmosphere as shown in Figure 2 did start increasing in 1800 from a fairly stable ≈ 280 ppm (volume) in historical times to the present 330 ppm. As in 1800, world population was expanding in number and in activity levels, the connection cause-effect is considered obvious, although the share of emissions between various forms of activity, like forest clearing and fossil fuel combustion and oth-ers, is still under discussion.

The last point is important for our purposes, because we will propose *active solutions for the control of* CO_2 *emissions linked to energy use*. Forest clearing would require other means of control. Looking at the past, the share can be calculated checking ^{14}C concentration in tree rings. Fossil fuels have no ^{14}C and appear as diluents. The results are reported in Figure 3 (Bolin, 1985) and show some interesting features, e.g., that land clearance was the largest contributor of CO_2 to date (265 10^9 tons) and fossil fuels generated a little more than half that value (170 10^9 tons). The situation, however, has drastically changed after World War II, with fossil fuels becoming the dominant emitters.

The cure then has to be applied to fossil fuels. The next question is the share of the different fossils, coal, oil, gas, in the consumption of fossils during the next 50 or 100 years, so that the cure can be applied to the dominant ones, to simplify the procedures and maximize the effect of the measures taken.

This forecast in terms of shares can be done in a very robust form by using the Volterra substitution model we did employ at IIASA to map the dynamics of primary energy markets during the last 100 years (Figure 4). The model is parsimonious and predictive, even long term, when the dates for new competitors (e.g., fusion) entering the market are available. This is possible using a more abstract model dealing with innovation (Marchetti, 1980), but in our case this is not necessary as we deal with fossil fuels only. Their future is sealed by nuclear penetration alone.

Nuclear systems (or solar for those who believe in it as a future large source of primary energy) will finally close the fossil fuel era, as the Volterra substitution model, applied to big envelopes: wood and renewables, fossils, nuclear and solar, shows (Figure 5). So the question boils down to the mix coal, oil and gas. This comes from Figure 4, but it will be shown in a more detailed way later on. It will be in any case clear that the *dominant primary fossil energy* in the next 100 years will be *natural gas*, and that most of the CO_2 will come out of its combustion, even if it is the fuel with lowest carbon content and highest efficiency in use. *The cure has to be applied there in order to get maximum effect.*

The model shows market shares. But we need absolute quantities for the CO_2 released. For that purpose we must make an hypothesis about the growth of energy consumption at world level. By smoothing from Kondratiev oscillation world energy consumption during the last 150 years, we find a steady increase by 2.3% annually (Figure 6). We may then take this *business as usual* as an *hypothesis* for growth in the next 150 years. This is in a sense arbitrary, but at this rate the final consumption per head will bring developing countries *then* at the same level of the U.S. *today.*

Having set the context, Figure 7a,b,c shows the careers of coal, oil and g. seen as *product cycles.* Figure 8a,b,c shows actual demand and integrated demand for each of them in absolute amounts, with the hypothesis of 2.3% growth of global energy markets. It is clear from the figures that *gas consumption will integrate at a level one order of magnitude larger than oil consumption.*

From Figure 8, the actual quantities (10^9 tons) of CO_2 emitted can be calculated for each of the primary energies (Figure 9). The share of CO_2 emitted from methane is reported in Figure 10. The analysis is done using smoothed data, i.e., the Kondratiev oscillation is taken away which may change somehow the fine print. However, the date when 50% of CO_2 emissions will come from natural gas will not be very different from the year 2010 that appears in the chart.

Another idea which is implicit to the competition game, as shown in the chart of Figure 5, is that in due time all primary energy will come from nuclear sources, be fission or fusion, and consequently also chemical fuels will be produced from them. We all know that hydrogen from water is candidate number one for that role, but we also know that the technologies to make it directly from water on the *scale* and the *cost* it will be required to be initially competitive with the fossil fuels, even in specialized market niches, will take time to develop.

However there are possible *short cuts for a transition period*, which may last for tens of years, and *I will talk about them, also in view of alleviating the problem of* CO_2 *emissions*. It comes out that doing the two things together the marginal cost of eliminating a substantial share of CO_2 emissions will be marginal if not nil, and that is why I did title this presentation as a "solution without tears".

The point is that in order to "inject" nuclear heat into the chemical fuel system, the easiest way to date is to steam-reform natural gas. The reaction is endothermic and absorbs heat at temperatures a High Temperature Reactor (HTR) can provide. The reaction and its heat balances are shown in Figure 11. Essentially two molecules of water are decomposed per molecule of methane, and the oxygen goes into the CO_2. All the energy of the original methane, plus about 30% deriving from nuclear heat, appear in form of hydrogen. The use of HTR nuclear heat to reform methane has been studied for twenty years at the Kernforschungsanlage Jülich in the FRG, and the pilot plants are called EVA (R. Harth et al., 1985; Kernforschungsanlage Jülich GmbH, 1985).

It we now look at world maps for production and transport of natural gas, we find that most of it travels long distance through relatively few *large pipelines*. It is then natural to think that *nuclear power complexes can be put some place along these pipelines and a share or the totality of the gas is processed to hydrogen*. The energy throughput as said would then increase by about 30%.

Naturally the hydrogen does not need to be pure. It could be an h: ogenous mixture, containing also CO, CO_2, N_2 and CH_4. The main scope of the operation would be to *introduce nuclear energy beyond electricity*, on one side, and on the other to reduce fossil fuel demand, beyond the fraction which is going into electricity where the use of nuclear reactors for primary energy is well established.

Penetration of nuclear energy in that area has been fast and the market may be basically exhausted around 2020 or so (Figure 12). In order to continue general penetration as depicted in Figure 5, the use of nuclear energy for these reforming processes should be started soon because these technologies have long induction times and slow (if exponential) initial growth. So one should start early to catch up in 2020 at a level that will keep nuclear industry growing smoothly as from the chart in Figure 5.

Let us come back to the process. EVA reforms CH_4 using nuclear heat and with an efficiency of about 50%. Larger and more developed processes may have higher efficiencies, but for the sake of the argument, let us stick to that round figure. In current EVA experimental set up, the reforming leads to almost equal amounts of CO and CO_2. For the purpose I have in mind, CO_2/CO should be maximized, because I want to extract CO_2. This will permit its disposal outside the atmosphere as we will see.

To have some simple figures to visualize, the Soviet Union is exporting about $50 \times 10^9 m^3$ of natural gas to Europe, coming through two sets of pipelines. This is roughly equivalent to a mean power of 50 GW. Reforming may add about 30% of that, i.e., 15 GW, and the nuclear reactors providing the primary energy should have double that power as the efficiency in the reforming is assumed to be 50%. This means these lines only may employ *30 GW(th) of nuclear power*. The system *could* produce $100 \times 10^9 m^3$ *per years of hydrogen from water*, an equal amount coming from the breaking down of CH_4. Also $50 \times 10^9 m^3$ of CO_2 would be produced.

The same operation at world scale would represent around the year 2000 a *potential market* for nuclear energy and hydrogen production *two orders of magnitude* larger. This means 3000 GW of primary nuclear heat. At present, nuclear plants produce about 1000 GW of primary heat. The potential for electricity production in the year 2000 is again about 3000 GW of nuclear heat, and the addition of the reforming market would certainly be welcome by the nuclear industry. Incidentally, the present almost stoppage in nuclear plants orders is related to the Kondratiev cycle (Marchetti, 1985).

The secondary, but possibly very important, sideline of this operation is that CO_2 *should be separated and disposed, away from the atmosphere*. Separation has to be done anyway, as transporting a CO_2 ballast for thousands of kilometers certainly does not pay. Disposal can be done in various ways, as I proposed in old papers of mine (Marchetti, 1977; Marchetti 1979). It can be done by injecting into the underground, or, in an appropriate form, into the oceans.

For the Soviet Union case I would choose *underground injection*. But where? The most obvious solution would be to reinject CO_2 in the fields were CH_4 comes from. Volume for volume. As CO_2 is a much less ideal gas than CH_4, it could even be injected as a liquid or a near critical gas. There is enough space underground, and an appropriate reinjection could even increase the final output of the fields.

Second case of *reinjection is in oil fields*. The technique starts to be practiced for tertiary recovery, as CO_2 dissolves in oil making it expand and reducing drastically viscosity, so that small clinging drops can flow out of the rocks.

However geologic traps are very abundant everywhere, if usually filled up with water, and oil techniques usually localize them first. This would decouple the reforming complex from oil provinces, but would require extra drilling and infrastructures. So it is to recommend only in very special cases. *The best compromise would be in my opinion to spot an oil province midway to the consumer and localize there all the processing* (see the Case History).

Another reason for putting it near an oil field is that the complex would release large amounts of heat, at temperatures interesting for tertiary recovery. Availability of cooling water otherwise requires a river or a medium size lake.

Coming to gross estimates about the cost of this operation, I would say that *if it pays to make hydrogen by reforming natural gas with nuclear heat*, then the marginal cost of reinjecting CO_2 may be very small or even negative. U.S. oilmen pay up to \$3 per million cubic feet of CO_2 for their tertiary recovery, which is more than the price of an equal volume of CH_4.

The last question is how fast such technology would penetrate if one finds it fits the basic requirements for a commercial start which are usually much more intricate than bare economic break even and profits. The normal penetration times for such things can be 40 years. This means 50% of the market could be covered in 2030. At that time CO_2 emission from CH_4 combustion will be about 75% of total CO_2 emissions, and the effect of our initiative would be to reduce them by 40%. In 2040 the reduction would be above 50%.

This is a magic figure, because *at present about 50% of the CO_2 emitted into the atmosphere is reabsorbed by the ocean sink*. This sink has an extremely large capacity, and the bottleneck is in the kinetic of CO_2 absorption and transport. The amounts depend on the concentration of CO_2 in the atmosphere and at current rates of concentration increases, it is likely that it will remain the same in the future (Bolin, 1985).

This means our remedy can stop CO_2 increase in the atmosphere around 2040, without any sacrifice in energy consumption. And lead to a production of hydrogen in the order of 10^{14} m^3/year.

166

Case History

A Reforming Complex in West Ukraine

The Soviet Union is a large gas exporter to Europe. At present, the amount is in the range of 50×10^9 m/year. Although speculative, the maximum *possible* level of export has been estimated to be an order of magnitude larger.

The location of the large oil and gas provinces in the USSR is shown in Figure 13. The most important, from the point of view of gas, is the West Siberian province (Tyumen). A set of large gas pipelines moves from there to the Moscow area and to Europe (Figure 14). Gas is exported also from the south with lines going north through Kiev, and meeting the other ones in northwestern Ukraine (Figure 15).

There is an oil province, indicated as "West Ukraine" on the map of Figure 13, which could be a very promising site for the installation of a large nuclear gas reforming complex. The fields in fact have passed their prime production life, and are the natural target for tertiary recovery. This means they can be very good customers for CO_2 to be used for flooding. Furthermore, the complex would produce large amounts of heat at temperatures interesting for thermal stimulation. The region has also a system of large rivers, in the Dnjestr basin, and cooling water could be available when necessary.

Complete reforming of the 50×10^9 m^3 of natural gas exported, would require the installation of about 30 GW(th) of nuclear power on the site. The Soviet Union has much experience in graphite moderated reactors, but a limited one in HTR which is the variety very useful for that job. They have been developed mostly in the FRG. Also very large HTRs have not been designed yet, the tendency being toward a modular fail safe system.

The complete reforming would lead to a production of 200×10^9 m^3/year of H_2 and 50×10^9 m^3 of CO_2 to be injected for oil recovery *or* for permanent storage. The cost of the reactors should be reabsorbed in the extra caloric value of the gas sold, if not in the higher commercial value of the hydrogen.

CO_2 for oil recovery however has a premium value. Shell has recently completed the construction of a 500 mile CO_2 pipeline from wells in Colorado to a West Texas field in order to recover an expected 40 million tons of oil through CO_2 flooding. The pipeline operates at a pressure of 140 atm and can transport 2.5×10^9 m^3 of CO_2 per year. Its cost has been 3.3 billion dollars.

Assuming the cost of operation, amortization and interest rates on capital to be around 20%, this brings the indicative value of CO_2 to about 20 US cents/m^3. Current international prices of natural gas in the order of 3\$/MBtu correspond to a round price for natural gas of 10 US cents/m^3.

CO_2 may be assumed to be worth the double of natural gas, obviously in the appropriate location and amounts. This is certainly true for the first GWs of reforming installed, and could be *the driving force for a start*. Furthermore, adding 5% or 10% hydrogen to natural gas changes only marginally its characteristics, in the sense that the consumer would use the same equipment and get the same effect.

This may be a golden opportunity for the West Ukrainian area to get a large reforming complex that can develop into a large petrochemical complex (CO + H_2 available!). The fields would get the CO_2 to enhance their production. The Germans would get the extra energy coming from reactors they may probably produce, having for the moment the best know how in that area.

A mega joint venture in the spirit of perestroika?

MATHEMATICAL APPENDIX

The equations for dealing with different cases are reducible to the general Volterra–Lotka equations

$$\frac{dN_i}{dt} = K_i N_i + \beta_i^{-1} \sum_{n}^{i=1} a_{ij} N_i N_j \quad , \tag{1}$$

where N_i is the number of individuals in species i, and a, β, and K are constants. The equation says a species grows (or decays) exponentially, but for the interactions with other species. A general treatment of these equations can be found in Montroll and Goel (1971) and Peschel and Mende (1986). Since closed solutions exist only for the case of one or two competitors, these treatments mainly deal with the general properties of the solutions.

In order to keep the analysis at a physically intuitive level, I use the original treatment of Verhulst (1845) for the population in a *niche* (Malthusian) and that of Haldane (1924) for the competition between two genes of different fit. For the multiple competition, we have developed a computer package which works perfectly for actual cases (Marchetti and Nakicenovic, 1979), but whose identity with the Volterra equations is not fully proven (Nakicenovic, 1979).

Most of the results are presented using the coordinates for the linear transform of a logistic equation originally introduced by Fisher and Pry (1970).

The Malthusian Case

This modeling of the dynamics   population systems started with Verhulst in 1845, who quantified the Malthusian case. A physically very intuitive example is given by a population of bacteria growing in a bottle of broth. Bacteria can be seen as machinery to transform a set of chemicals in the broth into bacteria. The rate of this transformation, *coeteris paribus* (e.g., temperature), can be seen as proportional to the number of bacteria (the transforming machinery) and the concentration of the transformable chemicals.

Since all transformable chemicals will be transformed finally into bacterial bodies, to use homogeneous units one can measure broth chemicals in terms of bacterial bodies. So $N(t)$ is the number of bacteria at time t, and $\bar{N}$ is the amount of transformable chemicals at time 0, before multiplication starts. The Verhulst equation can then be written

$$\frac{dN}{dt} = aN(\bar{N} - N) \quad , \tag{2}$$

whose solution is

$$N(t) = \frac{\bar{N}}{1 - e^{-(at+b)}} \quad , \tag{3}$$

with b an integration constant, sometimes written as t_0, i.e., time at time 0; a is a rate constant which we assume to be independent of the size of the population. This means that there is no "proximity feedback". If we normalize to the final size of the system, $\bar{N}$, and explicate the linear expression, we can write equation (2) in the form suggested by Fisher and Pry (1970).

$$\log \frac{F}{1 - F} = at + b \quad , \quad \text{where } F = \frac{N}{\bar{N}} \quad . \tag{4}$$

Most of the charts are presented in this form. $\bar{N}$ is often called the *niche*, and the growth of a population is given as the fraction of the niche it fills. It is obvious that this analysis has been made with the assumption that *there are no competitors*. A single species grows to match the resources ($\bar{N}$) in a Malthusian fashion.

The fitting of empirical data requires calculation of the three parameters $\bar{N}$, a, and b, for which there are various recipes (Oliver, 1964; Blackman, 1972; Bossert, 1977). The problem is to choose the physically more significant representation and procedure.

I personally prefer to work with the Fisher and Pry transform, because it operates on *ratios* (e.g., of the size of two populations), and ratios seem to me more important than absolute values, both in biology and in social systems.

The calculation of $\bar{N}$ is usually of great interest, especially in economics. However, the value of $\bar{N}$ is very sensitive to the value of the data, i.e., to their errors, especially at the beginning of the growth. The problem of assessing the error on $\bar{N}$ has been studied by Debecker and Modis (1986), using numerical simulation.

The Malthusian logistic must be used with great precaution because it contains implicitly some important hypothesis:

- That there are no competitors in sight.
- That the size of a niche remains constant.
- That the species and its boundary conditions (e.g., temperature for the bacteria) stay the same.

The fact that in multiple competition the starts are always logistic may lead to the presumption that the system is Malthusian. When the transition period starts there is no way of patching up the logistic fit.

The fact that the niches keep changing, due to the introduction of new technologies, makes this treatment, generally speaking, unfit for dealing with the growth of human populations, a subject where Pearl (1924) first applied logistics. Since the treatment sometimes works and sometimes not, one can find much faith and disillusionment among demographers.

One-to-One Competition

The case was studied by Haldane for the penetration of a mutant or of a variety having some advantage in respect to the preexisting ones. These cases can be described quantitatively by saying that variety (1) has a reproductive advantage of k, over variety (2). Thus, for every generation the ratio of the number of individuals in the two varieties will be changed by $\frac{1}{(1-k)}$. If n is the number of generations, starting from $n = 0$, then we can write

$$\frac{N_1}{N_2} = \frac{R_0}{(1-k)^n} \quad , \quad \text{where } R_0 = \frac{N_1}{N_2} \text{ at } t = 0 \quad . \tag{5}$$

If k is small, as it usually is in biology (typically 10^{-3}), we can write

$$\frac{N_1}{N_2} = \frac{R_0}{e^{kn}} \quad . \tag{6}$$

We are then formally back to square one, i.e., to the Malthusian case, except for the very favorable fact that we have an initial condition (R_0) instead of a final condition ($\bar{N}$). This means that in *relative terms* the evolution of the system is not sensitive to the size of the niche, a property that is extremely useful for forecasting in multiple competition cases. Since the generations can be assumed equally spaced, n is actually equivalent to time.

As for the biological case, it is difficult to prove that the "reproductive advantage" remains constant in time, especially when competition lasts for tens of years and the technology of the competitors keeps changing, not to speak of the social and organizational context. But the analysis of hundreds of cases shows that systems behave exactly *as if*.

Multiple Competition

Multiple competition is dealt using a computer package originally developed by Nakicenovic (1979). A simplified description says that all the competitors start in a logistic mode and phase out in a logistic mode. They undergo a transition from a logistic-in to a logistic-out during which they are calculated as "residuals", i.e., as the difference between the size of the niche and the sum of all the *ins* and *outs*. The details of the rules are found in Nakicenovic (1979). This package has been used to treat about one hundred empirical cases, all of which always showed an excellent match with reality.

An attempt to link this kind of treatment to current views in economics has been made by Peterka (1977).

172

References

Blackman, Jr, Wade A. (1972), A mathematical model for trend forecasts, *Technological Forecasting and Social Change, 3*:441-452).

Bolin, B. (1985), How much CO_2 will remain in the atmosphere? Chapter 3 in *The Greenhouse Effect, Climatic Change, and Ecosystems*, B. Bolin et al. (eds.). Chichester-New York: John Wiley & Sons.

Bossert, R.W. (1977), The logistic curve revived, programmed, and applied to electric utility forecasting, *Technological Forecasting and Social Change, 10*:357.

Debecker, A. and T. Modis (1986), *Determination of the Uncertainties in S-Curve Logistic Fits*. Geneva: Digital Equipment Corporation.

EPRI (1988), The politics of climate. Cover story. *EPRI Journal 13*(4):4-15.

Fisher, J.C. and R.H. Pry (1970), A simple substitution model of technological change. *Technological Forecasting and Social Change 3*:75-88.

Guiot, J. (1987), Reconstruction of seasonal temperatures in central Canada since A.D. 1700 and detection of the 18.6- and 22-year signals. *Climatic Change 10*(3):249-268.

Haldane, J.B.S. (1924), The mathematical theory of natural and artificial selection, *Transactions, Cambridge Philosophical Society, 23*:19-41.

Harth, R., H-F. Niessen, V. Vau, H. Hoffmann, and W. Kesel (1985), *Die Versuchsanlage EVA II/ADAM II – Beschreibung von Aufbau und Funktion*. Jül-1984. Jülich, FRG: Kernforschungsanlage Jülich Gmbh (in German).

ISI Inc. (Institute for Scientific Information), Bibliography search on *Science Citation Index*, Philadelphia, Penn.

Kernforschungsanlage Jülich GmbH (1985), *Nukleare Fernenergie. Zusammenfassender Bericht zum Projekt Nukleare Fernenergie (NFE)*. Jül-Spez-303 (in German).

Lotka, A.J. (1925), *Elements of Physical Biology*. Baltimore: Williams and Wilkins.

Lotka, A.J. (1956), *Elements of Mathematical Biology*. New York: Dover Publications, Inc.

Maier-Reimer, E. and K. Hasselmann (1987), Transport and storage of CO_2 in the ocean – an inorganic ocean-circulation carbon cycle model. *Climate Dynamics 2*(2):63-90.

Marchetti, C. (1977), On geoengineering and the CO_2 problem. *Climatic Change 1*:59-68.

Marchetti, C. (1979), Constructive solutions to the CO_2 problem. In *Man's Impact on Climate*, W. Bach, J. Pankrath and W. Kellog, eds. New York: Elsevier.

Marchetti, C. (1980), Society as a learning system: discovery, invention and innovation cycles revisited. *Technological Forecasting and Social Change 18*:267-282.

Marchetti, C. (1985), Nuclear plants and nuclear niches: on the generation of nuclear energy during the last twenty years. *Nuclear Science and Engineering 90*:521-526.

Marchetti, C. and N. Nakicenovic (1979), *The Dynamics of Energy Systems and the Logistic Substitution Model*. Research Report RR-79-13. Laxenburg, Austria: International Institute for Applied Systems Analysis.

Montroll, E.W. and N.S. Goel (1971), On the Volterra and other nonlinear models of interacting populations, *Rev. Mod. Phys., 43*(2):231.

Nakicenovic, N. (1979), *Software Package for the Logistic Substitution Model*. Research Report RR-79-12. Laxenburg, Austria: International Institute for Applied Systems Analysis.

Nakicenovic, N. (1984), *Growth to Limits: Long Waves and the Dynamics of Technology*. Laxenburg, Austria: International Institute for Applied Systems Analysis.

Neftel, A., E. Moor, H. Oeschger, and B. Stauffer (1985), The increase of atmospheric CO_2 in the last two centuries. Evidence from polar ice cores. *Nature 315*(6014):45-47.

Oliver, F.R. (1964), Methods of estimating the logistic growth function, *Applied Statistics, 13*:57-66.

Pacific Power & Light (1987), Partners eye using power plant CO_2 for enhanced oil recovery. *Electricity Utility Week*, January 26, p. 12.

Pearl, R. (1924), *Studies in Human Biology*. Baltimore: Williams and Wilkins Co.

Pearl, R. (1925), *The Biology of Population Growth*. New York: Alfred A. Knopf, Inc.

Peschel, M. and W. Mende (1986), *The Predator-Prey Model*. Springer Verlag: Berlin-Heidelberg-New York.

Peterka, V. (1977), *Macrodynamics of Technological Change – Market Penetration by New Technologies*. Research Report RR-77-22. Laxenburg, Austria: International Institute for Applied Systems Analysis.

Verhulst, P.F. (1845), in *Nouveaux Memoires de l'Academie Royale des Sciences, des Lettres et des Beaux-Arts de Belgique 18*:1-38.

Volterra, V. (1931), *Lecon sur la Theorie Mathematique de la Lutte Pour la Vie*. Paris: Gauthier-Villars.

174

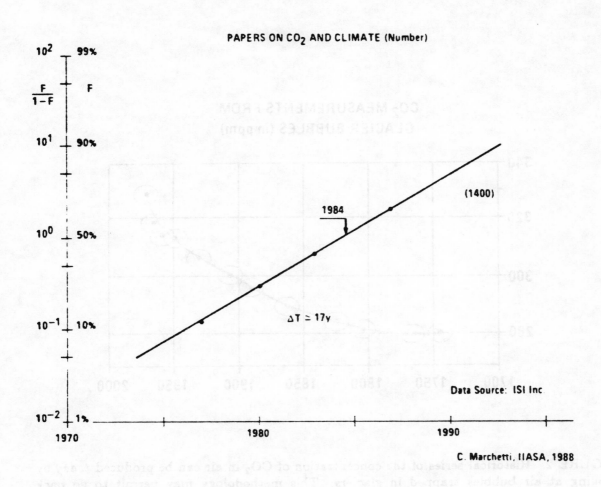

PAPERS ON CO$_2$ AND CLIMATE (Number)

C. Marchetti, IIASA, 1988

FIGURE 1. Cumulative number of papers on CO$_2$ and climate as reported in ISI Inc. (1988). Their number is fitted to a logistic equation (see Mathematical Appendix) and the best fit leads to a calculated saturation level of 1400 papers. The point of maximum production of papers, i.e., the flex of the logistic, was passed already in 1984 and we are now on the ebb side of the wave. The time constant being 17 years means in ≈1992 90% of the papers on the subject will have been written. This logistic shape of publication waves applies to all sort of subjects.

175

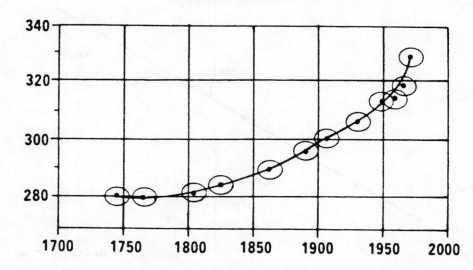

CO$_2$ MEASUREMENTS FROM GLACIER BUBBLES (in ppm)

FIGURE 2. Historical series of the concentration of CO$_2$ in air can be produced *today* by looking at air bubbles trapped in glaciers. This methodology may permit to go back perhaps 100,000 years and compare CO$_2$ levels with prevalent climatic situations that can be evaluated by various types of analyses of sediments (and tree rings for the last 1500 years). This reconstruction may help calibrating the climatic models over which much of the CO$_2$ controversy is based.

ANNUAL CO₂ PRODUCTION
(10⁹ tons of C   )

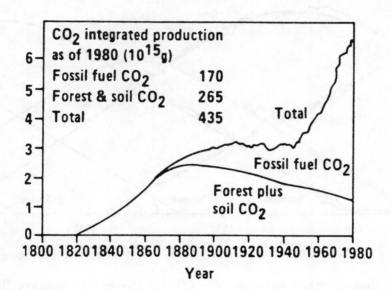

FIGURE 3. Knowledge of the fossil CO_2 emissions and analysis of tree rings for ^{14}C and ^{13}C permits a reasonable reconstruction of the amounts of CO_2 put into the atmosphere by changes in the level of carbon storage in standing forests and soil. From these calculations it appears as the integrated amount of CO_2 that burdens CO_2 levels in air, is due mostly to activities related to agriculture and forests. Only after World War II emissions from fossil fuels have become dominant.

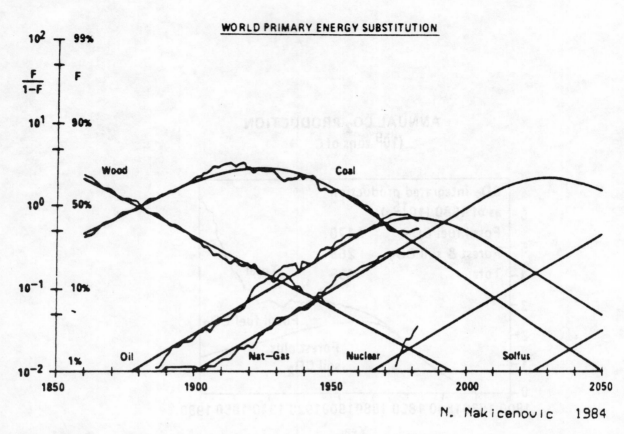

FIGURE 4. The idea that primary energies compete for the energy market like the varieties of a species for the resources of a niche, give a conceptual framework and a mathematics to deal with the evolution of the energy markets. The excellent fitting of the equations (smooth lines) with the statistical data for more than one hundred years give much weight to their use for forecasting. The fast rise of nuclear by respect to a business as usual market penetration equation is probably due to the fact that nuclear sells wholesale and has not the necessity of laying its own distribution grid. A similar phenomenon did occur when natural gas started diffusing in countries where city gas distribution nets did exist already (Marchetti and Nakicenovic, 1979).

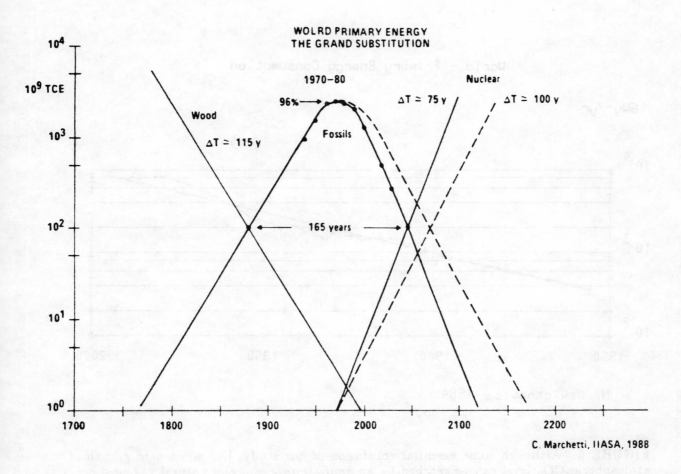

WOLRD PRIMARY ENERGY
THE GRAND SUBSTITUTION

C. Marchetti, IIASA, 1988

FIGURE 5. Fossil fuels can be lumped together by summing their energy contribution to the energy market. We obtain then a line for phasing out wood and other renewable energies. Fossils have a "product life cycle" of about 400 years, after which they will be substituted by nuclear energy in various forms. We gave two time constants for the penetration of nuclear energy to show their effect on the phase out of the fossil fuels.

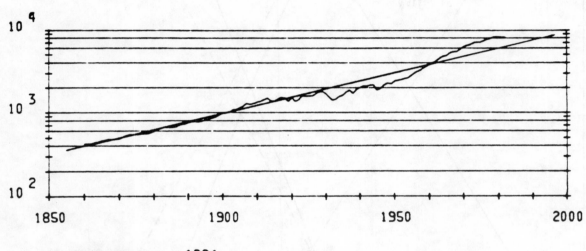

World – Primary Energy Consumption

GWyr/yr

N. Nakicenovic 1984

FIGURE 6. Although some essential conclusion of our study, like when zero growth of atmospheric CO_2 level can be reached by appropriate processing of natural gas need only shares, an insight into quantities is necessary to get a physical feeling of the size of the equipment to be put in place. Here the *primary world energy consumption* is reported including renewable energy and it is fitted with a 2.3% per year growth line. It can be shown that the deviations can be reduced to Kondratiev oscillations with a period of about 55 years. For the following consideration the 2.3% per year growth line will be used.

Figure 7a

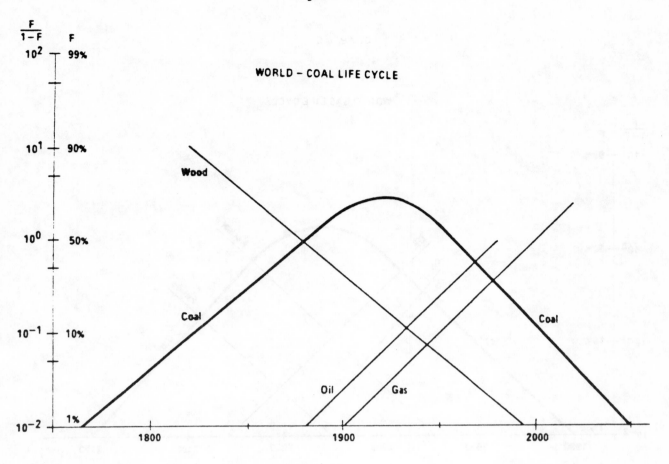

WORLD – COAL LIFE CYCLE

Figure 7b

WORLD OIL–LIFE CYCLE

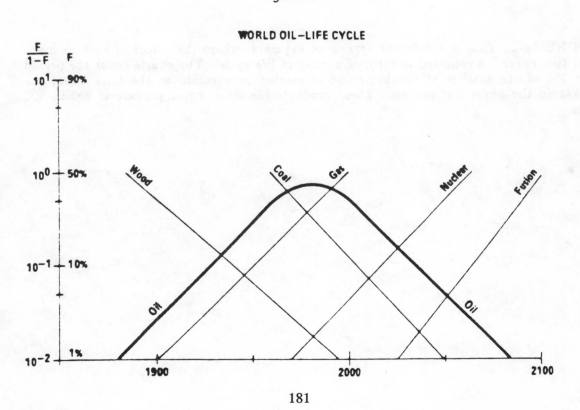

Figure 7c

WORLD GAS LIFE CYCLE

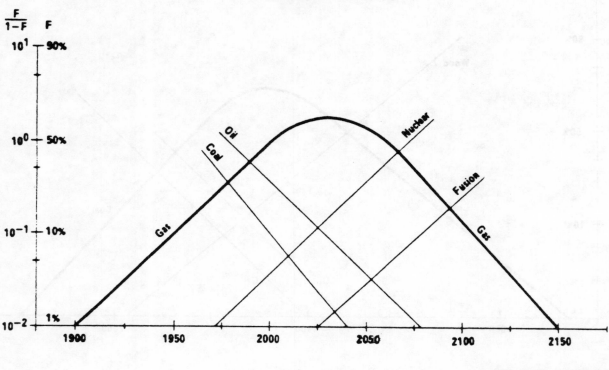

C. Marchetti – IIASA '86

FIGURE 7a-c. This is a different version of Figure 4, where the career of each primary fossil fuel energy is reported in form of a product life cycle. The charts cover the period from 1% of the market at the beginning of market penetration to the final 1% of the market in the phase out process. These products life times cover periods of 200 to 300 years.

Figure 8a

WORLD COAL CONSUMPTION (10^9 Tons)

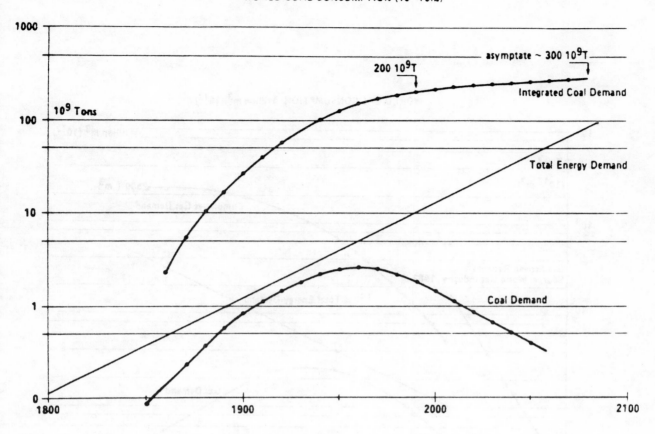

Figure 8b

WORLD OIL CONSUMPTION 10^9 tons

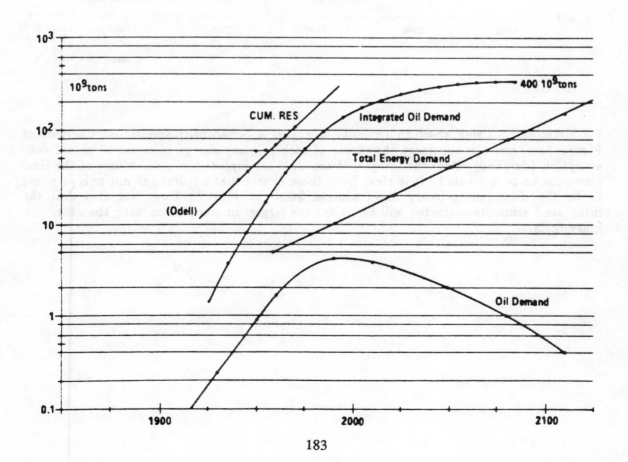

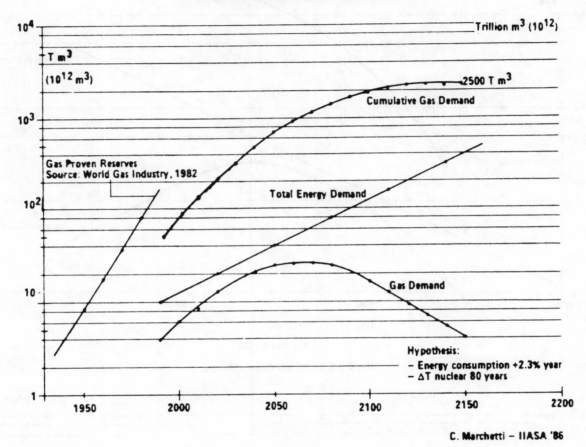

FIGURE 8a-c. Using absolute amounts of Figure 6 (smoothed) and actual fractions of Figure 7a-c, one can construct the career of each primary energy in terms of actual consumption (demand). Also the integrated amounts are reported to give an idea of the final resources to be activated. It is clear from these charts that natural gas not only is going to be the dominant primary fossil sources from the year 2000 on, but also that the integrated amounts extracted will be by far the largest in comparison with the other primary fuels.

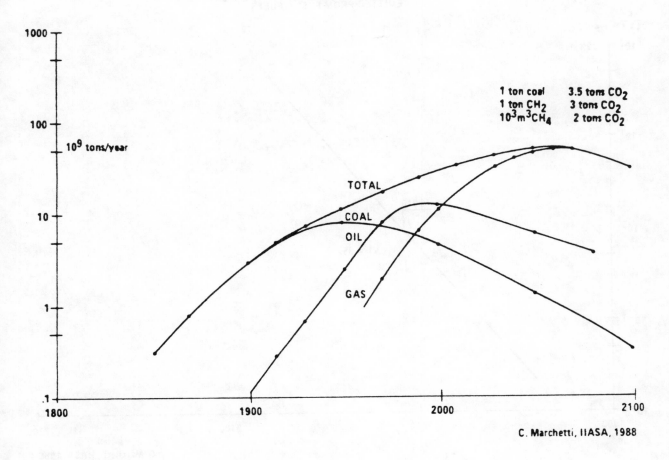

FIGURE 9. The evolution of CO_2 emissions can then be calculated from Figure 8 a-c, applying an appropriate emission factor to the fuels. As the fuels are fairly inhomogeneous, these factors are only indicative. A finer analysis is not worthwhile in view of the use we are going to do of the results.

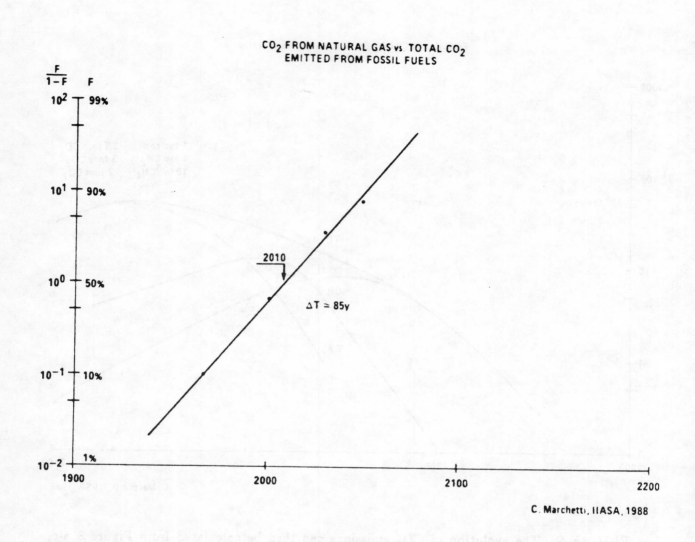

CO$_2$ FROM NATURAL GAS vs. TOTAL CO$_2$
EMITTED FROM FOSSIL FUELS

C. Marchetti, IIASA, 1988

FIGURE 10. We can use again the concept of substitution to the "penetration" of CO$_2$ emitted by burning natural gas by respect to the total CO$_2$ emitted by burning fossil fuels. Because of the increasing dominance of this fuel, 50% of the CO$_2$ emitted will come from it already in 2010. This shows that processes for controlling CO$_2$ emissions to the atmosphere should concentrate on natural gas.

REFORMING PLANT
Gross Energy Balances

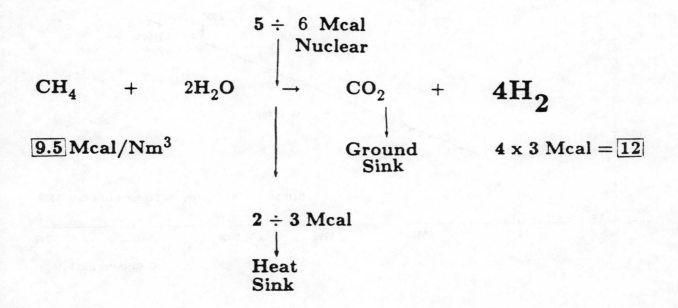

$$CH_4 \quad + \quad 2H_2O \quad \rightarrow \quad CO_2 \quad + \quad 4H_2$$

5 ÷ 6 Mcal
Nuclear

9.5 Mcal/Nm^3

Ground
Sink

4 x 3 Mcal = 12

2 ÷ 3 Mcal

Heat
Sink

FIGURE 11. A skeleton description of the steam reforming process with the help of nuclear heat is here given to show the energy balances. Basically, reforming adds about 30% to methane's energy input. This extra energy obviously comes from the nuclear heat, with an efficiency of 50% or more. This process appears relatively simple and very suited to introduce large amounts of nuclear energy into the fuel system.

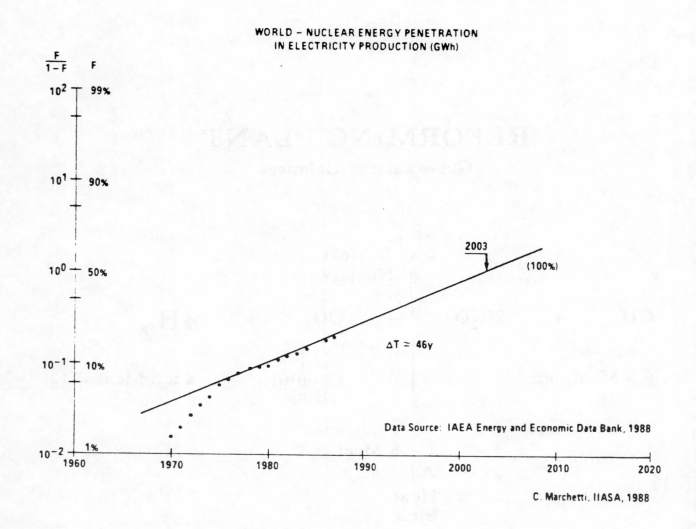

FIGURE 12. Nuclear energy penetration in the market of electricity production (GWh) is here reported. In spite of the very moody descriptions, the situation does not seem to be bleak. Penetration proceeds at a slow but consistent pace and has reached (1987) about 18% of all electricity produced (including hydro). The fitting is done assuming a 100% penetration as a maximum level. The low penetration level reached to date impedes the calculation of a more realistic saturation point (75%?). The analysis is at world level. The chart shows that by 2020 the "conquest" of the electrical system will be substantially concluded, and that, if penetration has to follow the lines of Figure 4, new very important uses have to be found in the meantime.

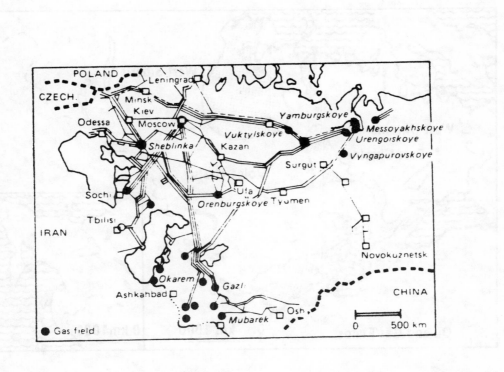

FIGURE 13. This map shows the main gas pipeline system in the Soviet Union.

SOVIET UNION
OIL AND NATURAL GAS PROVINCES

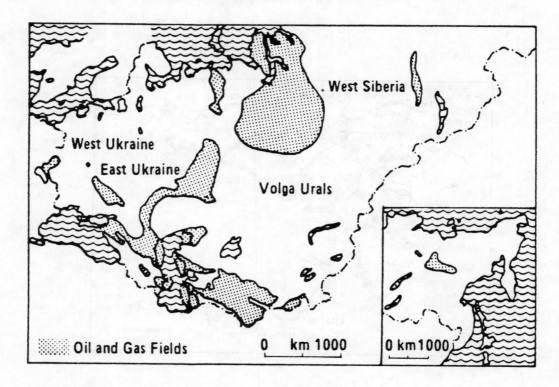

FIGURE 14. This map shows the main oil and gas provinces of the Soviet Union. Attention to the area here marked "West Ukraine" at the triple point of the Soviet Union, Czechoslovakia and Poland.

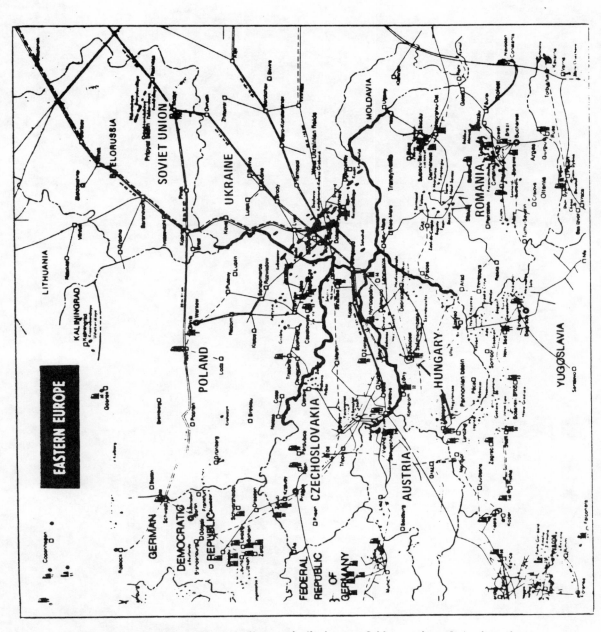

FIGURE 15. This map shows pipelines and oil plus gas fields together. It is clear that most of the exported gas is passing through or near the west Ukrainian fields, where a large reforming plant can be situated as explained in the text.

The Greenhouse Effect and Its Connection with the Area of Fossil Fuels: the Concept of CO_2 Emission Intensity

J. C. Balacéanu, A. Bertrand, J. J. Lacour
Institut Français du Pétrole

Correlations (11) between climatic evolution and the carbon-dioxide rate over the last 160,000 years (Figure 1), together with a recent and constant increase in the carbon-dioxide rate on Earth since the start of the industrial era, have led to the fear that the Earth is warming up in general as a result of the greenhouse effect, in which carbon dioxide lets sunlight pass through but absorbs the infrared rays reemitted by the Earth (1-5). Moreover, carbon dioxide is not the only gas of industrial origin having the same effect. Trace gases such as methane, nitrogen oxides, Freon and ozone have absorption properties that give them a similar impact even at low concentrations (4-6). However, carbon dioxide plays a specific role in that it is involved in cycles and a series of natural balances. Likewise, it is an inevitable product of industrial development or, more basically, of the improvement of the standard of living.

Because of its solubility in seawater, which decreases with the temperature, carbon dioxide is subjected to a series of physicochemical states of equilibria that regulate any increase in its concentration (Figure 2). Furthermore, it is transformed by chlorophyllous assimilation into vegetable matter, including plankton, while breathing and combustion emit it into the atmosphere. These phenomena also have a controlling effect, because a higher concentration of CO_2 increases the development of plants and plankton. For plants, these effects are linked to the nature of the species growing on the ground. Any deforestation, even for the planting of crops, brings about a reduction in CO_2 consumption, for example.

The effects of the warming of the Earth are not apparent enough at present, and pertinent theories are not obvious. In particular, there is no evidence of any accelerating effect that warming by the greenhouse effect leads to a desorption of the CO_2 dissolved in oceans and to a saturation of the atmosphere with water vapor, resulting in an increase in the greenhouse effect and a further increase in temperature (5). Any increase in temperature would also cause evaporation and rainfall, reducing the CO_2 content as well as the growth of plankton. Ice would probably also play an important and unelucidated role, with opposing effects. The "lid" effect would stem from ices making up a barrier to diffusion between the atmosphere and cold oceans

(7).   When ice melts, it causes a redissolution of carbon dioxide in cold water and inhibits the greenhouse effect, amplifying temperature fluctuations linked to changes in the Earth's orbit around the Sun. On the other hand, the melting of ice reduces the Earth's power of reflection and contributes to an increase in the average temperature of the Earth.   Released volcanic carbon dioxide increases the volume of CO_2 and accentuates the greenhouse effect. However, dust particles and aerosols bring down the temperature by increasing reflecting power.

The development of an unquestionable model satisfactorily describing the way the Earth is evolving must take important anthropogenic and complex processes into consideration. In any case, all the considerations that have been made lead us to raise with acuity a problem which has now reached an appreciable magnitude, that of carbon dioxide released through human activity, mostly from energy consumption and human influence on nature.

We will approach this problem as energy specialists. From 1860 to the present, the atmospheric CO_2 content has increased from 275 to 350 ppm (8), with CO_2 released by combustion rising from less than 100 million tons to nearly 20 billion tons per year. The world has been entering an industrial era, leading to population growth and a rise in the standard of living, but involving an increase in energy consumption, with fuels playing and continuing to play a major role.

Let us cite a few figures compiled by the Institut Français du Pétrole on the basis of research made available by the Conservation and Studies Committee of the World Energy Conference (9,10), and with the sole aim of giving orders of magnitude:

	1960	2060
. World population (Ginb)	3.02	9.69
. North	1.00	1.67
. South	2.02	8.02
. World GDP [US 1$ (1980)]	4.05	41.85
. North	3.46	25.55
. South	0.59	16.30
. World Energy Demand (GtOE)	3.50	20.75
. North	2.72	11.91
. South	0.78	8.84
. World Fossil Fuel Demand (GtOE)	2.84	12.70
. North	2.47	7.48
. South	0.37	5.22
. World CO_2 released (Gt)	11.8	54.1
. North	10.2	34.2
. South	1.6	19.9

In such a scenario, the percentage of fossil fuels has dropped from 80 percent in 1960 to 60 percent in 2060 of total energy consumption, but the absolute amount released has been multiplied by 5, reaching 54 Gt, which should be compared to natural transfers between oceans and the atmosphere and between the continents and the atmosphere, estimated at 400 Gt/year apiece.

Careful attention should be paid to the North/South breakdown and to the weight of the North countries (OECD + Eastern Europe) which account for more than 60 percent of the carbon dioxide released in the world in 2060 by fossil fuel consumption (Figures 3 and 4).

What can be done to ensure the indispensable rise in the standard of living and the concomitant growth of energy consumption, and yet still limit CO_2 emission, knowing that any reduction or stagnation in the standard of living and any underdevelopment lead to such social evils as hunger, malnutrition, epidemics and the like, all of which are themselves veritable pollutions of the human species?

Can energy consumption be reduced? Or within energy consumption, can the share of fuel energy be reduced, knowing that 1 tOE of oil releases 3.14 tons of CO_2, 1 tOE of natural gas releases 1.98 tons of CO_2, and 1 tOE of coal releases 5.92 tons of CO_2?

The industrial era was born because mankind learned to discover, produce and use these types of fossil energy, which will continue to be available for a long time.

WORLD FOSSIL FUEL RESOURCES

FOSSIL FUELS	PROVEN RESERVES (R) (GtOE)	1986 PRODUCTION (P) (GtOE)	"DEPLETION" RATIO R/P (Years)	RESOURCES
OIL	95	2.88	33	350 - 450 (1)
NATURAL GAS	95	1.51	60	250 - 400
COAL (2)	900	2.31	400	7000 - 9000

Notes : (1) Of which : Enhanced Oil Recovery: 100 - 150 GtOE
Deep Offshore and Arctic zones: 50 - 100 GtOE

(2) Lignite included

Source : IFP/Department of Economics/1988

Forecasts covering nearly a century (from 1980 to 2060) show that world industrial production will increase by 1.8 percent per year, while energy consumption will increase by only 1.3 percent and that of fossil energy sources by 0.9 percent, thus indicating appreciable progress in energy conservation. These forecasts, however, call for a 1.1 percent per year growth in the release of carbon dioxide, which will increase from 22.8 Gt in 1980 to 54 Gt in 2060, despite considerable advances in reducing emissions.

An analysis of energy independence leads to a definition of the concept of energy intensity or oil intensity. Hence, let us introduce a new fundamental concept for an unavoidable energy pollutant, i.e. the number of metric tons of CO_2 pollutant released per unit of industrial product or CO_2 emission intensity, so as to analyze the degree of pollution associated with the demand for fossil fuels linked to the development of industrial activities. In this context, the number of tons of carbon dioxide released per U.S. $1000 (1980) will decline from 2.2 in 1980 to 1.29 in 2060, thus decreasing by 0.7 percent over this period as a whole. This concept, moreover, does not apply solely to CO_2 but also includes CO and NO, all of which are inevitable pollutants for fossil fuels that use air as an oxidiser.

Can we do better, and further reduce CO_2 emission on a world-wide scale?

Of course, and the example of France is very significant since, for economico-political reasons and because of the availability of its own energy resources, France has undertaken a very goal-oriented energy conservation program since the oil crisis of 1973 and has been concentrating on reducing its consumption of oil and coal, and especially on the possibility of developing its nuclear energy. From 1970 to 1987, for a gross domestic production growth of 63 percent, the increase in French energy consumption was only 30 percent, and consumption of oil and coal decreased respectively by 10 and 48 percent. The share of fossil energy sources in the national energy balance dropped from 90 to 65 percent. Hence, according to its economic plan, by the end of the century France will achieve an intensity of carbon-dioxide release of 0.37 ton per $1000 (1980 dollars) of GDP (Figure 5), whereas for the world as a whole this figure will be 1.77 tons and will still be 1.29 tons in the year 2060.

To what extent can this example be generalized, and is this sufficient?

The effort can involve only the developed countries in the North (OECD and Eastern Europe), the leading CO_2 polluters (73 percent of the emissions in the year 2000 and about 63 percent in 2060). They are developed, rich and large-scale energy consumers. Furthermore, it is easier to orient or even plan their industrial policy.

With oil and gas becoming largely depleted toward the end of the next century, the contribution by coal will become preponderant, and it can be compared with the French balance for 2000, where coal accounts for 8 percent, oil for 33 percent and nuclear energy for 41 percent (Figure 6). The estimated balance for the "North" in 2060 is for coal to contribute 41 percent, oil 6 percent and nuclear energy 19 percent. Hence it is hardly conceivable that the intensity of French CO_2 emissions in 2000 can be imposed in the North, even though this would reduce CO_2 emissions by 25 Gt in 2060. A twofold reduction of emissions by the North by 2060 would reduce global emissions to 37 Gt, which is the order of magnitude of the year 2000, but in the North this would imply the conversion of 3 approximately GtOE/year of coal into nuclear energy (rising to 43 percent of the balance) and/or energy conservation. All these figures show that the efforts aiming to reduce CO_2 emissions to a level equivalent to 5% of the 800 Gt/year of natural exchange fluxes are extremely difficult and costly, and that their even partial implementation would have repercussions on economic development. There are still technical solutions such as the chemical processing of effluents from coal-fired thermal power plants, in particular with reinjection of CO_2 into cold waters or even partly into oil fields to improve recovery. Even though the technologies exist (ethanolamines), their costs (more than $100 per ton of CO_2) and the energy consumption required seem exorbitant.

Within the current economic context, it is obviously important to give priority to improving models for assessing the global carbon-dioxide content in the light of anthropogenic and natural emissions. These include volcanism and natural regulatory effects such as photosynthesis and dissolution. Climatic forecasting models must also be improved in the light of the influence of different factors on the Earth's thermal equilibrium, i.e. the greenhouse effect, astronomic effects, variations in the Earth's albedo, etc. Only a sufficient degree of accuracy in these models will enable the impact of anthropogenic CO_2 to be assessed, thus justifying technical and industrial measures capable of modifying the rate and of orienting the developments of our civilization.

REFERENCES

1   DEAGUE T.K.,
"Global atmospheric consequences of the combustion of fossil fuels"
Journal of the Institute of Fuel, p 153-162, September 1975.

2   WOODWELL G.M.,
"The Carbon Dioxide Question"
Scientific American, p 34-43, January 1978.

3   LAMBERT G.,
"Le gaz carbonique dans l'atmosphère"
La Recherche, p 778-787 (in French), June 1987.

4   "Le gaz carbonique et l'effet de serre"
Etude CEA-CITEPA, 144 p (in French), July 1987.

5   BERTRAND A.,
"L'effet de serre par le CO_2 et les gaz traces"
Revue de l'Institut Français du Pétrole, p 655-669 (in French), October 1987.

6   RAMANATHAN V.,
"The Greenhouse theory of climate change : a test by an inadvertent global experiment"
Science, p 293-299, 15 April 1988.

7   FAURE H.,
"Mécanisme d'amplification du cycle climatique global : l'effet de couvercle de la glace de mer contrôle le CO_2 atmosphérique"
Compte Rendu de l'Académie des Sciences de Paris, p 523-527 (in French), t 305 - série II.

8   KEELING C.D. et al.,
"Atmospheric Carbon Dioxide Variation at Marina Loa Observatory"
TELLUS, p 538-551, 1976, vol 28.

9   Oil Substitution : world outlook to 2020.,
    <u>Conservation Commission/World Energy Conference/New Delhi</u>,
    Graham & Trotman, 1983.

10   FRISCH J.R.,
    "Future Stresses for Energy Resources/Energy Abondance: myth or
    reality"
    <u>Conservation Commission/World Energy Conference</u>, Cannes,
    Graham & Trotman, 1986.

11   JOUZEL.J. et al.,
    "Vostok ice core: a continuous isotope temperature record over
    the last climatic cycle (160,000 years)"
    <u>Nature</u>, p 403-408, 1 October 1987.

FIGURE 1

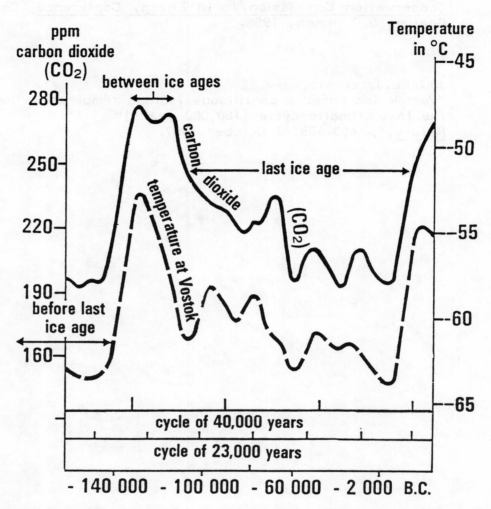

SOVIET CAMPAIGN AT VOSTOK, ANTARCTICA (1980 -1985)

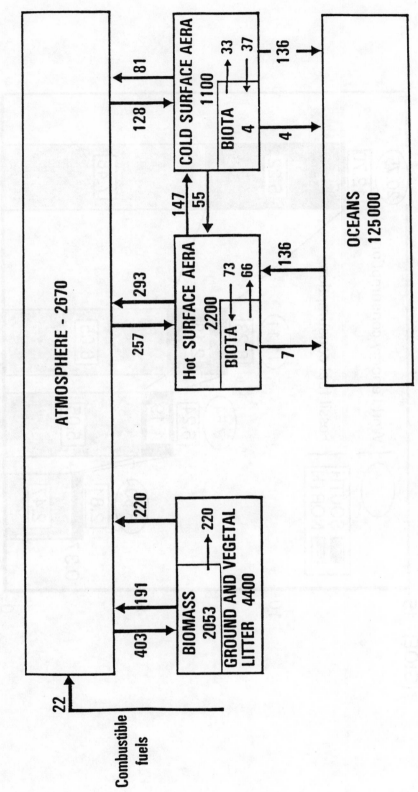

FIGURE 2

GLOBAL CARBON CYCLE (in Gt of CO₂)

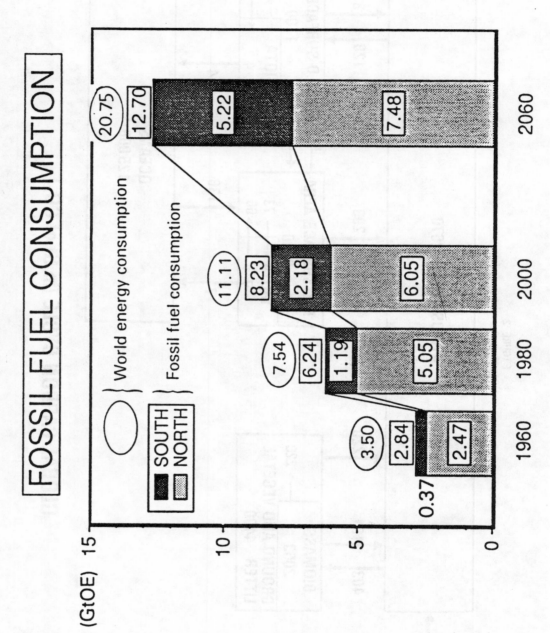

FIGURE 3

FOSSIL FUEL CONSUMPTION

(GtOE)

Source : Conservation and Studies Committee /World Energy Conference/1983 and 1986

202

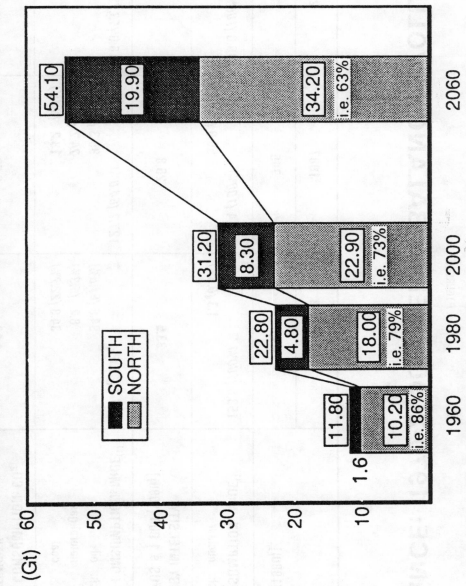

FIGURE 4

FOSSIL FUEL CONSUMPTION : CO2 EMISSIONS

Source : IFP/Department of Economics/1988
from Conservation and Studies Committee /World Energy Conference/1983 and 1986

FIGURE 5

FRANCE: 1970-2000 ENERGY BALANCE EVOLUTION

	1970	1987	2000[2]
GDP [US G$(1980)]	459	745	1 000
ENERGY CONSUMPTION: MtOE[1] — of which: nuclear	153.6 (100%) 1.3 (0.8%)	199.4 (100%) 59.0 (29.6%)	216.0 (100%) 89.0 (41.2%)
↑ ENERGY INTENSITY [tOE/US $ 1 000 (1980)]	33.5	26.8	21.6
FOSSIL FUEL CONSUMPTION: MtOE[1] — of which: oil natural gas coal	137.7 (89.6%) 94.6 (61.6%) 8.2 (5.3%) 34.9 (22.7%)	127.7 (64.0%) 85.3 (42.8%) 24.2 (12.1%) 18.2 (9.1%)	116.0 (53.7%) 71.0 (32.9%) 28.0 (12.9%) 17.0 (7.9%)
FOSSIL FUEL CONSUMPTION: CO_2 EMISSION INTENSITY [t/US $ 1 000 (1980)]	1.13	0.56	0.37

Notes: (1) Corrected for weather.
(2) Median values; from «Energy forecast to 2000» (DGEMP - Observatoire de l'Energie)/April 1987.

FIGURE 6

WORLD: 2060 ENERGY BALANCE

	NORTH	SOUTH	WORLD	NORTH 2060 (%)		FRANCE 2000 (%)	
GDP [US T $ (1980)]	25.55	16.30	41.85				
ENERGY CONSUMPTION (GtOE)	11.91	8.84	20.75	100		100	
— of which: nuclear	2.24	0.56	2.80		18.8		41.2
FOSSIL FUEL CONSUMPTION (GtOE)	7.48	5.22	12.70	62.8		53.7	
— of which: oil	0.75	1.55	2.30		6.3		32.9
natural gas	1.87	1.56	3.43		15.7		12.9
coal	4.86	2.11	6.97		40.8		7.9
FOSSIL FUEL CONSUMPTION: CO₂ EMISSION INTENSITY [t/US $ 1 000 (1980)]	1.34	1.22	1.29	1.34		0.37	

(Values rounded off)

Source: IFP/Department of Economics/1988
from Conservation and Studies Committee/World Energy Conference/1983 and 1986
and DGEMP/Observatoire de l'Energie/April 1987.

205

A SYSTEMS APPROACH TO A TECHNOLOGY-BASED RESPONSE TO THE GREENHOUSE GAS ISSUE[*]

L.D. Hamilton
Biomedical and Environmental Assessment Division
Brookhaven National Laboratory
Upton, NY 11973 U.S.A.

SUMMARY

A systems approach to a technology-based response to the greenhouse gas issue is one in which technologies are evaluated within the context of the energy, economic, and environmental systems of which they are part. We can learn from the systems approach that has gradually evolved to address the acid rain problem in Europe:

- Search for a single set of technology characteristics applicable to all countries is futile.

- Preoccupation with precise cost data is unrewarding.

- In evaluating emission-control technology-by-technology, the problem of allocating joint costs appears as soon as more than one type of emission is considered.

- Close comparison of technologies that are near-substitutes is irrelevant; there are too many omitted criteria influencing the choice.

Systems analysis helps determine which data are important; co-evolution of system models and improved data define is the preferable approach. The goal is improved technology; systems analysis can point to where improvements would pay.

For emission policy decisions, results should be presented within an economic context at the national level. National energy system models already exist for many nations. The EFOM and MARKAL models are being used to evaluate emission-control technologies and strategies for SO_x, NO_x, and CO_2. These national models evaluate the possibilities of fuel switching, technology substitution, and energy conservation as well as direct emission abatement. Macroeconomic effects are also properly determined at the national level.

We therefore propose an international effort to extend or develop national system models of energy and energy-related emissions of greenhouse gases for the industrialized and industrializing nations.

---

[*]Prepared for presentation at IEA/OECD Expert Seminar on Energy Technologies for Reducing Emissions of Greenhouse Gases, Session I, Technology Options and Strategies for Immediate or Short-term Action, Paris, 12-14 April 1989.

The near-term task is to use national models to improve the input data to existing global-climate models. The long-term task is to develop a global model set that capitalizes on the capabilities of national EFOM/MARKAL models. While it is now recognized that there is limited value in comparing technologies on the basis of one kind of emission, it doesn't seem to be as well recognized that there is limited value in a model that treats only one kind of emission. An important criterion for second-generation global emission control models is that tradeoffs between different kinds of emissions can be made within the set of models.

Emission tradeoffs can only be made by changing the nature and the mix of technologies that produce the various kinds of emissions. Therefore, the tradeoff analysis must be made within the entity where these technology/economy decisions are made: the nation.

The technology data-base for greenhouse gas emission control needs to be generated, not simply collected. IEA collaborative R&D agreements should be taxed to develop and report relevant data. Collaboration between system modelers and technologists will serve the goal which is not just better models or better data but better technology.

INTRODUCTION

Increasing atmospheric concentrations of CO_2 and other greenhouse gases have raised concerns among scientists, the public, and government. The United States policy goals in this area include identification of technologies and activities that can limit mankind's adverse effect on global climate, slowing the rate of increase of greenhouse gases in the atmosphere, and, over the long term, stabilizing or reducing atmospheric concentrations of greenhouse gases (U.S., 1987). Anthropogenic sources of greenhouse gases, especially CO_2, are well known; the problem is to evaluate feasibility, cost, market penetration, and effectiveness of emission reduction opportunities.

Fossil and biomass fuel combustion are the principal source of anthropogenic CO_2 emissions and are an important source of other greenhouse gases. They are thus one of the few keys to reducing anthropogenic emissions. They are also the world's principal fuels. Global policies changing world energy use patterns can have great socioeconomic, political, and developmental effects. Such effects are much more certain, and more near-term, than potential climate change.

Strategies to reduce emissions of greenhouse gases from fossil and biomass fuel use must consider the ramifications of any emissions-reduction policy, and the total scope of policy (Table 1) and technical (Table 2) options available. Also, strategies for reducing emissions of greenhouse gases from fossil and biomass fuel combustion must be considered in light of strategies for emission control of other pollutants. In most cases the two are supportive; controls aimed at one purpose will help to achieve the other (Koyama and Ihara, 1988). In some cases, however, the two may conflict; an example is increased CO_2 emissions from use of limestone to control SO_2 emissions.

A systems approach is needed that will include all such options with consideration of costs, state of development, market-penetration timing, public perception, national energy development plans, and available resources at national, regional, and global levels.

A systems approach to a technology-based response to the greenhouse gas issue is one in which technologies are evaluated within the context of the energy, economic, and environmental systems of which they are part.

A systems approach can be contrasted with the commonly taken pollutant-by-pollutant, technology-by-technology approach. Figure 1 diagrams how environmental assessment -- along with the direct costs of energy production and other societal considerations -- influences energy policy.

USING ACID RAIN EXPERIENCE AS AN EXAMPLE

The response to the acid rain issue in Europe is a good example from which to learn. Acid rain also is heavily linked with fossil-fuel combustion emissions; it shares a similar history in that modeling efforts and control technology development expanded from simple considerations of SO_2 to include NO_x, volatile organics, and other pollutants, and, although effects are regional rather than global, multinational solutions emerge as necessary.

A systems approach to the acid rain issue has evolved gradually. The need to evaluate technologies within the context of national energy systems was recognized quite early (UNECE, 1984). A cooperative international program has now been defined (UNECE, 1988) that would coordinate macroeconomic models, integrated assessment models (e.g., RAINS), and technology-oriented energy emission models (e.g., EFOM and MARKAL). Nonetheless, much of the acid rain effort has concentrated on one pollutant (SO_2) and technology-by-technology comparisons.

Lessons learned from the acid rain experience should be considered in planning strategies for reducing emissions of greenhouse gases:

* Cross-country comparisons of technologies can be useful in reaching internationally consistent characterizations, but the search for a single set of technology characteristics applicable to all countries is futile (OECD, 1986).

* Preoccupation with precise cost data is unrewarding. Precision in part of the problem is unrewarding where gross uncertainties prevail in other parts. Systems analysis helps determine which data are important; co-evolution of system models and improved data is the best approach.

* In evaluating emission control technology-by-technology, the problem of allocating joint costs appears as soon as more than one type of emission is considered (OECD, 1987). One cannot omit considerations of whole classes of strategies (e.g. energy conservation) because they lead to benefits beyond their effect on climate change.

* Close comparison of technologies that are near-substitutes is irrelevant; there are too many ignored criteria that affect choice. Instead, focus should be on differing system characteristics of different types of technology, i.e., those that have different combinations of emissions characteristics. In a preliminary analysis of CO_2 constraints on the Netherlands energy system using MARKAL, for

example, Kram and Okken (1988) classify CO_2 emission reduction measures as follows: CO_2 removal, fuel-mix changes, reduction by recycling, reduction by savings in conversion, and reduction by savings in end-use (Figure 2).

* <u>For emission policy decisions, results should be presented within an economic context at the national level</u>. The IIASA RAINS model provides an example.

RAINS, the Regional Acidification Information and Simulation model developed by the International Institute for Applied Systems Analysis (IIASA), exemplifies the first-order systems approach. RAINS uses national-level emissions reduction cost curves (Figure 3) in combination with a long-range atmospheric transport model and models of the impact of sulfur deposition on lakes, forests, and soils to aid international SO_2 emissions control decisions (Amann and Kornai, 1987; Shaw and Amann, 1988). These national cost curves can then be used in conjunction with transport and effects models to estimate the distribution of environmental effects throughout Europe resulting from individual national energy use scenarios or direct inputs of sulfur and nitrogen emissions. In its optimization mode, the model can determine least-cost national emissions requirements to meet predetermined maximum sulfur-deposition limits. However, the level of aggregation required in such integrated assessment models probably cannot adequately reflect country-specific conditions. At present, for example, such models do not take into account the full energy conservation potential in individual countries.

GREENHOUSE GAS EMISSIONS: NATIONAL, REGIONAL, OR GLOBAL MODEL?

The atmospheric model for greenhouse gas build-up and climatic effects must be global, but emissions estimates must be subglobal because of variation in technology and density of energy use. Whether to estimate emissions on a regional or national basis is a key decision.

Early studies -- Global 2000, for example -- used emissions estimates from regions that reflected economic classes of countries (e.g., OECD, centrally-controlled economies, less-developed nations). At the other extreme, useful insights on greenhouse gas emissions can be produced with models even at the community level (Larsson and Wene, 1988). Although there are areas in the world that may be adequately modeled regionally (e.g., much of Africa), <u>we urge an international effort to develop national system models of energy and energy-related emissions of greenhouse gases for industrial and industrializing nations</u>.

There are several reasons why greenhouse gas models should be formulated at the national level:

• National energy systems are distinct.

• Differences in culture, legal apparatus, economic systems, industry, and technology may be more distinct nationally than regionally.

• Data are collected nationally.

• Energy planning is done nationally.

- Environmental control regulations vary nationally; see, for example, IEA (1988).

- Final action on control of emissions of greenhouse gases will be taken by individual sovereign nations.

- <u>National energy models already exist for many nations</u>.

The EFOM and MARKAL models of some nations are used already to analyze emission control measures for CO_2 as well as SO_2 and nitrogen oxides (Koyama and Ihara, 1988, 1989; Kram and Okken, 1988; Yasukawa et al., 1988; Wagner, 1988). No significant changes are needed in the models; only data on alternative emission controls need to be added.

A technology-based response to the greenhouse gas issue is not a novel problem. Measures to control greenhouse gases are not distinct from other emission control decisions. The same set of technologies emits CO_2, SO_2, nitrogen oxides, and other pollutants. Choosing future energy and emission control technologies in the light of restrictions on many kinds of emissions is best accomplished within a single model.

In the language of the recent conference on air pollution in Europe (UNECE, 1988), the integrated assessment model needs to be global in scope for greenhouse gases, but the same technology-oriented energy-emission models can be used nationally.

Following the example of the IIASA RAINS model and the IEA Energy Technology Systems Analysis Project (ETSAP), emissions and emission cost curves should be developed for each of the industrial and industrializing nations.

Examples of such results from the use of the MARKAL model in Japan are given in Figures 4 and 5. Figure 4 shows the effect of increasingly stringent emission limits on SO_x and NO_x through 2025. In this scenario, the limit on NO_x is binding; the measures to reduce NO_x by the required amount also reduce SO_x to below the required linear decrease over time. These same measures also reduce the growing emissions of CO_2 (Koyama and Ihara, 1988).

Figure 5 shows the increasing cost of the Japanese energy system as greater value is placed upon reduced CO_2 emissions for two scenarios (Koyama and Ihara, 1989). These curves correspond to the national cost functions of the RAINS model shown in Figure 3. They differ in that they represent the present value of cost accumulated over 45 years due to a variety of changes in the energy system -- including fuel switching and technology substitution -- rather than the levelized cost in a single year only for direct emission abatement.

National emissions estimates would be based on mutually developed global scenarios, using national technological and policy constraints. The national models or their results (see below) would then be coupled with global atmospheric and climatic change models to explore effects of different strategies and to develop equitable and cost-effective strategies that would become the basis of international agreement on reduction of greenhouse gases.

National models would vary in complexity as do national energy systems. The amount and quality of data available would also vary, but the availability of data from many countries should overall improve data quality. One source of

211

data improvement would be the opportunity to investigate why the discrepancies among nations with similar economic and social characteristics; this would provide quality control as well as a better understanding of the factors that contribute to differences in emissions. Another source of data improvement is that data gaps could be filled by appropriately tailored modification of data from similar countries rather than relying on potentially inappropriate data because of data availability, as was done in the past.

A program of assistance to implement MARKAL/EFOM would be required for less-developed countries. Even if all countries did not participate, however, participation of enough countries would provide sufficient scope of data to improve greatly the ability to estimate emissions and emission-reduction cost curves in nonparticipating countries.

EFOM/MARKAL FAMILIES OF MODELS AND THEIR APPLICATION TO THE GREENHOUSE GAS ISSUE

MARKAL and EFOM are two similar dynamic linear programming models used to describe the energy systems of a number of countries. MARKAL (an acronym for "Market Allocation") was originally designed to evaluate possible impacts of new energy technologies on national or regional systems. MARKAL is demand driven; end-use energy demands are exogenously specified for each demand sector and for each time period. The number and character of the demand sectors and the number and duration of time periods are specified by the user. The interactions between MARKAL and the natural and anthropogenic environment are shown in Figure 6.

MARKAL was developed within the framework of the International Energy Agency (Fishbone et al., 1983). It is available at computer centers in at least 18 locations around the world (Table 3) and has been applied in more than 20 countries, regions, and localities. It has been used in industrialized nations of Europe, North America, and the Far East, and in developing nations in Asia, South America, and the Pacific (Figure 7). EFOM (Energy Flows Optimization Model) is a similar model developed by the Commission of the European Communities (CEC) (Van der Voort, 1984). EFOM is currently being used for energy-environment studies by the nations of the European Community (Morgenstern, 1988; Patzak, 1988).

The widespread use of EFOM/MARKAL and their demonstrated international acceptance as an energy and energy-environment planning tool argue for their application to global study and planning of energy-related emissions of greenhouse gases. This is not a new application for MARKAL. As early as 1980, 15 industrialized nations[*] completed a systems analysis using MARKAL that included preliminary analysis of CO_2 emissions in each nation. In the analysis, several hundred new and existing energy technologies competed for market shares over the 1980-2020 time span. CO_2 emissions were produced for each country under a variety of assumed policy options and global constraints. All scenarios studied predicted a steady increase in CO_2 emissions throughout the 45-year time span with no tendency to level off. Only one scenario achieved a reduction in

---

[*]Participating countries included Austria, Belgium, Canada, Denmark, F.R. Germany, Ireland, Italy, Japan, New Zealand, Norway, Spain, Sweden, Switzerland, the United Kingdom and the United States. The Commission of European Communities also contributed and conducted a separate analysis for the aggregated European Economic Community.

the rate of increase to less than 1% per year. Since the scenarios bracketed existing national energy policies of the time, it was clear that a global objective to reduce or even level off CO_2 emissions would require major revisions in national energy policies (Sailor, 1980).

More recently, the IEA program has continued with participation of nine countries.[+] This effort focused explicitly on expanding the environmental relationships of the energy model, including emissions of CO_2 and acid-rain related gases. Preliminary products of the effort were presented in a workshop jointly sponsored by the IEA Energy Technology Systems Analysis Project and the IIASA Acid Rain Project (ETSAP 1, 1988; ETSAP 2, 1988).

MACROECONOMIC INTERACTIONS WITH THE ENERGY SYSTEM

Macroeconomic effects are properly determined at the national level.

Although pollution-control programs have had small macroeconomic impacts to date (UNECE, 1988), they may become important with larger emission reductions. Energy systems models (predecessors of MARKAL) can be successfully linked with economic input-output models (Behling et al., 1976). A similar linkage was made between MARKAL and an econometric model and applied to energy-environmental analyses in Japan. The structure of this model is shown in Figure 8. Figure 9 shows how the changing economic structure changes the real cost of energy technologies over time (Sato and Yasukawa, 1988). Adjustments in the economic system will almost certainly reduce the cost of emission controls below that estimated by an energy-system model alone (Bergman and Carlsson, 1988).

Were the earth's climate to change significantly, important feed-back processes would become important to long-range predictions. Some of these processes are a direct part of the global climatic system and should be included in the global climatic model. Others are natural ecological process, such as faster plant growth with increasing temperature and increasing CO_2 levels in the atmosphere. Of importance to the energy-related anthropogenic emissions of greenhouse gases are changes in end-use energy demand resulting directly from climate change (e.g., increased air conditioning loads) and indirectly, through climate-mediated social change. As climate changes, incentives to conserve energy or to make energy technology changes to reduce greenhouse gas emissions may well intensify. These changing energy demands and social patterns will be reflected in the energy and economic systems. The operation of these feedback loops are diagramed in the coupled energy-economic-global climate model shown in Figure 10.

The solid lines on this figure show the linking of an energy systems model to a global climate effects model in its simplest form, based on the example of national cost-curves as input to a continental environmental model in the IIASA RAINS model for acid rain emission controls (Amann and Kornai, 1987). The feedback from environmental effects through an econometric model to the national energy system model is shown in broken lines.

[+]Participating countries are Canada, F.R. Germany, Italy, Japan, the Netherlands, Sweden, Switzerland, the United Kingdom, and the United States.

STRATEGIES FOR SHORT-TERM ACTION

In the short term, MARKAL results can be used to improve the data input to first-order global CO_2 models.

Assumptions about future energy systems are fundamental to the results of the first generation of global CO_2 models. Nordhaus and Yohe (1983), for example, found that the atmospheric concentration of CO_2 in 2100 calculated with their model was most sensitive to "ease of substitution between fossil and nonfossil fuels."

Table 4 shows that many of the required major input assumptions of the Edmonds-Reilly model can be provided by MARKAL. Clearly, several of the inputs that can be specified independently by the user -- e.g., technological change in energy production, environmental costs of energy production, and market penetration of supply technology -- are interrelated. As outputs of MARKAL, these inputs to the global CO_2 model would be more assuredly feasible and consistent.

In a Monte Carlo test of the sensitivity of CO_2 emissions to each of 79 parameters in their model, Edmonds and Reilly (1985) found that "end-use energy efficiency" was second in importance. End-use energy efficiency is defined as the rate at which energy use per unit output would decline over time as a consequence of technological change, process improvements, and changes in the mix of goods produced if population, GNP, and prices were constant. End-use energy efficiency by energy carrier and demand sector is specified for 1975 with no geographical breakdown, as shown in Figure 11 . The user is able to define improvement over time from zero to 3% per year with a default value of 1% (as shown), also for the entire globe (Edmonds and Reilly, 1986).

Clearly a sample of the national model results showing plausible trends of end-use efficiency over time should be an improvement over such simplified input assumptions.

STRATEGIES FOR LONG-TERM ACTION

While it is now recognized that there is limited value in comparing technologies on the basis of one kind of emission, it doesn't seem to be as well recognized that there is limited value in a model that treats only one kind of emission.

RAINS in its present version was described as a first-order model (UNECE, 1988). It should be pointed out that any such model, regional or global, will never be more than a first-order model if it treats only one kind of emission; that is to say, if it cannot address the question of tradeoffs in the control of different kinds of emissions. The ability to track different emissions is not enough.

If the near-term task is to use national models to improve the input data to an existing global CO_2 model, the longer-term task is to develop a global model that capitalizes on the capabilities of the national EFOM/MARKAL models.

The first step in producing a second-generation global model is to establish the guidelines for what it should be able to do.

214

<u>An important criterion is that tradeoffs among different kinds of emissions can be made within the set of models</u>.

The acid rain problem won't go away. In certain countries, acid rain emissions may be the limiting factor; in others, greenhouse gas emissions. The global model set should be able to analyze these tradeoffs among countries. Besides, the global model set should be able to treat the issue of different emission standards among countries either for reasons of economics (stricter emission standards for acid rain, for example, where the damaging emissions originate) or for reasons of equity (developing countries making the case that they should not abide by the same standards as the developed countries).

<u>Tradeoffs can only be made among the set of technologies that produce the various kinds of emissions. Therefore, the tradeoff analysis must be made within the entity where these technology/economic decisions are made: the nation</u>.

The preference for treating macroeconomic effects at the national level has been pointed out above. National economies may be linked (maybe even tightly linked), but they are distinct. Economic decisions are made nationally in the end, not globally. Moreover, coupled national energy-econometric models are presently in being.

<u>Modeling frameworks exist for a global model that exchanges data with national models, specifically in the form of price information</u>.

Algorithms have been developed that would permit emission tradeoffs to be made at the national level in consonance with global as well as regional or national constraints. For example, the Dantzig-Wolfe decomposition algorithm provides a modeling mechanism for coupling national linear programming models with a global model (Dantzig and Wolfe, 1960). The implementation of this concept was developed under sponsorship of the Commission of the European Communities for coupling the EFOM models of the EC nations (Ho et al., 1979; Ho and Loute, 1983).

Another approach is use of the "cobweb" algorithm to link linear programs in which individual submodels are linked without a master model. This approach makes it possible to use exchange prices that are not marginal costs generated by the program itself (Berger et al., 1988a). MARKAL is currently being used to evaluate electricity exchanges between Quebec and New York State (Berger et al., 1988b).

BETTER DATA, BETTER TECHNOLOGY

<u>The technology data-base needs to be generated, not simply collected; IEA collaborative agreements should be taxed to develop and report relevant data</u>.

A systems approach to a technology-based response to the greenhouse issue begins with an inventory of the technologies that affect greenhouse gas emissions. As indicated in Figure 12, the systems view extends from fossil fuel extraction through fuel treatment and energy conversion to end use. A better technology data base needs to be generated, not simply collected. A potential source of the broad range of expertise needed to develop such a data base exists in IEA collaborative R&D agreements.

215

Generating a CO_2 reduction data base has begun at Brookhaven National Laboratory. Investigations of advanced technologies conducted at Brookhaven have considered three technological applications: (1) removal, recovery, and disposal of CO_2; (2) improved energy technology; and (3) hydrogen economy based on fossil fuel (Steinberg and Cheng, 1987). A scheme for fossil energy system and CO_2 emission control technologies is shown in Figure 13.

Removal, Recovery and Disposal of Carbon Dioxide

A systems study for the removal, recovery, and disposal of CO_2 from fossil fuel power plants in the U.S. was made (Steinberg and Cheng, 1984). The technologies defined in this study are among those evaluated with MARKAL models of the Netherlands (Figure 2) and Japan (Figure 5).

The U.S discharges about 30% of the world's total man-made CO_2 emissions. Utilities, which constitute large central sources of CO_2, account for about 30% of the total U.S. emissions. Estimates indicate that a reduction in the annual incremental atmospheric CO_2 content of up to 10% would be achieved by application of CO_2 control systems to the U.S. utility industry.

The projected removal and recovery system employed at each power plant site is based on an absorption/stripping system using an improved stack gas scrubbing solvent. The recovered CO_2 is liquefied for transmission to the ultimate disposal site. The CO_2 control process is integrated with the power plant operation in that low back-pressure steam from the power-generating turbines is used to regenerate the solvent. In this manner, for 90% removal of CO_2, the power plant efficiency is reduced by only 3 percentage points from conventional plant efficiency of 38% to a value of 35% with the control system.

Three methods of ultimate disposal were considered: (1) injection into 300-meter-deep ocean below the thermocline, located approximately 100 miles off the shores of the U.S. (No CO_2 exists in the waters below the thermocline and, thus constituting an unlimited sink or storage site for CO_2); (2) storage in depleted oil and gas wells; and (3) storage in excavated salt domes.

Improved Energy Technology

Brookhaven has explored the potential for better use of energy through improvements in energy technology to reduce future buildup of atmospheric CO_2 (Cheng et al., 1986). The efficiencies of end-use technologies vary widely among countries. The evolution of these efficiencies and differences among countries indicate the magnitude of the potential for global energy savings through development and implementation of improved end-use technologies. Energy technologies were grouped by evolutionary stage as (1) now in use; (2) emerging; and (3) advanced. The energy-efficiency levels associated with each technology were evaluated in terms of (1) present; (2) achievable; and (3) theoretical maximum values. The improved technologies were considered with regard to the potential energy savings achievable by their introduction into the residential and commercial, industrial, and transportation end-use sectors.

The possible energy savings due to improved end-use technologies were estimated first in terms of efficiency improvements applicable to specific energy services for each end-use sector. Then the use of the electricity-generating technologies was assumed to determine the additional savings in fuel. This approach allowed estimating of the possible minimum fuel requirements for each energy end-use service, and it therefore allowed determination of the potential savings of fossil fuels and associated reduction of CO_2 emissions to expected from technology improvements that might be adopted globally.

Hydrogen Economy Based on Fossil Fuel

A new technology process being developed at Brookhaven, called HYDROCARB, offers the possibility of generating and using hydrogen from fossil fuel sources without generating any CO_2 whatsoever. A two-step process was devised that consists of (1) hydrogenating coal (or any other fossil fuel) to produce methane, followed by (2) thermal decomposition of methane to pure carbon and hydrogen (Steinberg, 1987). The process can be made highly energy and mass efficient and thus economical. The net effect is a cracking of the fossil fuel to its elements while the oxygen and sulfur impurities are recovered with a small loss of carbon or hydrogen. The net energy that can be recovered in hydrogen generation ranges from 56% for natural gas to 24% for coal. This is not inconsequential, considering that the so-called "hydrogen economy" based on solar or nuclear energy must generate electricity first at approximately 30% efficiency and then use the electrical energy in an electrolytic cell to decompose water to hydrogen and oxygen -- a process, at best 80% efficient, resulting in an overall efficiency of 24%. Moreover most of the energy is in the clean carbon produced that can be placed into "monitored retrievable storage." Should it be found in the future that the CO_2 greenhouse effect is not materializing, then this clean carbon can be readily extracted and used as a clean source of energy.

IEA Collaborative Agreements as a Source of Data

A technology-based response to the greenhouse gas issue may call upon virtually all branches of energy technology either for efficiency improvements or as replacements for other technologies. Developing technology improvements and developing data to describe them can go hand in hand.

The participants in the IEA collaborative R&D agreements constitute a unique international resource for providing the data needed to assess potential improvements in energy systems as well as to further technology development. Their expertise should be drawn into the system for making these evaluations. With proper feedback of the system model results, this process can help establish the direction for further improvement.

The ultimate goal is not simply better data or better models but better technology.

CONCLUSIONS

We recommend a systems approach to guiding anthropogenic emissions reduction of greenhouse gases from the energy system. In developing a systems approach, one draws on the experience gained in energy systems planning and in planning emissions reductions for acid rain control -- an environmental problem with many similarities. The key lesson is that, the problem is global, yet emissions estimates need to be made on the national level, the reasons being to improve input data quality and to facilitate implementation. Technology and energy-demand characteristics are likely to depend on national characteristics and policies; assumptions based on technology characteristics, costs, energy demand, or energy policy may not hold from one nation to the next. From the standpoint of implementation, final implementation must be at the national level; international negotiations on emissions reduction will be facilitated if realistic, specific estimates of ramifications at the national level are available.

Because of their wide availability, acceptance as a national energy planning model, and experience in coupling them with CO_2 emissions data, the EFOM/MARKAL family of models is the choice for the national energy systems model to couple with a global climate change model.

ACKNOWLEDGEMENTS

I thank Drs. Douglas Hill, Samuel Morris, and Meyer Steinberg for help in preparing this paper.

REFERENCES

Amann, M. and G. Kornai. 1987. Cost functions for controlling SO_2 emissions in Europe (WP-87-065), International Institute for Applied Systems Analysis, Laxenburg, Austria.

Behling, D.J., Jr., R. Dullien, and E. Hudson. 1976. The relationship of energy growth to economic growth under alternative energy policies (BNL 50500), Brookhaven National Laboratory, Upton, NY.

Berger, C., A. Haurie, E. Lessard, R. Loulou, and A. Paskievici. 1988a. MARKAL: Development of materials, financial and multiregional capability, in ETSAP 2, pp. 133-154.

Berger, C.R., R. Dubois, A. Haurie, E. Lessard, and R. Loulou. 1988b. Assessing the dividends of power exchange between Quebec and New York State: a system analysis approach, GERAD, Montreal, Canada.

Bergman, L., and A. Carlsson. 1988. Emission control cost estimating using a model of the Swedish energy markets, in ETSAP 2, pp. 1-12.

Cheng, H.C., M. Steinberg, and M. Beller. 1986. Effect of energy technology on global CO_2 emissions (DOE/NBB-0076, TRO30), U.S. Department of Energy, Washington, DC.

Dantzig, G.B., and P. Wolfe. 1960. Decomposition principle for linear programs, Operations Research, Vol. 8, 1960, pp. 101-111.

Edmonds, J.A., and J.M. Reilly. 1985. Future global energy and carbon dioxide emissions, in J.R. Trabalka (ed.), Atmospheric Carbon Dioxide and the Global Carbon Cycle, DOE/ER-0239, U.S. Department of Energy, Washington, DC.

Edmonds, J.A., and J.M. Reilly. 1986. The IEA/ORAU long-term global energy-CO_2 model: personal computer version A84PC, ORNL/CDIC-16, CMP-002/PC, Oak Ridge National Laboratory, Oak Ridge, TN.

ETSAP 1. 1988. Toward estimating national energy emission control costs, Proceedings of a joint workshop held by the Energy Technology Systems Analysis Project, International Energy Agency, and the Acid Rain Project, International Institute for Applied Systems Analysis, Vol. 1, Summary Record, Brookhaven National Laboratory, Upton, NY.

ETSAP 2. 1988. Toward estimating national energy emission control costs, Proceedings of a joint workshop held by the Energy Technology Systems Analysis Project, International Energy Agency, and the Acid Rain Project, International Institute for Applied Systems Analysis, Vol. 2, ETSAP Papers, Brookhaven National Laboratory, Upton, NY.

ETSAP 3. 1988. Annual report of the Energy Technology Systems Analysis Project to the Committee on Energy Research and Development, International Energy Agency, June 1988.

Fishbone, L.G., G. Giesen, G. Goldstein, H.A. Hymmen, K.J. Stocks, H. Vos, D. Wilde, R. Zulcher, C. Balzer, and H. Abilock. 1983. User's guide for MARKAL [BNL/KFA version 2.0], a multi-period, linear-programming model for energy systems analysis, (BNL 51701), IEA Energy Technology Systems Analysis Project, Brookhaven National Laboratory, Upton, NY, USA, and Kernforschungsanlage Jülich, Jülich, FRG.

Ho, J., E. Loute, Y. Smeers, and E. Van der Voort. 1979. The use of decomposition techniques for large-scale linear programming models, in A. Strub, (ed.) Energy models for the European Community, Commission of the European Communities, Brussels.

Ho, J.K. and E. Loute. 1983. Computational experience with advanced implementation of decomposition algorithms for linear programming, Mathematical Programming, 27 (1983), 283-290.

IEA. 1988. Emission controls in electricity generation and industry, International Energy Agency, Organisation for Economic Co-operation and Development, Paris.

Koyama, S., and S. Ihara. 1988 A preliminary energy-environment analysis of Japan, Part I (ETL), in ETSAP 2, pp. 27-58.

Koyama, S., and S. Ihara. 1989. Preliminary analysis on energy system structure for reducing carbon dioxide emission, presented to the Institute of Electrical Engineers, April 5, 1989, Tokyo. (In Japanese)

Kram, T. and P.A. Okken. 1988. Preliminary results of CO_2-constraints on the Netherlands energy system using MARKAL, presented at the Fifth workshop, IEA Energy Technology Systems Analysis Project (Annex III), Nov. 1988, Tokyo.

Larsson, T. and C.-O. Wene. 1988. Effects of limits on emissions of greenhouse gases on the energy system, in ETSAP 2, pp. 83-92.

Morgenstern, T. 1988. CEC energy and environmental project optimal control strategies for reducing emissions from energy conversion and use, in ETSAP 1, pp. 65-69.

Nordhaus, W.D., and G. Yohe. 1983. Future carbon dioxide emissions from fossil fuels, in Changing Climate, National Academy Press, Washington, DC.

OECD. 1986. Understanding pollution abatement cost estimates, Environment Monograph No. 1, Organisation for Economic Co-operation and Development, Paris, March 1986.

OECD. 1987. Energy and cleaner air, costs of reducing emissions, summary and analysis of Symposium Enclair '86, Organisation for Economic Co-operation and Development, Paris.

Okken, P.A. 1988. Impacts of environmental constraints on energy technology in the Netherlands, in ETSAP 1, pp. 50-52.

Patzak, R. 1988. CEC energy and environment project: extension of the CEC-EFOM model with special regard to environmental problems, in ETSAP 1, pp. 70-72.

Sailor, V.L. 1980. Future CO_2 emission levels for fifteen IEA countries, Annex II in V.L. Sailor and S. Rath-Nagel, MARKAL, A computer model designed for multi-national energy systems analysis, presented at the Seminar on Modelling

Studies and their Conclusions on Energy Conservation and Its impact on the Economy, 24-28 March, Washington, D.C.

Sato, O. and S. Yasukawa. 1988. Combined E-I/O model and its application to energy-environment analysis, in ETSAP 2, pp. 115-131.

Shaw, R.W. and M. Amann. 1988. IIASA RAINS model, in ETSAP 1, pp. 26-32.

Steinberg, M., H.C. Cheng, and F. Horn. 1984. A systems study for the removal, recovery, and disposal of carbon dioxide from fossil fuel power plants in the U.S., (BNL 35666), Brookhaven National Laboratory, Upton, NY.

Steinberg, M. 1987. Clean carbon and hydrogen fuel from coal and other carbonaceous raw materials, (BNL 39630), Brookhaven National Laboratory, Upton, NY.

Steinberg, M. and H.C. Cheng. 1987. Advanced technologies for reduced CO_2 emissions, (BNL 40730), Brookhaven National Laboratory, Upton, NY.

UNECE. 1984. Air-borne sulphur pollution: effects and control, United Nations Economic Commission for Europe, Air Pollution Studies 1, United Nations, NY.

UNECE. 1988. Conference on air pollution in Europe: environmental effects, control strategies, and policy options, incorporating a workshop on the application of cost-effective control strategies, executive report, Norrtalje and Stockholm, Sweden.

U.S. 1987. Global Climate Protection Act of 1987, 101 Stat. 1331.

Wagner, H. J. 1988. CO_2 Emissions in the Federal Republic of Germany, presented to the Fifth Workshop, IEA Energy Technology Systems Analysis Project (Annex III), 31 October - 4 November 1988, Tokyo.

Van der Voort, E., E. Donni, C. Thonet, E. Bois d'Enghien, C. Dechamps, and J.F. Guilmot. 1984. Energy supply modelling package EFOM-12C, Mark 1, Commission of the European Communities, Brussels.

Yasukawa, S., O. Sato, Y. Tadokoro, Y. Nakano, T. Nagano, and H. Shiraki. 1988. A preliminary energy- environment analysis, part II (JAERI), in ETSAP 2, pp. 59-70.

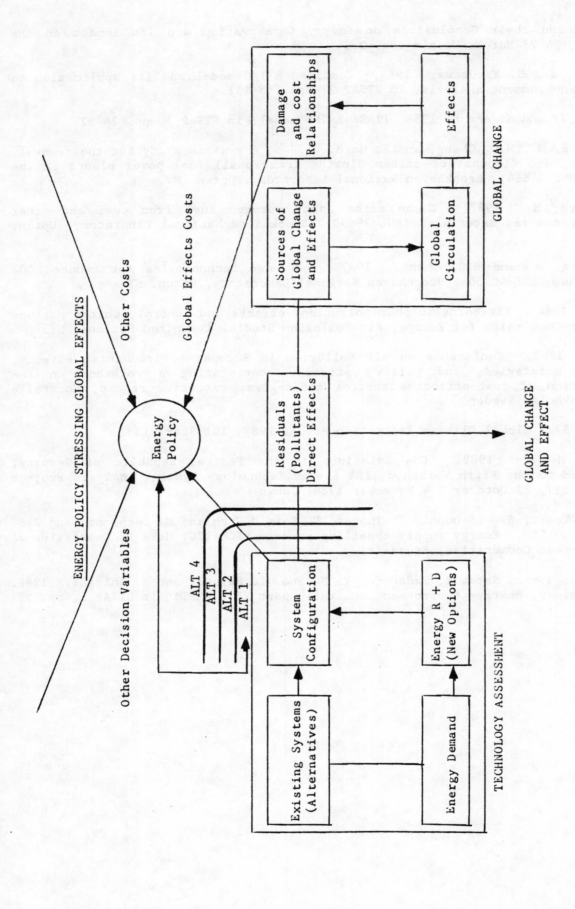

Figure 1. The position of global change in overall energy assessment.

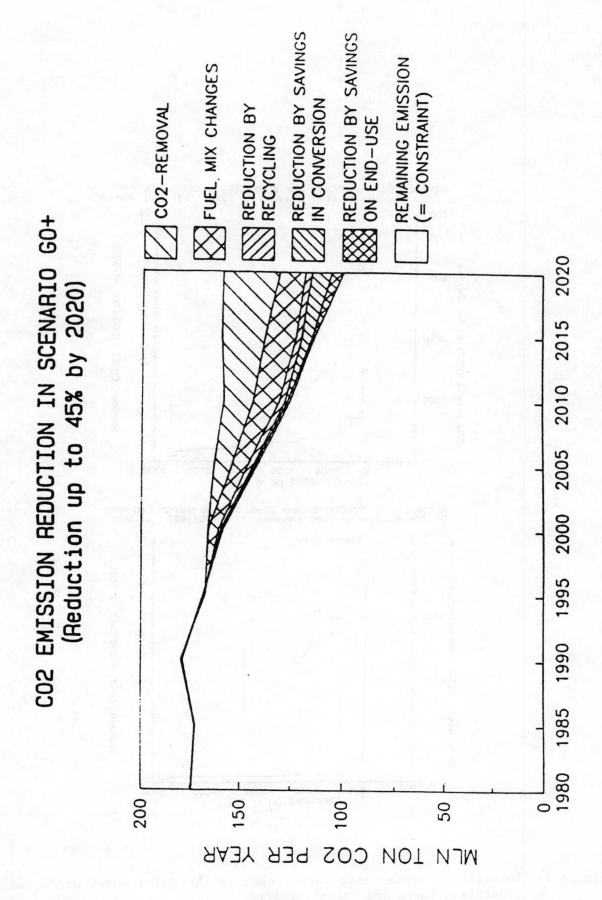

Figure 2. Effect of CO_2 emission reduction measures in the Netherlands (Kram and Okken, 1988).

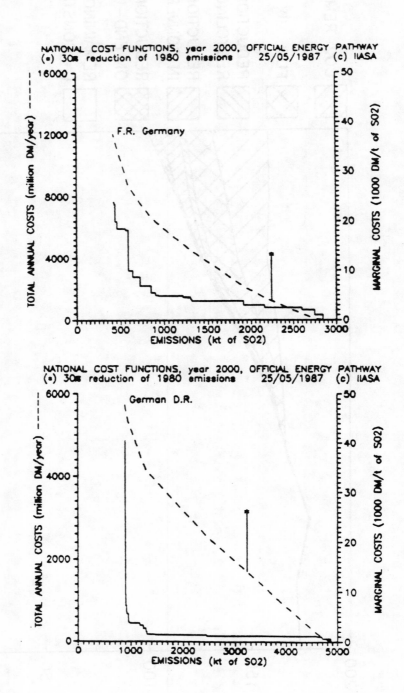

Figure 3.   Emission reduction cost curves used in the RAINS model differ among countries (Amann and Kornai, 1987).

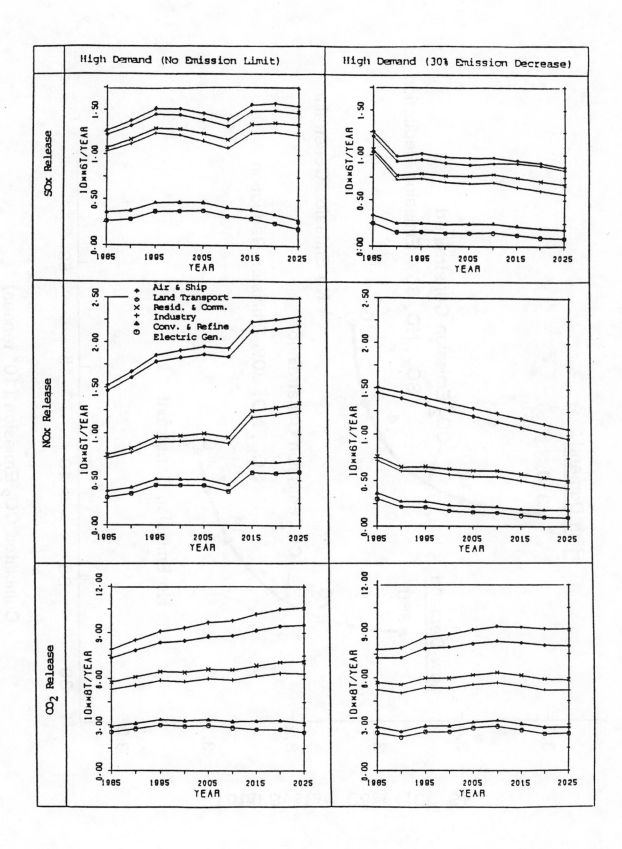

Figure 4. NO$_x$ emissions in Japan are limiting rather then SO$_x$ in a scenario requiring a linear decrease in future emissions, also reducing CO$_2$ emissions (Koyama and Ihara, 1988).

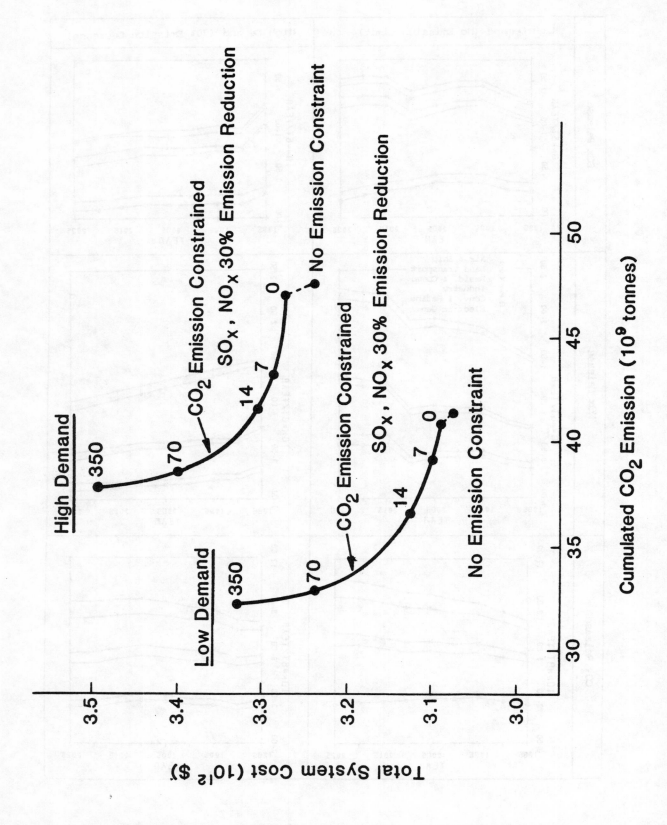

Figure 5. Increasing cost of the Japanese energy system as greater value is placed upon reducing CO2 emissions in two 45-year scenarios (Koyama and Ihara, 1989).

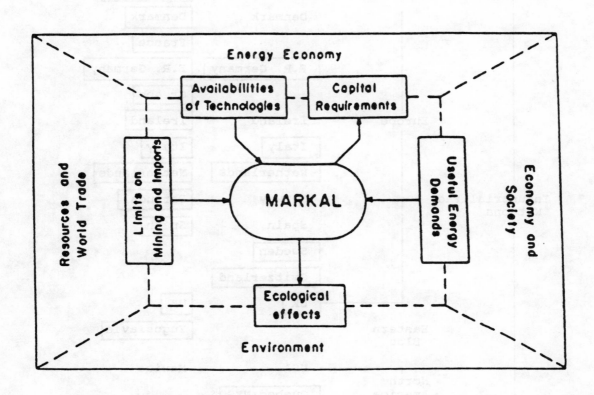

Figure 6.   Interfaces of MARKAL model (Fishbone et al., 1983).

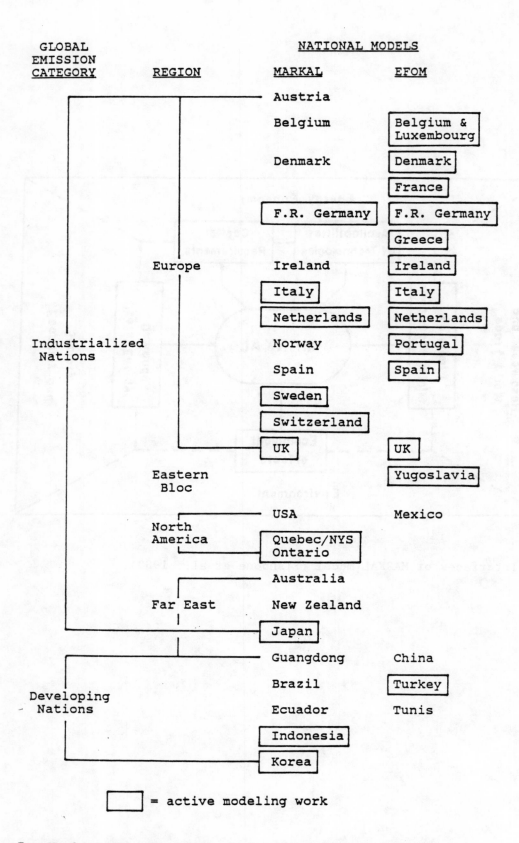

Figure 7. Worldwide application of the EFOM and Markal models.

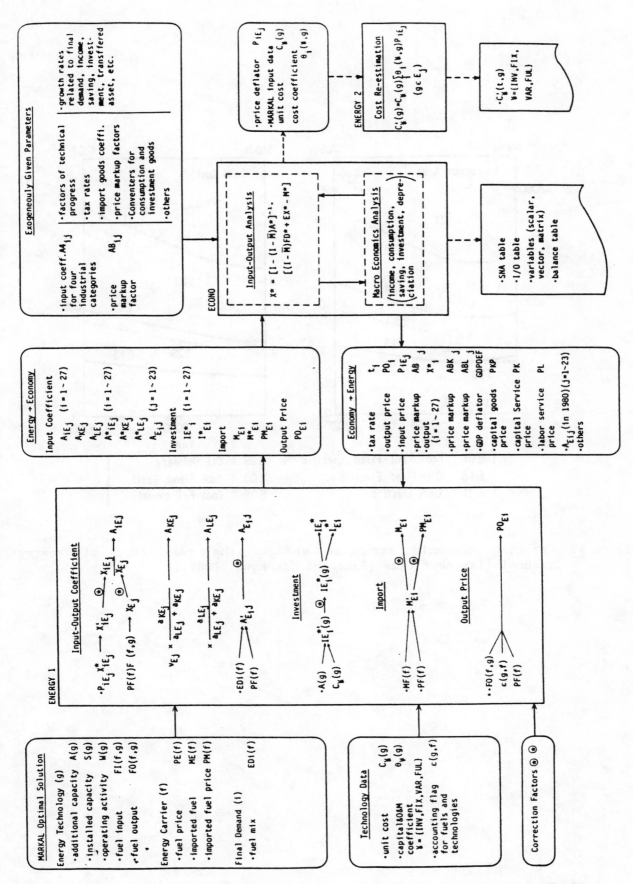

Figure 8. Linkage of the MARKAL model with an econometric model of Japan (Sato and Yasukawa, 1988).

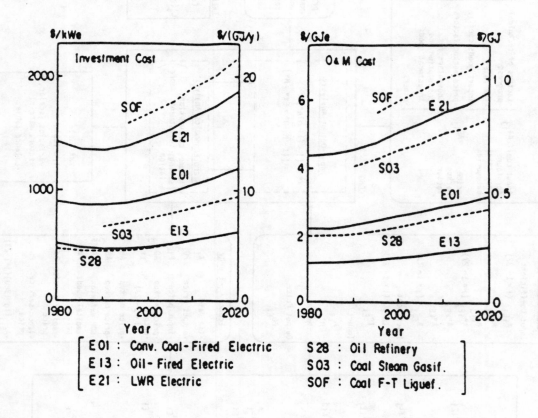

Figure 9. Changing economic structure changes the real cost of energy technologies over time (Sato and Yasukawa, 1988).

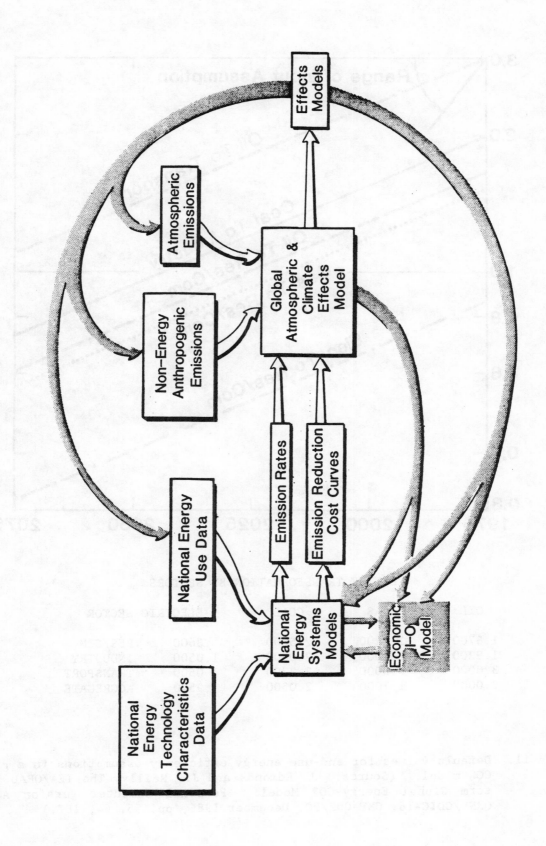

Figure 10.   A coupled energy-economic-global climate model.

231

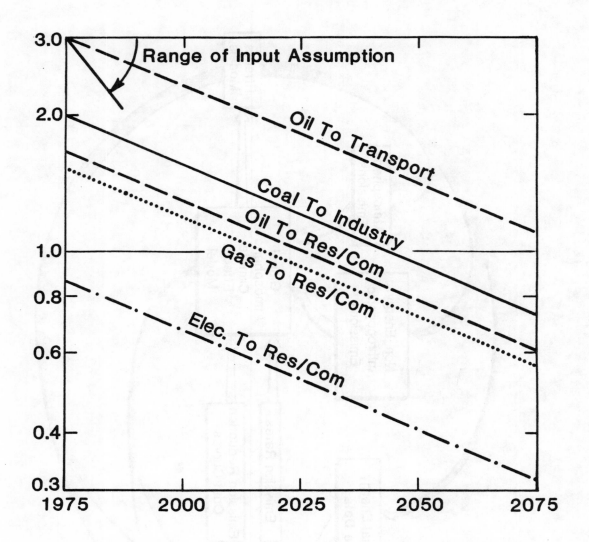

ENERGY TRANSFORMATION BY SECTORS

OIL	GAS	COAL	ELECTRIC	SECTOR
1.6700	1.5400	2.5000	.8600	RES/COM
1.9200	1.9000	2.0000	1.0500	INDUSTRY
3.0000	3.0000	3.3300	1.0500	TRANSPORT
2.0000	1.7000	2.0500	.9500	AGGREGATE

Figure 11. Default values for end-use energy efficiency assumptions in a global CO_2 model. (Source: J. Edmonds and J.. Reilly, The IEA/ORAU Long-term Global Energy-CO2 Model: Personal Computer Version A84PC, ORNL/CDIC-16, CMP-002/PC, December 1986, pp. 55, 64, 167.)

232

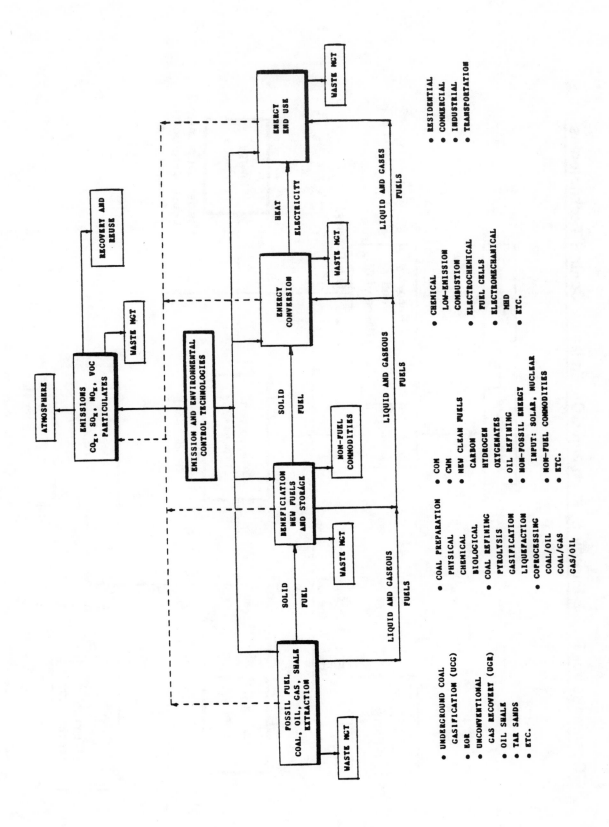

Figure 12.   System for emission and environmental control technologies applied
to fossil energy operations (Steinberg).

233

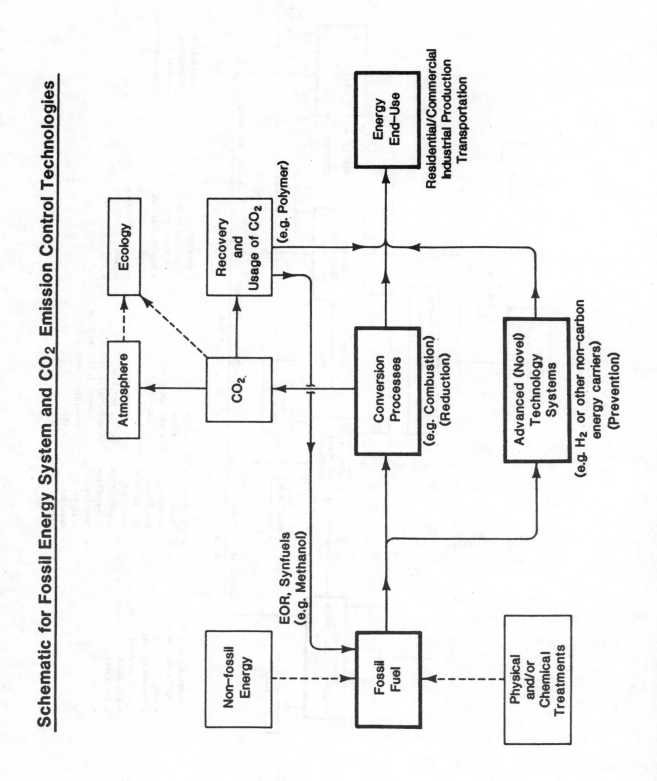

Figure 13. Schematic for fossil energy system and CO_2 emission control technologies (Steinberg).

Table 1. Options for reducing emissions of CO_2 and other greenhouse gases from fossil and biomass fuel use.

* conservation,

* increased efficiency,

* techniques to remove carbon or the constitutes of other greenhouse gases from the fuels,

* techniques to remove gases from the effluent gas stream,

* switches among fossil and biomass fuels (e.g., from coal to natural gas),

* switches to other energy sources (e.g., solar and nuclear power).

Table 2. Technical options for reducing emissions of CO_2 and other greenhouse gases from fossil and biomass fuel use.

* Removal

* Recovery

* Disposal

* Storage

* Reuse of energy

* Non-fuel material use

* Improved energy technology efficiency

* Coal-refining, including:

  - Hydrogen gas

  - Carbon black

  - Oxygenates, methanol, polyalcohols

  - Changing carbon-to-hydrogen ratio in solid and liquid fuels

  - Addition of non-fossil energy components, solar and nuclear

  - Monitored retrievable storage of discarded fuels.

Table 3. List of computer centers at which MARKAL is available [ETSAP annual report, 1988, Appendix C]

1. Commonwealth Scientific and Industrial Research organization (CSIRO). Sydney, Australia.

2. Energieverwertungsagentur (EVA), Austria.

3. CORE, Louvain-la-Neuve, Belgium.

4. Cia Auxiliar de Empresas Electricas Brasileiras (CAEEB), Centro de Processamento de Dados, Brazil.

5. École des Hautes Études Commerciales, Montréal, Canada.

6. Commission for Science and Technology of the Province of Guangdong, Kanton, P. R. China.

7. Kernforschungsanlage Jülich GmbH (KFA), Jülich, Federal Republic of Germany.

8. State Oil Company PERTAMINA, Jakarta, Indonesia.

9. ANSALDO Spa, Genoa, Italy.

10. ENIDATA Spa, Milan, Italy.

11. Electrotechnical Laboratory, 1-1-4, Umezono, Tsukuba-Shi, Ibaraki, Japan.

12. Japan Atomic Energy Research Institute, Tokai Research Establishment, Tokai-Mura, Ibaraki-Ken, Japan.

13. Department of Energy Conversion, Chalmers University of Technology,, Gothenburg, Sweden.

14. Paul Scherrer Institute, Würenlingen, Switzerland.

15. Netherlands Energy Research Foundation (ECN), Petten, the Netherlands.

16. Energy Study Group, The New University of Ulster, N. Ireland, United Kingdom.

17. Harwell Laboratory, Oxfordshire, United Kingdom.

18. Brookhaven National Laboratory, Upton, New York, U.S.A

Table 4. Major input assumptions of global CO_2 model[*] that are provided by MARKAL output.

Major assumption	ASSUMPTION DISAGGREGATION													
	None	Geographic	Aggregate price elasticity	Rate of interfuel substitution	Rate of end-use interfuel substitution	Energy-GNP feedback elasticity	Conventional oil	Natural gas	Coal	Shale oil	Nuclear power	Solar electric power	Biomass	Other
1. Population	○													
2. Labor productivity	○													
3. End-use energy efficiency		●												
4. Income effect						○								
5. Price effect				●	●									
6. Resource base							○	○	○	○	○			
7. Technological change in energy production							●	●	●	●	●			
8. Environmental costs of energy production							●	●	●	●	●			
9. Market penetration supply technology							●	●	●	●				
10. Solar and biomass energy costs												●		
11. Synfuel costs													○	
12. Number of forecast periods														○

○ Required input assumption.   ● Required input assumption provided by MARKAL output.

* Source: J. Edmonds and J. Reilly, The IEA/ORAU Long-Term Global-CO_2 Model: Personal computer Version A84PC, ORNL/CDIC-16, CMP-002/PC, December 1986, Table 4.1.

238

IEA/OECD Expert Seminar on Energy Technologies
for Reducing Emissions of Greenhouse Gases,
April 12-14, 1989, OECD, Paris

Basic Points of View on the Evaluation of Energy Technologies
towards the Global Changes: Economics and Energy Security

Akira Kinoshita
Electric Power Development Company (EPDC)

(Abstract)

Social preference influences the choice of technology.
Social and international support for technology assessment
cannot be predicted without comprehensive understanding of the
complex interactions between human activities and the global
environment in the context of sustainable development. The
intertemporal social preference is still an important element.

From the long term view, first of all, the security of a
durable energy supply has dominant importance and the provision
of a balanced energy mix remains our basic strategy. An energy
technology should be evaluated in terms of cost-effectiveness to
be realized by system-oriented enhanced energy use efficiency,
and where assessment of total inputs vs. total outputs is
important.

The aspect of poverty and equity should not be neglected
in insight of interactions between development and conservation
of global environment. With today's prevailing uncertainties,
strategies to deal with global environment issues have to be
flexible on step by step basis. Although issues of education
and ethics are also involved, no international consensus on
public policies concerning global environment could be created
without assessment of social cost-benefit.

Introduction

"The Limits to Growth" postulated by the Club of Rome was
a warning that development on the conventional basis would bring
the predicament of mankind based on the assumption that
resources of the earth are limited. Later on, as a result of
successful economic recovery, more optimistic arguments were
raised against this modern Malthusianism by the contention that
such limitation can be overcome through the price mechanism and
technological progress. Today, the global environmental
constraints being stressed again are likely to be a world
political issue while there seems to be no adequate policy
science capable of handling this issue.

Meanwhile, the status of the developing nations in the
world economy still remains vulnerable in their increasing

burden of accumulating debts. Their economic dependence on the
OECD industrialized nations is still high. The economic growth
of leading industrialized nations, which may be undergoing
certain qualitative changes, remains to be the key requisite for
accelerating the development of developing nations, which is
today the central issue in our attempt to keep the world economy
functioning normally. Here, we can define the qualitative
change in economic growth as the one more oriented to the
optimum use of resources with constraints. As world energy
consumption is rising again, the policy priority for
quantitative assurance of durable energy supply environmentally
acceptable, as well as prevention of economic turmoil as caused
by energy supply disruption, is quite high.

Recent world economic operations including energy supply
have been oriented toward deregulation and encouragement of free
competition. In the energy scene, the basic principle has been
to provide a balanced and diverse energy supply options by which
both stability and economics in energy supply are to be
secured. Those who advocate today the urgency of overcoming
global environmental constraints demand to establish a new
policy science based on their critiques on "market failures" and
"political failures".

What is being suggested is to take the immediate action,
which is based on the urgent concern on possible change of
global environment, but not on sufficient scientific knowledge
or proof. Nevertheless the facet of intertemporal social
preference is still situated as a fundamental factor in such
policy issue. Cost of uncertainties in a social decision for
such action would be extremely high if it is done based on
uncertain facts. It would not be easy to form a social
consensus on an issue which is not fully supported by scientific
analysis and evaluation. It is serious that such decision may
lead to another face of crisis for the human, or decline and
disturbances in socio-economy. At the same time, it could lead
to new confrontation between such nations centering on different
possible effects in spite of ongoing transformation of the
concept for national border which has been brought up through
expanded economy.

The technologies to meet global environmental change
should be evaluated based on the scientific understanding of
interactions between human and environment on the basis of
comprehensive socio-economic systems analysis of the production
systems which are defined by quality of human life.

Environmental Technology and Social Preference

More attentions are being paid in recent years to the
relation between technological progress and social needs.
Selection of technology is often closely related to social
preference. Summarizing the general direction of social
preference, which is becoming more complicated by reflecting
diversified value consciousness of the social constituents, is a

basic process in assessment of technologies. At the same time, the social preference is matured by level of the information feedback systems. It is difficult to assess technology on the assumption of the attainment of sustainable development without correctly understanding the complex interactions between human and environment. Not only impacts on interactions between production/consumption with emissions/waste and nature but also the withdrawal of conservation inputs can be pressure of economic society on environment. Analysis of the environmental management process depend on interdisciplinary systems analysis on structural issues in complexity of the interaction systems between ecosystem and living condition of human on the scale from local level to global level. What is important in such an analysis is to clarify the process of dynamic adaptation through structural changes, which generate sychronically and diachronically according to the cost-effectiveness.

When cost and benefit is to be evaluated over a long time frame, the difference of values at present and in future has to be adjusted by a certain discount rate. This discount rate is to reflect the intertemporal social preference. However, when global environmental problems are discussed, new philosophy that the costs and benefits occurring over different generations are expected to be evaluated based on the same value standard avoiding consideration of intergenerational difference, might come into some socio-economic decision. On the other hand, if it is expected that the economic growth can be sustained without diminishing efficiency in resources use through sophistication of production/consumption systems enbodied by advanced knowledge, or the efficiency in use of resources must be furthermore improved in future, application of the lower discount rate may not be allowed. If uncertainty in global change is higher due to the limited knowledge stock, the discount rate to be applied to net benefits related to such actions must be necessarily higher. If the importance is attached to prior preventiveness of environmental actions for possible adverse effects without clearance of uncertainties, there is greater possibility to increase contingency in outcomes of the control measures. Unless the rate of intertemporal social preference is extremely small, the present worth of the future economic loss related to the possible climatic change is by no means critical or the current incentive for investment on items other than prevention of uncertain future climatic change may be much larger, such investment opportunity may realize larger social benefit in future than the uncertain economic loss to be caused by the climatic change.

Economic system is directly linked with the ecosystem by science and technology given sustainable development. On what value consciousness human behaves in economic system, and what science and technology are adopted for development and use of resources is closely related to the value system of our society. The environmental policies of economic society today is directed toward internalizing the environmental control costs and treating them within the framework of economic policies. In attempting to extend this orientation to global change, the

241

fundamental problem is that the external economic factors are so complex for identification that they can not be internalized easily. In internalizing such costs, there must be the process that the interactions of the economic activities with the environment have to be evaluated precisely by scientific tools so that they are amenable to economic justification. The reason why environmental discussions get entangled so often is such impacts and interactions can not be clarified quantitatively. It would be very difficult to form the balanced policy mix with social consensus by neglecting this process.

In this view, it would be an attractive idea to build a holistic model by which the "natural systems" and "human systems" and their interactions can be described comprehensively and autonomously, including representations of the "hard systems" in the human systems comprising production, transportation, consumption, emissions, waste disposal and land use, as well as "soft systems" for socio-economic transactions such as decision making, cultural and institutional activities. It is fundamental proposition for the holistic policy systems to secure organic balances in various policy priorities. In building such a holistic model, diverse and interdisciplinary ideas and approaches will be required in addition to replenishment of data base in which point of view on costs and benefits of information is also indispensable.

For natural systems model, it is expected to develop scientific verification based on unified statistical standards concerning interactions between climate systems which contain considerable uncertainties in the climate model such as interactions between ocean and biosphere, heat exchange process between atmosphere and ocean, cloud effects, local temperature deviation and solar activity as well as interactions with other important factors to ecosystem in addition to greenhouse gases including CO_2. It means elucidation of the balanced measures for interactions with: effects of acid depositions on forests and lakes; destruction of tropical forests with accompanying species extinction and genetic erosion; increasing desertification and accelerated soil erosion; and chemical pollution.

Analysis of technological systems in human systems involves elucidation of potential technological changes centering on emission source reduction, recycling/reuse, disposal and containment as well as possible shifts in agriculture, forestry and land use patterns in addition to possibility of non-conventional strategic environmental control technologies.

Energy Technologies and Cost-Effectiveness

Various scenarios can be produced for the long term world energy demands in future depending on the assumptions on the scale of economy, economic structure, energy prices, technology level, living styles, population structure, population distribution, habitation/settlement and international policy cooperations. It is, first of all, crucial to secure sufficient amount of energy supply under any assumption, and such requirements can never be realized by a single energy supply source.

The fundamentals of energy supply strategy under uncertainty are to compose a balanced and diversified fuel mix, to make economic mechanism function through inter-fuel competition, and to assure suitable stability and flexibility in energy supply. In implementing such strategy, an international cooperation, designed to realize a balanced energy supply/demand structure on the world basis through coordinated policies inspired by globalism, is indispensable. Should environmental constraints prohibit development of a diversified energy supply structure including use of coal, this will not only disturb the supply/demand balance and bring disruption upon world economy but place serious burdens on the trade balances of developing nations through jumping energy prices.

In evaluating the value of a technology, cost is an important parameter as it represents the gross resource consumption and resource use efficiency. The cost of use technology as assessed in technology must be evaluated not only by the energy cost but the total cost including the capital cost, operating cost, environmental control cost, refurbishment cost, and research and development cost.

Further, it is important to evaluate a component technology by a comprehensive assessment of the total system, and a specific component technology must be assessed by the marginal utility of the system. This is natural as the improvement in the energy consumption efficiency by a specific technology can be realized only when it is incorporated into the system. At the same time, it is important to make provisions for assuring flexibility within the system-oriented structuring, so that the cost of adapting a technology to uncertainty is minimized. Reflecting the recent trend of evaluating multi-phase values of energy options under variety of possible situations, it would be important to place emphasis on the strict examination of contingency involved by the total cost of a technology. And also, it will be requested to secure objectivity in costing in a situation when the social preference on the energy technology plays important role in an energy option.

With globally arising interest on environmental problems, environmental control measure should be evaluated on the line of the optimum resource allocation problem, in accordance with universal perception for importance of overall resource consumption efficiency. That is, the problem we face is the optimization of resource consumption including that for environmental control, where the importance is attached to the overall evaluation of whole time history of environmental impacts as caused by overall inputs of the total system comprising a specific technology in the inter-industrial context. Such an evaluation includes not only the environmental impacts caused at the emission phase of energy production employing a specific technology, but those caused, directly and indirectly, throughout the life of that technology from investment to decommissioning including its whole fuel cycle impacts. An energy technology which may be recognized as being clean in its energy production phase, but for which the total resource inputs required throughout the process from investment to decommissioning exceeds its total output, is more likely to have negative impact on the global environment.

With this point of view of life-cycle total energy balance, many energy technologies having higher energy efficiencies can realize superior resource use efficiencies in terms of its impact on the global environment. Further, it should be noted that the global environmental emission can be saved by system oriented approaches such as cogeneration. Even if CO_2 fixation technology becomes technically feasible, more energy consumption and cost are incurred by such fixation process unless there is a market for the fixed CO_2.

The relation between the greenhouse gases and the global temperature has not been scientifically established. If the marginal effect of each greenhouse gas is different, it is natural to set priority on the emission control measures based on the cost-effectiveness from the point of view of efficient deployment of environmental control resources.

Global Environment and Policy Analysis

Today, the establishment of policy analysis to initiate studies on global change and possible solutions is demanded against the insufficient knowledge stock. This would be the first intellectual challenge never experienced. Considering the complexity and scope of this problem, as well as the insufficiency of knowledge, we may require a new paradigm beyond the existing rules of social science to deal with this problem. Since our knowledge is inadequate, the strategy selected would have to be flexible as it would require mid-course modification as new information is added. Such a strategy must involve efforts to minimize the cost of uncertainty related to decisions step by step basis and the alteration of decisions. On the occasion of social choice on new dimension, under these circumstances, not only phase of natural science is focused but also the research on social sciences should be initiated. Accordingly, it is required, first of all, to establish a policy to start research for the collection of basic information concerning the functions of the global environment and elucidation of the overall interactions between human activities and possible global environmental changes.

In such research, the study of aggregate social costs and benefits accompanying the prevention of global environmental change must addressed in examination of the public policy for rational resource allocation correctly set forth, despite the non-conventional nature of our theme as the global environment. Since the effects of any decision making will be also global, such study is indispensable, and no international agreement could be created without such efforts.

As we are dealing with the interactions between development and environmental conservation, the problem of deforestation and desertification, which is being caused by poverty in certain areas of the world, must be discussed based on a balanced point of view. This issue is one of equity between the rich and the poor, and also the equity between current and future generations. This problem shows us that, even if sophisticated scientific tools are developed for the conservation of the global environment, it is not worthwhile unless it is globally assured that everyone on this earth has access to it.

Remaining on equity, we cannot forget the importance of the long term viability or investment effect in developing areas, if we are concerned with the effective utilization of limited resources. It is very clear that regional distribution in population growth, which has a critical relationship with the problem of global environment. This leads us to think that the global environmental issue also covers the domain of human education and ethics. Questions may be raised as to whether it is sufficient to discuss global environmental issues within the framework of conventional cost-benefit concept.

In starting our research, we must analyze the possible social support for the allocation of research and development resources by sector and function based on long term macro-trends of the actual inputs for science and technology. In this analysis, the balance between research efforts in natural science and social science as well as "research on research" must be addressed. Also useful would be an analysis of decision making processes in science and technology policies and review of the processes for the distribution of research/development resources as well as implementation strategy of R/D policies. In the analysis of implementation strategy, the key factors are the method of resource procurement, regulation, private sector activities, the process of creating international consensus and the motivation of international cooperation. As long-term world energy supply demand depends on coal and its effects on global change are concerned, it is suggested that resource investment in research and development of coal use technology is drastically increased.

Conclusion

i) An international consensus on global environmental control measures cannot be created without research efforts for comprehensive evaluations.

ii) Any strategy adopted would have to be flexible under existence of the present uncertainties. Technology development should be focused on the improvement of energy consumption efficiency and the enhancement of cost-effectiveness.

iii) The basic energy strategy remains a balanced energy mix, and resource inputs for the development of coal use technology should be drastically increased.

TROPICAL FORESTS AND GREEN HOUSE GASES

Peter S. Ashton
Professor of Dendrology
Harvard University

Tropical Forests and Greenhouse Gases.

Forests may mediate the flux, by acting as sources and sinks, of greenhouse gases notably CO_2 and nitrous oxides, but also ozone, CO methane and neoprene.

1. Forests fix carbon, from gaseous CO_2, through photosynthesis, but also release it through plant respiration and soil litter decomposition. They likewise release and absorb oxygen. Mature forests in nature are in approximate dynamic equilibrium for these gases, releaving approximately as much as they absorb. There is some evidence that there is a small net uptake of CO_2 through soil solution in forest compared with non-forest soils, the long term fate of which is unknown. This is unlikely to be important. Therefore, the most effective use of forests as a sink for CO_2 is as fast-growing timber plantations: Plantations are net sinks of carbon, and the timber, provided it is being harvested for purposes which do not involve release of carbon, can potentially act as a continuously expanding sink. Sustainment of maximum production rates would require continuous application of fertilizer.

2.   A major fraction of global atmospheric photo-chemistry occurs in the tropics. Changes in photochemical processes in the tropics may disturb global balances for important greenhouse gases including CO and ozone. Tropical forests and soil represent globally significant sources of reactive hydrocarbons, notably isoprene, which enhance ozone production and, in seasonal forests, nitrogen oxides during the dry season; while they act as a major sink of ozone through solution onto leaf surfaces. Swamp forests also produce significant methane emissions. Atmospheric concentrations of methane, CO, nitrogen oxides and ozone above natural forests are universally low, although there is a seasonal fluctuation in nitrogen oxides which effects parallel fluctuations in ozone.

3. Currently, it is estimated that two thirds of wood produced in the tropics is burned as fuel, thereby releasing the carbon fixed by it into the atmosphere. The use of wood and other biomass, such as cassava and sugarcane, can create a sustainable means of energy generation, and generates a steady state carbon cycle without net accumulation of atmospheric CO_2. But biomass plantations require fertilizer and herbicide application, and the cost of energy derived from them will not compete with that from fossil fuel in the foreseeable future unless policies are implemented which artificially raise the price of the latter. Fertilizer would also increase soil emissions of nitrogen oxides.

4. It is estimated that there is currently a net increase of forest area in the temperate world. But this is far offset by deforestation in the tropics.

Strict deforestation, that is conversion of rain forest to other land use in which the vegetation is less than 5 m tall and with crown cover of less than 40%, has been estimated at less than 60,000 Km2 per annum and at 115,000 Km2 per annum (Lanly and Clement, 1979 and Myers, 1982 in Jacobs 1987 where there is an excellent review of estimates on pp. 10-12). Myers adds that a further 85,000 Km2 per annum are grossly disrupted by logging and other exploitation. It is now estimated that only 45% of the original area of tropical forest is now covered by forest growth of any kind, and that 200,000 Km2 amounts to approximately 2% of the residual tropical forest.

Net deforestation at world scale is estimated to be contributing between 20-40% of current increases in atmosphere CO_2 or 0.5-1.0 billion tons.

Keller and his colleagues have constructed simulation models to predict the affect of conversions of tropical forest to agriculture, including pasture, on the exchange of greenhouse gases over the Amazon basin. The predictions of their model are supported by NASA measurements undertaken with the Brazil-U.S. Amazon Boundary Layer Expedition. The causes of change includes burning of the biomass, decrease of Leaf Area Index (leaf area per unit area of ground beneath) from an estimated 10 to 2, and change in soil emissions through increase in surface day temperature and daily and seasonal variation in water content, and increase in nitrogen from burning and fertilizer application.

They estimate that fires, including annual burning for pasture reinvigoration and weed control, lead to methane, CO and nitrogen oxide emissions several orders of magnitude greater than from forest. The nitrogen oxides will be converted partially to nitric acid. This will replace organic acids in rain water which will in turn likely become sufficiently acid, with pH down to 4, that lakes may become acidified and forest growth reduced. There is some possibility, when a certain stage of deforestation is reached, that acid rain may kill residual forest, inducing a positive feedback effect.

Decrease in leaf area will lead to decrease in biogenic gas emission, hence also ozone production, but this will be more than offset by the decrease in area available for absorbtion of soluble gases including ozone and nitrogen oxides, while increased emission of nitrogen oxides will lead to increased ozone production. Keller and colleagues estimate that there may be a threefold increase in nitrous oxide emissions from

herbicide application, and the cost of energy derived from them
will not compete with that from fossil fuel in the foreseable
future unless policies are implemented which artificially raise
the price of the fossil fuel. Fertilizer would also increase
soil emissions of nitrogen oxides.

4. It is estimated that there is currently a net increase of
forest area in the temperate world. But this is far offset by
deforestation in the tropics.

Strict deforestation in the tropics , that is conversion of
rain forest to other land use in which the vegetation is less
than 5 m tall and with crown cover of less than 40%, has been
estimated at less than 60,000 Km² per annum (Lanly and Clement,
1979), and at 115,000 Km² per annum (Myers, 1982 in Jacobs 1987
where there is an excellent review of estimates on pp. 10-12).
Myers adds that a further 85,000 Km² per annum are grossly
disrupted by logging and other exploitation. It is now estimated
that only 45% of the original area of tropical forest is now
covered by forest growth of any kind, and that Myers' estimate of
Myers total annual disturbance, 200,000 Km², amounts to
approximately 2% of the residual tropical forest.

Net deforestation at world scale is estimated to be
contributing between 20-40% of current increases in atmosphere
CO_2 or 0.5-1.0 billion tons.

Keller and his colleagues have constructed simulation models
to predict the affect of conversion of tropical forest to
agriculture, including pasture, on the exchange of greenhouse
gases over the Amazon basin. The predictions of their model are
supported by NASA measurements undertaken with the Brazil-U.S.
Amazon Boundary Layer Expedition. The causes of change include
burning of the biomass, decrease of Leaf Area Index (leaf area
per unit area of ground beneath) from an estimated 10 to 2,
change in soil emissions through increase in surface day
temperature and daily and seasonal variation in water content,
and increase in nitrogen from burning and fertilizer application.

They estimate that fires, including annual burning for
pasture reinvigoration and weed control, lead to methane, CO and
nitrogen oxide emissions several orders of magnitude greater than
from forest. The nitrogen oxides will be converted partially to
nitric acid. This will replace organic acids in rain water which
will in turn likely become sufficiently acid, with pH down to 4,
for lakes to become acidified and for forest growth to be

reduced. There is some possibility, when a certain stage of deforestation is reached, that acid rain may kill residual forest, inducing a positive feedback effect.

Decrease in leaf area resulting from deforestation will lead to decrease in biogenic greenhouse gas emission, hence also decrease in tropospheric ozone production, but this will be more than offset by the decrease in area available for absorption of soluble gases including ozone and nitrogen oxides, while increased soil emission of nitrogen oxides will lead to increased ozone production. Nitrogen oxides in tropical ecosystems are derived solely from soil emissions. Keller and colleagues estimate that there may be a threefold increase in nitrous oxide emission from agricultural land compared with forest, which will be further increased with fertilizer application. Keller and colleagues point out that, though nitrous oxide has been measured, NO has not but is expected to increase consequent to deforestation in the same manner. They further estimate doubling of ozone production over agricultural land with significant periodic regional episodes favored by climatic conditions. They describe evidence, for instance, of substantial tropospheric increases of ozone and other greenhouse gases offshore in the southern Atlantic during and following the dry season burns in Amazonia and southern Africa. Keller's group emphasize evidence for the wide spreading of ozone and CO from their source in the tropics.

Methane emission, mostly from swampy soils, is also significantly increased as a consequence of deforestation, in both pasture and arable land and, especially, in irrigated lands such as rice fields.

5. It is difficult to estimate what the capacity of forest plantation at global scale would be to act as a net sink of CO_2, and possibly other greenhouses gases. Biomass production rates are greatly influenced by soil water availability and, in the case of secondary growth and plantations, by available soil nutrients. Necessary fertilizer application will affect the soil as sink and source of greenhouse gases notably methane and nitrogen oxides. George Woodwell (in litt.) has estimated that 2 million Km^2 of fast growing industrial wood plantation, grown under favorable conditions, could absorb 1 billion tons of carbon annually, on a ten year harvesting cycle; but it is unclear to what use the aggregate biomass would be put, other than re-oxidized as fuelwood or forage. A reasonable rough estimate of the rate at which the original tropical forest biomass could be regrown, under optimal management conditions, would be 2% per annum for each hectare planted, if light hardwood timbers were planted. On this estimate, it would take in the order of fifty

251

years of plantation, on a scale equivalent to the current rate of deforestation, before the world forest carbon balance could be achieved if present deforestation rates continued during that time. Put another way, it would take about 50 years before forest timber plantation in the tropics could act as a sink for one quarter of the total atmospheric carbon released annually into the atmosphere on present levels.

One must conclude that implementation of a drastic plan for more efficient fossil fuel use has more promise than reforestation of the tropics as a rapid means to reduce net accumulation of atmospheric carbon, and other greenhouse gases, though reforestation of the tropics would have a significant longer term affect and would simultaneously generate other important benefits.

These other benefits are diverse, and in some cases differ according to the nature of the forest. The unique attribute of many tropical evergreen forests is their species diversity of plants and animals. It has been estimated (NRC, 1980 Chapter 2) that two thirds of all species on earth are confined to the tropics. Of these, about three quarters are confined to evergreen forests. The majority are restricted in geographical distribution, and many are quite limited in range ('endemic'). Areas of exceptional importance for species diversity and endemism, in approximate order of regional diversity, include the forests of the Pacific slopes of the Andes from Ecuador to Panama; the eastern Andean slopes; northwestern Borneo north of the line from Pontianak to Kota Kinabalu; the western Guyana Highlands; the southwestern foothill forests of central New Guinea (Digul to Fly catchments); the north-eastern hillslopes of Madagascar; the forests of western Cameroon; and the Atlantic forests of Brazil. There are others of exceptional importance, but some of the most extensive rain forests, including those of much of the basins of the Amazon and Congo rivers, are relatively uniform and low in local endemism.

The economic value of this species diversity must not be underestimated. The population of the forested tropics will more than double in the next thirty years, while that of the industrial nations, main consumers of tropical tree commodities, may only increase 15-20%. Most forest land in the tropics is only suitable for growing tree crops. Plainly, new crops must be developed, both to diversify traditional markets and to develop the expanding regional markets in the tropics. Germplasm for these crops must be largely sought from the forest.

So far, tropical forest products have been marketed at prices well below replacement value. An extreme case is the many pharmaceutically active plant chemicals that have acted as precursors for a significant proportion of modern industrially produced drugs, for which little or no rental has been paid for germplasm conservation in the forest.

Rain forests possess a research and educational value beyond calculation, and have a potential so far hardly realized for nature tourism.

Tropical rain forest will only be conserved if opportunity costs turn in its favor. For this to happen, the nations holding tropical forests will have to devise effective means to collect sufficient rent from the totality of goods and services yielded by these forests, which they have so far failed to do.

All forms of destructive harvesting of tropical rain forest are incompatible with conservation of biological diversity as a whole, and strict preserves must therefore be included as part of integrated development planning. This is because the majority of the species are concentrated in the mature phase of the forest. This old growth is reduced to fragments even by selective logging, increasing risks of local extinction, and reducing population densities which in turn can reduce successful mating and hence fecundity rates. Though few extinctions are likely following a single logging, extinction rates are predicted to accelerate in subsequent felling cycles.

Forest cover of any other type, though lacking the products and biodiversity characteristic of tropical evergreen forest, does simulate it more or less closely in its climatic and edaphic influences. I am not aware though of comparative studies of the exchange of greenhouse gases in tropical tree plantations.

All forests in full leaf evaporate water from their canopy in the order of 80-90% of an open water surface. This, combined with their albedo, is an important ameliorating influence on local weather. Cloud formation is enhanced. There is unpublished evidence from Malaysia that deforestation does not lead to local reduction of mean annual rainfall, but that rain comes to fall in heavy storms at longer intervals, and is thus less effective and more of an erosion hazard. In continental areas such as West Africa and northern South America there is evidence that rain is increasingly derived from upwind forest canopy evapotranspiration with increasing distance from the sea

during seasons of moist trade wind (Salati). Deforestation near
the coast is a likely major cause of the southern spread of the
western Sahara, and deforestation in the Amazon is expected to
have an analogous affect. Such changes will likely have a
significant major influence on regional, and even world,
emissions of greenhouse gases, directly and also indirectly
through their possible influence on broader climatic systems.

Much land in the tropics can only sustain tree crops. This
is because pasture and other low vegetation does not protect the
soil, which is either compacted or eroded or both. Day
temperature at the soil surface is much higher (up to 45°C, as
opposed to 24°C in forest), and humidity lower leading, with
lower litter fall, to lower surface humus and lower capacity to
exchange nutrients.

There are therefore multiple reasons why major investment in
reforestation in the tropics is urgently justified. The form
which it should take must take into account both socio-economic
and environmental factors, and could include commodity
plantations and social forests, producing both timber and non-
timber goods, integrated with conservation areas of residual
virgin forest.

Further reading.

Jacobs, M. 1987. The Tropical Rain Forest. A first encounter.
    Springer (particularly see chapter 1.)

Keller, M., J. Jacob, S.C. Wofsy and R.C. Harris. Effects of
    tropical deforestation on global and regional atmospherical
    chemistry. (Submitted to the journal Climatic Change).

National Research Council (U.S.) Committee on Research
    Priorities in Tropical Biology, 1980. Research Priorities
    in Tropical Biology. National Academy of Sciences,
    Washington. (Particularly chapter 2).

ENVIRONMENTAL EMISSIONS FROM ENERGY TECHNOLOGY SYSTEMS: THE TOTAL FUEL CYCLE

By Dr. Robert L. San Martin
Deputy Assistant Secretary for Renewable Energy
U.S. Department of Energy
Washington DC, U.S.A.

April 1989

ENVIRONMENTAL EMISSIONS
FROM ENERGY TECHNOLOGY SYSTEMS:
THE TOTAL FUEL CYCLE

Abstract

To accurately quantify and compare environmental emissions from energy technologies, each phase of the fuel cycle, including resource extraction, facility construction and facility operation, must be evaluated. Meaningful comparisons among the various technologies should also be based on a common measure of each technology's useful output. This analysis establishes a framework for conducting a comparative evaluation of the total fuel cycle of different energy technologies. Environmental considerations for each technology and each phase of the fuel cycle, categorized by major types such as air emissions, water emissions, solid waste emissions and materiel requirements, are evaluated individually for different environmentally significant substances.

The result is a comparative analysis of 14 electric generating technologies using the total energy cycle framework and metric tons per gigawatt hour (GWh) as a consistent unit of measurement for comparison.

Introduction

The analysis presented in this paper examines environmental factors by building on a previous study conducted for the U.S. Department of Energy's Office of Renewable Energy, *Energy System Emissions and Materiel Requirements,*[1] which developed an overall methodology for direct comparison of electric power technologies. That assessment viewed all environmental impacts associated with a technology as part of a total system designed to extract and produce energy over a specified operating life.

By relating environmental emissions from the resource extraction, facility construction, and facility operation phases, a basis was established for comparing electric technologies that have different capital, fuel, and operating characteristics. The five electric power technologies evaluated were:

- a conventional pulverized coal plant
- an Atmospheric Fluidized Bed Combustion (AFBC) plant
- an Integrated Gasification Combined Cycle (IGCC) plant
- a boiling water nuclear reactor
- a central station photovoltaic plant

The earlier work evaluated more than 30 environmental factors including atmospheric emissions such as carbon dioxide (CO_2) and nitrogen oxide (NO_x); water emissions such as dissolved solids; solid waste; and land and water requirements; all reported on the basis of quantities per unit of electric output (e.g. tons/GWh).

This paper builds upon the earlier report by expanding the number of energy technologies compared. Fossil fuel technologies included in this analysis are:

- a conventional pulverized coal plant
- an Atmospheric Fluidized Bed Combustion (AFBC) plant
- an Integrated Gasification Combined Cycle (IGCC) plant
- an oil-fired steam electric plant
- a gas-fired steam electric plant

The non-fossil energy technologies examined include:

- a boiling water nuclear reactor
- a wood-fired steam electric generating station
- an open-cycle Ocean Thermal Energy Conversion (OTEC) plant
- a dry-steam hydrothermal geothermal power station
- a large hydropower plant
- a small hydropower plant
- a wind energy conversion system
- a central station photovoltaic plant
- a distributed receiver solar thermal electric plant

Among the types of emissions analyzed, carbon dioxide represented one of the most significant quantities of emissions on a per gigawatt-hour basis. Therefore, to illustrate how comparative analyses can be conducted using a total energy cycle methodology, data for CO_2 emissions from each of the above technologies will be the focus of this presentation. Also, some studies suggest carbon dioxide, from a combination of fossil fuel combustion and deforestation, accounts for nearly 50% of the "enhanced" greenhouse effect resulting from increasing concentrations of greenhouse gases.[2]

Analysis Concept

Because this analysis attempts to take a detailed, directly comparative view of emissions from power production technologies, only limited data were readily available. For the most part the literature on emissions of electric technologies tends to focus on power production. Emissions associated with extraction and transportation of fuel, or associated with plant construction, have been less fully documented and the available literature is limited with respect to the relationship to point-of-use characterizations. The National Acid Precipitation Assessment Program (NAPAP) has made important progress in addressing integrated fuel cycles and identifying data gaps. Most do not address the effects of fuel extraction and facility construction. As a result of these limitations this analysis is restricted to examining major issues using data from available sources.

This analysis does not seek to recommend one technology over another. Rather, it is intended to provide a useful comparison of each technology's emissions profiles, which is only one factor that should be considered in their deployment. Without information on costs, the suitability of a technology to particular sites and energy demand situations, and other environmental impacts associated with particular projects, it is impossible to say one technology is preferable to another.

Study Approach

The analysis used in this paper is based on two fundamental considerations. First, the environmental effects of energy production at all stages of the energy production cycle must be viewed as a direct function of generating the final energy product. Only by analyzing the complete energy cycle can these effects be fully and consistently evaluated. The second consideration requires that a common measure of the environmental factors be established such that the total energy cycle for different technologies can be cross-compared within specific categories of emissions, while controlling for variation in energy output, materiel requirements, fuel demand, etc.

By investigating the impact of each stage of the energy production process, the analysis attempts to normalize differences between materiel- and fuel-intensive technologies in order to provide a fair basis for comparison. When emissions are normalized in terms of each facility's useful power output, the association between electricity production and emissions for each technology becomes clearer.

CO_2 emissions are rarely expressed in terms of quantities as a function of useful power output, largely because CO_2 has never been regulated or measured as an air pollutant. Raw tonnages of CO_2 only indirectly show the impact of the product society actually consumes -- watt-hours of electricity.

258

This study estimates CO_2 emissions associated with each stage of energy production for each technology as part of one system designed to produce energy, from fuel extraction through construction, operation, and decommissioning. The goal of this approach is to make the impact of a technology like photovoltaics, which has practically no emissions during operation, but requires significant one-time inputs of raw materiel, comparable to emissions from a technology like a coal plant, which produces its most significant emissions during operation.

By necessity the comparisons presented are generalizations. Each energy facility is to some extent unique. For example, the amount of steel and concrete used in a PV facility will vary with site conditions and the type of equipment used. Coal mining impacts depend on the extent and depth of deposits, site conditions, and mining methods. Combustion emissions from coal are impacted by both generating equipment and coal chemistry, which varies from mine to mine. Some issues, such as the impact of iron ore mining associated with the steel used in plant construction, were not addressed. The following section discusses and compares the impact of resource extraction, facility construction and plant operation for the fossil fuel, nuclear, and renewable energy technologies examined.

Emission Analysis and Comparison by Energy Production Stage

Comparing nuclear and renewable energy to the coal technologies confirms the generally accepted belief that non-fossil technologies represent an advantage from the standpoint of CO_2 emissions. The results also clearly show, however, that their contribution is not zero when all the elements of their fuel cycle are considered. No technology is completely environmentally benign. The CO_2 emissions from the power production technologies examined are shown in Table 1.

Fuel Extraction

Fossil Fuel Extraction -- The fuel extraction stage for fossil fuel includes the impacts of mining, processing, and transporting fuel to the site where it will be converted to energy. Emissions associated with fuel extraction and transportation for the fossil fuel technologies were scaled to the fuel demands of each fossil fuel technology by dividing the annual fuel demand of the power plant by the capacity of the fuel extraction, processing, and transportation facilities. This demand/output ratio was multiplied by the emissions from each fuel supply facility to derive the share of emissions from the facility attributable to the final generating plant. For coal it was assumed that the fuel supplied to each technology was mined and transported under the same conditions, so variations in emissions from fuel extraction are mainly a function of each plant's relative efficiency in generating electricity.[3] Oil and gas fuel extraction data were not complete so the impact of fuel extraction activity could not be assessed.

Renewable Energy Fuel Extraction -- Most of the renewable energy technologies, including photovoltaics, solar thermal, wind, hydropower, and ocean

Table 1. Carbon Dioxide Emissions: Electric Technologies

Technologies	Emissions by Energy Production Stage (Metric Tons per GWh)			
	Fuel Extraction	Construction	Operation	Total
Conventional Coal Plant	1.0	1.0	962.0	964.0
AFBC Plant	1.0	1.0	960.9	962.9
IGCC Electric Plant	1.0	1.0	748.9	750.9
Oil Fired Plant	-	-	726.2	726.2
Gas Fired Plant	-	-	484.0	484.0
Ocean Thermal Energy Conversion	NA	3.7	300.3	304.0
Geothermal Steam	0.3	1.0	55.5	56.8
Small Hydropower*	NA	10.0	NA	10.0
Boiling Water Reactor	1.5	1.0	5.3	7.8
Wind Energy	NA	7.4	NA	7.4
Photovoltaics	NA	5.4	NA	5.4
Solar Thermal	NA	3.6	NA	3.6
Large Hydropower	NA	3.1	NA	3.1
Wood (sustainable harvest)	-1509.1	2.9	1346.3	-159.9

(-) Missing or inadequate data for analysis, estimated to contribute ≤1%.

(NA) Not Applicable

*This analysis considered construction of new dams. According to a recent Federal Energy Regulatory Commission report there is 8,000 MW of small hydropower under construction or projected, much of it involving refurbishing or refitting existing dams, which would substantially reduce small hydropower's CO_2 impact.

thermal energy conversion (OTEC) have no direct fuel extraction impacts. Geothermal field development and well drilling activities emit minor amounts of CO_2 as a result of gas released from wells.

Biomass energy can produce net reductions in CO_2 over the life of the facility assuming that fuel is extracted from a sustainable, managed source of biomass such as a short-rotation, intensive-culture wood plantation, which is examined here. Sustainable biomass energy production will fix CO_2 equal to the amount of CO_2 released through combustion over the life of the plant. Sources of CO_2 emissions external to this cycle, notably from inputs of fertilizers and pesticides and the use of fossil fuels in cultivating, harvesting, and transporting the fuel, were evaluated and included in the analysis as net contributors to CO_2 emissions. However, these emissions, are offset by the carbon storage capacity of the roots and other unharvested portions of the biomass that remain in place (and growing in the case of coppiced species). Over the life of a generating plant this harvest/regrowth cycle can yield a net *reduction* in CO_2 emissions over all stages of biomass-fired electricity production.

A Scenario of Biomass Fuel Regrowth

A managed, short-rotation forest fixes or sequesters 45 metric tons of carbon per hectare per year (165 metric tons CO_2 per hectare per year) during its growth period. If trunks and branches from the short-rotation forest are harvested and used in a power plant, 82.5 metric tons of CO_2 per hectare remain in the forest, stored in the root system and soil. Therefore, even with harvesting and energy production, substantial carbon remains fixed after the first harvest. Subsequent harvests and wood utilization would be balanced from a CO_2 standpoint in that growth of new trunks and branches would offset CO_2 released back to the atmosphere during combustion.

In 1986, U.S. fossil-fired electricity production emitted 1.62 billion metric tons of CO_2. Thus, it would require 9.88 million hectares (approximately the size of the state of Virginia) of newly planted forest to offset all 1.62 billion metric tons of CO_2. Each tree offsets approximately 338 kilograms (0.34 metric tons) of CO_2 per year during its annual growth cycle. At that recapture rate, 4.8 billion new short-rotation trees would need to be planted to absorb the 1.62 billion metric tons of CO_2 in 1986. With a population in the U.S. of 240 million, each person would have to plant approximately 20 trees to achieve this offset. (At CO_2 offset rates for natural forests, 26 million hectares would be required to uptake 1.62 billion metric tons of CO_2 or each person would plant over 50 trees.)

Nuclear Fuel Extraction -- The nuclear calculations were made in the same general manner as the coal calculations, with fuel demand at the power plant traced back through fuel fabrication, enrichment, processing, and mining in order to allocate the emissions from each stage of fuel manufacture in proportion to each stage's contribution to final power production.[4] An additional increment to emissions was added to the source values based on each fuel processing facility's electricity demand. A coefficient for CO_2 emissions as a function of the electric generating fuel mix in the U.S. was calculated and then applied to the electricity demand of the nuclear plant.

Construction

The construction phase includes the indirect impacts of the technologies in terms of CO_2 emissions associated with manufacturing the raw materiel inputs. Steel and concrete are the major materiel inputs examined and the major sources of CO_2 emissions.

The construction stage accounts for the greatest differences between materiel- versus fuel-intensive technologies, with the former producing the highest environmental impacts at this stage. The estimates in this analysis focus exclusively on emissions from final manufacture of major materiel used in construction; it does not represent a comprehensive estimate of all emissions. There are secondary emissions associated with the mining of raw materiel (such as iron ore, bauxite, etc.) and actual in situ assembly of materiel and components, but these types of impacts were not addressed.

The emissions associated with materiel manufacture were divided by the annual output of the technology times the operational life of the technology to derive CO_2 emissions per unit of output over plant life. The CO_2 emission factor for steel was derived by examining fuel demand as a function of industry output, and then multiplying the resultant estimate of fuel use per ton of output times a CO_2 emission coefficient to derive an estimate of CO_2 per ton of output. This estimate was then used to calculate the CO_2 emissions associated with steel demands. Electricity as an energy input to steel was converted to CO_2 inputs by calculating the fuel mix for electricity in 1987, multiplying the quantities by their respective coefficients, and then allocating the gross CO_2 emissions over the total number of gigawatt-hours produced in 1987.[5, 6, 7] The CO_2 coefficients for steel, concrete, and for the various fuels considered are based on data from industry data bases or global climate investigations, respectively.[8, 9, 10]

Fossil Fuel Construction -- In the case of the IGCC plant and the AFBC plant, direct estimates of materiel requirements were unavailable. Therefore the values were derived by adjusting the materiel used in a conventional plant by the proportionate capacity associated with the AFBC and IGCC plant.[11] It is acknowledged that this assumption ignores the significant technology differences and the effects of differing economies of scale between technologies. Data were unavailable for the oil and natural gas plants, and no estimates of their impacts were made. In general, emissions from fossil fuel plant construction are small relative to the output over the operating life of the plant.

Renewable Energy Construction -- Like conventional technologies, the materiel requirements for renewable energy plants can vary widely depending on specific site conditions and technical requirements. The different technologies vary widely in their materiel intensity and CO_2 emissions per GWh. For each renewable energy technology, the Department of Energy Renewable Energy Program has estimated materiel requirements per MW of capacity, given an "average" or typical facility.

The steel and concrete estimates of a PV plant are for a conceptual utility-scale design developed by the Electric Power Research Institute. The PV plant is assumed to employ flat-plate, thin-film arrays with 15% efficiency located in Barstow, California. Plant size was 100 MW, with 209 GWh of annual energy output.[12] Geothermal plant construction requirements are basically equivalent to a conventional fossil fuel plant with comparable materiel requirements. The wood combustion generating plant also has construction materiel requirements similar to a comparable fossil plant.

Nuclear Construction -- Construction-related CO_2 emissions from nuclear energy are quite low when considered over the life of the plant. Although they require a considerable amount of materiel initially, nuclear plant impacts are spread over a high lifetime power output.[13]

Operation

The values for emissions and materiel inputs associated with operating the technologies were taken from source documents and Renewable Energy Program inputs. The annual value for emissions was then divided by the annual GWh of output for each technology to derive emissions per unit of output. Values for the IGCC and AFBC plants were assumed to be similar to the conventional plant in terms of the rate of emissions, and thus were only adjusted for the increased efficiency and power output per ton of coal input gained from each technology (if any).

Fossil Fuel Plant Operation -- Impacts at the operation stage are measured in terms of emissions produced while the plants are actively generating energy. The conventional coal plant in the assessment is assumed to be a 500 MW facility producing 3500 GWh of electricity annually. It represents a new plant built to meet or exceed existing environmental standards, and to maximize performance. The plant lifetime is assumed to be 30 years, prior to major refurbishment, repowering or retirement.

The AFBC plant examined is rated at 500 MW with annual energy production of 3500 GWh. Its useful life is 30 years. Nearly 2 million tons of Illinois coal is required to fuel the plant annually. The IGCC plant is rated at 945 MW and produces roughly 6700 GWh annually. The assumed heat rate for the plant is 9,410 kJ/kWh. Its useful life is 30 years. The plant consumes roughly 3 million tons of coal annually.[14]

The oil-fired plant is rated at 800 MW and produces 3850 GWh annually using 954 million liters of #6 residual fuel oil. The gas-fired plant is rated at 800 MW and produces 3850 GWh annually using 1.05 billion cubic meters of natural gas annually. Both are conventional steam turbine plants. A combined cycle gas plant would be much more efficient and thus produce lower emissions per useful unit of energy production, but data for an assessment of a combined cycle plant were not available. In general the fossil-fired emissions of CO_2 during operation are 962 metric tons per GWh for conventional coal, over 740 metric tons per GWh for IGCC, 725 metric tons per GWh for oil, and 484 metric tons per GWh for natural gas.

Renewable Energy Plant Operation -- Hydropower, wind, photovoltaic, and solar thermal technology emissions during plant operation are essentially zero. The wood-fired generating facility has the highest CO_2 emissions of any technology during operation but it is important to note that this is offset by fuel regrowth, so that net CO_2 emissions are zero, or slightly negative. Among the renewable energy technologies, the OTEC plant has the next highest emissions during operation and the highest overall emissions at 304 tons per GWh. This represents only one OTEC technology option. A closed-cycle system would dramatically reduce the release of entrained gas in the seawater as it is flashed, thus bringing OTEC CO_2 emissions in line with the other renewable energy technologies. Similarly, the geothermal dry-steam system is also an open-cycle, which allows venting of CO_2 trapped in the hydrothermal steam that powers the turbine generator. This open-cycle hydrothermal system produces 56 tons of CO_2

per GWh. Closed-cycle flash steam systems and binary-cycle plants would eliminate the majority of these emissions. Binary technology is especially suited to the most abundant moderate temperature resources, and so is likely to play a larger role in future development of geothermal energy.

Nuclear Plant Operation -- The nuclear plant is a boiling water reactor design rated at 1000 MW, producing 6130 GWh annually over a useful life of 30 years.[15] The CO_2 emissions during nuclear plant operation should be viewed as the high end of a possible range of emissions, since they are based on the assumed operation of fossil fuel backup generators and boilers during normal operation. Under actual operating conditions a nuclear plant can be expected to operate with less reliance on fossil-fired auxiliary systems. It is estimated that a Pressurized Water Reactor (PWR) will have a similar ($\pm 5\%$) CO_2 profile. Although the PWR requires somewhat less fuel per gigawatt hour, it uses a more highly enriched fuel concentration.

Carbon Dioxide Emissions Summary

Summary by Technology

The total CO_2 emission profile of each of the technologies is shown graphically in Figure 1.

Fossil Fuels -- Conventional coal provides a baseline for comparison of CO_2 emissions from electric generating technologies; it is an established technology with well-known characteristics that provide a benchmark for alternatives. The Atmospheric Fluidized Bed Combustion (AFBC) plant represents an innovative alternative to conventional coal combustion and scrubbers.[16] The Integrated Gasification Combined Cycle (IGCC) plant represents an emerging advanced technology which offers significant improvements in coal combustion.[17] Oil and particularly gas are attractive for their lower CO_2 emissions profile, and gas is an increasingly important component of the U.S. electric generating system.

For fossil-fired generating technologies, most CO_2 emissions occur during operations. CO_2 emissions per ton of coal combusted are assumed to be basically similar for each technology, but the gross emissions are spread over a higher GWh output per ton of coal for the IGCC plant, which accounts for its improved emissions profile. Oil and gas have much lower CO_2 emissions per unit of energy output, but still have significantly higher emissions than renewable energy technologies.

Renewable Energy -- CO_2 emissions from the hydropower, wind, photovoltaic and solar thermal plant are primarily related to the construction of the generating station and the emissions from the steel and concrete plants. For these technologies, air emissions related to construction are higher than the emissions related to construction for the other technologies because of the materiel-intensive nature of the technology. But overall their emissions are a very small fraction of the emissions from coal

264

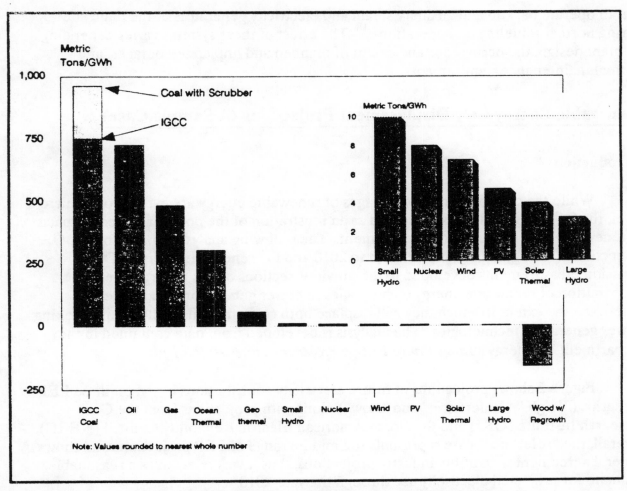

Figure 1. Carbon Dioxide Emissions: Electric Technologies

technologies and are for the most part less than or comparable to nuclear. Biomass, OTEC, and geothermal have relatively higher emissions during operation. Biomass in particular has higher emissions than a coal plant during operation, but when a managed biomass fuel cycle is considered, which includes regrowth of the feedstock, utilization of wood to produce power can minimize or eliminate net CO_2 emissions.

An open-cycle OTEC plant has CO_2 emissions comparable to a gas-fired plant during operation, although these emission levels are not inherent in the technology since a closed-cycle could substantially reduce CO_2 emissions. Geothermal's emissions during operation are large in comparison to the solar, wind and hydropower technologies, but far less than gas-fired generation. Like OTEC, geothermal CO_2 emissions are not inherent in the technology, and could be substantially eliminated through the use of closed-cycle systems.

Nuclear -- CO_2 emissions from the nuclear reactor should be viewed as a range, since a portion of the emissions are associated with fossil fuel combustion required to produce electrical and other inputs to uranium processing operations and the occasional use of fossil fuel boilers and generators during operation. There is also an input of fossil

265

fuel to operate backup and auxiliary steam and electricity generators at the plant site during normal refueling and operations.[18] The effect of these systems varies depending on plant design, the occurrence and extent of planned and unplanned outages, and normal maintenance requirements.

Renewable Energy CO_2 Displacement Projections (A Sample Case)

Introduction

While the environmental advantages of renewable energy are evident on a micro level, the following analysis is presented as an illustration of the potential macro impacts of renewable energy technology deployment. The following analysis is based on a DOE projection of energy supply and demand to 2010 and the renewable energy CO_2 emission measurements developed in the previous sections of this report. The contribution of renewable energy technologies in power generation is examined to determine the extent to which they will displace both conventional baseload and peaking power generation technologies. The analysis is developed from data contained in the Department of Energy's *Long-Range Energy Projections to 2010 (LEP)*.[19]

Figure 2 shows projections of future electricity contributions based on three LEP scenarios ("High," "Reference," and "Low") along with projections from the Gas Research Institute (GRI) and the North American Electric Reliability Council (NERC). Overall, the "reference" case represents the middle range of LEP projections and shows general agreement with utility industry projections, thus it was selected as a reasonable estimate for projected electricity use through the year 2010.

LEP Assumptions

Electricity consumption is projected to grow in every sector, averaging just over 3% per year between now and 1990 and between 2.4% and 2.7% per year thereafter. The projected growth in electricity consumption is due to a number of factors, including its inherent flexibility, the continuing increases in the efficiency of its end uses and, perhaps most important, the increasing relative prices of oil and natural gas. LEP assumed oil prices in the range of $18 and $22 per barrel ($1986) by 1990. Beyond 1990, price projections are much more uncertain, but are projected to be between $29 and $37 by 2000 and between $44 and $61 by 2010.

The electricity consumption projections imply that significant new capital expansion will be required starting in the early 1990s. By 2000, according to LEP projections, at least 50 gigawatts (GW) of new generating capacity in addition to the approximately 70 GW currently under construction or announced will be needed. In the LEP "reference" projection, over the near term the bulk of new capacity coming into operation will be coal and nuclear, as plants currently under construction are completed. Much of the new and as yet unplanned generating capacity, anticipated in LEP, is for low-emission coal-fired technologies, with newer "clean coal" technologies such as coal

266

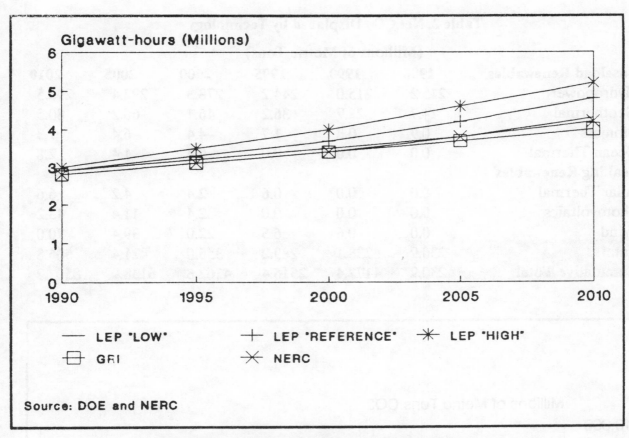

Figure 2. Comparisons of Electricity Projections

combined cycle and fluidized bed combustion making a growing contribution. Oil use in the electric utility sector is projected to rise, but existing excess oil capacity may negate the need for significant quantities of new conventional oil capacity. Natural gas consumption is also expected to rise, with small amounts of new gas turbine and gas combined-cycle capacity expected. However, by the late 1990s, oil use is expected to decline due to rising fuel costs, while nuclear expansion is assumed to diminish due to the lack of new plant orders over the past fifteen years. Small hydro, geothermal, wind, and photovoltaic renewable energy facilities are projected to produce moderate but growing amounts of electricity.

In order to determine the potential contribution of renewable energy technologies in displacing future fossil-fired CO_2 emissions, fossil-fired electric generating systems were compared with renewable energy systems with similar operating characteristics. It was assumed that gigawatt-hours from hydropower, geothermal, biomass, and ocean thermal production would displace a mix of baseload fossil- and nuclear-generated electricity. Gigawatt-hours generated by wind, photovoltaics, and solar thermal technologies were assumed to displace a mix of intermediate/peaking oil- and gas-fired electricity.

Table 2. Net CO$_2$ Displaced by Technology

(Millions of Metric Tons)

Baseload Renewables	1986	1990	1995	2000	2005	2010
Hydropower	215.2	215.0	244.2	278.3	292.4	296.5
Geothermal	14.7	21.7	36.2	46.1	66.2	80.5
Biomass	0.9	0.8	1.7	4.4	6.4	7.3
Ocean Thermal	0.0	0.0	0.0	0.4	1.4	2.8
Peaking Renewables						
Solar Thermal	0.0	0.0	0.6	2.4	4.2	6.6
Photovoltaics	0.0	0.0	0.0	2.4	11.4	43.2
Wind	0.0	0.6	6.5	22.0	39.4	70.0
Total	230.9	238.1	289.2	356.0	421.4	506.8
Cumulative Total	230.9	1172.4	2516.4	4162.8	6138.9	8502.2

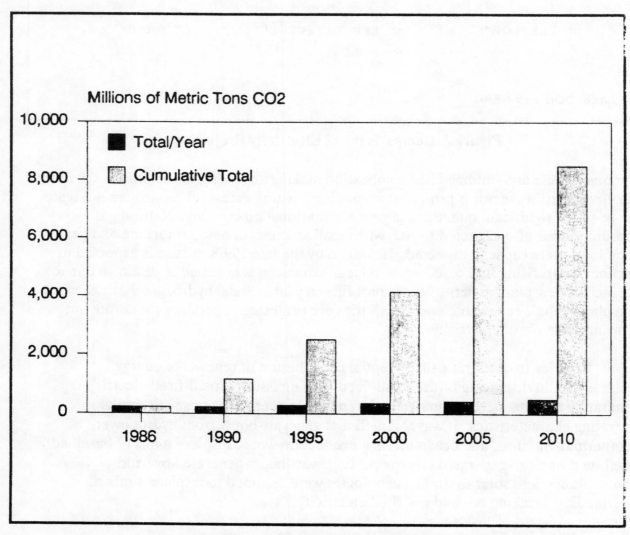

Figure 3. Renewable Energy CO$_2$ Displacement

Net CO₂ Displaced by Renewable Energy Technologies

By the year 2010 renewable energy technologies taken collectively are projected to displace over 8.5 billion metric tons of CO_2, on a 24-year cumulative basis, as shown in Table 2 and Figure 3, and would continue to expand substantially beyond 2010. The scenario considered here is based on conservative estimates of future energy use and renewable energy contribution. As the authors of the LEP point out, their scenario(s) should be interpreted simply as points of departure for understanding possible future energy development. The same is true for this analysis of CO_2 displacement potential.

Scenarios for renewable energy's contribution in the U.S. could significantly exceed the projections by the LEP, depending on the future price of conventional energy, the overall competitiveness of renewable energy technologies in the future, and the nature and aggressiveness of U.S. and international policy initiatives for addressing global climate change. Advances in renewable energy technology research could greatly accelerate their overall contribution to mitigating CO_2 emissions from conventional electric generation technologies.

Conclusion

In order to compare measures of emissions and materiel requirements for power production technologies, they must be examined in their entirety, taking into account each stage of the energy production process. This comprehensive approach provides a cumulative view of emissions that focuses on quantities of emissions as a function of energy supplied; a measurement convention that facilitates comparisons between different technologies.

From a historical perspective, the mix of fossil-fired electric power generation in the U.S. in 1986 produced an average of 874 metric tons of CO_2/GWh, while renewable energy technologies produced an average of approximately 18 metric tons of CO_2/GWh. Thus each GWh from renewable energy displaced approximately 856 metric tons of CO_2, or a 98% reduction. From a future perspective, projections to 2010 indicate that renewable energy electric technologies could reduce CO_2 emissions by 519 million metric tons per year in the U.S., or an 18% displacement of CO_2 related to an equivalent electrical output from fossil-fired power facilities.

Endnotes

1. Meridian Corporation. *Energy System Emissions and Materiel Requirements*. Prepared for the U.S. Department of Energy Office of Renewable Energy. February, 1989.

2. Rind, Hansen, et al. "A Character Sketch of Greenhouse Gases." *EPA Journal*. January/February 1989.

3. U.S. Department of Energy. *Technology Characterizations: Environmental Information Handbook*, DOE/EP--0028. June, 1981, p.76.

4. Ibid., p.2

5. U.S. Department of Energy, Energy Information Administration. *Electric Power Monthly*. DOE/EIA--0226 (88/06). June, 1988, p. 21.

6. Cheng, H.C., M. Steinberg, and M.Beller. *Effects of Energy Technology on Global CO_2 Emissions*. U.S. Department of Energy, Office of Energy Research. DOE/NBB--0076. April, 1986 p. 52.

7. Marland G., and R.M. Rotty. *CO_2 Emissions From Fossil Fuels: A Procedure for Estimation and Results for 1950-1981*. TN: Oak Ridge Associated Universities Incorporated. DOE/NBB--0036. June 1983, p. 13.

8. "Industrial Sector Energy Analysis: The Steel Industry." *Gas Energy Review*. VA: American Gas Association. vol. 16, no. 11, November, 1988, pp. 6-12.

9. Marland G., and R.M. Rotty. *CO_2 Emissions From Fossil Fuels: A Procedure for Estimation and Results for 1950-1981*. TN: Oak Ridge Associated Universities Incorporated. DOE/NBB--0036. June, 1983, p. 13.

10. Cheng, H.C., M. Steinberg, and M.Beller. *Effects of Energy Technology on Global CO_2 Emissions*. U.S. Department of Energy, Office of Energy Research. DOE/NBB--0076. April 1986, p. 52.

11. U.S. Department of Energy. *Technology Characterizations: Environmental Information Handbook*, DOE/EP--0028. June, 1981, p.72.

12. Electric Power Research Institute. *Integrated Photovoltaic Central Station Conceptual Designs*. Black and Veatch Engineers - Architects for Electric Power Research Institute, EPRI AP-3264, June 1984, p. 3-24.

13. U.S. Department of Energy. *Technology Characterizations: Environmental Information Handbook*, DOE/EP--0028. June, 1981, p. 2.

14. U.S. Department of Energy, Office of Fossil Energy. *The Role of Repowering in America's Power Generation Future*. DOE/FE--0096. December, 1987, p. 9, 15.

15. U.S. Department of Energy. *Technology Characterizations: Environmental Information Handbook*, DOE/EP--0028. June, 1981, p. 16.

16. U.S. Department of Energy, Office of Fossil Energy. *The Role of Repowering in America's Power Generation Future*. DOE/FE--0096. December, 1987.

17. U.S. Department of Energy. *Technology Characterizations: Environmental Information Handbook*, DOE/EP--0028. June, 1981, p. 76.

18. U.S. Department of Energy. *Technology Characterizations: Environmental Information Handbook*, DOE/EP--0028. June, 1981, appendix p. 29.

19. U.S. Department of Energy, Office of Policy Planning and Analysis. *Long Range Energy Projections to 2010*. DOE/PE-0082. July, 1988.

IMPACT OF SPECIES SELECTION AND CO$_2$ CONCENTRATION ON GLOBAL PHOTOSYNTHESIS.

Mirta N. Sivak
Robert Hill Laboratory,
Department of Animal and Plant Sciences,
University of Sheffield,
26 Taptonville Road,
Sheffield S10 5BR,
United Kingdom.

ABSTRACT

Plants in terrestrial or acqueous environments are a source of biomass and a sink for CO_2 emissions. Increased CO_2 and other greenhouse gases will affect plants in an unknown manner, adding to the difficulty of predicting short- and long-term climatic changes. Increased [CO_2] will affect plants *via* photosynthesis and other aspects of plant metabolism, but the degree of change brought about will depend on the plant species and on environmental conditions, e.g. temperature and rainfall, likely to change in response to increased emission of greenhouse gases. Relative ignorance in some areas of plant sciences is limiting the application of genetic engineering, plant breeding and other resources for the manipulation of global photosynthesis to stabilize the balance of atmospheric CO_2.

INTRODUCTION

In photosynthesis, light energy is used to reduce CO_2 into CH_2O (this being a simple representation of carbohydrate) and O_2 is formed. Over milenia, photosynthesis has progressively increased O_2 concentration and decreased CO_2 concentration in the atmosphere. Plants evolved different strategies to adjust to these changes. Some groups of plants developed additional metabolic pathways, associated with changes in cell or leaf anatomy, optimizing photosynthesis under these unfavourable conditions.

During this century atmospheric carbon dioxide concentration has been increasing steadily as a result of human activities, mainly because of the combustion of carbon-based fuels. Although the full extent of the changes occurring in the earth's atmosphere is still a matter for investigation, it is certainly far reaching. The expected changes in CO_2 concentration will have important and very complex effect on plants (CO_2 being a substrate of phosynthesis). The information available on effects of CO_2 on plants has been extensively reviewed (see, for example, Lemon, 1983; Strain and Cure, 1985). The mechanisms that plants evolved in response to *decreasing* [CO_2] will strongly influence their response to the expected *increase* in [CO_2] and, for this reason, I shall summarise them below. The main purpose of this paper is to describe the basic resources available for the management of global photosynthesis, how they could be used to increase its stabilizing effect on the balance of atmospheric CO_2, and to discuss the areas of ignorance limiting the usefulness of those resources.

STRATEGIES EVOLVED IN RESPONSE TO DECREASING CO_2 CONCENTRATIONS

Photorespiration

Like animals, plants respire, oxidising carbohydrates and other metabolites, to CO_2 and H_2O. In the light, C_3 plants also lose newly fixed carbon to the atmosphere, as CO_2, in a pathway termed photorespiration. Carbon dioxide fixation in the Benson—Calvin cycle (or reductive pentose phosphate pathway, RPPP) is catalysed by the enzyme Rubisco. In the presence of excess CO_2, only carboxylation occurs, but because Rubisco also has oxygenase activity (i.e., O_2 can compete with CO_2) in the presence of air concentrations of CO_2 and O_2 oxygenation occurs, leading to the loss of newly fixed carbon and energy.

The massive concentration in the chloroplast stroma of Rubisco, the most abundant protein in nature, may be regarded as an evolutionary response to decreasing CO_2 in the atmosphere. The photorespiratory pathway is now considered a way to bring some of the carbon "lost" by oxygenation back into the cycle, forming some useful by-products.

C_4 metabolism

The C_4 dicarboxylic acid pathway of photosynthetic carbon assimilation (the C_4 pathway) is a complex biochemical and physiological elaboration of the common photosynthetic carbon reduction cycle, the RPPP. The immediate end-product of the RPPP is a three-carbon compound. All green plants operate this sequence of carbon assimilation and most temperate species depend upon it entirely and are, therefore, called C_3 plants. Others operate an additional sequence, which yields a four-carbon compound as the first product of CO_2 fixation, and are therefore called C_4-plants. In order to provide a high, saturating CO_2 concentration to the RPPP, these plants have evolved a metabolic CO_2 pump which is energy-driven. It functions to trap atmospheric CO_2 in an outer layer of photosynthetic cells and shuttle it into a inner layer of the leaf where it is assimilated in the RPPP. Many of the worst weeds of the world are C_4 species, whilst most crops are C_3. The most economically important C_4 crops are sorghum, maize and sugar-cane, millets, and a number of pasture grasses.

CAM metabolism

Crassulacean acid metabolism (CAM) is a pathway which is found mostly in succulent species growing in arid environments. In C_4 photosynthesis, the first and second carboxylation occur in outer and inner compartments, but in CAM the reactions are separated in time instead of in space. CAM plants normally close their stomata (the valves that control gas exchange) during the day preventing water loss. This would severely limit photosynthesis through limiting CO_2 if its fixation by another carboxylating enzyme did not occur during the night when the stomata are open. Some CAM plants can switch to a C_3-like photosynthesis under certain environmental conditions. In the "CAM mode", the stomata are open by night and closed by day, while the reverse is true in the C_3 mode.

CAM is best described, ecologically, as an adaptation to water stress and water conservation. CAM plants have a higher water-use efficiency than C_3 or C_4 plants but this is accompanied by a lower productivity.

Mechanisms for the accumulation of inorganic carbon in algae

The mechanisms for the accumulation of inorganic carbon (IC, i.e. CO_2 or HCO_3^-), different to C_4 metabolism, have attracted the attention of plant biologists. This is because they also decrease photorespiration under natural conditions. There is a group of cyanobacteria (blue-green algae), which possesses the ability to concentrate IC via an energy requiring pumps. These photosynthetic microorganisms can change their relative affinities for external IC depending on the level in the external medium, and there is some active accumulation mechanism which allows CO_2 to be concentrated inside the cell. Accumulation in excess of 10^3 that of the external medium has been demonstrated (Canvin *et al*, 1987).

THE TIMESCALE

Global photosynthesis has been unable, so far to prevent atmospheric CO_2 from rising. This is partly the result of man's activities, which oxidise biomass back to CO_2. We must assume that global phosynthesis contributes significantly to slow down what would othwerwise be out of control, and this is why this process is so important.

The predicted changes in the composition of earth's atmosphere are relatively small when compared with those that plants had to face during evolution. If plants have adapted successfully to such major changes in the past, why should we worry about the predicted future increases in $[CO_2]$? The expected changes in $[CO_2]$ may look slow and incremental on a historical scale, but in geological terms, and in comparison with the previous decreases in $[CO_2]$ to which plants had millions of years to adapt, they are very fast, almost instantaneous. There is no reason to believe that the resulting changes in the productivity and composition of natural and man-managed ecosystems should be beneficial. Directed intervention may be required to preserve stability.

Plants can adapt to changes in the environment, in the sense of acquiring heritable characteristics enabling them and their progeny to survive and reproduce better than other plants lacking those characteristics. But we cannot expect adaptation in all species to occur over brief period. Plants with shorter growth cycles will have more generations than those with longer cycles, and a better chance to adapt.

Acclimation, a non-heritable modification in response to the new environment, will be more important in the short term. Different species vary in their capacity to acclimate to environmental changes.

Can we accelerate these adaptative processes? Can we influence the genetic changes so they have the desired outcome? What resources are at our disposal? In some specific cases, it may be possible to manipulate the genetic composition of the plant population in order to optimise yield and increase the potential of plants to act as stabilizers of atmospheric $[CO_2]$. To achieve these aims, it may be necessary to use plant breeding and genetic engineering to compress into a few decades changes the changes that, if left to natural evolution, would require thousands of years. Agricultural and forestry management could be adjusted to ensure that maximum yield is obtained within the environmental and economic constraints.

PLANT RESPONSES TO CO_2.

At present, photosynthetic rates and growth of plants are limited by ambient CO_2 when other factors (e.g. light, water, nutrients, temperature) are optimal. If CO_2 limits photosynthesis for most plants in most circumstances, a simple minded extrapolation leads to the conclusion that more CO_2 in the atmosphere should mean more yield and more carbon accumulated in plants. Some greenhouse crops will respond well to the expected increase in $[CO_2]$. Indeed, CO_2 enrichment has been used for some decades now to improve yield in crops such as tomato. We also know that plants with C_3 metabolism will respond more than C_4 plants to added CO_2. But there are complications. Plant growth and yield are not controlled by photosynthesis alone. More carbohydrate produced by photosynthesis will affect metabolic and maintenance dark respiration, and $[CO_2]$ may affect the activity of specific enzymes. After the initial incorporation of CO_2 by photosynthesis, there is a hierarchy of increasingly complex processes controlling the production and allocation of proteins, starch, sucrose and lipids into the various sinks that contribute to leaf expansion, root growth, flowering and fruiting, and final yield. The timing of each part of the growth cycle is also important and CO_2 is known to affect many of the partial processes involved in growth and development in some species.

Drought resistance may become less important, because stomata are less open in higher $[CO_2]$, thus decreasing water loss. This factor will be more important in C_4 plants, in which stomata are more sensitive to $[CO_2]$.

Increased photosynthesis and plant growth should increase the plant demand for nutrients. The availability of nutrients already limits plant growth, and the ratio carbon/nitrogen is expected to decrease, unless more nitrogen is made available by fertlization. Increased photosynthesis will stimulate nitrogen fixation by the root nodules of the legumes and may increase migration of legumes into grassland. Photorespiration is involved in N metabolism, and the partial supression of the photorespiratory pathway by increased $[CO_2]$ will affect N assimilation.

CLIMATE CHANGE

Up to a point, plants photosynthesise better as the temperature increases. A broad optimum is soon reached, however, as the advantage derived from faster carbon assimilation is offset by various deleterious factors. Photorespiration is important amongst these and therefore C_4 plants tend to have higher temperature optima. C_3 species photosaturate at about 1/5 full sunlight (i.e. photosynthesis does not increase much as the light intensity is raised above this level). Conversely, photosynthesis by C_4 species (or by C_3 species in high CO_2 or low O_2) increases with increasing light intensity up to, and beyond, full sunlight. Again, the difference is principally attributed to photorespiration, and C_4 species are likely to be at an advantage over other species under these particular conditions.

Not only a change in the global temperature is predicted as a result of the greenhouse effect, but changes in the pattern of rainfall are also forecast. Some parts of the world will have an increase in rainfall whereas others will have significant decreases. Although atmospheric $[CO_2]$ is a limiting factor, weather (particularly extremes such as frost and drought) is and will continue to be, an important source of variation in yield.

NATURAL ECOSYSTEMS.

Interactions at this level become so complex that prediction of the effects of CO_2 enrichment is extremely difficult. Competition between species will be affected by any change in climate and $[CO_2]$. No two species are the same, and within a given species the responses will be different according to growth conditions. These differences, added to the expected climatic changes, should result in significant changes in the composition of the plant communities, and through a complex net of ecological relationships, will bring changes in the associated animal communities.

RESOURCES

Increased research efforts are directed to characterise the long term response of plants to an increase in $[CO_2]$ of the order predicted, and to understand the biochemical mechanisms underlying these responses. The challenge is to produce the scientific information required to model the extremely complex responses of the main crop species and of natural ecosystems to such changes; and to use genetic engineering, plant breeding and other means to advantage.

Can we circunvent the new bottlenecks?

Increased $[CO_2]$ will bring earlier development of leaf area, increasing early cover and the efficiency in the use of incident light in the initial stages of the crop. Increased photosynthesis will not necessarily result in higher yield. If utilization of the carbohydrates produced by photosynthesis is relatively slow, starch may accumulate in the leaves and photosynthetic rate may decrease.

The so called "inhibition of photosynthesis by sink–source inbalance" refers to decline in rate that accompanies the slow utilization of photosynthates by the "sinks", i.e. the organs that use the assimilates such as young leaves, flowers, fruits, and it can be caused by low temperature or any factor that slows down the growth rate of the sink. In some species (but not in others), increased $[CO_2]$ leads to starch accumulation in leaves and decreased photosynthetic rates, a response that has attracted much attention because of its potential impact on yield. Some progress has been made in the understanding of this problem.

The ortophosphate (Pi) requirement by the chloroplast depends on the supply of the other substrates of photosynthesis (i.e., light and CO_2), and on the rates of the partial processes in which these substrates are utilised. The plant cell can adjust its Pi supply up to a degree by increasing the rates of one or more of the processes which recycle Pi (Sivak & Walker, 1987). Measurement of different aspects of photosynthesis *in vivo* (CO_2 and O_2 exchange, light–scattering, chlorophyll fluorescence and its quenching components) led to the characterisation of a broad range of symptoms caused by experimental limitation of phosphate supply to the chloroplast. Inadequate phosphate supply was also recognised in several circumstances of physiological relevance, for example symptoms have been described in leaves of plants grown in warm conditions and tranfered to relatively low temperatures, in leaves photosynthesising in saturating light and $[CO_2]$, in plants infected by pathogens or illuminated for a long time in optimum conditions. Symptoms of inadequate Pi supply to the chloroplast have also been described in plants grown under high $[CO_2]$.

Decreased photorespiration in response to higher $[CO_2]$ should decrease the light and $[CO_2]$ thresholds at which Pi becomes limiting, firstly because less Pi will be made available through photorespiration and secondly because increased carboxylation will make higher demands on the supply of Pi, itself a substrate of photosynthesis. If limited Pi supply could be recognised as limiting photosynthesis and growth in plants growing in enriched CO_2, improved understanding of photosynthetic regulation could be used for the improvement of the crop.

Acclimation

The better acclimation to changes in the environment observed in some species of plants (including increased $[CO_2]$), may be related to their capacity to adjust more eficiently the Pi supply to the chloroplast. The role of Pi re–cycling in acclimation to $[CO_2]$ becomes evident when one of the partial processes involved is affected. When the response of photosynthesis to increased $[CO_2]$ was studied in a starchless mutant of *Arabidopsis thaliana*, it was found that very small increments in $[CO_2]$ elicited oscillations in photosynthesis and lead to a decline in rate (Sivak and Rowell, 1988). Feeding P_i through the petiole alleviated the symptoms of inadequate Pi supply brought about by the increase in $[CO_2]$. It was concluded that in this species the contribution of starch synthesis to P_i re–cycling is essential for efficient adaptation to changes in environmental conditions that, like increased $[CO_2]$, pose increased demands on photosynthesis.

If better acclimation of some plant species to changes in the environment is related to their ability to adjust the rates of the Pi–recycling processes, it should be possible to improve acclimation of other species by introducing new Pi–recycling mechanisms into them or improving on the existing ones. One interesting pathway is that leading to the accumulation of fructans, found in leaves of temperate grasses. It has been proposed that it plays an important role in the ability of temperate species to grow in northern latitudes, by providing storage of fixed carbon in a readily accesible state under conditions of low temperatures and short daylength (Pollock *et al.* 1980). Fructans are accumulated in the vacuole, and this compartmentation may provide added flexibility. Starch is accumulated inside the chloroplast, the site of the photosynthetic process, and it has been suggested that excess starch may lead, by itself, to decreased photosynthesis, by disrupting the fine structure of the chloroplast.

Changes in agricultural practices

As mentioned above, if the higher yields allowed by increased $[CO_2]$ are to be realised, increased inputs of N, P and K will be required. Increased nitrogen fixation by symbionts in leguminous crops and by free living bacteria could reduce to some extent the use of fertilizers.

As far as weeds is concerned, for the C_3 crops (e.g. wheat, potato, rice), because they will benefit more from increased $[CO_2]$, weeds are likely to be less of a problem. The reverse is true for the C_4 crops (the most important, corn, sorghum, sugarcane and millet). The climate changes will result in shifts in the geographical areas more suitable for each species. Developments in herbicide technology and pest control will improve the overall situation and the same is valid for insect and infectious diseases (see below, "genetic engineering"). Increased understanding of the action of growth regulators could improve commercial yield of many crops, because it will allow the manipulation of partition of carbohydrates in the direction of increased harvest index. If, as expected, increased $[CO_2]$ leads to higher efficiency in the use of water by many crops, their cultivation could be extended to drier climates.

Forestry has different requirements and problems, because it deals with long–lived perennial species which will have too few generations to adapt to higher $[CO_2]$ and accompanying climate changes. The long life–cycles make breeding and screening for good genotypes especially difficult, but because of the importance of forests in the stability of earth climate they must be attempted.

Plant selection by breeding

In spite of the complexity of plants response to increased $[CO_2]$, some generalizations can be made. For example, plants with different photosynthetic metabolism each respond differently to a given change in $[CO_2]$: plants with C_3 metabolism respond more than C_4 plants to added CO_2. Thus, it should be possible to modify the productivity of a system by changing, say, from corn to wheat. This is, however, a drastic and probably impractical solution. Breeding or genetic enginering could provide a more satisfactory alternative, by introducing an advantageous character into the desired species.

The current breeding practices select those varieties best adapted to the environmental conditions of the site during the last few years of testing. Thus, the plant breeding programs incorporate the responses to changing climate and $[CO_2]$ in the part of the world where the crop is being tested. Wheat is an interesting example of this characteristic inherent to plant breeding, and the modern varieties are more resistant to a gaseous pollutant, SO_2, than varieties that were common at the beginning of this century (A. Bolan, P.J. Lea & T. Mansfield, 1989, personal communication).

Once good genotypes have been selected by conventional breeding, if the biochemical and physiological mechanisms underlying the higher yield are understood, screening in laboratory rather than in field conditions becomes possible. Better understanding should also facilitate the transfer of the beneficial characters to other varieties by breeding, and even to genetically distant species, by genetic engineering.

Tissue culture

Tissue culture is a relatively new methodology which allows the production of many genetically identical plants from a single individual in a relatively short time, and in a limited space. This technique has been used to obtain virus-free individuals from an infected plant. If a good genotype is obtained by plant breeding or genetic engineering, tissue culture can shorten the time necessary to obtain the large number of plants required for commercial crop production. A problem inherent to this technique is that, as genetic diversity is diminished, the likelihood of a new disease decimating the crop, increases.

Genetic engineering

The application of recombinant DNA technology to plants requires input information from several disciplines, especially genetics, biochemistry, biophysics and physiology. That information is scarce and inadequate and because of this limitation, molecular genetic research in plants so far has been focused on a few areas where the necessary background exists. Beneficial characters are usually controlled by several genes, but application of genetic engineering to crop improvement is at present limited to single-gene traits. A foreign gene can be introduced into what becomes a transgenic plant leading, for example, to the synthesis of an antimetabolite, trypsin inhibitor that provides protection against attack by insects (Boulter et al., 1988). This kind of character could become even more useful if higher biomass, combined with warming, increased the risk of plant diseases and attack by insects. Current research into diverse mechanisms of resistance to herbicides, insects, viruses and fungus should produce the technology needed to introduce such resistance into sensitive species. Even the introduction of resistance to some pollutants seems feasible, and synthesis of a human protein confering resistance to heavy metals has been achieved in transgenic tobacco and *Brassica napus* plants (Misra, 1988).

Much of the efforts of molecular biologists is devoted to the improvement of photosynthetic performance at low $[CO_2]$, through the modification of the carboxilating enzyme (Rubisco), introduction of IC pumps similar to those found in some cyanobacteria, or modification of the photorespiratory pathway. In the future, emphasis may have to be shifted to the partial processes likely to become limiting or (co-limiting) under increased $[CO_2]$.

The obtention of qualitative changes in plants is, however, a far more ambitious objective. We can include in this category the reduction of photorespiration in C_3 plants (making them more like C_4 plants) or the incorporation nitrogen-fixing capacity into non-legume species. These adavantageous characteristics were achieved by some plants through evolution spanning thousands of years, and involve not just a single protein but a series of proteins, and a degree of compartmentation and regulation of the sort that is beyond the abilities of genetic engineering at present. From the point of view of the plant molecular biologist, the simultaneous modification of many proteins, or a major modification of leaf anatomy, is a very complicated matter. We can only guess how long it could take or whether it is at all feasible to achieve such objectives, but it is worth noting that what can be done now routinely in laboratories all over the world, was science fiction one or two decades ago. One of the essential requirements for the incorporation of new pathways, the re-targetting of a single protein into any membrane in the cell, is now possible (Ellis, 1985). And yet, for plant molecular biology to provide solutions to classical agricultural problems, a great deal more of understanding of basic plant biochemistry will be required. Plant physiologists, biochemists, molecular biologists and users will have to collaborate in the tasks of choosing the most desirable characteristics, and of evaluating the feasibility of their introduction and the effects of this genetic engineering on the overall performance of the plant.

CONCLUSIONS

The factors limiting yield at present are not the same factors that will be limiting in a future CO_2-enriched world. The areas of science constraining our capacity to manage global photosynthesis at present are not the same the will constrain it in one or two decades time. We need to keep flexibility in the way we allocate resources, human and material, to different areas of scientific research.

ACKNOWLEDGEMENTS

My work is supported by grants from the Agricultural and Food Research Council (U.K.) and by Shell Research (U.K.).

REFERENCES

Boulter, D., Gatehouse, A.M.R. and Hilder, V. 1988. Trypsin inhibitor as insect protectant. Abstracts, 2nd International Congress of Plant Molecular Biology, Jerusalem, November 13–18, 1988.

Canvin, D.T., Miller, A.G. and Espie, G.S. 1987. C_3 Photosynthesis in algae: The importance of inorganic carbon concentrating mechanisms. In: *Carbon dioxide as a source of carbon.* (M. Aresta and G. Forti, eds.) pp 199–212.

Edwards, G. and Walker, D.A. 1983. *C3, C4: mechanisms, and cellular and environmental regulation, of photosynthesis.* Blackwell Scientific Publications, Oxford.

Ellis, J. 1985. Eucaryotic proteins retargetted among cell compartments. *Nature,* 313: 353–354.

Hanson, A.D., Hoffman, N.E. and Samper, C. 1986. Identifying and manipulating metabolisc stress–resistance traits. *HortScience,* 21: 1313–1317.

Kramer, P.J. 1981. Carbon dioxide concentration, photosynthesis and dry matter production. *BioScience,* 31: 29–33.

Lange, O.L., Nobel, P.S., Osmond, C.B. and Ziegler, H. (eds.) 1981. *Encyclopedia of Plant Physiology, New series, Vol. 12A (Physiological Plant Ecology I, Responses to Plant Environment).* Springer, Berlin.

Lemon, E. (ed.) 1983. *CO_2 and plants: The response of plants to rising levels of atmospheric carbon dioxide,* AAAS Symposium Vol 84. Westview Press, Boulder, Colorado, 280 pp.

Misra, S. 1988. Heavy metal tolerant transgenic *Brassica napus* and tobacco plants. Abstracts, 2nd International Congress of Plant Molecular Biology, Jerusalem, November 13–18, 1988.

Pollock, C.J.; Riley, G.J.P.; Stoddart, J.L. and Thomas, H. 1980. The biochemical basis of plant response to temperature limitations. *Annual Report of Welsh Plant Breeding Station for 1979,* pp. 227–246.

Sivak, M.N. and Walker, D.A. 1987. Oscillations and other symptoms of limitation of *in vivo* photosynthesis by inadequate phosphate supply to the chloroplast. *Plant Physiol. Biochem.* 25: 635–648.

Sivak, M.N. and Rowell, J. 1988. Phosphate limits photosynthesis in a starchless mutant of *Arabidopsis thaliana* deficient in chloroplast phosphoglucomutase activity. *Plant Physiol. Biochem.* 26: 493–501.

Strain, B.R. and Cure, J.D. (eds). 1985. *Direct effects of increasing carbon dioxide on vegetation.* United States Department of Energy, Office of Energy Research, Office of Basic Energy Sciences, Carbon Dioxide Research Division, 286 pp.

PART B

GREENHOUSE GAS CONCERNS IN COAL

CONVERSION TECHNOLOGIES

The role of coal use and technology in the greenhouse effect

by

Irene M Smith

IEA Coal Research, London, UK

Prepared for the IEA/OECD Expert Seminar on Energy Technologies for Reducing Emissions of Greenhouse Gases

OECD Headquarters, Paris, 12–14th April 1989

Abstract

An introductory overview is presented of the contributions to the global greenhouse effect from different sources and energy use sectors. The rates of increase in the greenhouse gases and future projections are discussed. Data on CO_2 emission factors for different coals are presented, showing the effect of improved combustion efficiency on emissions.

Introduction

It is an established fact that the concentrations of the greenhouse gases: CO_2, chlorofluorocarbons (CFC), methane (CH_4), nitrous oxide (N_2O) and ozone (O_3), in the atmosphere are increasing. This is due to human activities, including the fossil energy, chemical and agricultural industries. The greenhouse gases, including water vapour, control the radiative heat balance of the earth. These simple facts are complicated by the complexities of the climatic system (atmosphere-ocean-ice-earth). Because of this, it is not yet possible to make definitive predictions about the response of the global climate to increased concentrations of the greenhouse gases. The results of climatic model experiments broadly agree that there will be an unprecedented global warming next century and observations fit this general picture. However, there is as yet no conclusive evidence that increasing concentrations of the greenhouse gases are resulting or will result in a climatic warming and no agreement about the regional effects of a warming.

Despite uncertainties in most aspects of the issue, the potential implications are so serious that precautionary measures to slow down the increase of greenhouse gases in the atmosphere are being considered by the governments of many countries. IEA Coal Research has been keeping a watching brief on the greenhouse issue since 1977 and in a recent review (Smith, 1988) ranked the precautionary measures as follows:

1. Substitute for CFC;
2. Improve efficiency of energy use;
3. Promote reafforestation and improve forest management;
4. Use energy sources with no net CO_2 emissions.

Emissions of CFC will probably be reduced as a result of the Montreal protocol to protect the ozone layer. If substitutes with little or no greenhouse effect are chosen, there will be a major reduction in this group of greenhouse gases. The second and fourth measures were recently addressed by the Toronto Conference on The Changing Atmosphere. The conference advocated a reduction in CO_2 emissions by approximately 20% of 1988 levels by the year 2005 as an initial global goal. One half of this reduction would be sought from energy efficiency and other conservation measures; the other half by modifications in supplies (The Changing Atmosphere: Conference Statement Committee, 1988). Attention is also focussing on the deforestation issue but with little success.

This paper addresses the role of coal as one of the contributors to the greenhouse effect and the potential for improving the efficiency of its use. Information is drawn both from the 1988 review and from preliminary material collected for a further review by IEA Coal Research (CO_2: Abatement, Control) which is being prepared.

Emissions

The contributions to the total greenhouse effect from human activities in 1980-85 were estimated by Warrick (1988) as follows:

- 56% CO_2;
- 24% CFC;
- 14% CH_4;
- 6% N_2O.

These data would be modified by the contribution from increases in O_3 in the lower atmosphere. However the net influence on the greenhouse effect of this increase with O_3 depletion in the stratosphere is quite uncertain and is discussed in depth by Bolle and others (1986). Both the rates of increase and the residence time in the atmosphere are greater for CFC than for most other contributors (see Table 1).

Looking more closely at sources of CO_2 the emissions from fossil fuels, cement manufacture, deforestation and land exploitation had reached approximately 6 to 8 Gt C/y in 1984. The portion from coal amounted to 2.1 Gt C or 27% to 34% of CO_2 emissions, depending on the range of estimates recently considered plausible for deforestation and land exploitation. Assuming that CO_2 was responsible for about 56% of the total greenhouse effect (Warrick, 1988), the contribution from CO_2 emissions from coal combustion to the greenhouse effect was about 15% to 19% (see Table 2). Although the proportion of coal used for power generation in OECD countries is about 72% (IEA, 1988), considerable amounts are being used in other ways in industrialising countries such as China. It was estimated that the proportion of coal used worldwide in power generation was 49% in 1986 (Matthews and Gregory, 1989). Hence, it can be concluded that the greenhouse effect from coal-fired power generation was about 8% of the total.

The overall greenhouse effect from fossil fuels is made up of the contribution from CO_2 which is about 43% (see Table 2) and that from the CH_4 and N_2O emissions associated with fossil fuel extraction and combustion respectively. Warrick (1988) assumed an additional 5% for these emissions, bringing the greenhouse effect from fossil fuels to about 48% of the total. In terms of energy use sectors, the global greenhouse effect due to fossil fuels was estimated by Warrick as follows:

- 14% electricity generation;
- 14% transportation;
- 10% industry;
- 8% residential/commercial;
- >1% miscellaneous losses.

Although electricity generation and transportation appear to be the largest single contributors to the greenhouse effect, the range for deforestation and land exploitation (see Table 2) suggests that this could be almost as significant, and possibly even more significant, a factor than either electricity generation or transporation. However, at the lower end of the range, CO_2 from both industrial and residential/commercial sectors could each be contributing more to the greenhouse effect than deforestation and land exploitation.

While CH_4 and N_2O are associated with coal use, CO_2 is the principal greenhouse gas produced from coal so the rest of this paper will concentrate on CO_2 emissions. One of the major areas of uncertainty is that of future emissions of CO_2 from fossil energy. For the future, there are much larger measured resources and reserves of coal than oil or gas and there is sufficient accessible coal in significant coal fields for reserves not to constrain coal utilisation. More reserves of all fossil fuels may become available in future.

The regional patterns of energy consumption and hence CO_2 emissions are changing as newly industrialising countries increase their use of fossil fuel. For example the regional patterns of CO_2 emissions (% of world total which was 1.6 Gt C and 5.2 Gt C in 1950 and 1980 respectively) are estimated by Rotty and others (1984) to have changed for specific regions as follows:

	1950	1980
North America, Western Europe, Japan, Australia	68	43
Centrally planned countries	19	33
Newly industrialising countries	6	12

Figure 1 shows the trends of CO_2 emissions in six countries. China exemplifies an industrialising country with a fast growth rate using mainly coal. Australia, Canada and Denmark represent industrialised countries with increasing use of coal and the Netherlands one with increasing use of natural gas. It is interesting that the emissions of the latter three countries and of Japan show reduced overall CO_2 emissions following the oil crises of the 1970s, with or without increasing use of coal. In addition to Denmark, the following countries also show reduced overall CO_2 emissions with an increasing use of coal: Finland, Spain and the USA.

None of the trends in Figure 1 can be extrapolated with any certainty into the future because the use of energy beyond the next twenty years is extremely uncertain. Projections of carbon emissions in the year 2050 made since 1981 were evaluated critically by Keepin and others (1986). The results from different authors are presented in a simplified form in Figure 2 to show their wide range. Keepin and others attributed the large variation here to a general ignorance about future cultural, political, social and technological developments. The upper and lower bounds in Figure 2 were intended to define informally a range of projections outside of which conditions and scenarios were regarded as extreme. The upper bound corresponds to a growth rate for carbon emissions of 2.3%/y (considerably greater than the average of 1.46%/y since 1973) and would be associated with relatively high fossil fuel consumption. The growth rate for the lower bound is -1.4%/y and would require the development of a suitable mixture of, for example, conservation, efficiency improvements, renewable or nuclear energy sources. Hence, the total CO_2 emissions from fossil fuel in 2050 might lie between 2 and 20 Gt C/y (compared with the current amount of about 5 Gt C/y). An additional contribution of up to 1 or 2 Gt C/y is assumed from deforestation and changing land use. The uncertainty factor in CO_2 emissions by the middle of the next century is therefore nearly an order of magnitude but is considered likely to lie in the middle range between the bounds. Hence future emissions of CO_2 to the atmosphere will depend on a choice of strategies involving political, technological and social developments, influenced by the results of many investigations of the issue worldwide.

Reduction of CO_2 emissions

Emissions of CO_2 from fossil fuels depend on their C content and heating value as well as on how they are burned. The emission factors (kg C/GJ) used generally (Marland, 1983; Marland and Rotty, 1984) and based on the higher heating value (HHV which is the heat of combustion when all of the product water remains liquid) are as follows:

- 24.1 bituminous coal;
- 19.9 crude oil;
- 13.8 natural gas.

The value for coal does not vary greatly with rank. Hence Hughes (1989) and Fortune (1989) concluded from the analyses of coals as received from Australia, Canada, China, Colombia, South Africa, UK and USA for power generation that the emission factors varied only from 23.2-25.4 kg C/GJ. Anthracites and semianthracites had higher values up to 26.4 kg C/GJ. Maude (1989) also found that Yallourn brown coal from Australia had an emission factor of 26.1 kg C/GJ similar to that of anthracite. Peat and wood were a little higher at 27.5 and 27.3 kg C/GJ respectively (Fortune, 1989). There is therefore little opportunity for reducing CO_2 emissions by choice between types of coal but an appreciable reduction by using oil or natural gas instead of coal.

On the other hand there is an opportunity for reducing CO_2 emissions from all fossil fuels by developing more efficient utilisation technologies. Energy efficiency improvements have already had a significant effect. For example the specific coal consumption for power generation in the Federal Republic of Germany (FRG) has decreased from about 950 grams of coal equivalent (gCE)/kWh in 1920 to only 330 gCE/kWh today (Zimmermeyer, 1988) (see Figure 3). This means that CO_2 emissions are about a third today compared with what they might have been without improvements in efficiency. Can this trend continue in future?

In the power generation sector the current efficiencies of conventional pulverised fuel combustion with flue gas desulphurisation (PF + FGD) and circulating fluidised bed combustion (CFBC) are 35-38% (325-350 gCE/kWh). Advanced power generation technologies however have the potential of realising much higher efficiencies (see Figure 4). The use of integrated gasification combined cycle (IGCC) and pressurised fluidised bed combustion (PFBC) are expected to increase the efficiency to 38-40% (310-325 gCE/kWh) (Harrison, 1988; Torrens, 1988). In the topping cycle the gases from PFBC would be heated by addition of a topping gas from a separate gasification system to feed a high temperature gas turbine. This technology is being developed in the UK and the USA and promises an improvement in efficiency up to 45%. This increase over PF + FGD represents a reduction in CO_2 emissions by 17% (Harrison, 1988). Efficiencies of 50-60% (205-245 gCE/kWh) are believed achievable with combined cycle methods and fuel cells (Torrens, 1988). Even with these advanced technologies there are still significant heat losses. These can be exploited in combined heat and power systems for all the power cycles (with a small reduction in electrical power) to achieve efficiencies exceeding 80% (150 gCE/kWh).

Conclusions

The largest contributors to the greenhouse effect caused by human activities are CO_2 at 56% and CFC at 24%. Three activities: electricity generation, transportation and deforestation (including land exploitation) each contribute significant proportions of around 14%, although emissions from industrial and residential/commercial installations are also important. The total CO_2 emissions from all coal use worldwide are responsible for about 17%, of which only about 8% of the greenhouse effect is due to coal-fired power generation.

There is a good deal of uncertainty regarding future emissions of CO_2. These will depend on a choice of strategies. Industrialising countries such as China and India, which are dependant largely on coal for their industrial development, are likely to contribute increasingly to the greenhouse effect in future, both absolutely and proportionally. However the experience in some countries such as Denmark, Finland, Spain and the USA shows that an increasing use of coal does not necessarily lead to overall increasing CO_2 emissions because of fuel substitution and improvements to efficiency.

Throughout this century, improved efficiency has been an important factor in reducing emissions and this will continue. Whilst it is not certain how fast advanced power generation technologies will reach commercial status it is theoretically feasible to meet the Toronto target of a 20% reduction in CO_2 emissions by the year 2005. Continued improvements are believed achievable.

At any given level of energy use, CO_2 emissions can also be reduced by substitution with other fossil fuels. The replacement of coal with natural gas and oil which produce less CO_2 per unit energy than coal may not always be desirable or convenient. A further option is the replacement of fossil fuels by nuclear power or various forms of renewable energy, any one of which gives rise to environmental problems of a different nature.

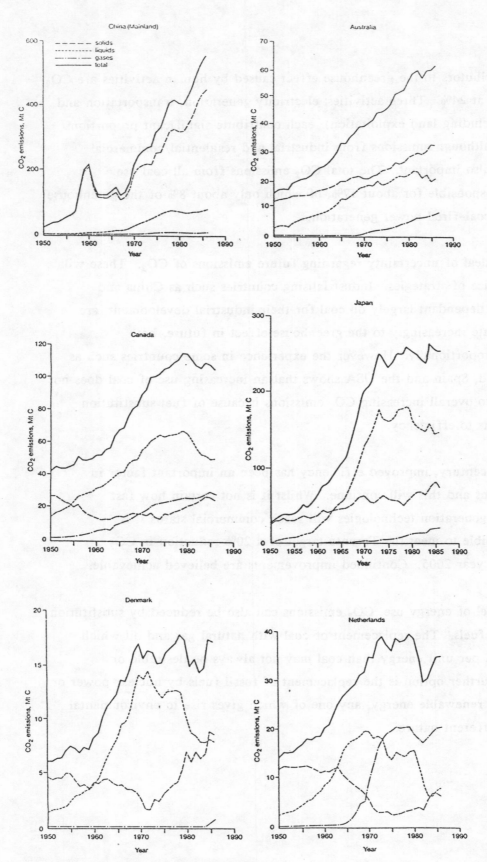

Figure 1 Production of CO_2 from fossil fuels (Marland and others, 1988)

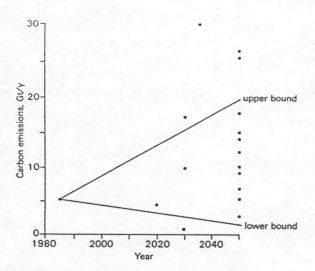

Figure 2 Projections of carbon emissions in 2050 (Keepin and others, 1986)

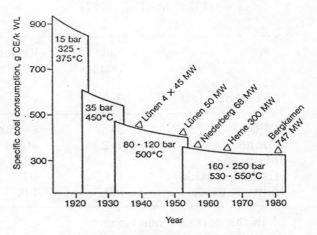

Figure 3 Specific coal consumption in power generation in the FRG (Zimmermeyer, 1988)

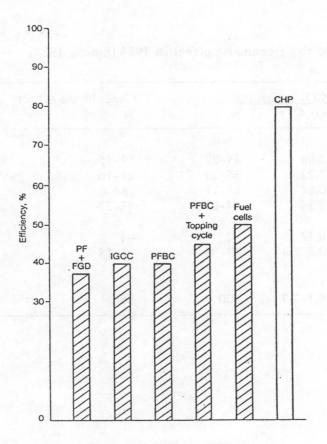

Figure 4 Efficiency of power cycles (Harrison, 1988; Torrens, 1988)

Table 1   Global increase in the concentration of the
          greenhouse gases in the lower atmosphere
          (Bolin and others, 1986; Bolle and others,
          1986; MARC, 1987)

Greenhouse gas	Rate of increase %/y	Residence time y
CFC	5-6	8->500
Ozone*	1-2	~0.25
Methane	1	10
Carbon dioxide	0.4	~7
Nitrous oxide	0.2	170

* in the northern hemisphere

Table 2 Contributions to the greenhouse effect in 1984 (Smith, 1988)

	CO_2 emissions Gt C/y	%	Greenhouse effect %
Coal	2.08	34-27	19-15
Oil	2.22	36-29	21-16
Gas	0.84	14-11	8- 6
Fossil fuel	**5.14**	**84-67**	**48-38**
Cement manufacture	0.13	~2	~1
Deforestation and land exploitation	0.8-2.4	13-31	7-18
Total CO_2	**6.1-7.7**	**100**	**56**

References

Bolin B, Döös B R, Jäger J, Warrick R A (eds.) (1986a) The greenhouse effect, climatic change, and ecosystems, SCOPE 29, Chichester, UK, John Wiley and Sons, 572 pp (1986).

Bolle H-J, Seiler W, Bolin B (1986) Other greenhouse gases and aerosols: assessing their role for atmospheric radiative transfer. In: The greenhouse effect, climatic change, and ecosystems, B Bolin, B R Döös, J Jäger, R A Warrick (eds.), SCOPE 29, Chichester, UK, John Wiley and Sons, pp 157-203 (1986).

Fortune D B (1989) London, UK, Shell Coal International Limited. Private communication (Feb 1989).

Harrison J S (1988) Innovation for the clean use of coal. Lecture presented at the British Coal Utilisation Research Limited, 1988 Robens Coal Science Lecture, London, UK, 3 Oct 1988. Cheltenham, UK, Coal Research Establishment, vp (Nov 1988).

Hughes I S C (1989) Carbon dioxide emission coefficients - variation with coal type, Cheltenham, UK, Coal Research Establishment. Private Communication (Jan 1989).

International Energy Agency (1988) Coal Information 1988, Paris, France, OECD/IEA, 515 pp (1988).

Keepin W, Mintzer I, Kristoferson L (1986) Emission of CO_2 into the atmosphere: the rate of release of CO_2 as a function of future energy developments. In: The greenhouse effect, climatic change, and ecosystems, B Bolin, B R Döös, J Jäger, R A Warrick (eds.), SCOPE 29, Chichester, UK, John Wiley and Sons, pp 35-91 (1986).

Marland G (1983) Carbon dioxide emission rates for conventional and synthetic fuels. Energy; 8 (12); 981-992 (1983).

Marland G, Rotty R M (1984) Carbon dioxide emissions from fossil fuels: a procedure for estimation and results for 1950-1982. Tellus; (36B); 232-261 (1984).

Marland G, Boden T A, Griffin R C, Huang S F, Kanciruk P, Nelson T R (1988) Estimates of CO_2 emissions from fossil fuel burning and cement manufacturing using the United Nations energy statistics and the U.S. Bureau of Mines cement manufacturing data. NDP-030, Carbon Dioxide Information Analysis Center, Oak Ridge, TN, USA, Oak Ridge National Laboratory, Environmental Sciences Division, vp (Oct 1988).

Matthews A K, Gregory K (1989) A global estimate of the proportion of coal used in the production of electricity (Draft). OR/951/20/1, London, UK, British Coal, 4 pp (Feb 1989).

Maude C (1989) London, UK, IEA Coal Research. Private communication (Feb 1989).

Monitoring and Assessment Research Centre (1987) United Nations Environment Programme: environmental data report. Oxford, UK, Basil Blackwell Ltd., 352 pp (1987).

Rotty R M, Marland G, Treat N (1984) The changing pattern of fossil fuel CO_2 emissions. DOE/OR/21400-2, Washington, DC, USA, US Department of Energy, 24 pp (Sep 1984).

Smith I M (1988) CO_2 and climatic change. IEACR/07, London, UK, IEA Coal Research, 52 pp (May 1988).

The Changing Atmosphere: Conference Statement Committee (1988) The changing atmosphere: implications for global security. In: The changing atmosphere conference, Toronto, Ontario, Canada, 27-30 Jun 1988. Downsview, Ontario, Canada, Environment Canada, 12 pp (1988).

Torrens I M (1988) Global climate change linkages; strategies for control of emissions. Paper presented at the conference on global climate change linkages; acid rain, air quality and stratospheric ozone, Washington, DC, USA, 15-16 Nov 1988. Palo Alto, CA, USA, Electric Power Research Institute, 28 pp (1988).

Warrick R A (1988) Norwich, UK, University of East Anglia, Climatic Research Unit. Private communication (Feb, Dec 1988).

Zimmermeyer G (1988) Climatic risks and the role of energy use. Prepared for the World Congress: climate and development; climatic change and variability and the resulting social, economic and technological implications conference, Hamburg, FRG, 7-10 Nov 1988, 41 pp (1988).

CO_2 and the Greenhouse Effect:

Actions and Strategies of German Hard Coal Mining Industry

Gunter Zimmermeyer, Essen

Friedrich H. Esser, Hückelhoven

Paper presented at the IEA/OECD Expert Seminar on Energy

Technologies for Reducing Emissions of Greenhouse Gases

12. - 14.4.1989 in Paris

Summary:

German hard coal mining industry started assessing a climate risk
due to the greenhouse effect at the beginning of the 70s. Despite
the various open questions and uncertainties in scientific knowledge which are presented in the paper, the opinion is advocated
that measures to reduce this risk have to be globally implemented within the concept of precaution. That means measures that
remain justified, should the greenhouse effect not turn out to
have the detrimental climatic effects currently predicted.

A high standard of energy efficiency and energy saving employed
globally might keep the CO_2 emissions on the present constant
level, thus delaying a CO_2 doubling for centuries. Technologies
and policies should be developed for further steps to be implemented in accordance with extended scientific knowledge and validity about the greenhouse issue.

To achieve common action an international convention has to be
set up and ratified by at least the fifteen most important CO_2

emitting countries. Some items of such a convention are discussed.

INTRODUCTION

To reduce the risk of changing global climate in an appropriate way turns increasingly out to be a challenge for all nations and especially for the industrialised countries. Increasing public perception and political pressure additionally calls for actions.

Mankind has got a chance to implement solutions. However, all known facts and aspects of the problem have to be taken into consideration. To be able to develop strategies for risk management it is at first necessary to assess the size and the nature of this risk.

Assessing climate risk was started in the German hard coal mining industry at the beginning of the 70s. Since then we have surveyed the literature, contacted leading scientists and discussed their findings and the reasons for their opionions with them and last not least we have financed scientific programs, some of them in cooperation with the German Environmental Protection Agency (Umweltbundesamt). We need and use those results of risk assessment as a basis for risk management. That means we endeavour to put our concepts for reducing risk appropriately on a sound basis.

296

RISK ASSESSMENT

In scientific knowledge about the greenhouse effect several facts are known but there are also a lot of unsolved questions and un- certainties.

The Facts

It is a fact that CO_2 acts as a greenhouse gas together with other gases, emissions and activities . Others are water vapour (H_2O), methane (CH_4), dinitrogen oxide (N_2O), the CFCs and ozone but even sulphur dioxide and nitric oxides may influence the radiation budget via atmospheric transformations. Anthropogenic emission of water vapour needs not to be taken into consideration because man's contribution to natural emissions is neglegible. However, water vapour emitted in high altitudes may produce cir- rus clouds absorbing infrared radiation significantly. Non-CO_2 IR-active gases may have an extremely high capability to absorb heat radiation. For example, compared to CO_2 methane is more radiatively effective by a factor of up to 40 and some CFCs even by a factor of about 10.000 or even more. Obviously their con- tinuing increases in atmospheric concentrations has to be taken into consideration primarily. This paper, however, specificly deals with questions relating to energy use and in particular with CO_2.

It is another fact that atmospheric concentrations of CO_2 and other greenhouse gases are increasing; CO_2 presently by about 0.5 % per year, the CFC's for example by nearly 5 % per year. The well known Mouna Loa curves show the trend in atmospheric CO_2 concentrations rising from 310 ppm in 1958 to about 350 ppm today (Fig. 1). Equally accepted are the data for former periods. Measurements from ice cores indicate that in the 18th century CO_2 concentrations had been around 270 ppm. So there is scientific evidence to designate the data as pre-industrial CO_2 concentrations (Fig. 1). For the future, however, no definite projection can be stated but it is likely that steadily increasing global CO_2 emissions will produce a CO_2 doubling within the next century. Estimated climatic consequences may occur even earlier taking the other greenhouse gases into consideration.

The contributions of the various countries and regions of the world to global CO_2 emissions are known to be different. Solely three countries (USA, UdSSR, China) share more than 50 % and about 15 countries sum up to more than 80 % (Fig. 2). The position of several countries in this sequence and so their importance has changed in the past and it is expected to change drastically in the future.

Regarding the development of CO_2 emissions since the 50s a striking change in the trend for industrialised countries is shown in 1973, the year of the oil crisis (Fig. 3).

From that time onwards, caused by higher energy prices, these countries have increasingly used efficient energy technologies that have been available and affordable for them. However, developing countries and centrally planned countries do not show this break. There, these technologies have not been available and/or have not been affordable.

In the past decade we have also learned much about the fate of atmospheric CO_2. The astonishing fact that only half of the emissions produced by burning fossil fuels could be monitored in the atmosphere induced scientists to discover this missing sink. Meanwhile there is much scientific evidence that the ocean plays the key role in this context. Recent ocean models by the Hamburg Max-Planck-Institute for Meteorology based on tritium measurements in the deep sea indicate that the rate of CO_2 uptake by the ocean via gas exchange, supported by a biological pump generated by assimilation of phytoplankton and taken up by zooplankton, will increase the more the lower the emission rate into the atmosphere will be. If mankind therefore succeeds in keeping CO_2 emissions constant at the present level, in the medium term, the ocean is capable of absorbing up to 70 % of those emissions, thus delaying a CO_2 doubling for two or three centuries (Fig.4) and reducing a possible temperature increase in such a way that mankind might be able to adapt to it.

Uncertainties and open questions

The main uncertainties and unsolved questions are the development of future emissions, the validity of the climate models, the observed global temperature change and sea level rise, and the expected regional climate change and its effects on man and on the environment.

Future emissions of CO_2 have been predicted by several institutions ranging from an increase by a factor of four till the year 2050 to a substantial decrease to about one half of today's global emissions (Fig. 5). The World Energy Conference estimates a growth in annual energy demand from about 12×10^9 t CE today to about 15×10^9 t CE in the year 2000 (Fig. 6). It is not foreseeable which estimate will turn out to be correct. To predict future CO_2 emissions it has additionally to be taken into account that an increasing energy demand may eventually be supplied by a specific lower fossil fuel use and so the increase in CO_2 emissions might be much smaller than the development of energy demand.

Furthermore the World Energy Conference estimates the Proved Recoverable Reserves of fossil fuels, i. e. the tonnage that can be recovered under present expected local economic conditions with existing available technology to about $1,050 \times 10^9$ t CE compared to the geological ressources of about $13,000 \times 10^9$ t CE and to the data used in the greenhouse issue of between 5,000 and

300

10,000 x 10 ^{9} t CE. So, assuming that only about half of the emitted CO_2 remains airborne, a doubling of the atmospheric CO_2 concentration would not be possible by burning all the reserves (Fig. 7). It has of course to be admitted, that this argument is not suitable to nullify the whole problem because the other greenhouse gases have to be considered and because the size of the proved Recoverable Reserves is a variable function of energy demand, energy prices and technology and may increase in future. However, the present value demonstrates that predictions of doubling the atmospheric CO_2 within the next five decades have to be put into perspective. This will presumably not be feasible.

It is said to be a common consensus among scientists that global temperature will increase by 3 ± 1,5° C due to a doubling of CO_2 concentration. Accurately, this is the result of the best available climate models. However, one has to take into consideration that major uncertainties exist with respect to the role of clouds and the role of the oceans and their impact on climate. More recent model calculations show evidence that temperature effects may be damped significantly even more than the lower bandwidth of the present model results of 1.5° C indicates.

Some scientists argue that a global temperature rise has occurred within the last decade indicating a greenhouse effect. Others question that and regard these changes as to be within natural fluctuations (Fig. 8). This suggestion seems to be appropriate as there are unexplained variations in the temperature curve and the abrupt rise in the last years can be explained by an extreme temperature rise in tropical and subtropical zones. In higher latitudes, the actual temperatures tend to be even lower than 30 years ago. These findings are in contrast to the model results and have to be explained.

At present models are not capable in predicting regional climate and sea level rise. Therefore, no politician is able to check whether the country he is responsible for might be a winner or a loser. Arguments are used to support the feeling that no country will turn out to be a winner. Nevertheless some important countries do believe that reducing CO_2 emissions might be more disadvantageous than adapting to a climate change.

CONCLUSION FROM RISK ASSESSMENT

Further global growth in the emissions of greenhouse gases turns out to be a significant risk causing climate changes.

Keeping global CO_2 emissions constant at the present level for the next decades helps to reduce this risk to a tolerable level. In this context, however, it has to be mentioned that the Toronto Conference in 1988 demanded a much more stringent reduction. However, as we noticed, this conference had not been aware of the recent results in oceanic CO_2 absorption and additionally the possibilities and the constraints in energy demand have not been fully taken into consideration.

Think globally, act nationally was a slogan presented at the Toronto-Conference too. This request might be acceptable if at least all important CO_2 emitting countries act that way. If, however, only a few countries fulfill that demand their economic risk thereby will be very high and for all that no reduction in global emissions of IR-active gases will be achieved.

Because of the ongoing scientific debate the greenhouse problem is a classical field of precaution (Vorsorge), a concept in German environmental law. It means that such measures should be implemented, that will remain justified, should the greenhouse effect not turn out to have the detrimental climatic effects currently predicted.

So the aim must be to achieve an international protocol or convention which has to be ratified by as many countries as possible, at least by the fifteen largest CO_2 emitting countries. This convention should consider that measures have to be implemented stepwise, intending the individual succeeding steps to demand more stringent actions to limit emissions and activities. Those succeeding steps should be put into force in accordance with extended scientific knowledge and validity about the greenhouse issue.

In any case, the important principle has to be satisfied that these measures must not generate greater risks - environmental or economic - for the community than the one that has to be expected by a climate change.

The first step has to be implemented according to the principle of precaution. For the CO_2 problem this means to increase energy efficiency and to save energy globally.

RISK MANAGEMENT

To get solid informations on what can be achievable in the global energy issue it is helpful to analyse the past development in energy demand and energy use.

Global energy demand has increased since the 50s from about 3 x 10^9 t CE to about 12 x 10^9 t CE per year in the second half of the 80s (Fig. 6). Moreover the contribution of the different energy sources to supply energy demand has changed markedly during this time.

As an example for the industrialised countries the development in the FRG is demonstrated, being more or less typical. Official energy programs and scenarios on energy consumption of the 60s have proven to overestimate the actual energy consumption by far. More recent scenarios made by energy companies expect future energy consumption not to increase further but to remain nearly at the present level. Scenarios made by alternative institutes indicate their ambitions of exhausting the whole potential of energy saving. They think it possible to reduce the energy demand by 30 % or even more within the next decades.

Even in the past decade the CO_2 emission total in the FRG has been reduced by about 10 % . Actually, the emissions are lower than in the 70s (Fig. 9). There are several reasons for this development:

- Efficiency in power plants

  Efficiency in power plants has increased drastically since the beginning of this century (Fig. 10) and even since the 50s. Only a few plants of that period are still in operation. Presently only about 330 g of coal are burned to generate 1 kWh in comparison with 1000 g at the turn of the century. Therefore, specific CO_2 emission today in power plants is only one third.

- Cogeneration of heat and power

  Since the 50s in the FRG the connected heat load and the heat input to district heating from cogeneration has continuously increased and amounts to about 35.000 MJ/s resp. 200.000 TJ/a today. Without district heating oil consumption would be higher and about 1×10^6 t/a CO_2 would have been emitted additionally.

- Energy demand in the production sector

  Aside from the increased multifold use of energy in industry, the specific demand of energy input to produce a product has also been reduced significantly .

- Energy demand in the consumption sector

  There are many examples that the energy demand of applicances, for instance in the household sector, has decreased significantly (Fig. 11). It is of course fair to mention that the number of such appliances like TV sets or washing machines has increased much quicker in that time so that the net energy consumption in this sector has grown. It would, however, have grown significantly higher if the specific energy demand had remained constant.

- Energy demand in traffic

  A most impressive example of energy conservation is given by the locomotives. Despite increasing kilometric performance the primary energy consumption decreased to less than 40 % because of using modern three-phase current motors.

Such improved energy efficiency contributed to the above mentioned reduction in CO_2 emission total in the FRG together with increased use of such energy sources that specifically emit less or no CO_2.

Only a few other industrialised countries had even more success

in reducing CO_2 emissions. Most others, however, have only a-
chieved a significantly lower standard in energy efficiency and
energy saving. Therefore, the resulting CO_2 emissions for different countries show a different trend (Fig. 12).

In the centrally planned economies and the developing countries
there was no development comparable to this. Despite globally
increasing energy prices in 1973, energy efficient technologies
could not be employed because they were not achievable or affordable. So these parts of the world expose different CO_2 trends
compared to the western industrialised countries not showing a
marked break after 1973 (Fig. 3) because they did not react to
higher energy prices.

THE FIRST STEP OF MEASURES

At the first step to be implemented as precaution to reduce global CO_2 emissions all countries, especially the centrally planned
economies, but also western industrialised countries with a relatively high specific energy consumption are to be led up to the
latest state of energy efficiency. This would save a significant
amount of CO_2 emissions. Therefore a net global reduction might
result even with constantly increasing emissions in the developing countries, occurring on a much lower level. To this end, an
enormous investment program has to be installed. This can only be
employed if western industrialised countries are willing to help

307

with technology, education and massive financial support.

Besides aspiring this ambitious target in the next years more advanced energy efficient technologies can be implemented in the industrialised countries in addition. These technolgies already under development will penetrate the market the easier the higher energy prices will be. In power plant technology, for example, the next generation might be a coal gasification combined cycle or a pressurized fluidized bed combustion, in any case a plant with an efficiency further increased by up to 25 %.

Many densely populated areas with heavy individual traffic suffer from its various detrimental impacts. Extending underground and rail mass transit systems could make a large contribution in increasing energy efficiency and thus reducing CO_2 emissions. This would have additional local benefits, too.

And above, flanking measures in the CO_2 issue are necessary and possible for instance in agriculture, forestry and land use. They may serve to absorb CO_2 from the atmosphere and on the other hand to reduce emissions of the various greenhouse gases from these activities. It goes without saying that preventive measures to curb the use of CFCs with the prime aim of protecting the stratospheric ozone layer as envisaged by the so called Montreal Protocol have to be implemented and even tightened.

PROSPECTS FOR THE FUTURE

For the more distant future, measures have to be thought of and developed that should be implemented if scientific evidence of the greenhouse effect increases. Therefore, it is necessary to enhance research funds to increase efforts for research and development. Even energy systems that seem to be highly speculative today should have a chance of being looked at thoroughly and developed if there is a feasibility to realise it. To accelerate the development and implementation of technical progress political measures like subsidies and governmental incentives or even laws may be necessary.

Should a more stringent energy policy be necessary in the long run to cope with a greenhouse effect there will presumably be options available like photovoltaics or nuclear fusion and even other renewables that will penetrate the market anyway. Energy technologies especially for developing countries have to be developed that are in accordance with local / regional environmental, economical and technical conditions and requirements. In any case we can have confidence that man's intelligence and creativity will be able to provide any solution that is necessary.

In the very long run it is not imperatively necessary to banish fossil fuels completely even if a detrimental greenhouse effect would turn out to be true. There is much scientific evidence that the ocean as an enormous carbon store will absorb as much CO_2, that about 2×10^9 t C as CO_2 can be emitted without increasing the atmospheric CO_2 content.

THE COMMON ACTION

It is not automatically assured that all important CO_2 emitting countries think and act in a common agreement. Some of those countries have already indicated that a possible greenhouse effect is regarded as a net benefit. Such countries cannot be convinced to curb CO_2 emissions by exaggerating possible effects of the greenhouse gases and painting an overly gloomy picture of climate risks in general. The only chance to make them join the line is to demonstrate the advantages of efficient energy use and to support their interest by offering technical and financial assistance and education to implement these technologies. This is the most delicate and ambitious task for the western industrialised countries in this issue.

A possible way for cooperation could be an international convention that will have to be ratified by at least the fifteen most important CO_2 emitting countries if it is to have a chance to be successful in reducing these emissions. It is, however, still an open question if such a strategy can really avoid or mitigate a climate change because nobody knows the effect of a specific CO_2 reduction. This convention to reduce the risk associated with the greenhouse gases should allocate a quota to reduce radiatively effective gases or climatically relevant activities for each country, leaving it to the independent countries to find their own way in achieving this quota. To meet this quota it should be allowed, that highly developed countries with only a low additional potential for reductions could finance and support technologies in other countries to reach an equivalent reduction instead.

Efforts to control SO_2 and NO_x emissions as precursors of climate relevant airborne particulates which were already undertaken in some countries have to be taken adequately into account.

Such a procedure will have the highest probability to achieve significant reductions without increasing other risks. In any case, mankind has to bear in mind that all actions undertaken must not generate risks or detriments of other kind that are greater than the climate risks that have to be expected otherwise.

The greenhouse effect, might it be true or not, might it cause significant detrimental effects or not, should be viewed as a challenge to industry and energy producers and not as a motive for hostility to technology. Mankind must and is able to act immediately to increase global energy efficiency and thereby reduce climate risk. Energy producers should be at the top to implement well balanced global strategies for future energy use to reduce risk of climate change appropriately.

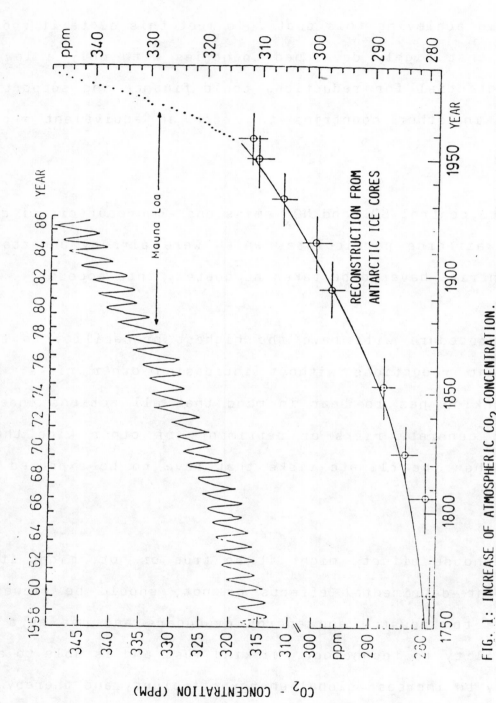

FIG. 1: INCREASE OF ATMOSPHERIC CO_2 CONCENTRATION.
DIRECT MEASUREMENTS (UPPER LEFT, MONTHLY AVERAGES; DOTTED CURVE, UPPER RIGHT
YEARLY AVERAGES) AT MAUNA LOA, HAWAII ACCORDING TO KEELING; THE CIRCLES AND
UNCERTAINTY RANGES INDICATED BY CROSSES ARE FROM RECONSTRUCTIONS BY NEFTEL,
OESCHGER AND CO-WORKERS (KEELING, 1986; OESCHGER, 1985)

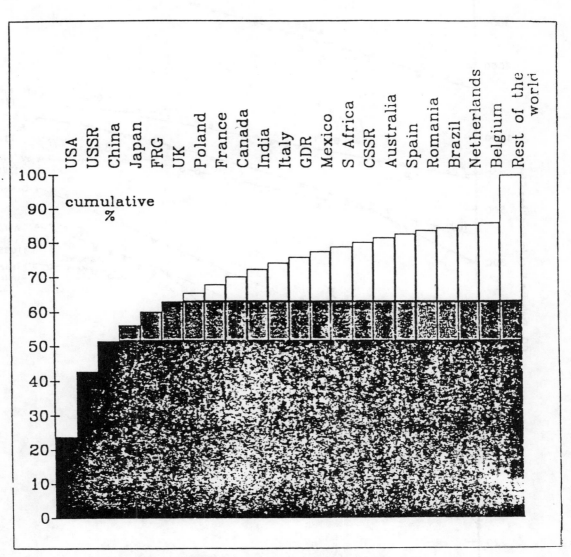

FIG. 2: CO_2 EMITTING COUNTRIES 1984 (ROTTY, 1987)

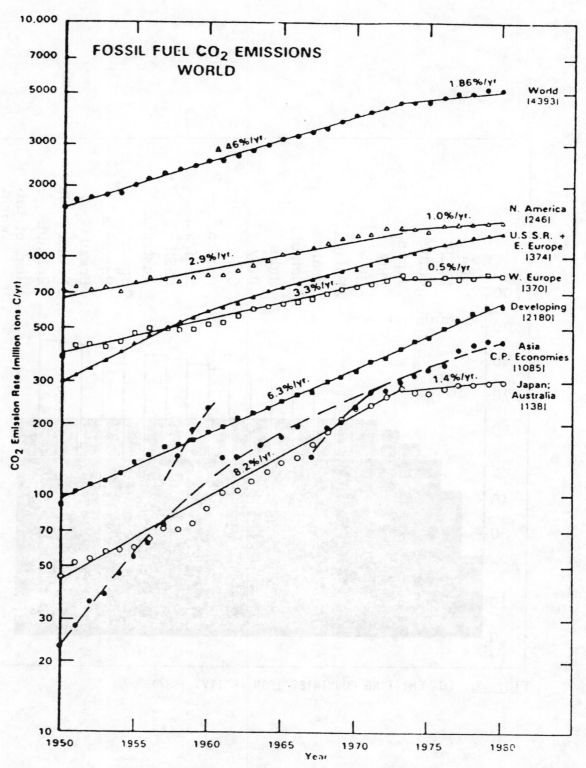

FIG. 3: GLOBAL AND REGIONAL CO$_2$ EMISSIONS 1950 - 1980 (ROTTY, MARLAND, 1984)

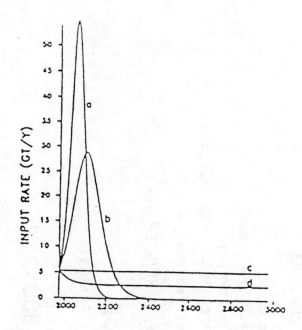

EMISSION SCENARIOS FOR THE FUTURE

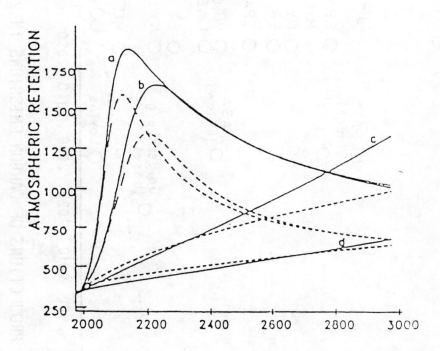

FIG. 4: INCREASE OF ATMOSPHERIC CO_2 CONCENTRATION
DEPENDING ON THE EMISSION SCENARIOS
(HASSELMANN, MAIER-REIMER, 1989)

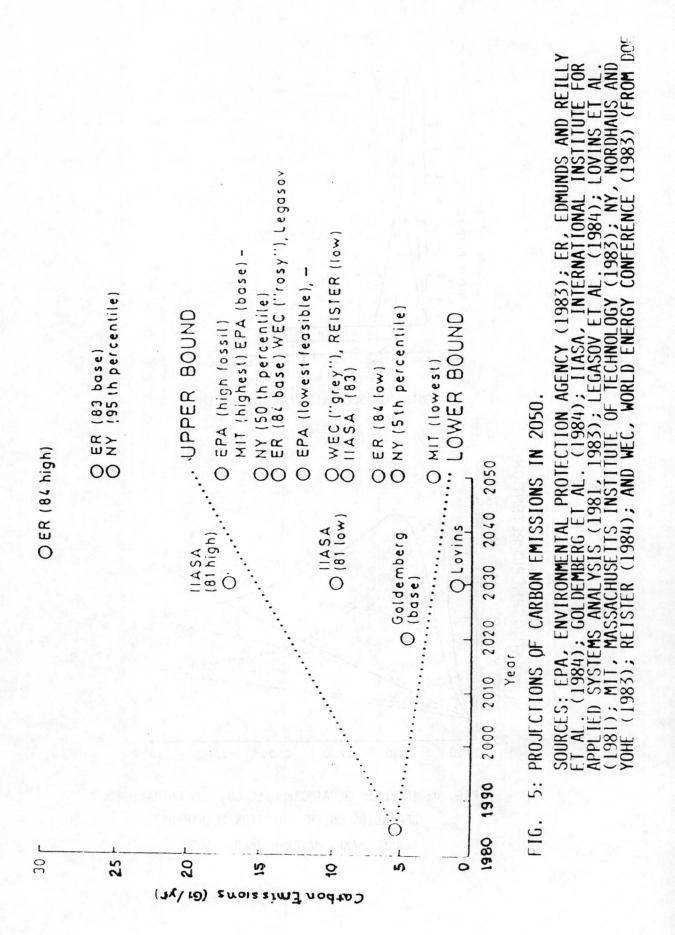

FIG. 5: PROJECTIONS OF CARBON EMISSIONS IN 2050.

SOURCES: EPA, ENVIRONMENTAL PROTECTION AGENCY (1983); ER, EDMUNDS AND REILLY ET AL. (1984); GOLDEMBERG ET AL. (1984); IIASA, INTERNATIONAL INSTITUTE FOR APPLIED SYSTEMS ANALYSIS (1981, 1983); LEGASOV ET AL. (1984); LOVINS ET AL. (1981); MIT, MASSACHUSETTS INSTITUTE OF TECHNOLOGY (1983); NY, NORDHAUS AND YOHE (1983); REISTER (1984); AND WEC, WORLD ENERGY CONFERENCE (1983) (FROM DOE

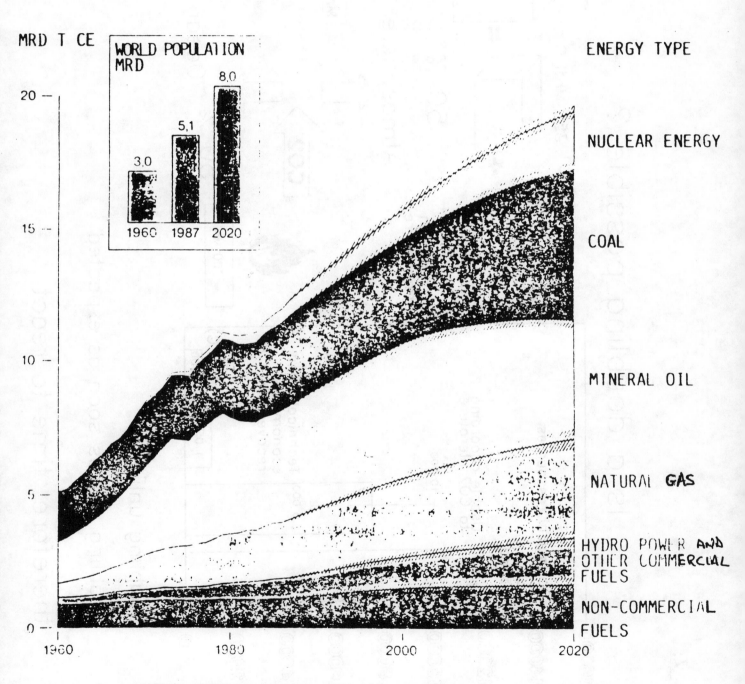

MRD T CE

20 —

15 —

10 —

5 —

0 —

1960   1980   2000   2020

WORLD POPULATION
MRD

8.0

5.1

3.0

1960 1987 2020

ENERGY TYPE

NUCLEAR ENERGY

COAL

MINERAL OIL

NATURAL GAS

HYDRO POWER AND
OTHER COMMERCIAL
FUELS

NON-COMMERCIAL
FUELS

FIG. 6: WORLD ENERGY DEMAND AND ITS SUPPLY

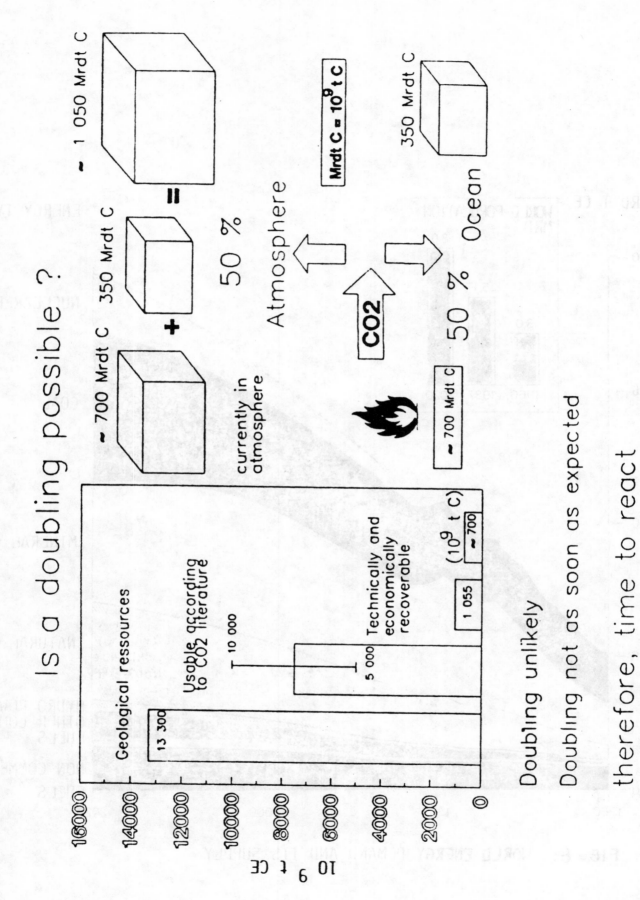

FIG. 7:

Is a doubling possible ?

Mrdt C = 10^9 t C

Atmosphere

CO2

50 %   Ocean

~ 1 050 Mrdt C

=

~ 700 Mrdt C + 350 Mrdt C

currently in
atmosphere

50 %

350 Mrdt C

~ 700 Mrdt C

Geological ressources

Usable according
to CO2 literature

Technically and
economically
recoverable

(10^9 t C)

13 300

10 000

5 000

1 055

~700

16000
14000
12000
10000
8000
6000
4000
2000
0

10^9 t CE

Doubling unlikely

Doubling not as soon as expected

therefore, time to react

318

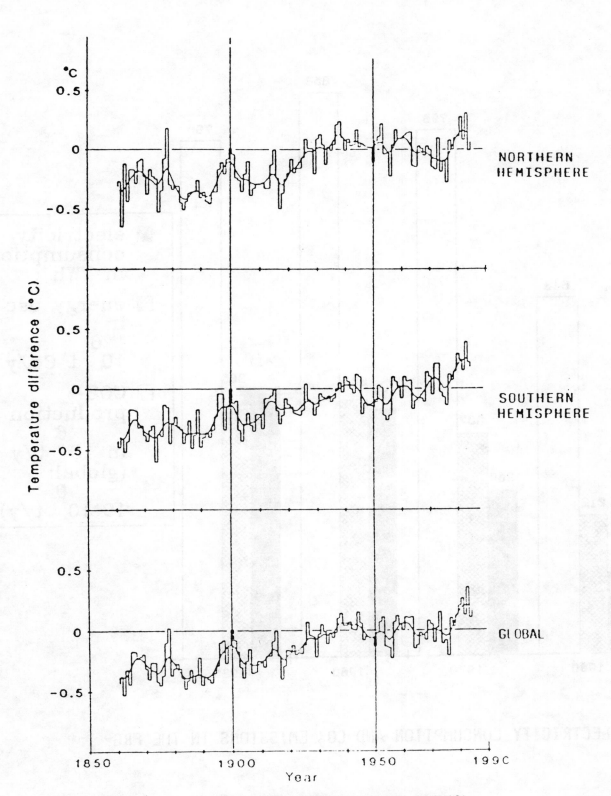

FIG. 8: GLOBAL AND HEMISPHERIC ANNUAL
MEAN TEMPERATURE VARIATIONS SINCE 1861
(JONES AND OTHERS, 1986)

319

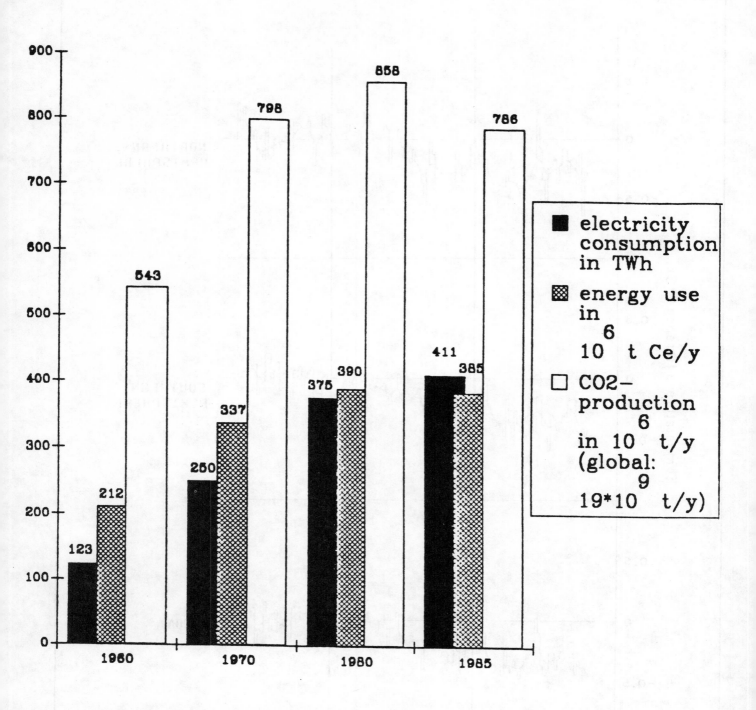

FIG. 9: ELECTRICITY CONSUMPTION AND CO_2 EMISSIONS IN THE FRG

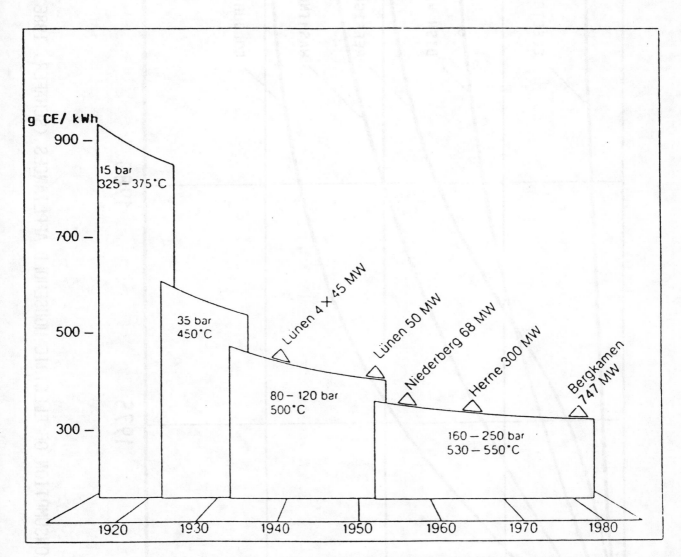

FIG. 10: SPECIFIC HEAT CONSUMPTION OF GERMAN COAL FIRED POWER PLANTS (STEAG, 1988)

321

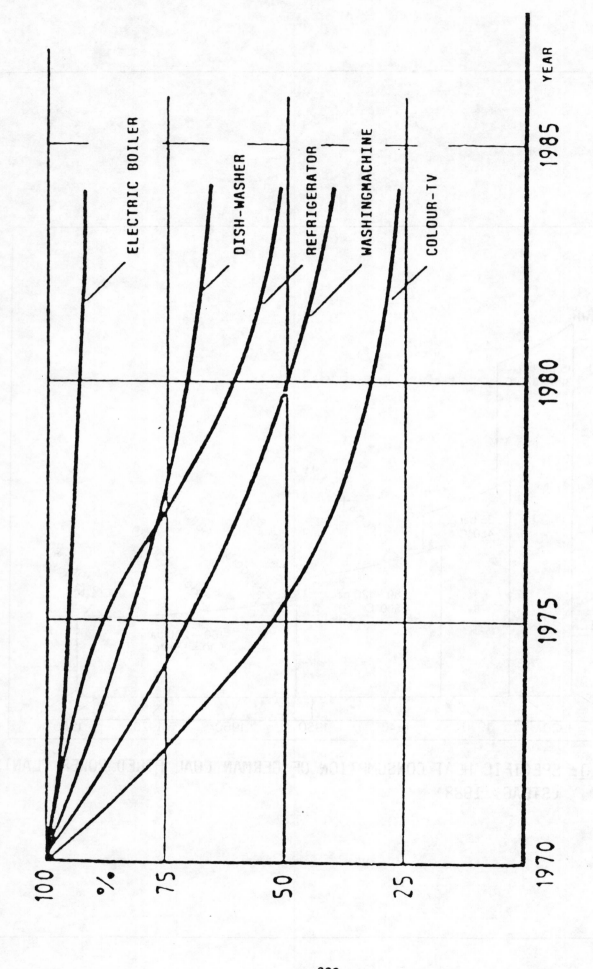

FIG. 11: SPECIFIC CONSUMPTION OF ELECTRIC HOUSEHOLD APPLIANCES (SCHÄFER, 1986)

322

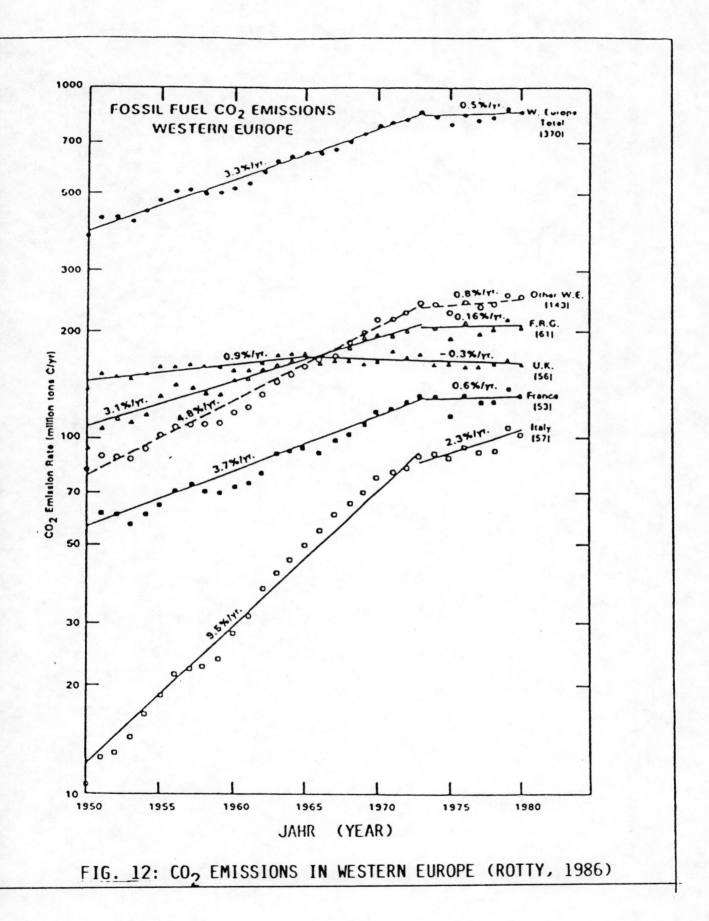

FIG. 12: CO₂ EMISSIONS IN WESTERN EUROPE (ROTTY, 1986)

COAL TECHNOLOGIES AND THEIR IMPACT ON THE GREENHOUSE EFFECT

K. M. SULLIVAN*

INTRODUCTION

A number of gases including CO_2, when emitted to the troposphere allow the sun's rays to penetrate to the earth, but due to specific properties of the gas, do not permit all of the heat to re-radiate back to the atmosphere. As a result, the earth's surface is warmed by the radiative characteristics of these gases. This characteristic, which is referred to as the "greenhouse effect" is fundamentally essential for the earth to exist with its present temperate climate. Up until recent times, CO_2 was thought to be the main contributor, but it is now known that other gases contribute about 50% to the effect, resulting from man's activities.

Man is able to exist in a relatively comfortable fashion throughout the world because of the greenhouse effect. History indicates that man has survived considerable climatic change, due to natural causes. However, there is concern that due to man's own activities the quantity of greenhouse gases will increase in the troposphere, resulting in an increase in the earth's surface temperature. This in turn may affect climate, agricultural response, human and animal reactions, ocean levels and land use. Impacts of this nature would have an effect on national wealth distribution and many aspects of life of all people.

The increase of greenhouse gases in the troposphere is a direct result of the world's increasing population, increased agricultural activity, deforestation and increased energy use.

Since CO_2 results from the combustion of all fossil fuels, then it is important to examine the contribution that the world use of coal makes to the greenhouse effect.

RADIATIVE GASES

A number of gases, such as carbon dioxide, methane, water vapour, nitrous oxide, ozone and chlorofluorocarbons are transparent to incoming short-

* Dr Ken Sullivan, Technical Advisor,
  Coal & Allied Industries Limited, Sydney, New South Wales.

wave radiation, but are relatively opaque to outgoing longwave radiation and as a result, are said to be radiatively active. Variations in the concentration of these gases in the troposphere can alter the thermal balance of the earth's atmosphere. Outgoing terrestrial radiation which would otherwise escape to space, is trapped within the inner layer of the atmosphere, resulting in a greenhouse effect, due to a rise in the surface temperature and cooling of upper levels of the atmosphere.

Several base line monitoring stations exist in various parts of the world for determining both CO_2 and other radiative gases in the atmosphere. Records, commencing in 1958 at Mauna Loa, Hawaii, show that there is a continuing increase in CO_2 concentrations in the atmosphere and this has been confirmed at other stations, as far afield as the South Pole and Cape Grim, Tasmania. Similar increasing trends have been more recently observed for other trace gases, such as methane, nitrous oxide and the various chlorofluorocarbons (1). Table 1 summarises published data on recently observed rates of increases for a number of radiative gases.

Radiative Gas	Approximate Average Annual Increase
Carbon Dioxide	0.4%
Methane	1.3%
Nitrous Oxide	0.3%
Freon - 11	5%
Methyl Chloroform	10%

TABLE 1

Measured Average Annual Increase of Specific Radiative Gases.

At present, it is estimated that radiative gases other than carbon dioxide, contribute about 50% to the greenhouse effect (2,3). However, in the future, the contribution made to the greenhouse effect by gases other than CO_2 will become increasingly more significant as illustrated by Fig.1, which shows the cumulative near surface air warming for an adopted trace gas scenario for a period approximately 40 years hence. From these predictions, it can be seen that in the future, the carbon dioxide contribution will diminish in relation to the total quantity of man-made radiative gases emitted.

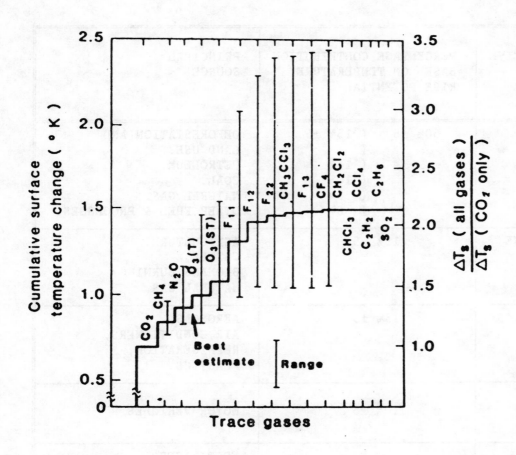

Predicted Influence of Various Radiative Gases
on Future Near-Surface Air Warming (from ref. 4).

FIGURE 1

Radiative gases arise from a wide range of sources and their escalating
increase is largely related to an increase in the world's population, an
increase in the standard of living of many areas and changes in
lifestyle.

The current relative contribution from anthropogenic sources of each of
the greenhouse gases to the potential earth's surface temperature rise,
is shown in Table 2.

RADIATIVE GAS	PERCENTAGE CONTRIBUTION BASED ON TEMPERATURE RISE POTENTIAL	PRINCIPAL SOURCE
CO_2	50% ±5 (15% ±5 ((14% (10% (5% (6%	DEFORESTATION AND LAND USE. PETROLEUM. COAL. NATURAL GAS. OTHER FUEL & PROCESSES.
METHANE	15% ±5	AGRICULTURE, CATTLE, BIOMASS BURNING, NATURAL GAS.
C.F.C.'S	13% ±3	AEROSOLS, AIR CONDITIONERS, REFRIGERATION, PLASTICS.
OZONE	10% ±5	MOTOR VEHICLES
N_2O	9% ±2	FERTILISERS, BIOMASS BURNING, MOTOR VEHICLES, FOSSIL FUEL BURNING.
OTHERS	3% ±2	

TABLE 2

Relative Contribution
of Greenhouse Gases Resulting
from Man's Activities.

CO_2 emissions occur as a result of all combustion processes from both stationary and mobile sources using liquid, gaseous and solid fuels.   In total, they are now estimated to approximate 5.4×10^{15} grams of carbon

328

per annum.   Possibly, the major single source of man-created CO_2 is
deforestation and erosion.   In addition forest fires, part of the
natural carbon flux, approach, but probably do not exceed the annual
release of carbon from the burning of fossil fuels (5).

Methane release into the atmosphere is increasing at an escalating rate,
largely as a result of man's increased agricultural activity. Examination
of data derived from ice-core studies, indicates that methane emissions
have remained comparatively static for most of the last thousand years.
However, in the present century methane emissions have almost trebled and
are increasing (6).

Chlorofluorocarbons   are   used   mainly   as   refrigerants,   solvents,
propellants in pressurised cans and for foam manufacture and have only
come into use during the last few decades. These gases are of particular
concern since they have a long life and in addition to being powerful
radiative gases have the ability to affect the stratosphere's ozone
layer.   As a result of the recent Montreal Protocol many countries will
agree to limit the manufacture and use of the most ozone destructive
chlorofluorocarbons.   However, this will not prevent, but will de-
escalate their increase into the atmosphere.

Nitrous oxides principally occur from natural soils, land use, the sea,
nitrogenous fertilisers and biomass burning, but a minor component (10%-
15%), is derived from fossil fuel combustion.

Ozone is derived from many sources as a result of atmospheric chemistry.
It is formed in the troposphere from hydro-carbons, carbon monoxide and
oxides of nitrogen, which occur as a result of combustion, mainly from
mobile sources.

In addition, there are a number of hydro-carbons and organic liquids
which all give rise to radiative gases in the atmosphere.   These largely
arise from various industrial processes and mobile combustion.

THE GLOBAL CARBON CYCLE

Prior to the 19th century, man appeared to have caused little impact on
the global carbon cycle.   However, with the development of transport
and communications systems, plus equipment for both industrial and
agricultural production, the world's consumption of energy expanded
from   500 x 10^6 tonnes   of   oil equivalent in 1900 to more than

329

6,900 x 10⁶ tonnes of oil equivalent in 1986.   During the same period,
the world's population increased threefold from 1.6 x 10⁹ to
4.8 x 10⁹.

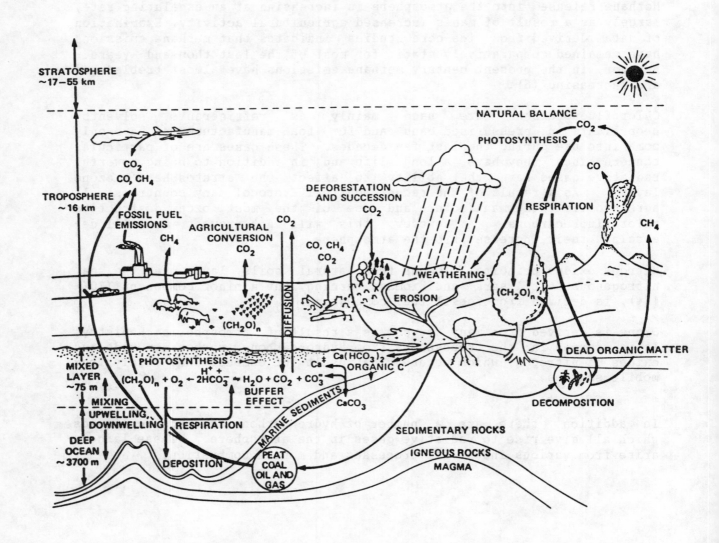

Schematic Diagram of the Global Carbon Cycle   (from ref. 7).

FIGURE 2

This vast increase in both population and energy consumption, associated with increased industrialisation and agricultural development, meant that man's activities had begun to play a small role in the carbon cycle.

This is confirmed by the fact that the amount of CO_2 injected into the atmosphere annually from fossil fuel combustion is approximately 2.5% of the combined amount exchanged between the atmosphere and the terrestrial biosphere (including deforestation and land use) and between the atmosphere and the oceans (8).

Figure 2 is a schematic diagram of The Global Carbon Cycle, illustrating the natural exchanges that occur between the oceans, the atmosphere, vegetation and human activity and super-imposing on these, the emissions arising from deforestation, from agriculture and from the combustion of fossil fuel from both mobile and stationary sources.

Although the processes that give rise to carbon dioxide and the exchange of carbon in the atmosphere are well-known, the rate of exchange and potential for these exchanges to occur with various levels of CO_2 in the atmosphere are not known. Therefore, if the effects of increased levels of CO_2 in the atmosphere are to be adequately modelled, improved knowledge of the carbon transfer mechanisms is required.

The fact that the carbon cycle is not at a static equilibrium condition, is illustrated from observation that annual increase in carbon content of the atmosphere is less than the cumulative input from fossil fuel combustion and other sources. This is because of an increase in CO_2 uptake by other reservoirs. The dynamics of these exchanges will determine the future CO_2 concentrations that result from additional fossil fuel combustion.

Net carbon exchange between the atmosphere and oceans is controlled by the temperature dependent equilibrium between carbonates and bi-carbonates, by biological processes, by turbulent mixing and by circulation of water masses that transport carbon from the surface into deeper waters. Many uncertainties in the atmosphere-ocean flux remain and even though improved data collection and modelling have occurred in recent times, even the most sophisticated analyses indicate that only 30% to 40% of the carbon admitted to the atmosphere has been taken up by the oceans. However, as much as 40% of the carbon admitted to the atmosphere is unaccounted for.

Estimates of the release of carbon to the global cycle from the burning of fossil fuels during the period 1860 to 1982 were 0.17×10^{18} grams. Whilst over the same period net releases of CO_2 to the atmosphere from forest clearing and other disturbances in the terrestrial carbon system, resulted in a greater contribution than that of fossil fuel and has been injected into an estimated world carbon reservoir of approximately $75,000 \times 10^{18}$ grams.

There is considerable uncertainty concerning the annual contribution to the atmosphere of deforestation and erosion and this has led to various investigators reporting widely varying estimates of the net annual flux of carbon from this source. The major reasons for the differences between the estimates seems to be in data used for carbon stocks in vegetation, the changes in carbon content of soils following disturbance, for rates of deforestation and for rates of decay of organic matter. In addition, differences in basic assumptions and methodologies used for evaluation account for some of the variations.

It has been reported (9) that deforestation can contribute quantities of CO_2 to the atmosphere that would range from 0.8 to 2.4×10^{15}g C per annum. A further study (10) indicates a release in the range $0 - 4 \times 10^{15}$g C per annum, whilst another report covering a comprehensive review (11) gives a range for annual net biotic flux from -2 to $+20 \times 10^{15}$g C.

THE CONTRIBUTION OF COAL TO THE GREENHOUSE EFFECT

About 90% of the world's primary energy requirement is derived from fossil fuels, whilst the remainder mainly comes from hydro electricity generation or the use of nuclear fuel. Of the total primary energy consumed, coal contributes approximately 30% (12), which represents an annual carbon use of 2.25×10^{15}g. Coal's use may be divided into three main areas:

1. Electricity generation which accounts for approximately 60% of the total.

2. General industry and domestic use which accounts for approximately 15% of the amount of coal used annually,

3. The remaining 25% is used for metallurgical purposes, where solid, liquid and gaseous by-products are produced. In addition an amount of coal is used for gasification or liquefaction, where other products are produced. During these conversion processes, only a small proportion of the coal is used for heat generation, with the result that less than 6% of the energy input

332

is converted into CO_2 emissions. The resultant by-products are then used for both non-combustion and combustion processes. Where these products are used for combustion then CO_2 emissions will occur, which will be included in the inventories of gaseous fuel, liquid fuel and solid fuel other than coal.

Based on present usage of coal, direct combustion results in approximately 1.7×10^{15} g per annum of carbon being emitted as CO_2. Since CO_2 makes up approximately 50% of the radiative gases contributing to the greenhouse effect and since deforestation and land use may contribute the order of one third of the annual total input of carbon dioxide to the atmosphere. Then the contribution of CO_2 from direct coal combustion to the greenhouse effect is approximately 10%. Included in this is the contribution of the world's coal fired power stations which approximates 8% of the total man-made radiative gas input to the troposphere.

These values are of a sufficiently low order to indicate that the contribution from the total combustion of coal to the greenhouse effect is small and that envisaged projections for an increased use of coal for combustion purposes will have only a small impact on the greenhouse effect.

STRATEGIES FOR THE FUTURE

With current or envisaged technology the control of CO_2 emissions from combustion is neither practical nor economic. The majority of emissions are generated from small mobile and stationary sources and no means of control can be imagined. A number of processes have been proposed for capturing the CO_2 from fossil fuel fired power plants. However, it has been concluded that although some are technologically feasible, none is practical (10). They would add substantially to plant costs and operating expenses and would consume a considerable portion of the plant's energy output, resulting in a considerable fuel increase. In addition, uncertainties exist concerning the practicability of long term storage of the collected CO_2.

Since energy use is directly related to both population and standard of living, then the most effective means of future control of CO_2 emissions from the combustion of fossil fuel, including coal, would be to effectively control the world's population and improve the efficiency of energy conversion and utilisation.

333

Improved internal combustion engineering and changes in attitude to motor vehicle size have demonstrated that a considerable reduction in energy use in transportation can be achieved. However, these achievements need to be extended to all parts of the world. Improvements in the efficiency of energy use have been demonstrated to the extent that although the world's energy use is expanding, primary energy consumption is being maintained and obviously this improvement in utilisation efficiency needs to be continued since by and large, it has mainly occurred in the developed western world.

Modern coal fired power stations have achieved peak conversion efficiencies for present power generation procedures. However, these current efficiencies of the order of 36% can be anticipated to improve to the order of 46% when present research and demonstration have been developed to the stage of full scale production. Currently, pressurised fluidised bed combustion, pressurised pulverised coal combustion and coal gasification coupled with combined cycle plants, are at pilot stage to operational stage and their expanded use is largely dependent on the development of high pressure, high temperature, gas clean-up equipment. When these new generation developments are available, coal use and associated CO_2 emissions will be reduced by 20% - 25%.

Emissions as a result of deforestation and land use need to be better identified and means of effecting control, while still permitting the optimum use of land, need to be devised and adopted internationally.

Forest burning although considered to be a natural phenomena and adding to the natural biotic flux, may contribute as much as the combined total of fossil fuels. Since a percentage of their occurrence and the severity of the event can be controlled, then international means of achieving this should be examined.

The role of other trace gases now accounts for roughly half of the warming from greenhouse gases in the atmosphere and their role is increasing at a greater rate than the effect from carbon dioxide, in addition, the control of greenhouse gases other than CO_2 may be technically, economically and practically preferable (13).

Methane released into the atmosphere is largely as a result of man's increased agricultural activity. Methane emissions have almost trebled in the past century and are increasing at an ever escalating rate. Since control of agricultural development seems in its own right to be difficult, then the obvious means of effecting control will be on population growth.

The major anthropogenic source of nitrous oxide is fertilisers. Combustion of coal is a minor contributor of emissions of N_2O (14). Combustion sources of nitrous oxide may be reduced by NOx control and by improved efficiency of energy utilisation.

Ozone is increasing in the troposphere due to photochemical processes involving hydro-carbons, nitrogen oxides and carbon monoxide. The main contributor being liquid fuels used for transport and biomass burning. Control on transport sources can be effected by improved emission control, a change to smaller more efficient units and conversion to gas as a fuel source. International control on biomass burning should also be identified as an objective. Pursuit of these control aims will not only minimise ozone, but will significantly reduce other man-made radiatively active gases. In particular, the use of gas in motor vehicles could have a profound effect on not only the reduction of greenhouse gases, but also on the most significant air pollution problem of the world's major cities.

Chlorofluorocarbons are of major concern because they have the dual effect of being a highly radiative gas and influence the stratospheric ozone layer. Some countries of the world have banned the use of chlorofluorocarbons as propellants in spray cans, and recently, the Montreal Protocol indicated that an international control will be effected on use for some purposes and on manufacture. The use and disposal of chlorofluorocarbons can be controlled for many applications, they can be replaced by substitutes which would reduce their effect on the ozone layer and may reduce their effect as radiative gases in the troposphere.

Both non-methane hydro-carbons and organic materials that contribute to radiative gas build-up in the atmosphere are, by and large, used in industrial processes and their application and use could be reviewed and potentially controlled.

Each of the foregoing control measures needs to be evaluated in relation to the need for control, the overall effect of control and the cost to the community of the control. Table 3 shows the potential reduction in greenhouse gas warming that may be achieved by adopting practical strategies that would have minimum impact on the community.

SOURCE	CONTRIBUTION	POTENTIAL REDUCTION
C.F.C.'S	13 ±3%	SUBSTITUTION AND ———> 13 ±3% CONTROL.
DEFORESTATION & LAND USE	15 ±5%	TREE PLANTING, CONTROL, ———> 15 ±5% ECONOMIC ASSISTANCE.
MOTOR VEHICLES CO_2 + O_3	12 ±3%	IMPROVED EFFICIENCY 3 - 4% USE OF GAS 6 - 7%
COAL-FIRED POWER STATIONS	8%	IMPROVED CYCLE ———> 2% EFFICIENCY.

TABLE   3

Control Potential for Reducing Greenhouse Gas Warming Effect.

336

CONCLUSIONS

The future expanding use of fossil fuel combined with increased deforestation and land use means that CO_2 emissions to the atmosphere will continue to increase, but at a lesser rate than the escalating contribution from other more active radiative gases. Hence in the future, the relative influence of CO_2 to the greenhouse effect will reduce from its present 50% contribution.

The result of increasing these radiative gases is unable to be exactly predicted. Although, there are scientific opinions that an increase in the magnitude of greenhouse gases will result in a warming of the earth's atmosphere, with an associated climate modification, which would have both positive and negative effects, impacting on most countries and people of the earth. However, at this stage of knowledge based on both models and historical data, there is no means of accurately predicting whether such a change will occur and if it does occur, the ultimate effect is unknown, due to the uncertainty of the magnitude of the negative and positive feed-back into the earth's system.

To address these uncertainties, there is an increasingly important role for science to play in ascertaining the consequences of increasing radiative gases in the atmosphere. This should largely be in relation to developing data and models that can be used to both evaluate and predict the consequence of past and future actions in this area, so that policy decisions can be made, which will be in the overall interests of the world at large.

The nominal 10% contribution that CO_2 emissions from the combustion of the world's coal makes to the greenhouse effect is small and even with the anticipated growth of coal in this area, will continue to be small having a minor impact on both the present and future total greenhouse effect.

REFERENCES

1. Tucker, G.B. "Trace Gas Trends in the Southern Atmosphere", Proc. The Seventh World Clean Air Congress, V 1, p.3, Australia, August 1986.

2. Edmonds, J.A. and Scott, M.J., "Energy and Future Climate Forcing" 87th Annual Meeting of APCA, U.S.A., June 1987.

3. Testimony of Donna Fitzpatrick, Under Secretary of Energy, House Committee on Energy and Commerce, U.S.A., September 22, 1988.

4. Ramanathan, V., Cicerone, R.J., Singh, H.B. and Kiehl, J.T., "Trace Gas Trends and their Potential Role in Climatic Change", J. Geophys. Res.D., V 90, p.5547, 1985.

5. Olson, J.S., Garrels, R.M., Berner, R.A., Armentano, T.V., Dyer, M.I. and Yaalon, D.H., "The Natural Carbon Cycle", United States Department of Energy publication DOE/ER-0239, December 1985.

6. Khalil, M.A.K., Rasmussen, R.A., "Trends of Atmospheric Methane Over the Last 10,000 Years", 87th Annual Meeting of APCA, U.S.A., June 1987.

7. Solomon, A.M., Trabalka, J.R., Reichle, D.E., and Voorhees, L.D., "The Global Cycle of Carbon", United States Department of Energy publication DOE/ER-0239, December 1985.

8. Smith, Irene M., "Carbon Dioxide - Emissions and Effects" IEA Coal Research Report No. 1 CTIS/TR18, June 1982.

9. Smith, Irene, "CO$_2$ and Climate Change", IEA Coal Research Report No. IEA CR/07 May, 1988.

10. Shepard, Michael, "The Greenhouse Effect: Earth's Climate in Transition", EPRI Journal, U.S.A., June 1986.

11. Houghton, R.A., Schlesinger, W.H., Brown, S. and Richards, J.F., "Carbon Dioxide Exchange Between the Atmosphere and Terrestrial Ecosystems", United States Department of Energy publication DOE/ER-0239, December 1985.

12. British Petroleum Company (1987) "B.P. Statistical Review of World Energy", London, June 1987.

13. National Academy of Sciences, "Changing Climate", U.S.A., 1983.

14. Berge, Niklas, "Nitrous Oxide Emission from Combustion" Kolsektione Globala Miljofragor, Sodertalje, Sweden, October, 1988.

Paper for IEA/OECD Expert Seminar
on Energy Technology for Reducing
Emissions of Greenhouse Gas
April 12-14, 1989 in Paris

Coal Utilization Technologies on Japanese Electric Power Companies

by Masashi Hatano

Electric Power Development Co., Ltd. (EPDC)

1. Introduction

Since the Toronto Summit held in June 1988, global environmental issues
have become objects of study and investigation as problems common to
all human beings. The global environmental issues include three
different problems of acid rain, ozone depletion and greenhouse
effect.

As for the problem of acid rains, there has been a great improvement in
Japan, since the enforcement of the Air Pollution Control Act, in
reduction of SOx and NOx emissions through application of environmental
control technologies (as represented by the use of DeSOx and DeNOx
plants).

In developing nations, however, few measures have been taken on SOx and
NOx emissions, and such emissions are increasing at present with
practically no regulations on them. What we have to think on acid
rains is that SOx and NOx emissions travel the distance of 500 km or
1000 km easily on wind and environmental measures taken in one country
might be far from the solution for the problem of global scale.

It can be said in other words that all nations including developed and
developping nations must be united in the form of international
cooperation to solve this type of issues, not limited to the problem of
acid rains.

As for the issue of the greenhouse effect, it has been pointed out that
the level of carbon dioxide (CO_2), a substance which is suspected to be
causing the greenhouse effect, is increasing in the atmosphere year by
year along with the increase in consumption of fossil fuels, and
electric utilities depending much on thermal power generation in
advanced nations are thinking of it as the most serious problem. As
for the greenhouse effect, however, it would be necessary to cope with
it prudently because its causal relations are still unclear in many
points and its effects are far reaching to the point that it might
change our social structures.

339

It is not an easy matter to switch from fossil fuels to other energy sources without putting the world economy in turmoil for the purpose of reducing the amount of CO_2 generation. It should be allowable to generate CO_2 to some extent, not greatly disturbing the natural recycling of CO_2, within a range where the amount of CO_2 in the atmosphere is controllable below a certain level, and to this effect, it would be important to pursue the best mix of available energy sources in the total energy requirement.

The coal reserve in the world is several times as large as those of LNG or uranium in energy. In addition, coal is distributed widely in the world and contributing much to stable supply of energy in many parts of the world. The effective use of coal therefore would remain to be very important for stable supply of energy for long.

This report introduces DeSOx and DeNOx technologies, employed in Japan, that would be helpful in solving the problem of acid rains. In addition, we also discuss high-efficiency power generation technologies, which in one way lead to suppression of CO_2 generation, from the viewpoint of coal utilization.

2. For Solving the Problem of Acid Rains

Normal rains are weakly acidic with pH (hydrogen-ion concentration) on
the order of 5.6 due to carbon dioxide dissolved in them. Generally,
rains of pH 5.6 or lower are referred to as acid rains. Natural
phenomena such as volcanic eruptions can also cause acid rains, but
acid rains we discuss here are those which are caused by artificial
pollutants such as SOx and NOx emitted from boilers. Such pollutants
turn rains to dilute solution of sulfuric acid and nitric acid in the
atmosphere and cause acid rains.

DeSOx and DeNOx technologies help protect forests from acid rains and
minimize effects on the ecosystem including fish in lakes and ponds and
human beings, and they are those environment control technologies which
are now in the greatest demand. Current statuses of DeSOx and DeNOx
technologies in Japan are described below.

(1) DeSOx Technologies

The DeSOx process employed widely for DeSOx plants of utility
boilers in Japan is the wet limestone-gypsum process. The
principle of the process is illustrated in Fig. 1. The SO_2 gas
generated by combustion comes into contact with limestone-water
slurry and is absorbed in the slurry in the absorber. The
desulfurization efficiency of this process is 90% or greater.
DeSOx plants of this process have been installed not only to coal
fired utility boilers but also to oil fired boilers in Japan, and
about 90% of SO_2 gas generated by combustion is eliminated. The
SO_2 gas thus eliminated is recovered in the form of gypsum as
byproduct, and gypsum is used for making gypsum board or as a
cement additive.

This wet DeSOx process uses much industrial water, and a
relatively complicated wastewater treatment plant is required for
treating the waste water. In order to cope with such
disadvantages, R&D activities are going on to develop a dry DeSOx
technology which requires less industrial water and produces pure
sulfur as byproduct.

The process flow of such a dry DeSOx system is shown in Fig. 2.
This system uses activated charcoal to adsorb sulfur components
contained in the flue gas and produces element sulfur as its
byproduct. The molecular weight of sulfur is much smaller than
that of gypsum, and the system is also advantaged in the point
that the amount of byproduct is minimized.

341

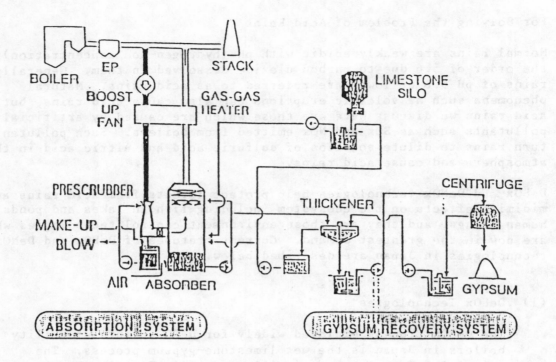

PRINCIPLE

SO2 ABSORPTION	SO2 + H2O + CaCO3 $\rightarrow$ CaSO3 $\cdot$ 1/2 H2O + CO2
GYPSUM PRODUCTION	CaSO3 $\cdot$ 1/2 H2O + O2 + H2O $\rightarrow$ CaSO4 $\cdot$ 2H2O

Fig. 1 Principle of the Wet FGD System

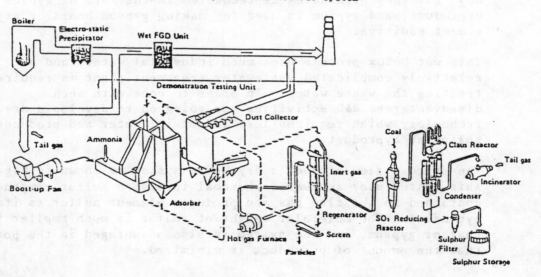

Fig. 2 Flow Sheet of Demonstration Unit for Dry FGD

(2) NOx Reduction Technologies

NOx is generated when fossil fuels are burnt. Such NOx comes from nitrogen contained in the fuel and that contained in the air. NOx thus generated is often responsible for acid rains and photochemical smogs as in the case of SOx, and its reduction is in great demand. NOx reduction technologies are classifiable into two major categories of the low-NOx combustion technology applied to combustion of fuel and the DeNOx technology applied in post flame.

a. Low-NOx Combustion Technology

One way for NOx reduction is to employ low-NOx combustion technologies. Major low-NOx combustion technologies include the following:

o Two-stage combustion (TSC)

o Low-NOx burner (LNB)

o Furnace volume expansion

Both of the TSC and LNB technologies lower the combustion temperature by controlling the combustion air or by mixing the flue gas to achieve slow and low-NOx combustion. Problems such as increase in unburnt carbon were encountered in applying these technologies to coal fired power plants, but such problems have been resolved through employment of various measures. The history of development of these technologies is shown in Fig. 3. By the use of these technologies, the NOx level of combustion gas at outlet of coal fired boilers is now as low as 150 to 200 ppm (O_2=6% equivalent).

b. DeNOx Technology

Japan has a large population in a small country. In addition, the size of industry is also large, and strictest environmental regulations have been enforced. For NOx emissions from power plants, it became impossible to meet such regulations only by improved combustion. The technology which was developed to cope with this situation is the flue gas DeNOx technology. Many different processes were proposed in the course of developing the technology, but the SCR (Selective Catalytic Reduction) system is now employed widely at electric utilities.

The principle of the SCR system is illustrated in Fig. 4. As seen in the figure, ammonia is injected in flue gas to transform NOx in the gas into N_2 and H_2O. The DeNOx efficiency of this process is on the order of 80%.

With the combination of the SCR system and the low-NOx combustion technologies, most NOx is eliminated from the flue gas, and it may not be too much to say that the problem of NOx has mostly been solved. A view of a typical SCR reactor is shown in Fig. 5 for reference.

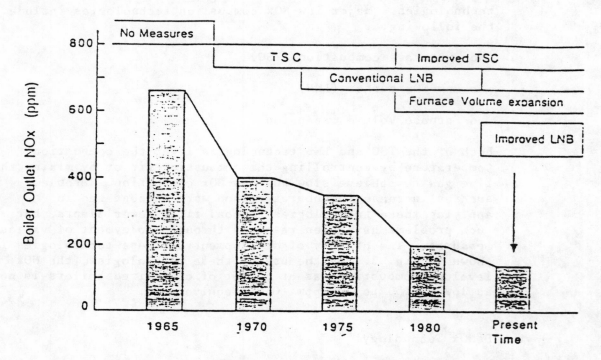

F i g. 3   History of Low NOx Combustion Technologies in Japan

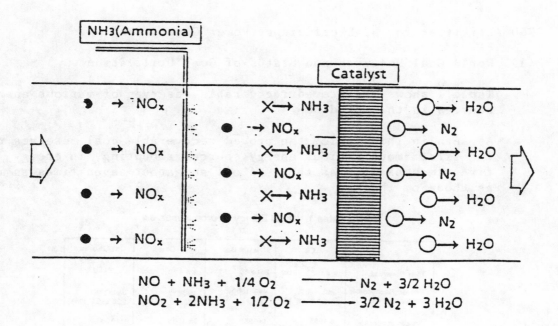

$$NO + NH_3 + 1/4\,O_2 \longrightarrow N_2 + 3/2\,H_2O$$
$$NO_2 + 2NH_3 + 1/2\,O_2 \longrightarrow 3/2\,N_2 + 3\,H_2O$$

F i g . 4 Principle of SCR System

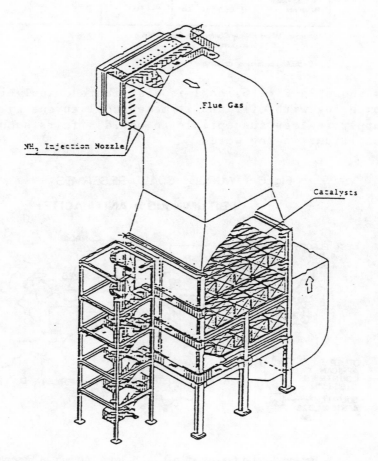

Fig. 5 Artistic View of a Typical SCR Reactor

3. R&D Activities for High-efficiency Power Generation

(1) World Coal Reserves and Status of Coal Utilization

Table 1 shows the proved recoverable reserves of various energy sources in the world.

As seen in this table, the proved recoverable coal reserves rank the first among various energy sources amounting, in energy, about five times as much as those of oil and about seven times as much as those of LNG.

Table 1. World Energy Source Reserves

	Oil	Natural Gas	Coal	Uranium
Total Reserves	2×10^{12} bbl	204×10^{12} m^3	8.4×10^{12} tons	N.D.
Proved Recoverable Reserves (R)	Jan. 1986 700.1×10^9 bbl	Jan. 1986 79×10^{12} m^3	1986 730.8×10^9 tons	Jan. 1983 226×10^6 tons
Annual Production (P)	In 1985 19.5×10^9 bbl	In 1985 1.28×10^{12} m^3	In 1985 3.18×10^9 tons	In 1983 37,000 tons
Recoverable Period (R/P) years	36	56	230	61
Oil Equivalent Reserves	95.9×10^3 tons	70.7×10^3 tons	485.3×10^3 tons	—

Sources: World Energy Conference, OECD/IAEA
Oil & Gas Journal, etc.

· Excluding communist countries

As shown in Fig. 6, coal is distributed widely in the world comparing with oil, and it makes coal an energy source of which supply is less susceptible to world affairs such as political conditions in the world.

Fig. 6   WORLD COAL RESERVES

(BITUMINOUS a ANTHRACITE)

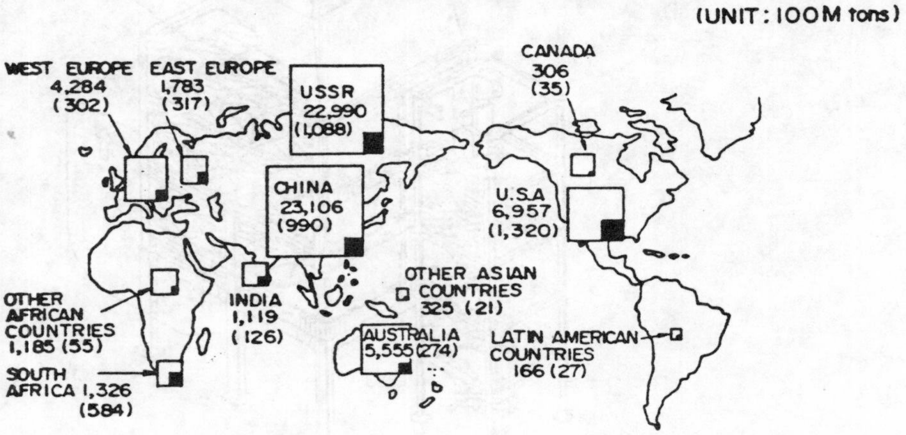

(UNIT: 100M tons)

WEST EUROPE 4,284 (302)
EAST EUROPE 1,783 (317)
USSR 22,990 (1,088)
CANADA 306 (35)
CHINA 23,106 (990)
U.S.A 6,957 (1,320)
OTHER AFRICAN COUNTRIES 1,185 (55)
INDIA 1,119 (126)
OTHER ASIAN COUNTRIES 325 (21)
SOUTH AFRICA 1,326 (584)
AUSTRALIA 5,555 (274)
LATIN AMERICAN COUNTRIES 166 (27)

Rectangle and figures without ( ): Total Amount of Reserves
Block Rectangle and figures with( ): Proved Recoverble Reserves

346

Figure 7 shows the recent trend in electric energy production in Japan.

As seen in this figure, most of its growth has been supplied by nuclear energy. Thermal energy, however, accounts for about two-thirds of the total electric energy production, and it remains to be the major source of electric energy production in Japan.

The total annual thermal power generation has remained almost the same for the last ten years or so, but its breakdowns by fuels have been changing drastically.

As shown in Fig. 8, the consumption of oil at electric utilities in Japan has been decreasing much through the two oil crises of 1973 and 1979 and those of coal and LNG have been increasing in place. Percentages which respective energy sources occupied in electric energy production in 1986 are as listed below.

Nuclear	28%
Hydraulic	14%
Coal	10%
Gas	25%
Oil	23%

As described above, coal is an important part of electric energy sources for Japan in the context of achieving the best mix of energy sources for power generation.

It is a fact, on the other hand, that coal produces more CO_2 per generated electric power than other fossil fuels, and the first thing to be done for coping with the greenhouse effect is to establish technologies for high-efficiency power generation.

Figure 9 shows how the efficiency of thermal power plants in Japan has improved in the last few decades.

As seen in the figure, the thermal efficiency improved rapidly up to 1970. It became necessary around that point to use high-grade materials to achieve higher efficiency. Savings in fuel cost that could be expected from higher efficiency matched economically with the increase in equipment cost which could be incurred due to the necessity of using high-grade materials. Because of this economic balance, the steam condition, or plant efficiency in other words, has improved little in the last twenty years. In Japan, however, the ratio of new power plants is high due to a rapid expansion of power generating capacity which occurred recently, and the overall thermal efficiency of Japanese thermal plants is now at the top level in the world.

347

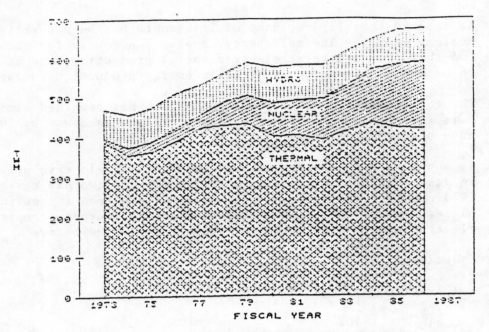

F i g. 7   Electric Energy Production in Japan

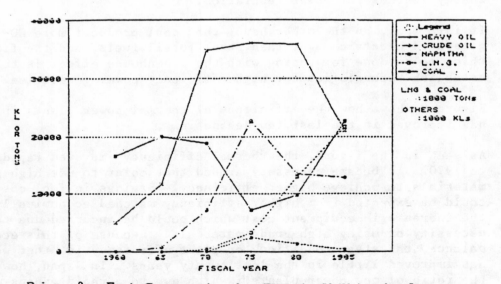

F i g. 8   Fuel Consumption for Electric Utilities in Japan

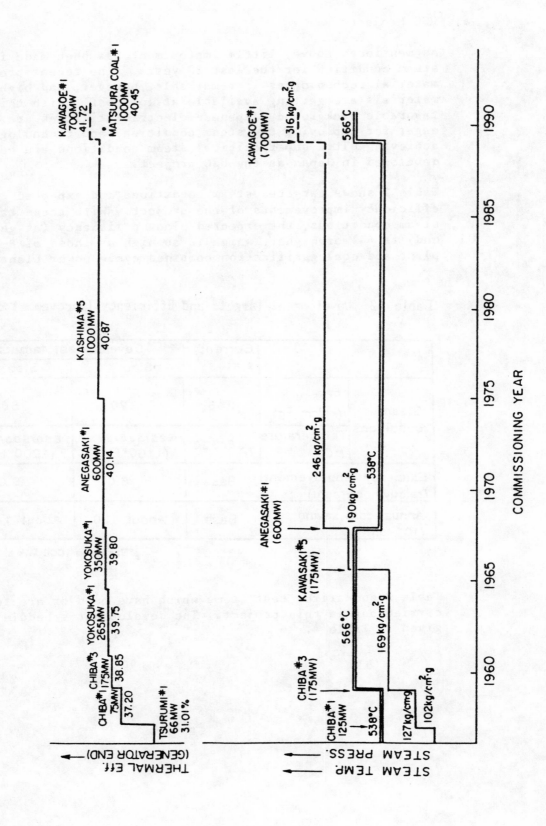

Fig. 9 HISTORY OF THE IMPROVEMENT OF STEAM CONDITION & EFFICIENCY

(2) High-efficiency Coal Fired Power Generation

a. USC Project

As mentioned above, little improvement has been made in the steam condition for the last 20 years. The recent progress in material technologies is remarkable, however, and high-grade materials are getting available at low cost. With this favorable background, Japanese electric utilities are getting eager for improving the steam condition, and technologies for achieving ultra super critical steam conditions are being developed in Japan as the USC project.

Table 2 shows targeted steam conditions and expected efficiency improvements of the project. With these targeted steam conditions, the expected plant efficiency (at generating end) is 44% or higher, which is as high as those of P-FBC plant and coal gasification combined cycle power plant.

Table 2 Development Targets and Efficiency Improvement of USC

		Current status	Development targets	
			Step 1	Step 2
Steam conditions	Pressure (kg/cm^{2}g)	246	320	350
	Temperature (°C)	538/538	595/595/595 (1100°F)	650/595/595 (1200°F)
Efficiency improvement (relative percent)		Base	6.0	8.0
Annual coal saving (10^3 ton)		Base	About 130	About 170

*Based on 1000 MW x 1 u

Table 3 summarizes test items which have been or are to be carried out in this project. The development schedule is given in Table 4.

350

Table 3   Outline of USC Tests

Test		Place of test	Test period	Development items
Actual boiler element test of high temperature materials	(1) Boiler element test (EPDC, MHI)	Takasago No.2 unit Coal-fired 250 MW Hyogo Prefec.	July 1981 to 1988	Development of heat resistant boiler tube materials / Development of heat resistant thick materials (for valves and pipes)
	(2) Turbine element test (EPDC, MHI)		Oct. 1982 to 1987	Development of vane, blade, casing and main valve materials
	(3) Corrosion test of overseas coals (EPDC, MHI)	Matsushima No.2 unit Imported coal-fired 500 MW Nagasaki Pref.	Oct. 1982 to 1987	Development of heat resistant boiler tube materials for overseas coals
(4) High-temperature turbine rotor test (EPDC, HITACHI, TOSHIBA)		Takasago No.2 unit Hyogo Pref.	June 1982 to 1988	Demonstration test (mainly on the test turbine rotor)

Companies in parentheses are cooperating companies.

[USC Turbine Demonstration Test]

Super high-temperature turbine demonstration test	Wakamatsu thermal power plant Fukuoka Pref.	June 1982 to 1991	Demonstration test of 1,000 MW class full size turbine rotor

Table 4   USC Test Schedule

Test \ FY	1981	1982	1983	1984	1985	1986	1987	1988	1989	1990	1991
(1) Boiler element test	595°C		620°C		650°C			analysis			
(2) Turbine element test			620°C		650°C			analysis			
(3) Corrosion test by overseas coal			595°C · 650°C					analysis			
(4) High-temperature turbine rotor test			Detail design and manufacture			595°C	650°C	material test 595°C		650°C	
(5) Super high-temperature turbine demonstration test		Conceptual design	Detail design and manufacture					Installation test	operation		

351

b. Integrated Coal Gasification Combined Cycle Power Generation Technology

The integrated coal gasification combined cycle power generation technology is one of the promising coal-fired power generation technologies of the future. This technology is expected to allow to have highly efficient and economical plants with characteristics favorable for environmental preservation, enlargement of coals usable for plants and plant operability. With such expectations, this technology is being developed in Japan under the leadership of the government. A test of 5 ton/day began in 1974. The gasification test amounting 6,900 hours, the gas refining test amounting 2,500 hours and the gas turbine element test amounting 1,100 hours were then carried out using a fluidized bed gasification plant of 40 ton/day to obtain various data. Based on such achievements, a pilot plant of specifications below is now under construction.

Gasifier

Fluidized bed coal gasification furnace

Coal processing capacity	200 ton/day
Gas generation capacity	42,900 m^3N/h (on standard coal base)

Gas refinery

Dry de-SOx and de-dusting equipment

Gas turbine

Combustor outlet temperature	1,260 °C
Generating end output	12,500 kW

The development targets of this project are as follows:

(1) Actual application of this technology shall be in the early part of the 21st century.

(2) The unit capacity shall be 250 MW and greater.

(3) The thermal efficiency shall be 43% or greater.

(4) The plant economy shall be better than that of pulverized coal fired plants.

(5) The environmental performance of the plant shall be on the order of that of oil fired plants. In addition, the ash treatment for the plant shall be easier than that for pulverized coal fired plants.

(6) Coal of various kinds and that of various ash melting points shall be usable for the plant.

(7) Operating performance of the plant shall be as good as or better than that of pulverized coal fired plants.

(8) The plant reliability shall be as high as that of pulverized coal fired plants.

The development schedule and a bird's-eye view of the pilot plant are given in Fig. 10 and Fig. 11, respectively.

Fig. 10 RESEARCH & DEVELOPMENT SCHEDULE

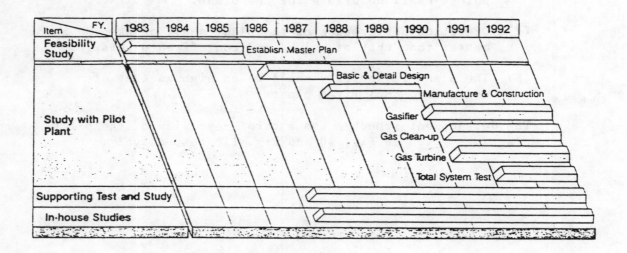

Item	FY.	1983	1984	1985	1986	1987	1988	1989	1990	1991	1992

Feasibility Study — Establish Master Plan

Study with Pilot Plant — Basic & Detail Design, Manufacture & Construction, Gasifier, Gas Clean-up, Gas Turbine, Total System Test

Supporting Test and Study

In-house Studies

Fig. 11   A bird's-eye view of the completed pilot plant

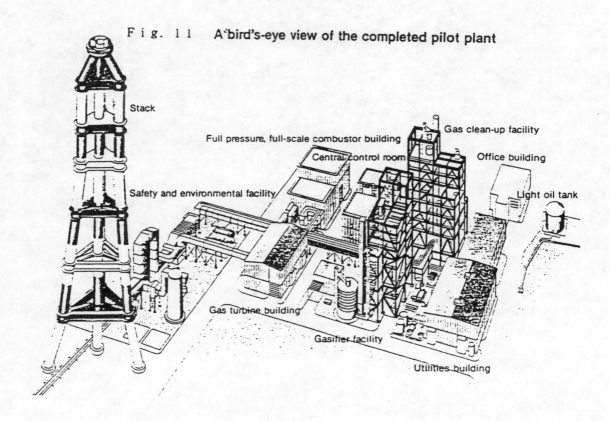

Stack

Full pressure, full-scale combustor building

Central control room

Gas clean-up facility

Office building

Safety and environmental facility

Light oil tank

Gas turbine building

Gasifier facility

Utilities building

354

c. Development of Fluidized Bed Combustion Boiler Technology

The pressurized fluidized bed combustion (P-FBC) technology is
one of the promising technologies for high-efficiency power
generation. Activities for developing this technology is yet
to begin in Japan, but those for developing the atmospheric
fluidized bed combustion (A-FBC) technology are under way. A
50-MW demonstration plant is being operated, and there is a
plan for converting the boiler of a 350-MW commercial plant to
an A-FBC boiler.

The development schedule of the A-FBC technology is shown in
Fig. 12.

Fig. 1 2 A-FBC Development Schedules

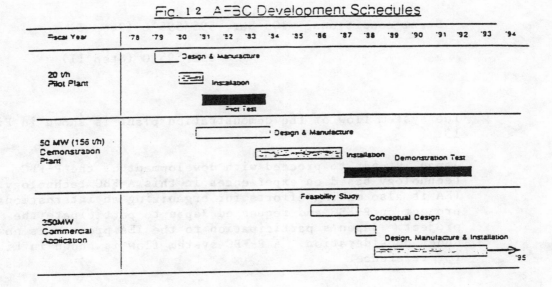

Major specifications of the A-FBC demonstration test boiler are as follows:

A-FBC Boiler

Boiler type: Atmospheric Fluidized Bed

                        Bubbling

                        Multi-staged Stacked

                        Under Bed Feed

Evaporation rate: 156 ton/h

Steam conditions: 105 bar, 595/595°C (Step I)

                        652/595°C (Step II)

The system flow of the demonstration plant is shown in Fig. 13.

Japan intends to proceed with development of the P-FBC technology based on experiences in this A-FBC technology. The IEA is also making efforts for organizing an international project on P-FBC and requested Japan to participate the project. Japan's participation to the IEA project is now under consideration. A P-FBC system flow is shown in Fig. 14 for reference.

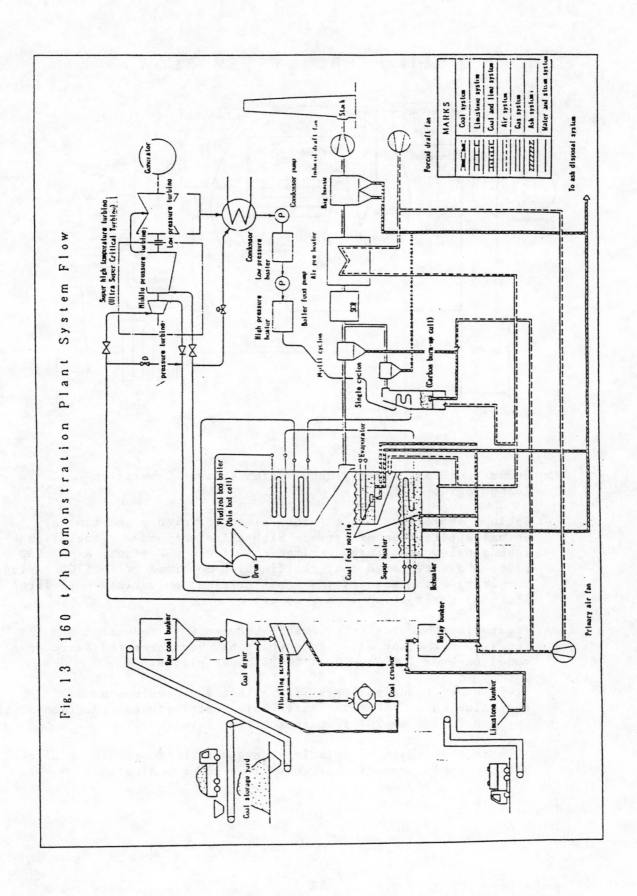

Fig. 13 160 t/h Demonstration Plant System Flow

357

Fig-14 P-FBC SYSTEM FLOW

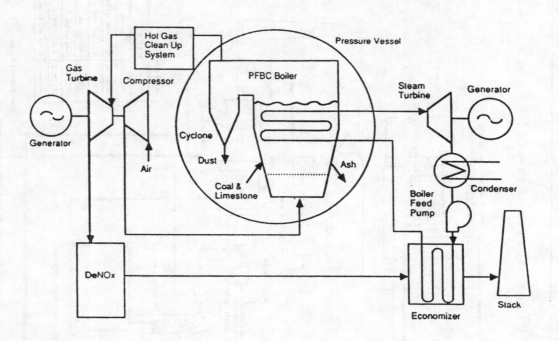

(3) Summary of Coal Fired High-efficiency Power Generation Technologies

Table 5 summarizes the expected plant efficiency and time of actual application of various high-efficiency power generation technologies such as pulverized coal fired USC plant, A-FBC USC plant and integrated coal gasification combined cycle (IGCC) plant comparing with those of the super critical pulverized coal fired (PCF-SC) plant which is now in use.

Figure 15 shows the relationship between CO_2 reduction and required net plant efficiency which has been prepared based on a model case of 1,000-MW coal fired power plants.

It is understood from this figure that the development of technologies of 45% efficiency would contribute to reduction of CO_2 generation by 20% from the current level.

This is the reason for placing the greatest importance on higher efficiency in promotion of coal fired power generation.

Table 5. Generation Technologies for Coal-Fired Power Plant Efficiency

Technology Development	Expectation for efficiency %		Period of commercialized utility (year)
	Plant efficiency (gross)	Transmit plant efficiency (net)	
PCF - SC	41.5	38.2	1981 (commercialized)
PCF -USC	44.8	41.2	1991 (expected start)
PFBC-SC	43.3	42.0	2000 (expected start)
PFBC-USC	46.1	44.2	2000 (expected start)
IGCC 1300 C	46.8	43.5	Since 2000
1500 C	48.6	45.6	

Fig. 15   Relation between CO2 Reduction and Required Net Plant Efficiency

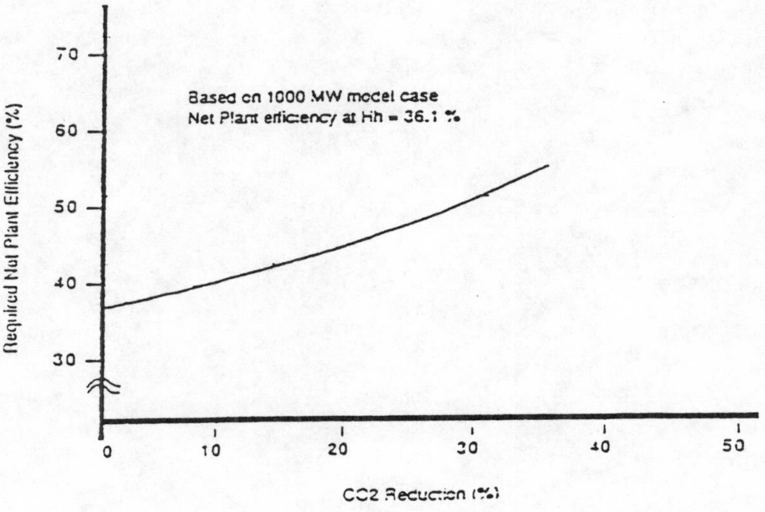

359

6. Concluding Remarks

As mentioned above, Japan is making continued efforts for high efficiency operation of conventional thermal power plants and for development of high efficiency systems through development of new technologies for coal utilization. Coal is ranked as one of the important energy sources for stable supply of energy even for the future.

Coal is a fossil fuel, and its combustion necessarily generates CO_2. It can be possible, however, to utilize coal without causing deterioration of the global environment through wasteless use of coal by high efficiency systems in accordance with optimal plan for consumption of all energy sources. That is, coal consumption can be reduced in this way, and the CO_2 generation can be suppressed.

Japan is making continued efforts not only for developing high efficiency systems for using coal but also for protection of environment from acid rains and other troubles through development and application of DeSOx and DeNOx technologies. It may not be too self-conceiting to say that some of such technologies are at the highest level in the world.

It would be our pleasure if Japanese technologies are utilized at many places in the world for solving the global environmental problems, and we would like to add that Japan is ready for accepting any requests for cooperation in this field.

Energy Technologies for the Use of Natural Gas
to Reduce CO_2 Emissions
Including Gas Reburn Technology for Coal Firing

Paper

presented by

Dr. Roland Pfeiffer,
Head of the Research and Development Division,
Ruhrgas AG, Essen,
Federal Republic of Germany,

in Paris
on 14 April 1989

Energy Technologies for the Use of Natural Gas
to Reduce CO_2 Emissions,
Including Gas Reburn Technology for Coal Firing

Summary

Carbon dioxide (CO_2) released to atmosphere by the combustion
of fossil fuels is held to be one of the main causes of the
anthropogenic greenhouse effect.

CO_2 emission control by post-formation clean-up techniques
which have been developed for the removal of sulphur dioxide
and oxides of nitrogen from flue gases does not seem feasible.
For practical purposes, carbon dioxide emissions can only be
reduced by burning less fossil fuels.

For this reason, efficient energy utilization and the substitution of low-CO_2 fuels for high-CO_2 energy sources must be
given priority.

By discussing typical examples, this paper shows how natural
gas can help to mitigate the anthropogenic greenhouse effect,
as the energy conservation potential associated with the use
of this fuel is substantial and CO_2 emissions per unit energy
consumption are lower than for any other fossil energy source.

1. Introduction

Politicians, engineers and environmentalists around the world are increasingly concerned about possible changes in the world's environment. Carbon dioxide (CO_2) released to atmosphere by the combustion of coal, oil products and gas is held to be one of the main causes of the anthropogenic greenhouse effect, which threatens to increase the Earth's temperature.

In its interim report published in November 1988, the Special Committee set up by the West German parliament to study atmospheric pollution came to the following conclusion:

"Theories developed on the greenhouse effect and the under-lying principles that have been identified are so conclusive that comprehensive action must be taken as soon as possible to reduce trace gas emissions."

To evaluate the impacts of CO_2 pollution by the combustion of fossil fuels, it is essential to grasp the scope of the problem:

- Fossil fuels presently cover some 90 % of world energy demand.

- Fossil fuel consumption differs remarkably in the various regions of the world. The United States, the Soviet Union and the People's Republic of China together emit some 50 % of the carbon dioxide released to atmosphere.

- The West German share in world CO_2 emissions is about 3.5 %. Slightly less than 50 % of the West German CO_2 emissions are attributable to oil products, some 25 % to hard coal, 13 % to lignite and another 13 % to natural gas.

- World energy demand is anticipated to grow further. Increased quantities of energy will be used above all in the developing nations and in the countries on the threshold of industrialization where fossil fuels will remain the most important energy sources. CO_2 emissions will, therefore, continue to rise.

- The CO_2 emission rates of fossil fuels differ (see Fig. 1) and will have to play an important part in the definition of CO_2 control strategies.

2. CO_2 Emission Control Possibilities

CO_2 emission control by post-formation clean-up techniques which have been developed for sulphur dioxide and oxides of nitrogen do not seem to be feasible. For practical purposes, CO_2 emissions can, therefore, only be reduced by using less fossil fuel.

In principle, the following options exist to cut carbon dioxide pollution:

- Energy conservation and development of new techniques to enhance energy efficiency

- Substitution of low-CO_2 fuels, such as natural gas, for high-CO_2 energy sources, such as coal and oil products

-   Substitution of nuclear energy for fossil fuels

-   Increased use of regenerative energy sources, such as hydropower, wind energy and solar energy.

Energy conservation and enhanced energy efficiency must be given priority among these options.

3. CO_2 Emission Control by the Use of Natural Gas

To minimize CO_2 pollution, a strategy must be developed giving priority to the use of fossil fuels that emit less carbon dioxide per unit energy consumption than other energy sources while at the same time offering a high energy conservation potential.

Natural gas satisfies both criteria.

Its properties make it a fuel that helps to save energy.

The combustion of natural gas is very efficient, since the gaseous fuel blends easily with combustion air, minimizing excess air. High excess air without combustion air preheating impacts on efficiency, because the air cools the flame temperature.

Some typical examples show the contribution which can be made by natural gas to reducing carbon dioxide emmissions.

3.1 Increase in the Efficiency of Conventional Boilers

The average efficiency of a gas-fired boiler has been
increased from approx. 60 % to some 86 % over the course of
the last 20 years. The following improvments have cut fuel
consumption:

- Continuous control of water temperature and boiler heat
  output as a function of actual heat demand

- Enhanced heat exchangers and enhanced thermal insulation

- Modulating boiler control

3.2 Recovery of the Heat of Condensation

Condensing boilers have been developed to recover the heat of
condensation contained in the water vapour formed during
combustion (see Fig. 2). To recover the heat carried by the
flue gas, heat exchangers are integrated in the boiler. They
can reduce the flue gas temperature to less than 40 °C. Heat
of condensation can only be recovered, though, if combustion
is clean, as is the case for natural gas.

Condensing boilers use 15 % less energy than modern high-
efficiency boilers.

3.3 Gas-Fired Heat Pumps

The idea of recovering practically inexhaustable quantities of
environmental heat for space heating and water heating
purposes was at the origin of heat pump development.

To exploit this potential, the temperature level of
environmental heat must be raised by means of work input. In
the case of a gas-fired heat pump, the ratio between primary
energy input and useful heat output is particularly high.

Compression-type heat pumps (see Fig. 3) driven by gas engines
use as much as 50 % less energy than conventional space
heating systems, since the useful heat they deliver may be 1.5
times higher than the primary energy input.

In West Germany, a total of some 750 compression-type heat
pump installations driven by gas engines are presently in
operation, covering a heat demand of nearly 900 MW.

3.4 Cogeneration

Cogeneration facilities are plants for the decentralized
production of electricity and heat. Today, most of these
facilities are equipped with internal combustion engines which
drive generators for the production of electricity and feature
heat exchangers for the recovery of heat from the engine
exhaust and the engine cooling water to produce hot water for
space heating (see Fig. 4). Most of the engines are fuelled by
natural gas, since natural gas combustion does not produce any
residue and enhances engine service life.

Currently, some 350 cogeneration plants designed for electricity output in excess of 200 MW are in operation in the Federal Republic of Germany.

The efficiency of the simultaneous production of power and heat using natural gas feedstock is 85 %. 35 % are accounted for by the conversion of natural gas into electricity and some 50 % by hot water production. As cogeneration facilities are decentralized plants, they avoid the transmission loss penalties of central power and district heating stations. Even small-sized cogeneration facilities save as much as 40 % of energy input if they are compared with plants for the separate production of electricity and heat.

Highly-efficient, low-pollutant gas turbines rated 5 MW or less are an attractive alternative to internal combustion engines for cogeneration plants. One of the main advantages of the gas turbine is the high temperature of the turbine exhaust of some 500 °C which may be used for the generation of steam to drive a steam turbine. The energy-saving potential of such combined cycle plants is particularly high.

In West Germany, some 30 turbine-driven cogeneration facilities designed for an electricity output of approximately 260 MW are presently in operation.

Fuel cells for the direct conversion of hydrogen-rich gas, such as natural gas, into electricity and heat may be the cogeneration system of the future (see Fig. 5). The efficiency of a phosphoric acid fuel cell which will soon be marketed is 40 % to 45 % for power generation alone. The overall cogeneration efficiency is between 80 % and 85 %.

Second-generation fuel cells such as molten carbonate and solid oxide fuel cells operating at a temperature of 600 °C and 1000 °C even promise power generation efficiencies of 55 % to 60 %. However, substantial research and development work will be required to advance these cells to the standards already achieved today by phosphoric acid fuel cells.

3.5   Heat Recovery

Heat recovery is a relatively simply method of substantially improving the efficiency of industrial processes.

Ceramic regenerators and recuperators have, for instance, been developed to recover heat from hot waste gases from industrial furnaces. Ceramic heat exchangers will preheat combustion air to a temperature in excess of 1000 °C, reducing fuel consumption by as much as 50 %.

3.6   Use of Natural Gas at Coal-Fired Power Station

Natural gas co-firing is a promising alternative to complex catalytic flue gas treatment techniques for NO_x control, such as selective catalytic reduction. Studies in the United States and West Germany have shown that gas reburn technology, a method of injecting gas into the hot flue gas from coal-fired boilers, will reduce NO_x output by more than 50 %, even if the share of gas in fuel input is only 10 %.

Gas reburn technology is not only an effective method of controlling NO_x emissions, though. It will also reduce CO_2 output since the combustion of natural gas produces less carbon dioxide than the combustion of coal.

If 10 % of all coal fired at West German power stations were, for instance, replaced by natural gas annual CO_2 emissions would be cut by some 10 million tonnes, equivalent to approximately 1.5 % of the annual West German CO_2 emissions of about 750 million tonnes.

4. Conclusion

The examples discussed above have shown the substantial contribution which natural gas can make to reducing energy consumption for thermal processes.

Cogeneration facilities, gas turbines, gas-fuelled heat pumps and condensing boilers have been developed for efficient gas utilization and fuel cells will also be an attractive option in a few years.

Apart from the cut in CO_2 emissions achieved by efficiency improvements through the use of natural gas, reasonable substitution of gas for other fossil fuels also helps to reduce CO_2 pollution. Gas reburn technology is but one example.

In conclusion, more efficiency in energy utilization and the substitution of low CO_2 fuels for high-CO_2 sources of energy may lower CO_2 emissions substantially and hence help counteract the anthropogenic greenhouse effect. Natural gas is capable of making a major contribution to this development.

Fig. 1

Emission Factors of CO_2

	emission factor (kg CO_2/kWh)	relation (%)
natural gas	0,19	100
oil	0,29	153
hard coal	0,33	174
lignite	0,40	211

ruhr gas 1989

Fig. 2

Condensing Boiler

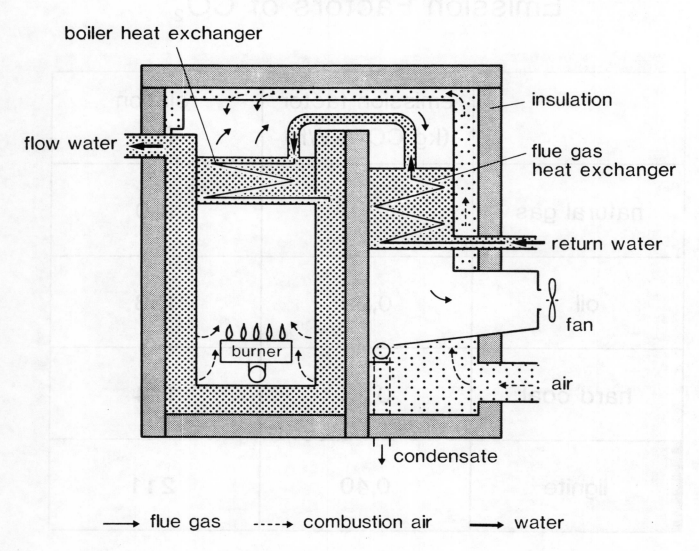

boiler heat exchanger

insulation

flow water

flue gas
heat exchanger

return water

fan

air

burner

condensate

→ flue gas ----→ combustion air → water

ruhr
gas 1989

Fig. 3

Schematic of a Gas-Fuelled Heat Pump

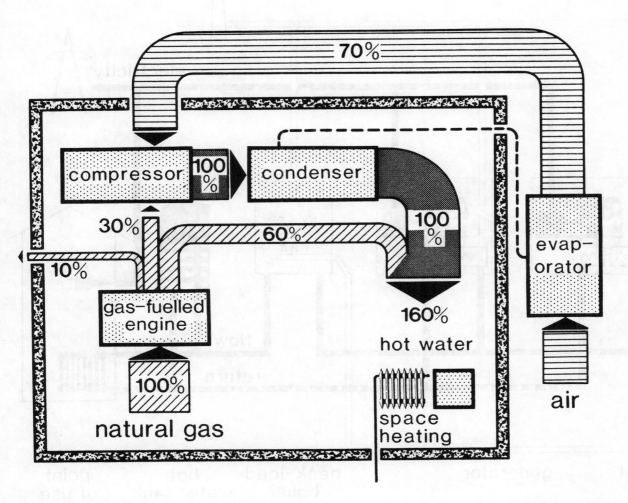

ruhr gas 1989

Cogeneration Plant

Fig. 4

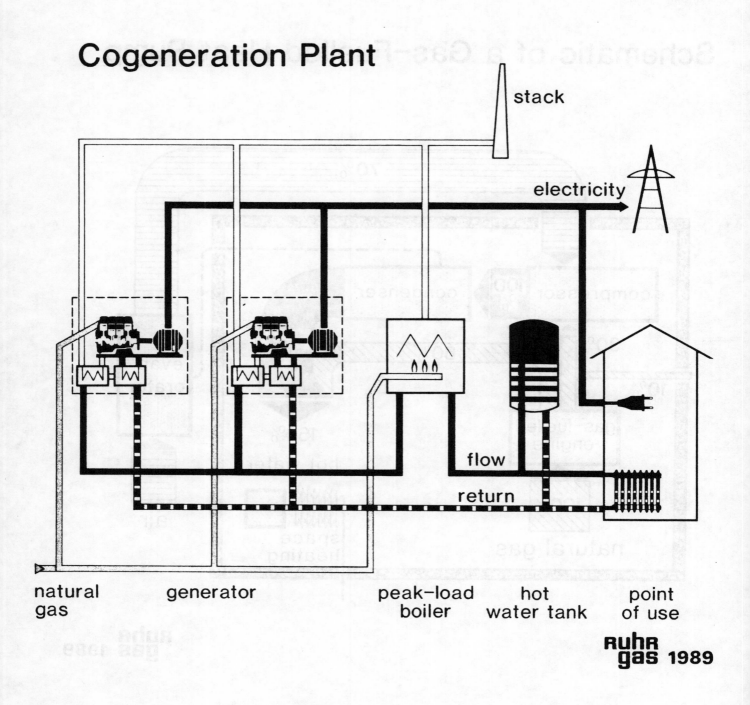

stack

electricity

flow

return

natural
gas

generator

peak-load
boiler

hot
water tank

point
of use

ruhr
gas 1989

374

Fig. 5

Fuel Cell Plant

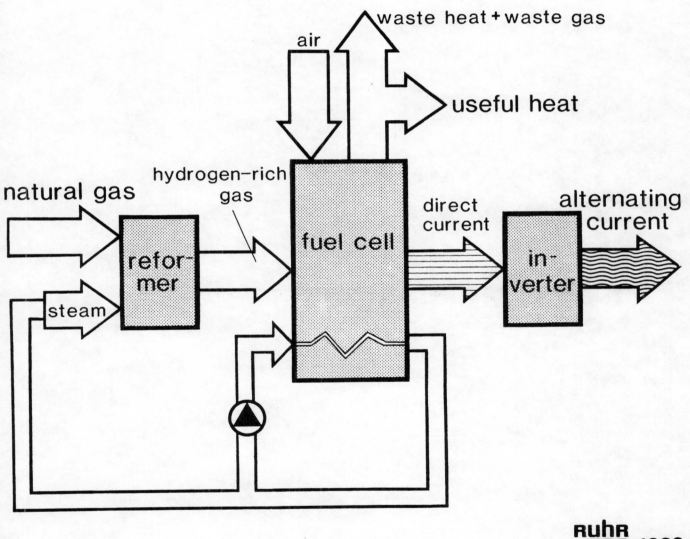

ruhr gas 1989

Natural Gas "Select-Use" Technologies: Opportunities for Emissions
Reductions Using Natural Gas in Conjunction with Coal

Mr. Lee Solsbery
Deputy Commissioner for Research and Fiscal Management
Texas General Land Office

Introduction

The problem of the depletion of the ozone layer and the
production of gases which lead to the "greenhouse effect" is a
critical one for the very future of our planet. The International
Energy Agency and other international organizations and major
governments must be commended for their concerted efforts to
address this complex problem in a systematic fashion.
Unfortunately, many of the potential solutions to reducing the
emission of "greenhouse gases" appear to be expensive,
controversial or in need of extensive research before full
commercialization can be obtained.

With all these obstacles in the path of possible solutions to
the global warming problem, we also confront the equally unsolved
problems of acid precipitation and ground-level air pollution
throughout our increasingly urbanized and industrialized world.

Immediate threats to human health from dangerous air pollutants, medium-term threats to forests and watersheds from acid rain, and longer-term threats to the balance of our global climate from greenhouse gases are very daunting technical, economic and political challenges for us all.

The Texas General Land Office, a state agency under elected Land Commissioner Garry Mauro, entered this difficult area of public policy with some trepidation. The agency's principal mandate is to manage over 22 million acres of public lands throughout the state of Texas, including over 14,000 oil and gas wells. A simultaneous mandate to protect and enhance the state's natural and environmental resources led us, however, to work with industry and the environmental community to identify and promote practical solutions to air quality problems that could enhance, instead of retard, economic development.

Benefits of Natural Gas

The solution we have chosen to promote is natural gas. As you all know, natural gas is by far the cleanest-burning fossil fuel. It produces far lower emissions of the most common energy-related greenhouse gases--carbon dioxide, nitrogen oxides, methane and CFCs. In fact, gas emits roughly half as much carbon dioxide per unit of energy as coal and 30 percent less than oil.[1] Gas is also a plentiful and cost-competitive fuel that can be handled safely and burnt efficiently without creating a secondary problem of solid or hazardous wastes.

The IEA/OECD sponsors of this seminar on greenhouse gases have wisely included a number of presentations regarding the emissions reduction benefits of natural gas. The most typical options discussed for expanding the use of natural gas in large industrial applications are fuel-switching or "repowering" with combined-cycle generation.

These natural gas repowering options can reduce emissions of sulfur dioxide, nitrogen oxides and carbon dioxide per unit of coal-fired electricity generated by 100, 90 and 64 percent, respectively.[2] Construction times for gas-fired combined-cycle power plants are only two to three years, at installed costs ranging from only $500 to $700 (U.S.) per kilowatt of capacity.[3]

For expansions and boiler replacements at existing power plants or the construction of new plants, these 100 percent gas-fired generators are very attractive economically and environmentally. Extensive literature exists on the various design options, cost factors and environmental characteristics of gas-fired generation.

Texas probably burns more natural gas in large utility and industrial boilers than any comparable region in the world. After ten years of Fuel Use Act mandates which spurred massive construction of coal-fired generation, most plant managers in Texas who are not burning natural gas wish they could because of the operational and environmental benefits of gas.

The Need for Natural Gas Select-Use Technologies

From a worldwide standpoint, however, the existing inventory of coal-fired generating plants, the vast reserves of coal available at competitive prices and the proven coal-fired technologies replicated and exported around the globe are fundamental factors to consider in any program to mitigate energy-related sources of greenhouse gases. We must not harbor unrealistic expectations about the ability of 100 percent gas-fired generation or other power options to supplant these coal markets through the end of this century and the beginning of the next.

Fortunately, for the coal plants and coal mines in the U.S. and elsewhere in the world that have years of useful life remaining, there are options to utilize natural gas in conjunction with coal for its operational and environmental benefits. They are the so-called "select-use" technologies and they are the principal focus of this paper.

Explanation of "Select-Use"

Natural gas select-use technologies involve the burning of natural gas with less environmentally attractive fuels in the same or separate combustion units in utility and industrial boilers.

"Cofiring" natural gas in a coal-fired boiler refers to the use of natural gas during start-up and normal boiler operations to provide a fraction of the boiler's total heat input and achieve a variety of boiler performance improvements due to the use of gas.[4]

Many coal-fired boilers already use natural gas as an ignition fuel. Cofiring goes a step further to expand the use of gas as a continuous source of heat input in the primary furnace region.

Cofiring just 10 percent natural gas in a high-sulfur coal boiler has been shown to reduce sulfur dioxide emissions by 12 percent; nitrogen oxide emissions by 25 percent; and carbon dioxide emissions by 4 percent. In a 500 Megawatt baseload powerplant, that would mean a reduction of 12,500 tons per year in SO_2; 4,000 tons of NOx and a substantial 140,850 ton reduction in CO_2 emissions.[5] Higher levels of gas cofiring can be used at 20 percent or more of heat input, with emissions reductions expected to track the additional input of gas by roughly a one-to-one proportion.

"Gas Reburn" technology involves the injection of natural gas into the upper furnace above the primary region of combustion to produce a fuel-rich zone which improves combustion efficiency and reduces emissions.[6] The gas reburn combustion process typically uses at least a 20 percent natural gas heat input over and above the input for gas cofiring in the primary region of the boiler.

The operation of gas reburn using 20 percent gas with 80 percent coal has been shown to reduce SO_2 emissions by 20 percent; NOx emissions by 60 percent; and CO_2 emissions by 9 percent. In the same example of a 500 Megawatt baseload powerplant, that would mean a reduction of 20,850 tons per year in SO_2; 9,600 tons of NOx and a very substantial 281,700 ton reduction in CO_2 emissions.[7] These reductions are all in addition to the reductions achieved by cofiring in the primary furnace region.

Gas reburn can be enhanced with sorbent injection to achieve SO_2 reductions of at least 50 percent through the use of a calcium-based sorbent in the gas reburning chamber. However, since the seminar is focused on greenhouse gases and not acid rain precursors, this paper will address cofiring and gas reburn costs and benefits without the addition of sorbent injection.

Cofiring Analysis

The first step in cofiring natural gas with coal is ignition and warm-up of the boiler. For boilers with fuel oil ignition and warm-up, natural gas igniters and guns can be added during routine annual maintenance without additional downtime. As a fuel oil displacement, gas provides cleaner startups and improved flame stabilization during ignition and warm-up.[8]

The second step in cofiring is to maintain the gas as a portion of the heat input in the primary combustion zone -- typically up to about 20 percent of the boiler rating capacity. (See Figure 1.) Because natural gas is a very homogeneous fuel, it will act as a buffer for inconsistencies such as sulfur and ash content, moisture content and even the heating value of non-homogeneous coal.[9]

Other operational benefits of cofiring gas with coal include reduced erosion, slagging and/or fouling, reduced maintenance requirements, lower particulate loading and/or reduced mill load

Figure 1

GAS/COAL COFIRING

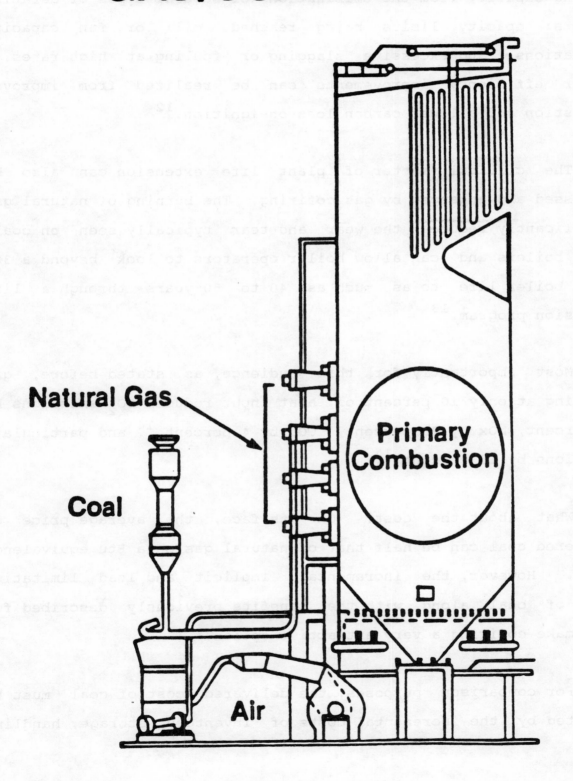

Natural Gas

Coal

Primary Combustion

Air

which can increase the unit's availability.[10]

A number of gas cofired boilers have experienced improved peaking capacity from the elimination of several types of derating effects: opacity limits being reached, mill or fan capacity limitations, and excessive slagging or fouling at high rates.[11] Boiler efficiency improvements can be realized from improved combustion and a lower carbon loss-on-ignition.[12]

The critical matter of plant life extension can also be addressed and enhanced by gas cofiring. The burning of natural gas significantly reduces the wear and tear typically seen on coal-fired boilers and can allow boiler operators to look beyond a 30-year boiler life to as much as 40 to 50 years through a life extension program.[13]

Most importantly for this audience, as stated before, gas cofiring at only 10 percent of heat input reduces SO_2 emissions by 12 percent, NOx by 25 percent, CO_2 by 4 percent,[14] and particulate emissions by 15 percent.[15]

What about the cost? On its face, the average price of delivered coal can be half that of natural gas on a Btu equivalency basis. However, the incremental, implicit and load limitation costs of coal, along with the benefits previously described for gas, make cofiring a very attractive option.

For comparison purposes, the delivered cost of coal must be inflated by the incremental costs of inventory, storage, handling

384

and drying. These costs do not accrue to natural gas. The implicit costs of ice, slag and tube leaks also add to the price of coal and not to gas. Finally, coal has load limitations from high ash moisture, air starvation and opacity or smoking.[16] In sum, at 5 to 20 percent of input, the operational and emissions benefits of gas, compared to the full costs of coal, make cofiring a very viable, cost-effective technology.

The American Gas Association has found that the incremental capital cost associated with cofiring with 10 percent gas is only $3 (U.S.) per kilowatt. The total levelized cost would be between 1.3 and 1.4 mills per kilowatt-hour in a 500 Megawatt coal-fired powerplant retrofit.[17]

A Cofiring Case Example

A 570 Megawatt coal-fired boiler at a major utility in the Eastern United States had been derated to 500 Megawatts due to slag opacity and excess air problems. In an operational test, cofiring only 5 percent gas, the utility was able to burn an additional 45 Megawatts of coal to bring the boiler back to a rated capacity of 550 Megawatts.[18]

A new market for 5 Megawatts of natural gas was created at the same time that a 45 Megawatt coal market was restored. This unit has not suffered a de-rate since the natural gas was put in almost three years ago.[19] Remarkably, total emissions were also reduced at the same time that energy output was increased.

Gas Reburn Analysis

Gas reburn is an add-on select-use technology to cofiring. Additional gas injectors are added to the boiler above the area of coal combustion. (See Figure 2.) Hydrocarbons in the gas react with NOx emissions to produce elemental nitrogen and ammonia. Above the gas, air is injected to complete the combustion of the natural gas.[20]

With only 20 percent gas in the reburn zone, NOx emissions can be reduced by an additional 60 percent, SO_2 by 20 percent, and CO_2 by 9 percent.[21] The efficiency of the boiler is improved by reducing excess oxygen and improving combustion efficiency through enhanced carbon burn-up.[22]

A particularly attractive feature of the reburn technology, since it involves an additional combustion region, is that it can be easily retrofitted and is compatible with all boiler types.[23]

According to the American Gas Association, the capital costs required for gas reburn are $12 per kilowatt. The total levelized cost would be between 3.5 and 3.9 mills per kilowatt-hour in a 500 Megawatt coal-fired powerplant retrofit.[24]

Pilot Projects in the United States

The Consolidated Natural Gas Service Company of Pittsburgh, Pennsylvania, currently has eight cofiring projects in its service region that are either in operation or will be shortly. They are

Figure 2

Gas Reburn – NO$_x$ Control

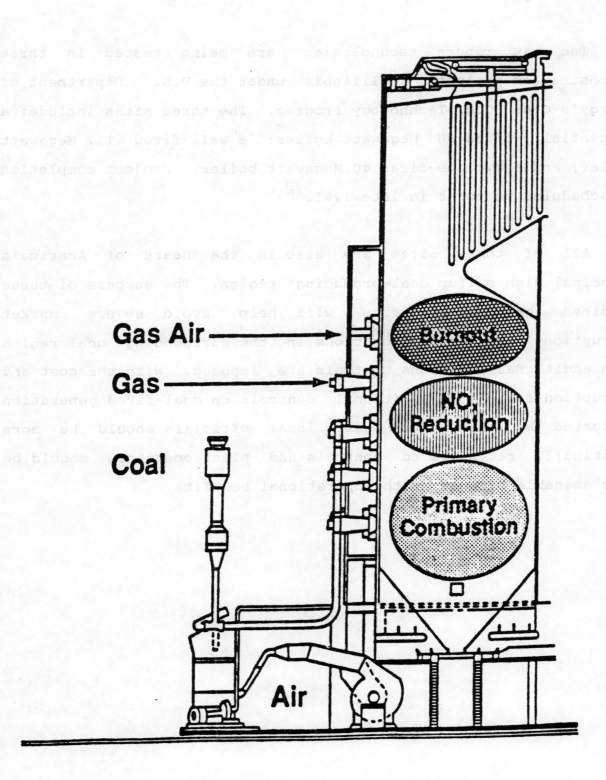

all coal-fired facilities that range from very small industrial boilers up to 550 Megawatt electric utility boilers. They also represent a wide base of coal combustion technologies, with different types and sizes of boilers.[25]

The gas reburn technologies are being tested in three demonstration projects in Illinois under the U.S. Department of Energy's Clean Coal Technology Program. The three sites include: a tangentially-fired 80 Megawatt boiler; a wall-fired 117 Megawatt boiler; and a cyclone-fired 40 Megawatt boiler. Project completion is scheduled to occur in late 1991.[26]

All of these sites are also in the heart of America's principal high sulfur coal-producing region. The success of these combined gas/coal projects will help avoid severe market disruptions and job dislocations in the surrounding coal region when additional emissions controls are imposed. With the cost and disruption factors of additional controls on coal-fired generation mitigated by the use of gas, local officials should be more politically receptive to controls and plant operators should be more amenable because of the operational benefits.

Conclusion/Policy Implications for Natural Gas Select-Use

The favorable policy implications of natural gas select-use technologies outlined in this paper are profound. In fact, gas cofiring and gas reburn offer a serendipity in emissions control that is very seldom achieved. The serendipity exists because natural gas select-use:

o   increases heat input and boiler output (instead of the energy penalty realized from scrubbers and most other pollution controls);

o   improves boiler operations and allows for plant life extensions (instead of the complications associated with scrubbers and most other pollution controls);

o   represents a retrofit approach with little or no additional downtime and the least-cost of the major retrofit options;[27] and

o   most importantly, provides the ability to reduce many pollutants simultaneously, including CO_2, NOx, SO_2, reactive hydro- carbons, particulates and carbon monoxide, without ash or sludge residue (unlike scrubbers and precipitators which, while reducing SO_2, can actually increase NOx and CO_2 emissions).[28]

The politics of coal-producing regions in America are not unlike those in other coal regions of the world. Fuel-switching, emissions taxes and expensive technological mandates are usually adamantly opposed by the interests in those regions.

Natural gas select-use technologies bring the serendipity mentioned earlier to a contentious battleground. They mitigate the threat to jobs and markets in the short run; they reduce the full range of emissions of concern for global warming, acid rain and ground-level pollution; and, they improve plant operations.

Select-use technologies are not a panacea. Most scientists believe that very fundamental changes in our energy production and consumption patterns worldwide must be made if we are truly going to affect the global warming trend in a significant manner.

However, for bodies like this which must tackle the very tough, long-range options for such fundamental changes, select-use technologies can be promoted and implemented now to mitigate short-term effects and offer a serendipitous transition to new approaches.

###

END NOTES

[1] "Global Climate Change and Emerging Energy Technologies for Electric Utilities: The Role of Natural Gas." American Gas Association. Arlington, Virginia. 1988. p.1, p.14.

[2] "An Evaluation of Alternate Control Strategies to Remove Sulfur Dioxide, Nitrogen Oxides and Carbon Dioxide at Existing Large Coal-Fired Facilities." Energy Analysis. American Gas Association. Arlington, Virginia. January 13, 1989. p.2.

[3] Henry R. Linden. "The Case for Increasing Use of Natural Gas in Generation." Public Utilities Fortnightly. Arlington, Virginia. February 18, 1988. p.2.

[4] "Assessment of the Potential Economics of Natural Gas Cofiring in Utility Coal-Fired Boilers." Gas Research Institute. Chicago, Illinois. November, 1988. p.1-1.

[5] Op.Cit., American Gas Association, p.3, p.13.

[6] Op.Cit., American Gas Association, p.10.

[7] Op.Cit., American Gas Association, p.3, p.13.

[8] "Natural Gas Cofiring for Coal Boilers." Gas Research Institute. Technology Profile. Chicago, Illinois. September, 1987. p.1.

[9] "An Update on Natural Gas Cofiring." Consolidated Natural Gas Service Company. Unpublished paper. February, 1989. p.1.

[10] "Assessment of the Potential Economics of Natural Gas Cofiring in Utility Coal-Fired Boilers." Gas Research Institute. Chicago, Illinois. November, 1988. p.1-1, p.2-10.

[11] Ibid., p.2-14.

[12] Ibid., p.2-18.

[13] Op.Cit., Consolidated Natural Gas Service Co., pp.2-3.

[14] Op.Cit., American Gas Association, p.3.

[15] Op.Cit., Gas Research Institute, p.2-6.

[16] Op.Cit., Consolidated Natural Gas Service Co., pp.6-8.

[17] Op.Cit., American Gas Association, pp.15-16.

[18] Op.Cit., Consolidated Natural Gas Service Co., p.8-9.

[19] Op.Cit., Consolidated Natural Gas Service Co., p.9.

[20] Op.Cit., Consolidated Natural Gas Service Co., p.13.

[21] Op.Cit., American Gas Association, p.3.

[22] Op.Cit., Consolidated Natural Gas Service Co., p.13.

[23] Op.Cit., American Gas Association, p.10.

[24] Op.Cit., American Gas Association, pp.15-16.

[25] Op.Cit., Consolidated Natural Gas Service Co., p.1-2.

[26] "Enhancing the Use of Coals by Gas Reburning and Sorbent Injection." U.S. Department of Energy, Office of Fossil Fuels. May, 1987. pp.3-5.

[27] Op.Cit., American Gas Association, pp.14-16.

[28] Op.Cit., American Gas Association, p.3.

#

Bibliography

American Gas Association. "An Evaluation of Alternative Control Strategies to Remove Sulfur Dioxide, Nitrogen Oxides and Carbon Dioxide at Existing Large Coal-Fired Facilities." Energy Analysis 1989-1. Arlington, Virginia. January 13, 1989.

American Gas Association. "Creative Energy Partnerships: Forging Solutions to National Problems." Panel Discussion on Developments in Natural Gas Cofiring Technology. AGA Annual Meeting. Arlington, Virginia. June 6, 1988.

American Gas Association. "Global Climate Change and Emerging Energy Technologies for Electric Utilities: The Role of Natural Gas." Arlington, Virginia. 1988.

American Gas Association. "Natural Gas and the Environment." Issue Brief 86-9. Arlington, Virginia. March 14, 1986.

American Gas Association. "Natural Gas for Efficient Electric Generation." Arlington, Virginia. December 10, 1986.

American Gas Association. "Use of Natural Gas for Environmental Purposes." Issue Brief 86-7. Arlington, Virginia. March 7, 1986.

ANR Pipeline Company. "Potential of Select Gas Use to Reduce Sulfur Dioxide Emissions in Wisconsin." Detroit, Michigan. October, 1985.

Consolidated Natural Gas Service Company. "An Update on Natural Gas Cofiring." Unpublished paper presented to the National Association of Regulatory Utility Commissioners. February, 1989.

Gas Research Institute. "Analysis of Natural Gas-Based Technologies for Control of SOx and NOx From Existing Coal-Fired Boilers." Topical Report. Chicago, Illinois. May, 1987.

Gas Research Institute. "Assessment of the Potential Economics of Natural Gas Cofiring in Utility Coal-Fired Boilers." Topical Report. Chicago, Illinois. November, 1988.

Gas Research Institute. "Co-Firing Gas in Coal-Fired Utility Boilers -- Summary of Benefits and Review of Operating Experiences." Topical Report. Chicago, Illinois. June, 1986.

Gas Research Institute. "Environmental Benefits Analysis Methodology and Model Users Manual." Topical Report. Chicago, Illinois. May, 1987.

Gas Research Institute. "Gas Reburning-Sorbent Injection (GR-SI): Development Status and Field Evaluation Plan." Topical Report. Chicago, Illinois. August, 1987.

Gas Research Institute. "Natural Gas Cofiring for Coal Boilers." Technology Profile. Chicago, Illinois. September, 1987.

Gas Research Institute. "Natural Gas/Pulverized Coal Cofiring Performance Testing at an Electric Utility Boiler." Topical Report. Chicago, Illinois. June-October, 1986

Henry R. Linden. "The Case for Increasing Use of Natural Gas in Generation." Public Utilities Fortnightly. Arlington, Virginia. February 18, 1988.

National Academy of Sciences, National Research Council. Changing Climate -- Report of the Carbon Dioxide Assessment Committee. National Academy Press. Washington, D.C. October, 1983.

National Academy of Sciences, National Research Council. Ozone Depletion, Greenhouse Gases and Climate Change. National Academy Press. Washington, D.C. 1989.

Tenneco Gas Marketing Company. "Select Use of Natural Gas in Partnership With Coal." Unpublished Presentation. September, 1988.

Texas General Land Office. Putting Together the Pieces -- The Recapitalization of the Texas Economy. Austin, Texas. January, 1989.

Texas General Land Office. Symposium Proceedings: Natural Gas and Clean Air, An Alliance for America's Future. Austin, Texas. April, 1988.

Texas General Land Office, Research Division. "Four Threats to the Air We Breathe... New Solutions Using Natural Gas." Austin, Texas. September, 1988.

Texas General Land Office, Research Division. "Natural Gas: A Fuel for the Future." Austin, Texas. November, 1988.

Texas General Land Office, Research Division. "Redefining the National Interest: Building Consensus on Energy & the Environment With Natural Gas." Austin, Texas. August, 1988.

U.S. Department of Energy, Office of Fossil Energy. "Enhancing the Use of Coals by Gas Reburning and Sorbent Injection." DOE/FE-0087. Washington, D.C. May, 1987.

SCENARIOS FOR REDUCING CO_2 EMISSIONS FROM ELECTRICITY GENERATION IN ENGLAND AND WALES BY FUEL SWITCHING AND ENERGY CONSERVATION

By A.J. Crane and P.L. Surman

Central Electricity Generating Board, U.K.

1. Introduction

In June 1988, the Toronto Conference on the Changing Atmosphere called for a reduction in CO_2 emissions to 80% of their 1988 level by 2005 as an initial global goal. This was the first time that a specific target for CO_2 emission reduction had been suggested at a major international meeting. It marked the beginning of a new phase in the greenhouse issue in which moves towards policy formulation are likely to acquire increasing prominence.

Whether or not understanding of the greenhouse effect has yet reached a stage at which particular emission reduction targets can be justified scientifically is an open question. Nevertheless, the Toronto Conference brought into focus the need to explore the technological means for achieving emission reductions.

This paper, based on a recent published submission by the CEGB to the House of Commons Select Committee on Energy (1), looks at possible ways in which CO_2 emissions from the electricity supply industry in England and Wales, which currently account for about one third of UK emissions, could be reduced. It first examines the potential for low- and non-CO_2 emitting technologies to increase their share of generation capacity by 2005. Based on this assessment, hypothetical scenarios are presented which would enable the electricity industry to meet its share of the Toronto target, assuming the 20% cut called for were applied *pro rata* across all energy sectors. Options for meeting a less severe target of holding CO_2 emissions constant at 1988 levels are also explored. The latter have significantly less onerous economic, political and social implications, demonstrating the importance of establishing a firm scientific foundation on which to base any energy policy. Without such a foundation, it will be difficult to evaluate the relative costs and benefits resulting from different limitation targets. The paper therefore concludes by indicating the extent of uncertainty in two important parameters that will govern the development of greenhouse gas-induced climatic change.

2. Technological options for reducing CO_2 emissions

(a) Improvement in energy conversion efficiency

Large improvements in conventional coal-fired plant efficiency have been achieved over the past 40 years, typically from about 25% thermal efficiency in the 1950s to 38% for a modern pulverised fuel (PF) plant. Table 1 shows past and present annual fuel consumption, electricity supplied and CO_2 emissions from power stations in England and Wales. Between 1950 and 1987, CO_2 emissions per kWh generated by fossil-fired plant decreased from 1.4 to 0.9 kg CO_2 /kWh. Introduction of nuclear power reduced the overall figure to 0.8 kg CO_2 /kWh with the result that, since 1970, CO_2 emissions have remained constant despite a 23% increase in electricity production. The thermal efficiency of conventional PF plant is now close to practical limits for this type of thermodynamic cycle. Further significant improvement in plant efficiency therefore requires a new technology.

Combined cycle generation using advanced gas turbines ahead of a conventional steam cycle and fuelled by natural gas can achieve 48% thermal efficiency with a consequent reduction in CO_2 emissions. An overall energy utilization of about 80% could be secured from a gas-fired combined heat and power (CHP) scheme. The potential for CO_2 savings by these technologies is considered below.

(b) Fossil fuel substitution

Typical carbon contents and relative calorific values of coal, natural gas and fuel oil are listed at the foot of Table 1. It can be calcualted from these figures that burning natural gas releases only 56% of the CO_2 produced by burning coal to produce the same amount of heat. The corresponding figure for oil is 82%. When advantage is taken of the higher thermal efficiency achieved by burning natural gas in a combined cycle plant, emissions of CO_2 can be reduced to 44% of those from an equivalent PF coal-fired plant.

If it assumed that natural gas supplies for electricity generation in England and Wales could reach a maximum of 20% of fossil fuel burn by 2005, some 10 GW of gas-fired plant could by then have displaced equivalent coal-fired generation (19 Mtce) and have led to a saving in CO_2 emissions of about 25 Mt. In principle, coal burn could be further displaced by burning oil in existing power stations. This could amount to a maximum of 40 Mtce per year, reducing CO_2 emissions by 16 Mt per year.

(c) Increased contribution from nuclear power

Although it is unlikely that with present planning processes more than the four PWRs currently envisaged by the CEGB could be commissioned before 2000, it is considered that thereafter the rate could be increased to two PWRs per year, or possibly more, depending on political determination, public acceptability and commercial considerations. By 2005, therefore, an additional 10 PWRs might be commissioned, abating 60 Mt CO_2 This would bring the total nuclear capacity to about 20 GW, equivalent to 50 Mtce.

(d) Renewable energy sources

It has been projected (2) that up to 18% of the UK's electricity demand might be met by renewable resources by 2030 if they become economic, technically viable and were publically acceptable. However, as set out in Table 2, the contribution by 2005 is unlikely to be more than 8% (11 Mtce), leading to a CO_2 abatement of 24 Mt. This total would require in particular the construction of the Severn and Mersey barrages and the successful introduction of wind farming in the UK. The CEGB and the Department of Energy expect to complete the construction of three 8 MW wind parks by 1993 to assess their viability. It is worth noting that some 500 square kilometres of land would be required to accommodate wind generators equivalent in output to one PWR.

Although the combustion of refuse produces CO_2, if it is burnt as a fuel in power stations it can displace the burning of coal. If the refuse is thereby diverted away from landfill, an additional benefit accrues in that it prevents the release of methane from refuse decomposition. Methane releases have a substantially larger greenhouse effect than CO_2, although the exact extent is difficult to quantify in view of uncertainty in the retention of methane in the atmosphere.

(e) Combined heat and power (CHP)

In the last 10 years, the CEGB has investigated some 150 potential CHP schemes. For the most part the overriding reason why these investigations have not resulted in practical, economic schemes has been the competition from direct heating, usually using natural gas, which has made CHP uneconomic without the injection of subsidies. The potential for CHP in the future is difficult to assess, although its prospects may be helped by the utilization of gas in combined cycle turbine plant. The emission abatement would occur across several energy sectors as the heat produced would displace other fossil fuel sources.

(f) Reducing energy consumption

Many assessments have indicated that a large economic potential exists for improvements in efficiency in the use of all forms of energy. A recent report by the Energy Technology Support Unit (3) suggests the potential savings in buildings are 20-30%, in industrial processes 20-25% and in transport 25-35%.

The extent to which this potential will be taken up is problematic however. Current electricity supply industry forecasts of electricity demand are based on an assessment that by 2005 efficiency in the use of electricity will be 12.5% higher than in 1986. The extent to which this rate of increase could be accelerated is most uncertain, as are the costs of the incentives which might have to be offered to electricity users.

Significant reductions in consumption would likely require intervention by government across all forms of energy use, whether by regulation, pricing or subsidy. Attempts to secure substantial savings through one energy sector alone can be frustrated if a cheaper alternative fuel is available to the consumer. In addition, in the domestic sector, the savings from improved energy efficiency may partially be lost by the choice of higher comfort standards.

3.   Hypothetical scenarios for electricity generation

To assess how the technologies described in the previous section might contribute to CO_2 reductions in the period to 2005, several, purely hypothetical scenarios of future fuel mixes have been constructed. These are purely illustrative and do not in any way constitute projections of future generation mix in England and Wales. The scenarios are set out in Table 3.

(a) Case 1: Unrestricted demand baseline

This scenario incorporates the 12.5% improvement in end-use efficiency referred to above and the construction of 5 PWRs, or equivalent non-fossil plant, but assumes the additional demand is otherwise met by increased coal and oil combustion. A 25% increase in CO_2 emissions above 1987 levels would occur.

(b) Case 2: 20% CO_2 reduction by changed fuel mix - demand unrestricted

To meet the 20% CO_2 reduction but with the same demand as in case 1, the contribution of renewables, nuclear and natural gas are set to the maximum likely potentials discussed in Section 2 above. Additionally, a substantial displacement of coal by oil is needed to meet the target with very severe consequences for the coal industry. The possible life-extention of some 9 GW of existing coal-fired plant has to be forfeited in this scenario. The major shifts in generation mix illustrated by this scenario indicates the need to consider demand restriction as a means of reducing CO_2 emissions to the extent required.

(c) Case 3: 20% CO_2 reduction by holding demand to 1988 level

This scenario maintains demand at 1988 levels. The same non-fossil and oil-fired generation as in case 1 is assumed with a more modest introduction of natural gas than in case 2. This allows coal burn to be reduced to 56 Mt, compared with 32 Mt in case 2. The continuation of reasonable rates of economic growth would be severely threatened by this degree of electricity demand reduction, which amounts to 37% in real terms.
It would involve major intervention in the energy and equipment markets and considerable social change.

(d) Case 4: Emissions held constant at 1988 level by demand restriction and changed fuel mix

If the Toronto target is judged to be impracticable or unnecessarily low, another hypothetical scenario would be to hold CO_2 emissions constant at their 1988 level. A 21% reduction in real demand is assumed, instead of the 12.5% assumed in case 1. Of the many ways in which fuel switching could then meet the balance of emission control, two examples involving increasing the nuclear or natural gas contributions are illustrated in Table 3. Although the fuel switches would appear to be feasible, it seems unlikely that the demand restriction could be achieved without intervention in the market place across all energy sectors.

4.   The scientific basis for CO_2 reduction targets

The previous section has illustrated the major implications, both in terms of changed fuel mix and the potential economic and social effects, of meeting the Toronto target. The consequences for other energy sectors and for countries with rapidly expanding populations and economies are likely to be even more profound. This highlights the need to establish a firm scientific basis for any particular target. Two aspects of the greenhouse effect are worth noting in this context which indicate that the level of understanding is probably not yet sufficient to justify the pursuit of the Toronto target, rather than a less onerous one.

Figure 1 (from ref 1) shows atmospheric CO_2 concentrations for different emission scenarios as projected by a carbon model. The Toronto target corresponds quite closely with the 1% per year emission reduction curve. It is seen that the less onerous target of holding emissions constant does not result in substantially higher CO_2 concentrations over the next several decades. It is arguable whether this difference is sufficient to warrant the choice of the more severe target at this stage. Both curves contrast strongly, however, with a scenario of accelerating emissions (2.5% per year), not uncharacteristic of global CO_2 emissions in the last few years. Growth in emissions at this scale will greatly advance the date at which a significant threat is posed. At present, therefore, it would be much easier to justify a policy aimed at slowing down this growth than one requiring absolute reductions in emissions.

The second aspect of the greenhouse problem to which attention is drawn is the continuing uncertainty surrounding the size of the climatic change likely to result from a given concentration of carbon dioxide. Projections of the average warming that may result from a doubling of the concentration, whenever that may occur, are highly dependent on quite subtle changes in the way various physical processes are approximated in the computer models. Different ways of representing clouds, for example, may lead to a warming of over 5 C or less than 2 C. If it is to address the issue effectively, the process of policy formulation will clearly have to take account of the large uncertainty that exists in the rate and scale of the climatic change expected. An acceleration of research to reduce these uncertainties must be a major priority at an international level. In the meantime, governments need to establish national and international mechanisms to reduce the present rate of increase in greenhouse gas concentrations. Principle aims should be the encouragement of more rapid penetration of improvements in energy efficiency and the conservation and restoration of the world's forests.

5.   References

1. CEGB (1989) Evidence to the House of Commons Select Committee on Energy. In: Energy Policy Implications of the Greenhouse Effect: Mempranda of Evidence. Energy Committee Session 1988-89, Vol 192-i

2. CEGB (1988) Prospects for alternative energy sources in electricity generation. House of Lords Select Committee on the European Community, Alternative Energy Sources. Session 1987-88 Report with Evidence. HL Paper 88, pp 17-24.

3. ETSU (1987) 1986 appraisal of UK energy research and demonstration ETSU Rpt R43, Harwell, p E3.

Table 1: Past, present and projected future annual fuel consumption, electricity supplied and carbon dioxide emissions by power stations in England and Wales. Note that two scenarios, one with and one without natural gas, are given for 2005.

Year	1950		1970		1987		2005			
	Mtce	Mt CO₂	Mtce	Mt CO₂	Mtce	Mt CO₂	Mtce	Mt CO₂	Mtce	Mt CO₂
Coal	31.61	68.6	68.5	148.7	79.7	173	98	213	79	171
Oil	0.07	0.13	19.7	34.9	5.2	9	9	16	9	16
Gas	—	—	0.4	0.5	—	—	—	—	19	18
Nuclear	—	—	7.0	—	14.2	—	33	—	33	—
Total	31.68	68.73	95.6	184.1	99.1	182	140	229	140	205
TWh supplied	48.9		186		228		332			
kgCO₂/kWh	1.4		1.0		0.8		0.69		0.62	

One tonne coal of 59.2 per cent carbon content produces 2.17 tonne CO_2
One tonne gas of 73.3 per cent carbon content produces 2.67 tonne CO_2
One tonne oil of 84.5 per cent carbon content produces 3.10 tonne CO_2
The net calorific value of gas is 2.20 times that of coal
The net calorific value of oil is 1.75 times that of coal
Mtce = million tonnes coal equivalent
Mt CO_2 = million tonnes CO_2.

Source: Reference 1

Table 2: Likely upper limit of carbon dioxide reductions from the increased use of renewable energy sources by year 2005 in England and Wales

	Installed Capacity MW	TWh/year	Mtce	% of total Mtce	MtCO₂/year avoided
Wind	1100	3	1.2	0.9	2.6
Tidal	9100	18	7.6	5.4	16.5
Geothermal	200	1.5	0.7	0.5	1.5
Wave	150	0.4	0.2	0.1	0.4
Hydro	10	0.02	—	—	—
Refuse	50	3.3	1.4	1.0	3.0
Totals		26.2	11.1	7.9	24.0

Source: Reference 1

Table 3: Hypothetical scenarios for carbon dioxide reduction from electricity generation in England and Wales

Scenario for 2005	TWh*	Mtce*	Mt CO_2	Renewable Mtce	Nuclear Mtce*	Gas Mtce	Oil Mtce	Coal Mt
1988 REFERENCE	246	106	182	—	21	—	5	80
CASE:								
1 Unrestricted demand, baseline	332	140	227	2	31	0	9	97
2 Target met by changed fuel mix: unrestricted demand	332	140	141	9	50	19	30	32
3 Target met by demand restriction	246	106	145	2	31	8	9	56
4 CO_2 held constant at 1988 level by conservation and (a) substitution of gas for coal	300	126	182	2	31	13	9	71
(b) substitution of nuclear for coal	300	126	183	2	38	0	9	77

*These figures include imports of electricity.

Source: Reference 1

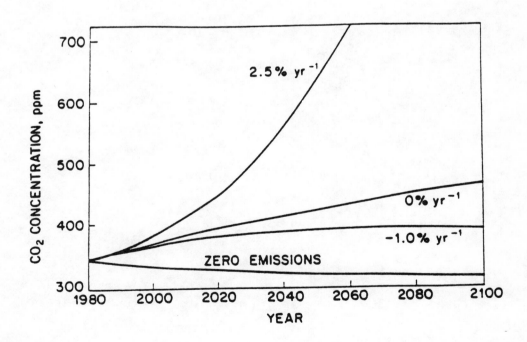

Figure 1: Projected atmospheric carbon dioxide concentrations for a range of future carbon dioxide input scenarios as derived from an atmosphere-ocean carbon cycle model (Source: Reference 1)

SUMMARY

**CO_2 EMISSIONS IN THE LAST DECADE AND ENERGY CONSUMPTION IN BELGIUM-
ELEMENTS OF CO_2 CONTROL STRATEGIES
(INCLUDING THE ECONOMICS OF TECHNOLOGIES FOR CO_2 RECOVERY FROM COAL
AND NATURAL GAS POWER PLANTS).**
Walter J. HECQ, Université Libre de Bruxelles

Between 1975 and 1985, the emission of CO_2 to the atmosphere from the burning of fossil fuels in Belgium decreased by 13% (15m. tonnes). During the same period, the consumption of primary energy increased by 6%. An analysis of the structure of energy consumption shows that what was mainly responsible for this result was the switch to nuclear energy, which now accounts for 60% as apposed to an earlier 6% of the electricity generated. So this strategy made it possible to avoid the emission of 5 m. tonnes of CO_2 in 1975, and up to 25 m. tonnes in 1985. During the same period the coefficient of emission per kWh produced passed from 617 gCO_2/kWhe to 311 g CO_2/kWhe; expressed in terms of MJe used, this means that the coefficient of emission decreased from 186 g CO_2/MJe to 94 g CO_2/MJe. All in all, the coefficient of emission per MJ (final) used in Belgium fell from 88 g CO_2/MJ to 78 g CO_2/MJ between 1975 and 1985.

Beside the nuclear programme, which could be expanded still further, there are other strategies which bear on the reduction of the amount of CO_2 emitted per unit of energy; an increase in the use of natural gas (51-56 g CO_2/MJ) is one possibility, but this solution runs into problems of availability regarding the future. What remains is CO_2 removal and recovery from flue gases, 90% efficient commercially avalaible technology which can be applied to fossil fuel-fired power plants. The cost and energy requirements for this process in place in two types of power plants, have been assessed. The results show that allowance must be made for a supplement of \$ 0.035 per kWhe for a coal-fired power station, and \$ 0.021 per kWhe for a power station equipped with gas and steam turbines. The energy requirements for such a control system reduce the efficiency of power generation from 39.6% to 32.7% (coal-fired) and from 48% to 42.7% (gas-fired). The coefficient of emission per kWhe generated is thus brought down from 950 g CO_2/kWhe to 115 g CO_2/kWhe (coal-fired) and from 440 g CO_2/kWhe to 49 g CO_2/kWhe (gas-fired).

Besides these strategies which bring about a decrease in the CO_2 emission per unit of energy, energy-saving policies must not be overlooked; after all, they showed their worth after the oil shocks. Although their potentiality for cutting CO_2 emissions is beyond doubt, this needs further studies to be evaluated.

****

RESUME

EMISSIONS DE CO$_2$ DURANT 10 ANS ET CONSOMMATION ENERGETIQUE EN BELGIQUE ELEMENTS DE STRATEGIES FUTURES DE REDUCTION DES REJETS DE CO$_2$ (INCLUANT L'EVALUATION ECONOMIQUE DE LA DECARBONATATION DES FUMEES DE DEUX TYPES DE CENTRALES ELECTRIQUES – CHARBON ET GAZ NATUREL.

Walter J. HECQ, Université Libre de Bruxelles

Entre 1975 et 1985, les émissions de CO$_2$ issues de la comsommation de combus-tibles fossiles en Belgique ont diminué de 13% (15 millions de tonnes). Au cours de la même période, la consommation d'énergie primaire a progressé de 6%. L'analyse de la structure de la consommation d'énergie montre que le principal responsable de ce résultat est le recours à l'énergie nucléaire dont la part dans la production d'électricité est passée de 6% à 60%. Une telle stratégie a ainsi permis d'éviter le rejet de 5 millions de tonnes de CO$_2$ en 1975 jusqu'à 25 millions de tonnes en 1985. Dans le même temps, le coefficient d'émission par kWhe produit est passé de 617 g CO$_2$/kWhe à 311 g CO$_2$/kWhe; il en résulte qu'exprimé par MJe consommé le coefficient d'émission a diminué de 186 g CO$_2$/MJe à 94 g CO$_2$/MJe. Au total en Belgique, le coeffi-cient d'émission par MJ (final) consommé est tombé de 88 g CO$_2$/MJ à 78 g CO$_2$/MJ entre 1975 et 1985.

A côté du programme nucléaire qui pourrait être encore étendu, il y a d'autres stratégies qui portent sur un abaissement des rejets de CO$_2$ par unité éner-gétique, l'extension de l'utilisation du gaz naturel (51-56 g CO$_2$/MJ) cons-titue une possibilité, mais cette solution se heurte à des problèmes de disponibilité future. Reste la décarbonatation des fumées, cette technique efficace (rendement 90%) et opérationnelle qui peut être appliquée aux cen-trales thermiques. Les coûts et les besoins énergétiques de ce procédé, installé sur deux types de centrales électriques, ont fait l'objet d'une évaluation. Les résultats montrent qu'il faut compter un supplément de 0,035 $ par kWhe pour une centrale de charbon et de 0,021 $ par kWhe pour une centrale équipée de turbines à gaz et à vapeur. Les besoins énergétiques du procédé entraînent une baisse du rendement de 39,6% à 32,7% (centrale au charbon) et de 48,0% à 42,7% (centrale au gaz). Le coefficient d'émission par kWhe produit est ainsi ramené de 950 g CO$_2$/kWhe à 115 g CO$_2$/kWhe (centrale au charbon) et de 440 g CO$_2$/kWhe à 49 g CO$_2$/kWhe (centrale au gaz).

A côté de ces choix stratégiques qui entraînent une diminution du coefficient d'émission de CO$_2$ par unité énergétique, les politiques favorisant les écono-mies d'énergie ne sont pas à négliger, elles ont porté leurs fruits après les chocs pétroliers. Leur potentialité sur un possible abaissement émissions de CO$_2$ est certaine mais des études supplémentaires s'indiquent pour qu'elle soit estimée.

EMISSIONS DE CO_2 DURANT 10 ANS ET CONSOMMATION ENERGETIQUE EN BELGIQUE ELEMENTS DE STRATEGIES FUTURES DE REDUCTION DES REJETS DE CO_2 (INCLUANT L'EVALUATION ECONOMIQUE DE LA DECARBONATATION DES FUMEES DE DEUX TYPES DE CENTRALES ELECTRIQUES - CHARBON ET GAZ NATUREL)

Walter J. HECQ - Université Libre de Bruxelles

La course à la croissance, source de richesse, que se disputent les économies, est inévitablement associée à une évolution croissante de la consommation de combustibles fossiles. Elle débouche sur un accroissement des quantités émises d'une série de produits de combustion.

Parmi ceux-ci, le dioxyde de carbone occupe une large place sur le plan quantitatif. Ce gaz s'accumule dans l'atmosphère avec des menaces de changement climatique bien connues. Devant les préoccupations soulevées par ce phénomène de large ampleur, il est utile d'examiner les responsabilités dans les rejets de CO_2.

Sur base de la relation étroite entre la consommation de combustibles fossiles et les rejets de CO_2, l'analyse de l'évolution structurelle de la consommation d'énergie permet de chiffrer les responsabilités des secteurs économiques et des catégories de combustibles dans les émissions de CO_2 et finalement de tirer des enseignements quant aux effets des politiques énergétiques.

C'est sur ce sujet que la première partie de cet article se penche. A côté de l'option stratégie énergétique, pour contrôler les rejets de CO_2, il existe l'option technologique.

La seconde partie de cet article s'y consacre au travers de l'étude d'un bilan énergétique et économique d'une technique de décarbonatation des fumées appliquée à deux types de centrales électriques classiques, grandes sources de rejets de CO_2.

1. Stratégie énergétique et rejets de CO_2 - Le cas de la Belgique

Chacun sait qu'à quantité égale d'énergie produite, le gaz naturel, puis les combustibles pétroliers rejetent moins de CO_2 que le charbon (tableau 1). Et les sources physiques d'énergie : nucléaire, hydraulique, solaire, éolienne, etc... n'en produisent pas.

En conséquence, si l'on adopte des stratégies énergétiques judicieuses quant à la répartition de ces formes d'énergie, on peut agir sensiblement sur les rejets de CO_2.

Tableau 1 : Coefficients d'émission du CO_2 (1)

GAZ NATURELS	:	51 - 56 g/MJ	1.795 - 2.245 g/Nm^3
GAZ DE COKERIE	:	± 46 g/MJ	± 880 g/Nm^3
GAZ DE HAUT FOURNEAU	:	± 242 g/MJ	± 750 g/Nm^3
GPL	:	± 30 g/MJ	± 3010 g/Nm^3
ESSENCE SUPER	:	± 75 g/MJ	± 2.400 g/l
GAS-OIL/FUEL-OIL	:	± 78 g/MJ	± 2.800 g/l
CHARBONS	:	85 - 105 g/MJ	2.590 - 2.835 g/kg

De telles stratégies associées à une augmentation de l'efficacité de l'utilisation (production, distribution, consommation) de l'énergie conduisent à des résultats positifs puisqu'elles ont produit une baisse des rejets de CO_2, ces dernières années, dans certains pays industrialisés et notamment en Belgique comme la suite le montre.

1.1. Evolution des rejets de CO_2 produits par la consommation de combustibles fossiles

La figure 1 indique comment ont évolué les rejets de CO_2 issus de la combustion de combustibles fossiles (2) en Belgique de 1975 à 1985 (figure 1). Ces rejets ont diminué de 110 millions de tonnes (1975) à 95 millions de tonnes (1985) avec une baisse drastique (plus de 25%) observée entre 1979 et 1983.

---

1) Communications : FIGAZ, Fédération Belge des Pétroles, Fédération charbonnière de Belgique.

2) Les combustibles pétroliers (liquides) et les charbons (solides) sont les principaux responsables des émissions de CO_2 (80%).
Parmi les autres sources, les fours à chaux et le cimenteries émettent grossièrement 3,5% des totaux mentionnés pour l'année 1985 (fig. 1). A titre indicatif, les rejets de CO_2 produits par la respiration des êtres humains se chiffreraient, en Belgique, à des valeurs comprises entre 3 et 5% des émissions de 1985.

Figure 1 : <underline>Evolution des émissions de CO_2 produites par la combustion de combustibles fossiles en Belgique</underline>

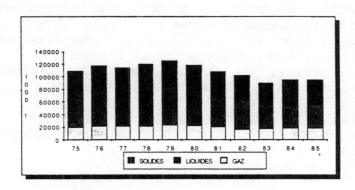

Cette diminution incombe essentiellement à l'abandon progressif
(- 200.000 TJ) de l'emploi de certains combustibles pétroliers
(liquides) (1) alors que la consommation des deux autres types de
combustibles fossiles (gaz naturel - 40.000 TJ et charbon + 50.000
TJ) (2) (3) a moins évolué (figure 2). Il en résulte une baisse de
la consommation des combustibles fossiles (- 190.000 TJ). Cette
baisse, très perceptible entre 1979 et 1983, est comblée avec
l'extension de l'utilisation de l'énergie nucléaire (+ 270.000 TJ)
(4). Il en résulte que de 1975 à 1985, le coefficient d'émission de
CO_2 par MJ primaire consommé a diminué de 73 g CO_2/MJ à 60 g CO_2/MJ.
Suivant les secteurs, des mesures d'économie d'énergie, de substitution de combustibles ou de fermeture et de rationalisation d'entreprises à haute intensité énergétique sont responsables de la
diminution de CO_2 comme le montre l'analyse reprise ci-après.

Figure 2 : <u>Evolution de la consommation d'énergie en Belgique</u>

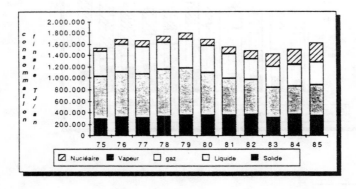

1) Statistiques Pétrolières Belges - Fédération Pétrolière Belge.
2) Annuaires Statistiques - Fédération de l'Industrie du Gaz.
3) Statistiques Administration de l'Energie - Service de l'Economie
   Charbonnière.
4) Rapports annuels - Fédération Professionnelle des Producteurs   et
   Distributeurs d'Electricité en Belgique.

1.2. Evolution des rejets de CO_2 issus de la consommation de combustibles fossiles par les secteurs économiques belges

La diminution totale des émissions de CO_2 après 10 ans est, bien sûr, la même que précédemment (figures 1 et 3).
On constate que l'effort est essentiellement supporté par deux secteurs : les producteurs d'électricité et l'industrie.
Dans une moindre mesure, le secteur domestique, tertiaire, et assimilé a contribué à la réduction des rejets de CO_2.

Figure 3 :   Evolution des émissions de CO_2 par les secteurs économiques belges (combustion de combustibles fossiles)

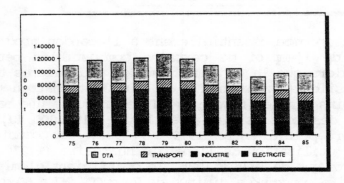

Figure 4 :   Evolution de la consommation d'énergie par le secteur des producteurs d'électricité (1)

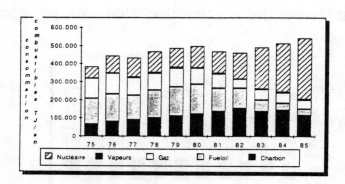

---

1) Annuaires Statistiques - Fédération Professionnelle des Producteurs et Distributeurs d'Electricité de Belgique.

i) <u>Producteurs d'électricité</u>
Les rejets de CO_2 issus de la production d'électricité sont passés de 24 millions de tonnes à 17 millions de tonnes entre 1975 et 1985, avec une pointe intermédiaire de 30 millions de tonnes en 1980. La diminution des rejets résulte de la part croissante prise par le nucléaire dans la production d'électricité (+ 270.000 TJ ; 60% en 1985) au dépens des combustibles pétroliers (-110.000 TJ) et du gaz naturel (- 59.000 TJ) (figure 4) alors que l'utilisation du charbon a crû (+ 51.000 TJ).

Endéans cette période 1975-1985, le coefficient d'émission de CO_2, et spécifique de ce secteur, est passé de 62,9 grammes de CO_2 par MJ consommé en 1975 à 31,2 grammes de CO_2 par MJ consommé en 1985. Grâce à cette évolution du coefficient d'émission, la croissance de la production d'électricité s'est accompagnée d'un effet bénéfique quant aux émissions de CO_2. Cet effet bénéfique est indirectement ressenti au niveau de la consommation (finale) d'électricité des autres secteurs puisque le coefficient d'émission spécifique a diminué de 670 g CO_2/kWhe (1985) à 340 g CO_2/kWhe (1985).

ii) <u>Industrie</u>
Premier secteur, par ordre d'importance, quant aux émissions de CO_2 provenant de l'utilisation des combustibles fossiles, l'industrie a contribué à la réduction des rejets de CO_2. Ils ont baissé de 42 millions de tonnes (1975) à 35 millions de tonnes (1985). Compte tenu des rejets indirects provenant de la consommation d'électricité, les émissions de CO_2 sont passées de 56 millions de tonnes (1975) à 44 millions de tonnes (1985). Cette diminution est imputable à la baisse de la consommation du gaz naturel et des combustibles pétroliers (- 108.000 TJ au total) alors que la consommation du charbon et d'électricité a parallèlement crû (+ 20.000 TJ,+ 8.000 TJe). Les mesures d'économie d'énergie (la consommation spécifique a chuté de 31% entre 1979 et 1983), les changements structurels intersectoriels (tertiarisation) et la récession qui a frappé les branches de la métallurgie et de la construction sont les principaux facteurs qui sont intervenus dans la baisse de consommation des combustibles fossiles (1) et finalement dans la baisse des émissions de CO_2. Il faut également ajouter l'effet positif de la baisse du coefficient d'émission spécifique de l'électricité (§ i) qui a largement inhibé l'effet de l'accroissement de la consommation de charbon (+ 20.000 TJ) puisque le coefficient d'émission global par MJ consommé dans le secteur de l'industrie belge est passé de 96 g CO_2/MJ à 86,5 g CO_2/MJ entre 1975 et 1985.

---

1) T. MOROVIC et Al., Energy Conservation Indicators - Springer - Verlag (1987) .

iii) <u>Domestique, tertiaire et assimilés (DTA)</u>
Deuxième secteur par ordre d'importance dans les rejets de CO_2, ceux-ci sont passés de 33 millions de tonnes en 1975 à 30 millions de tonnes en 1985, soit en valeurs corrigées (1) dues aux apports indirects de la consommation d'électricité, de 42 millions de tonnes à 37 millions de tonnes.

Si cette diminution est assez faible, il faut remarquer que ce secteur a, toutefois, consommé en 1985 un peu plus d'énergie qu'en 1975. La substitution d'énergie a joué ici un rôle déterminant dans la diminution des émissions de CO_2; l'accroissement de l'utilisation du gaz naturel (+ 93.000 TJ) et dans une moindre mesure de l'électricité (+ 23.000 TJe) s'est opéré au dépens des combustibles pétroliers (- 66.000 TJ) et du charbon (- 20.000 TJ). Le coefficient d'émission global par MJ consommé dans le secteur domestique, tertiaire et assimilés (DTA) est passé de 97,4 g CO_2/MJ à 74,5 g CO_2/MJ endéans la période 1975-1985. Il faut également mentionner le rôle significatif des mesures d'économie d'énergie (isolation, meilleure efficacité de chauffage) dans le batiment qui se sont traduites à raison de 50% dans la baisse de consommation des combustibles entre 1979 et 1983 et qui ont freiné l'accroissement de la consommation énergétique.

iv) <u>Transport</u>
Dernier secteur par ordre d'importance, le secteur des transports est le seul dont les émissions de CO_2 ont progressé entre 1975 et 1985 : de 11 millions de tonnes à 13,5 millions de tonnes et ce parallèlement à la consommation de carburant. Pourtant le kilométrage moyen annuel des véhicules a diminué de 15% et la consommation moyenne des véhicules au km parcouru a baissé, ceci en partie à cause de la part croissante prise par les véhicules "diesel" dans le parc. Mais c'est l'accroissement du nombre total de voitures (de 2,5 millions à 3,3 millions d'unités) en circulation qui a entraîné une évolution croissante de la consommation de carburant (2) qui se répercute sur les rejets de CO_2.

1.3. <u>Commentaire</u>

Une stratégie énergétique judicieuse peut porter des fruits en matière de limitation des rejets de CO_2.
Dans ce cadre, le recours à l'énergie nucléaire ou au gaz naturel ainsi qu'aux économies d'énergie constitue des stratégies attractives. Elles ont permis de réduire substantiellement les rejets de CO_2 (13% entre 1975 et 1985) en Belgique. Elles offrent, en outre, l'avantage de diminuer la dépendance énergétique du pays. Toutefois, de telles stratégies présentent des limites :

---

1)   Compte tenu de sa faible contribution (entre 1 et 2%), l'apport indirect dû à la transformation des produits pétroliers est négligé.

2)   Gasoil + 37.000 TJ; essence - 11.000 TJ.

- En ce qui concerne les économies d'énergie : en partie
réalisées à la suite des chocs pétroliers ou pour des raisons de
rentabilité, ces mesures pourraient être encore étendues notamment
dans le bâtiment puisque d'après les enquêtes, environ 50% des
logements seulement possèdent un ou plusieurs types d'équipements
d'isolation (1). Mais dans de nombreuses branches, les économies
d'énergie sont limitées aux rapports du poids des investissements
à consentir vis-à-vis des gains réalisés sous forme de diminution
des coûts des combustibles. Par ailleurs, dans le bâtiment encore,
se pose le problème du confinement des polluants intérieurs
lorsque l'isolation est trop poussée.

- En ce qui concerne le gaz naturel : l'extension de son utilisation se heurte à des problèmes de disponibilité future puisque
les experts prévoient un déficit de l'offre de ce combustible
vis-à-vis de la demande à la fin de ce siècle.

- En ce qui concerne le nucléaire : cette filière, qui ne produit
pas de CO_2, est vouée à la production d'électricité de base.
Elle n'est pas encore adaptée pour répondre demandes de pointe.
La part du nucléaire dans la production d'électricité reste donc
encore limitée et doit être complétée par le recours aux énergies
fossiles comme c'est le cas en Belgique. Il en résulte que, compte
tenu des pertes de transformation et de transport d'électricité,
le coefficient d'émission par kJe utile reste encore élevé en
1985 : ≈ 94 g CO_2/MJe (2). Cette valeur, corrigée en fonction
des écarts de rendement des appareils de chauffage, se rapproche
du coefficient d'émission des combustibles pétroliers (tableau
2). En conclusion, seule une extension de la part du nucléaire
toujours possible dans une certaine mesure et d'autres énergies
physiques dans la production d'électricité peut rendre cette
forme d'énergie plus attractive sur le plan des émissions de
CO_2; à moins de se tourner vers une option technologique, la
décarbonation des fumées des centrales thermiques classiques.

2. Technologie de réduction des rejets de CO_2 - la décarbonatation des fumées de deux types de centrales électriques

Le CO_2 est un gaz bien connu des milieux industriels et ce, pour
diverses raisons, notamment :
- lorsqu'il constitue une impureté qu'il faut éliminer : c'est le
cas dans les unités de production d'ammoniac, d'hydrogène et de
gaz de synthèse;
- lorsqu'il est destiné à certaines applications : usages industriels
(synthèse chimique, réfrigération, extraction du pétrole, ...) ou
alimentaires (gazéification des boissons, solvant d'extraction,
...).

---

1) CH. JAUMOTTE, Marketing de l'Utilisation Rationnelle de l'Energie -
Programme National R&D Energie (1987).

2) Alors qu'en 1975 : 186 g CO_2/MJe.

413

Différents procédés physiques et chimiques ont donc été mis au point pour extraire et concentrer ce gaz. Leur emploi varie suivant les caractéristiques du gaz à traiter, des spécifications portant sur le produit, etc.

Dans le cas particulier de la fixation de CO_2 contenu dans les fumées de combustion : plusieurs procédés sont mentionnés :
- absorption-stripping au moyen d'alkalonamines;
- absorption-stripping avec du carbonate de potassium;
- tamis moléculaires;
- cryogénation;
- etc....

La plupart de ces procédés (1) ont des rendements élevés : $\approx$ 90%. Mais les coûts énergétiques et en capital qui leur sont associés sont également élevés. Sur ce dernier plan, les études et l'expérience démontrent la supériorité des procédés utilisant des solvants à base d'alkalonamines.
De ce fait, un de ces procédés est proposé ici pour décarbonater les fumées de deux types de centales électriques.
Schématiquement, les équipements d'une unité de décarbonatation des fumées à l'aide d'alkalonamines comprennent :
- une tour d'absorption dans laquelle les fumées sont mises en contact avec un solvant à base d'alkalonamines qui fixent le CO_2;
- une tour de régénération dans laquelle le solvant, enrichi de CO_2, est distillé par entraînement à la vapeur; un séparateur installé en aval de cette tour isole le CO_2 du solvant qui, appauvri, est alors recyclé.

Le CO_2 peut ensuite être liquéfié dans une unité de compression et transporté vers un lieu de stockage (océan, mine de sels, gisement pétrolier épuisé, ...).

Le tableau 2 reprend les exigences énergétiques et les coûts (2) (3) associés à l'implantation d'une unité de décarbonatation des fumées sur deux types de centrales électriques qui seront programmées dans le futur en Belgique. Sur le plan énergétique, on peut constater que la perte de rendement des deux centrales reste relativement modérée, respectivement 6,9% et 5,8%. C'est parce que le procédé utilise de la vapeur basse pression soutirée aux turbines de ces centales pour la régénération du solvant.

---

1) J.A. LAGAS - Selective Chemical Absorption - Comprimo - Amsterdam 1985.

2) H.C. CHENG, M. STEINBERG - A Study on Systematic Control of CO_2 Emissions from Fossil-Fuel Pwer Plants in the U.S., Environmental Progress - Vol. 5, n° 4, pp. 245-255 (1986).

3) Dow Chemical - Données technico-économiques.

Tableau 2 : <u>Caractéristiques techniques des centrales de référence et coûts de la décarbonatation des fumées (fin 1986)</u>

Centrale (puissance)	Charbon (1500 MW$_{th}$)	Gaz naturel (625 MW$_{th}$)
Type de générateur	Chaudière à charbon pulvérisé - cendres sèches	Turbines à gaz + turbines à vapeur
Facteur d'utilisation (heures/an)	5.000	5.000
Rendement (%) . Avant décarbonatation . Après décarbonatation	39,6 32,7	48,0 42,7
Emissions de CO_2 (1000 t/an) . Avant décarbonatation . Après décarbonatation	2.860 286	660 66
Investissement (1) (millions $) . Décarbonatation (2) . Compression-liquéfaction (3) . transport-stockage (4)	242,5 55 88,5 99	97,5 45,5 27,5 24,5
Coûts annuels totaux dont : (millions $/an) . Capital (5) . Energie (6) . Solvant	87,5 50 34,5 3	28 20 7 1
Coûts par kWhe ($/kWhe)	0,035	0,021

1) En dollars fin 1985. Localisation Belgique - Centrale neuve.
2) Procédé DOW FT-1 sur cycle combiné, efficacité 90%.
3) Compresseur 4 étapes (140 kg/cm^2, 25°C).
4) Pipeline (6 inches jusqu'à un centre de collecte, puis 36 inches jusqu'au stockage dans l'océan.
5) Intérêt 8,6%; dépréciation 5%, taxes 3%, maintenance 3%.
6) Charbon : 2,12 $/GJ; gaz nat. : 2,73 $/GJ.

Sur certains plans, la décarbonatation des fumées souffre, toutefois, de sérieux handicaps, notamment :

i)   l'ampleur des coûts à consentir est importante. Elle varie suivant le type de centrale et de facteurs spécifiques : teneur en carbone du combustible, excès d'air, etc.... Au total, il ressort, qu'il faut consentir à une élévation de 0,035 $/kWhe pour la centrale au charbon et de 0,021 $/kWhe pour la centrale au gaz (turbine) ce qui correspond à une élévation de plus de 50% du prix de revient du kWhe fourni par ces centrales. De tels coûts ont trait à des centrales neuves et devraient être encore plus importants dans le cas de centrales existantes.
Il existe des possibilités de réduire les coûts : effectuer une décarbonatation moins sévère et pratiquer des politiques d'échanges d'émissions. Mais ceci doit faire l'objet d'études étendues.

ii)   le stockage du CO_2 implique des volumes importants. Dans l'évaluation présente, la solution se fonde sur l'hypothèse d'une participation de plusieurs centrales, et d'un rejet en mer à une profondeur de quelques centaines de mètres afin de bénéficier de l'effet de ralentissement de la diffusion de ce gaz vers la surface. Une telle pratique n'est pas sans soulever des inquiétudes quant à ses effets sur l'environnement marin. Une solution plus sûre pour le stockage du CO_2 consiste à l'injecter dans certaines mines désaffectées ou des gisements d'hydrocarbures épuisés. Mais pour mesurer les potentialités de cette dernière possiblité, un inventaire approfondi des lieux de stockage et de leur volume doit être dressé.

Malgré ces handicaps, il faut rappeler que la technique de décarbonatation des fumées est efficace, elle permet d'abaisser le coefficient d'émission de 950 g CO_2/kWhe à 115 g CO_2/kWhe (centrale au charbon) et de 440 g CO_2/kWhe à 49 g CO_2/kWhe (centrale au gaz). Cette simple constatation mérite qu'on prête à cette technique une attention particulière.

3. Conclusions

L'option stratégie énergétique et l'option technologique offrent des potentialités en tant que stratégies de contrôle des émissions de CO_2. Leur analyse montre qu'elles revêtent deux formes : un abaissement des quantités de CO_2 émises par unité d'énergie (coefficient d'émission) ou une réduction de la consommation énergétique.

(i)   Abaissement du coefficient d'émission de CO_2 par unité énergétique

Celui-ci, exprimé par MJ (final) consommé, est passé de 88 g CO_2/MJ à 78 gCO_2/MJ entre 1975 et 1985, malgré un accroissement de la part du charbon et une diminution de la part du gaz naturel dans la consommation énergétique. Principal responsable de ce résultat, le recours à l'énergie nucléaire dans le production d'électricité ce qui a permis au coefficient d'émission de CO_2 par kWhe produit de passer de 617 g CO_2/kWhe à 311 g CO_2/kWhe soit, par MJ électrique consommé, de 186 g CO_2/MJe à 94 g CO_2/MJe.

En 1975, le programme nucléaire a ainsi permis d'éviter un rejet de 5 millions de tonnes de CO_2; en 1985, ce chiffre est passé à 25 millions de tonnes de CO_2.

Quoique limité sur le plan de la production d'électricité (modulation de la demande), le programme nucléaire pourrait encore être élargi. Il a cependant été gelé il y a peu.

Moins efficace que le nucléaire, l'extension de l'utilisation de gaz naturel constitue un autre choix pour abaisser les émissions de CO_2, étant donné les coefficients d'émission relativement peu élevés qui lui sont spécifiques (51-56 g CO_2/MJ).

Cette solution a été globalement peu suivie entre 1975 et 1985.

(ii) <u>Réduction de la consommation énergétique</u>

Après le second choc pétrolier de 1979, la consommation énergétique a diminué en Belgique de 20% au total jusqu'en 1983 et ce grâce aux économies d'énergie réalisées dans le bâtiment (isolation, amélioration du rendement des chauffages, changement de comportement) et dans l'industrie (accroissement de l'efficacité énergétique, fermeture d'entreprises à haute intensité énergétique). La récession qui a frappé les entreprises sidérurgiques et du bâtiment a également pesé dans cette réduction de la consommation énergétique. Ces deux facteurs ont donné un sérieux coup de frein à l'élévation des émissions de CO_2 qui s'était poursuivie jusqu'en 1979.

Au total, les émissions de CO_2 ont diminué de 15 millions de tonnes entre 1975 et 1985 en Belgique. Avec le gel du programme nucléaire décidé il y a peu, le programme de conversion au charbon des centrales au fuel ainsi que les prix réduits des combustibles ce qui n'incite pas aux économies d'énergie, les émissions de CO_2 devraient s'accroître à l'avenir.

Pourtant, l'extension du programme nucléaire, qui a porté ses fruits, devrait encore permettre de réduire les rejets de CO_2. Il en est de même des économies d'énergie; dans le bâtiment notamment, des possibilités existent, puisque 50% des bâtiments seulement diposent de un ou plusieurs équipements d'isolation.

A côté de ces possibilités qui font partie de l'option <u>stratégie énergétique</u>, il y a l'<u>option technologique</u> qui consiste à décarbonater les fumées des centrales thermiques. Malgré les coûts assez élevés et des problèmes de stockage de CO_2 qui lui sont associés, cette méthode se révèle très efficace pour réduire les émissions de CO_2 puisqu'elle permet d'abaisser le coefficient d'émission de CO_2 par kWhe produit de 950 g CO_2/kWhe à 115 g CO_2/kWhe (centrale au charbon) et de 440 g CO_2/kWhe à 49 g CO_2/kWhe (centrale au gaz).

Toutes ces options examinées dans ce texte présentent donc à titres divers de réelles potentialités. Il serait utile, dans les perspectives d'un contrôle accru des rejets de CO_2, de mesurer l'impact de leur développement futur tant sur le plan des émissions de CO_2 que sur le plan socio-économique et ce éventuellement assorti du concours d'autres options : hydrogène, énergies solaire et éolienne, ...

PART C

TECHNOLOGY SCENARIOS FOR THE NEXT CENTURY

TECHNOLOGICAL OPTIONS AND POLICY STRATEGIES TO REDUCE THE RISK OF RAPID GLOBAL WARMING

Irving M. Mintzer
World Resources Institute
Washington, D.C.

DEVELOPING A GENERAL STRATEGY TO REDUCE THE RISKS

Global warming from the greenhouse effect presents the economic and political institutions of human societies with a profound challenge. Present levels of carbon dioxide and other greenhouse gases together are absorbing 50 percent more thermal radiation from the earth's surface than did pre-industrial concentrations of CO_2 alone.[1] Increasing concentrations of CO_2, methane, N_2O, CFCs or other greenhouse gases have a direct effect on the radiation balance of the atmosphere. Because our planet is a complex, coupled, non-linear, fluid-dynamic system, many feedback loops affect the ultimate temperature response to this change in the radiation balance. Some feedbacks, like the water-vapor feedback loop and the ice-albedo feedback, are likely to amplify the warming due to a greenhouse gas buildup. Other, negative feedbacks, including those which increase the overall reflectivity of clouds or of the Earth's surface, are likely to counteract somewhat the warming effect of a greenhouse gas buildup. (For a more complete discussion of feedback processes, see Dickinson, 1986 and Lashof, 1988.)[2] The lower values in the reported range (i.e., 3° F) assume that counteracting negative feedbacks will reduce the effect of the trace gas buildup. The upper bounds (i.e. 9° F) assume that positive feedbacks dominate the process. Nonetheless, despite the uncertainties about the magnitude of the temperature change, the direction of future change is clear and unambiguous. Scientists believe that even at the lower end of the range, today's activities are propelling the planet over a period of a few decades toward a climate that will be warmer than any experienced at least since the last glaciation. Unless actions are initiated soon to slow the rate of greenhouse gas buildup, national economies and natural ecosystems are likely to suffer severe consequences from this rapid alteration of the global climate.

Because the greenhouse problem has been analyzed principally within the scientific community, the debate has been framed largely in terms of the greenhouse gases themselves. While this approach is chemically correct, organizing the analysis gas by gas obscures the human activities that are the source of the problem. Rather than focus exclusively on the chemistry, policy analysis can be more profitably organized in terms of the relevant economic and political divisions of human activity. The complex linkages among global warming, ozone depletion, and ground-level air pollution suggest that the primary policy areas that need attention are energy, agriculture, forestry and industrial development.

Table 1 provides a perspective for qualitatively linking scientific and policy analysis. Each element in this science-to-policy translation matrix represents an estimate of the percentage of total future warming commitment from a greenhouse gas produced in the indicated policy or investment sector. Numbers in italics are rough estimates on the part of the authors, and indicate the need for additional scientific research to establish an improved basis for future policy choices.

The numbers around the periphery of the array are estimates of the total contribution to warming by the economic sector in that row or the greenhouse gas in that column. The arrows associated with "natural feedbacks" indicate that particular greenhouse gases increase by unknown amounts in natural and managed ecosystems as warming occurs. Even with the substantial uncertainties, this double formulation provides us with a much clearer sense of where the policy leverage is likely to be the greatest.

It is clear from this framework that the greatest potential for preventing warming lies in the energy and industrial sectors, but that nearly one quarter of the potential warming arises from the biological sectors of forestry and agriculture. However, each sector must be examined individually country by country in order to determine which strategies are most readily implemented and cost effective. Unfortunately, it may be impossible to conduct a complete cost-benefit analysis in the context of a single country since costs accrued locally may provide only distant benefits, separated from the costs by thousands of miles and, perhaps, by the passage of several decades.

The problem of global warming is intimately and inextricably linked to the problems of stratospheric ozone depletion and the problems of tropospheric chemistry, both oxidant buildup and acid deposition, that contribute to local and regional air pollution. The three problems are linked economically because it is often the same activities which release the pollutants that cause all three problems. They are linked chemically because, once released, the pollutants interact in complex and synergistic ways in the atmosphere. Finally, they are linked at the policy level, because policies designed to mitigate or reduce the effects of any one problem will unavoidably affect the timing and severity of the others. The implications for the economies and social institutions of industrialized nations, newly industrializing states and developing countries must be considered together in seeking strategies and policies to reduce future rates of emissions growth. The linkage between the greenhouse effect and other atmospheric problems require a systems approach in order to reduce the combined risks to human economies and natural ecosystems. This complex web of interconnections suggests that successful policy strategies will most likely be those that have multiple economic, social, and environmental benefits, and which are cost effective and minimally disruptive to the present organization of human societies.

The seriousness of the consequences of rapid global warming, the momentum of investments in the industrial system and of chemical reactions in the atmosphere, the essential irreversibility of the trend toward future warming and the very real possibility of accelerating positive feedbacks all argue for the urgent need to slow the rate of greenhouse gas buildup. The fact that the planet will be committed to substantial future warming well before such a warming is directly observable also supports the need to develop appropriate strategies sooner rather than later.

There are several general strategies for reducing the rate of greenhouse gas buildup that might be considered. One strategy involves reducing the rate of greenhouse gas emissions. The second involves removing the effluents after they are produced. Included in the first category are policy options to (1) increase the efficiency of energy supply and use, (2) shift the mix of fossil fuels to less carbon-intensive alternatives, (3) introduce smokeless technologies such as solar and other renewables and if it can meet appropriate criteria, nuclear, (4) limit deforestation and (5) phase out the production and use of the most dangerous CFCs. In the second category are policy options to remove greenhouse gases from the atmosphere; these include (1) scrubbing of CO_2 from stack gases, (2) reforestation efforts and (3) measures to enhance the natural processes for removing greenhouse gases already in the atmosphere.

Implementing such strategies will require identifying both long-term goals and near-term measures that shift the pattern of future economic development to a safer trajectory.

CONCLUSIONS FROM PREVIOUS STUDIES

Several important conclusions on what can be done to slow the rate of future global warming emerge from a review of previous studies. The first and most important conclusion is that emissions of a number of other trace gases will add to the future warming effect of any further buildup in the atmospheric concentration of CO_2. The second conclusion is that future rates of greenhouse gas emissions cannot now be predicted with certainty. Future emissions rates will be determined by the emerging pattern of industrial and agricultural activities as well as by the effects of feedback processes in the earth's biogeophysical system whose details are not fully understood at the present time. Third, it is too late to prevent all future global warming. Finally, and most important, policy choices and investment decisions made during the next decade that are designed to increase the efficiency of energy use and shift the fuel mix could slow the rate of buildup sufficiently to avoid the most catastrophic potential impacts of rapid climate change. Alternatively, decisions to expand rapidly the use of coal, extend the use of the most dangerous CFCs, and rapidly destroy the remaining tropical forests will advance the timing of global warming and increase the risk of disruptive and sudden climate changes.

Many uncertainties persist about the regional implications of global warming. Nonetheless, the stakes are incomparably high -- a warming of just a few degrees could seriously damage the economies of both developing and industrialized countries and could disrupt many natural ecosystems. The dimensions of the policy choices are straightforward: national governments can make policy decisions that slow the rate of increase in the risks of global warming and adapt to those climatic changes which are now unavoidable, or they can passively observe the continuing buildup and hope that some unforeseen scientific breakthrough will mitigate the impacts of future climate change.

REDUCING THE RATES OF GREENHOUSE GAS EMISSIONS

One important approach to forestalling rapid climate change involves reducing greenhouse gas emissions. Such strategies will involve policies to reduce emissions of CO_2, methane, N_2O and ozone precursors from the energy sector as well as approaches to limit biotic emissions and releases of other greenhouse gases (including CFCs and halons) from industrial activities.

ENERGY USE: THE KEY TO CONTROLLING ANTHROPOGENIC EMISSIONS OF CO_2

The global pattern of energy supply and use will substantially determine the rate of future emissions of CO_2. It will also affect the rate of increase in the atmospheric concentration of methane, nitrous oxide and tropospheric ozone. The rate of growth in emissions of all these gases will be affected by choices made in the next decade that reflect the relative prices of various fuels.

Annual emissions of CO_2 from the combustion of fossil fuels were approximately 5.6 billion metric tons of carbon in 1987. Emissions of CO_2 from the biota add another 1 to 3 billion metric tons of carbon to the atmosphere each year.[3] A large fraction of these biotic emissions result from the burning of biomass for energy and the destruction of tropical forests, often to clear land for agriculture. As a result of the emissions from all these sources, the atmospheric concentration of CO_2 is increasing by nearly 0.5 percent per year. (See Table 2.)

Not all fossil fuels contribute equally to these emissions. On a global basis, coal and other solid fuels contributed approximately 44 percent of total CO_2 emissions from burning fossil fuels in 1985. Combustion of liquid petroleum products contributed an additional 40 percent, and natural gas added about 15 percent.

Since 1973, the amount of primary energy supplied globally by natural gas and coal have increased by about 36 percent. Oil use in 1986 is back to approximately the same global consumption level as in 1973. Over the same period, the amount of gas vented and flared has fallen by 52 percent and remained stable over the last several years. Global CO_2 emissions from fossil fuels follow these patterns. (See Figure 1.) CO_2 emissions from the combustion of coal and natural gas have increased by about 2.4 percent annually, while those from burning petroleum have increased by less than 0.2 percent per year. For commercial fossil fuels, emissions declined slightly from 1979 to 1983 and have increased rapidly (i.e. an annual rate in excess of 2%) during the last 5 years. Since 1976, overall CO_2 emissions from fossil fuels have increased by about 17 percent, implying an average growth rate of 1.2 percent per year.

All nations have not contributed equally to global CO2 emissions. The United States, which currently consumes about 30 percent of the global fossil fuels used each year, produced about 25 percent of the fossil-fuel-derived CO_2 in 1985. The Soviet Union contributed about 20 percent of fossil-fuel-derived emissions in 1985, Western Europe about 15 percent, China about 10 percent, and Japan about 6 percent. In 1985, developing countries (excluding China) added about 15 percent.[4]

During the past 25 years, the historical balance has shifted. (See Figure 2.) The most dramatic and important shift is in the role of developing countries. In 1960, these countries contributed less than 8 percent of global CO_2 emissions from fossil fuel use while the United States produced almost 33 percent. At that time, the share emanating from Western Europe was 21 percent while the Japanese share of global emissions stood at 3 percent.

As the levels of population and energy use increase in developing countries during the next 25 years, their share of global emissions is likely to grow. The rate of increase in emissions will be determined largely by international oil prices, development trends, and by the national energy strategies which are chosen by these countries.

Deforestation and other land use changes have also contributed in recent years to the global emissions of CO_2. Most analysts now agree that the terrestrial biota are adding about 1 to 3 Gt of carbon as CO_2 to the atmosphere each year. Some think that the current rate may be even higher.

The terrestrial biota contribute to the atmospheric buildup of CO_2 when the amount of stored carbon that is burned for fuel or allowed to decay in a given year exceeds the amount of carbon dioxide converted to plant matter by photosynthesis. If the quantity of material harvested and burned each year is limited to the amount of new biomass grown during the same period, CO_2 is cycled through the atmosphere but no net change in concentration occurs. Excluding fossil fuel used to plant, fertilize and harvest biomass, fuel derived solely from the annual yield of agricultural crops or tree plantations does not add to the atmospheric burden of CO_2.

Another source of CO_2 emissions is soil erosion. Over 2 billion tons of soil is lost to erosion each year in Africa alone.[5] Erosion releases large quantities of soil carbon into the atmosphere. Once released, this carbon is readily oxidized to CO_2. By reducing the rates of soil erosion through increased tree planting, improved forest and watershed management, and protection of fragile and marginal areas from uncontrolled development, a significant source of

426

emissions growth would be controlled and carbon fixation by soil bacteria might even be increased.

STRATEGIES TO REDUCE EMISSIONS OF N_2O, CH_4, AND CFCS

The sources of nitrous oxide are relatively poorly characterized. Some authors suggest that "it is emitted almost totally by natural sources, principally by bacterial action in the soil and by reactions in the upper atmosphere." [6] Certainly the biological component is significant although little is known about the factors in the denitrification process that favor the release of nitrous oxide instead of molecular nitrogen. Some have suggested that the use of excess inorganic nitrogen fertilizer in agriculture may contribute to the bacterial production of nitrous oxide. If this is confirmed by future research, then more careful metering of nitrogen fertilizer (particularly that based on anhydrous ammonia) might reduce the release of this gas. It is difficult to envision direct trapping of the biological sources of nitrous oxide in the field.

A significant part of the increase in nitrous oxide concentration is due to fossil fuel combustion. The burning of coal and residual fuel oil contributes the most. Combustion of natural gas and other petroleum products adds a smaller and uncertain amount.

All fossil fuel combustion releases additional oxides of nitrogen (NO_x) that contribute to the formation of tropospheric ozone. In addition, releases of carbon monoxide (CO) from the combustion of biomass and fossil fuels increase the rate of growth in methane concentration by competing for methane's chemically reactive sink.

It is now possible to reduce emissions of methane from certain anthropogenic sources or to interrupt their passage before they reach the open atmosphere. A considerable amount of methane is released by bacterial action on moist organic garbage in landfills and on manure piles in centralized feedlot facilities and from sewage treatment plants.[7] It may be possible to enclose these facilities or otherwise to capture and store the methane emissions as a commercial-quality fuel. (Such a system might be capable of capturing the methane emitted by the cows in a feedlot, as well as the methane generated within the manure piles.) Similar approaches might be feasible for capturing methane which is now emitted from coal seams and partially depleted oil fields. If markets for the fuel can be made available within short distances of the site of methane production, no substantial investment in compression and pumping would be necessary. No estimates are presently available concerning the costs of these measures or their precise effect on the rate of future increase in methane concentrations.

A number of options exist for rapidly reducing the most widely used CFCs to the level specified under the Montreal Protocol. Rapid reductions in emissions can be achieved by improved management, better maintenance practices, and recycling. In the area of refrigeration and air conditioning systems, for example, it would be simple and cheap to require installation of service valves in the cooling loops of most devices so that the refrigerant could be removed and captured either during routine servicing or at the point of disposal. In the case of auto air conditioners, higher-quality hoses and fittings would reduce leakage during routine operations. Installation of service valves and requirements for recapture and recycling of the CFCs during regular maintenance would substantially reduce emissions from this application.

In the process of manufacturing foam products, substantial potential exists for recapture of the blowing agents. Enclosing process lines, ageing the product before shipment, and carefully controlling spray heads can substantially reduce emissions in manufacturing and may allow recapture of the dangerous gases.

A number of recent advances have occurred in the use of CFC-based solvents in the electronics industry. Hewlett-Packard Corporation, among others, has been highly successful in minimizing solvent losses by enclosing their production lines and passing the captured gases through a bed of adsorbent carbon. Similar opportunities exist throughout the electronics industry in solvent applications.

The ultimate solution to the CFC contribution to global warming is to phase these chemicals out completely. New findings about the seriousness of their threat to stratospheric ozone, have lead to calls for their complete elimination by the year 2000.

SEARCHING FOR A STABLE CLIMATE

In an attempt to identify combinations of policy strategies that can reduce the rate of growth in warming commitment to zero, The World Resources Institute (WRI) has been evaluating a number of scenarios using the WRI Model of Warming Commitment (MWC).[8] One of these is a "Stabilization Scenario" that, following a 70-year transition period (1980-2050), does not commit the planet to any additional future warming. The elements of this Stabilization Scenario include various policy measures and technological developments to improve dramatically the efficiency of energy supply and use; to eliminate completely, and by binding international agreements, the use of CFCs; to replace most fossil fuel uses gradually by geothermal, solar and other renewables, and nuclear supply options; to stop deforestation and increase global reforestation efforts; and to slow the rate of population growth. In addition, three less-radical scenarios, a "Current Trends Scenario", a "Low Population Scenario" and a "Partial Stabilization Scenario" were also evaluated using the WRI model.

The principal criteria for comparing such scenarios with respect to the global climate problem are their effects on the emissions of greenhouse gases and the resulting impacts on the timing and extent of commitment to future warming. The timing of future warming is evaluated in terms of the date at which the atmosphere would be committed, from the combined effects of all greenhouse gas emissions, to a warming of 3-9° F. The magnitude of future warming is measured by the amount of the warming commitment at the end of the simulation period.[*] The complex feedbacks that occur as a result of interactions among the atmosphere, ocean, and terrestrial biota are incorporated as factors that amplify or suppress the direct effects of a greenhouse gas buildup and are expressed as a range of temperature. The details of the model's structure and operations are summarized in the report <u>A Matter of Degrees: The Potential to Control the Greenhouse Effect</u>.[9]

These scenarios use two different projections of future global population increase. The Base Case, Current Trends and Partial Stabilization scenarios use the UN/World Bank mid-range population forecast in which global population stabilizes at about 10.5 billion souls in about 2075.[10] The Low Population and Stabilization Case use the UN/World Bank low population estimate in which world population grows more slowly and stabilizes at about 8 billion souls in 2070. (See Table 2.)

The five scenarios share similar rates of GNP growth for industrialized countries but incorporate different rates of income growth for developing countries. In all five scenarios, for the period 1975-2025, GNP growth in the industrialized world occurs at an average annual rate of 2.4-2.5% For the developing countries, income growth during this period ranges from an average annual rate of 4% in the Base Case to 5.5% in the Stabilization Case (See Table 2.).

Commitment to future warming is used here as a measure of impact (rather that the observable or realized warming) because it is the date at which the atmosphere is irreversibly committed to a future equilibrium temperature increase.

The reference case is the "Business-As-Usual" Base Case developed for A Matter of Degrees. In this scenario, current trends continue over the next century with no policy interventions designed to slow the rate of greenhouse gas buildup. CFC use grows steadily in this scenario while tropical deforestation continues at about the current rate. No systematic effort is made to increase the efficiency of energy use or accelerate the commercialization of solar technologies. Nuclear fission power makes a small and slowly growing contribution to electricity supply.

The first new case, a "Current Trends" scenario, assumes that the energy growth of the last decade (approximately 2% per year) continues through the first quarter of the next century. In addition, this scenario assumes continued increases in the rate of tropical deforestation (but somewhat more slowly than has occurred during the last two decades.) This scenario also assumes that all countries sign, ratify and implement the Montreal Protocol on Substances that Deplete the Ozone Layer. (It does not assume any further CFC reductions beyond the 50% cuts for industrialized countries or the 0.3 kg per capita limits for low consuming countries.)

The second test case, a Low Population scenario, incorporates a smaller global population (8 billion instead of 10.5 billion people in 2075), with the same technology, policies, end-use patterns and per capita rates of economic growth as the Current Trends case.

The third test case, a Partial Stabilization Scenario, incorporates the same population as the Base Case but changes many other assumptions. This case follows the general pattern of a low-energy future described by Goldemberg et al. in the WRI Research Report Energy for a Sustainable World (Goldemberg, Johansson, Reddy, and Williams, 1987). It also assumes that energy efficiency improves dramatically in both industrialized and developing countries. For the industrialized countries, energy use per constant dollar of GNP declines at an annual rate of about 1.7 percent over 40 years. For developing countries, this measure of the economic efficiency of energy use declines at an annual rate of approximately 3.25%.

The Partial Stabilization scenario also assumes that a carbon tax is applied to the combustion of fossil fuels, increasing their price at a rate proportional to the amount of carbon dioxide released per unit of energy supplied. Solar energy and renewable development accelerates in this scenario and reaches a constant dollar cost equal to about $0.06/kWh-- equivalent to the fuel cost of electricity generated in an oil-fired steam turbine burning $33 per barrel oil. Implementation of the Tropical Forest Action Plan and new, off-site planting of trees to mitigate CO_2 emissions from fossil fuel sources is also assumed to reduce the biotic contribution to CO_2 releases to about 0.5 billion tons of carbon by 2010 and to less than 0.01 billion tons of carbon in 2075.

In addition, the Partial Stabilization scenario assumes that the production and use of the most dangerous chlorofluorocarbons (CFC)s is controlled initially by the terms of the Montreal Protocol on Substances that Deplete the Ozone Layer (UNEP, 1988). This scenario assumes that the Protocol is implemented by all nations and that there is full compliance with its terms. It assumes that significant additional steps to control ozone-depleting chemicals beyond the terms of the current agreement are implemented in the early part of the 21st century. These include a 95 percent phaseout of CFCs by industrialized countries in 2010 and a staged reduction by developing countries, ultimately reducing their per capita consumption to 0.05 kg per person per year in 2075.

The final policy scenario, the Stabilization Scenario, assumes that an aggressive effort is made now to reduce greenhouse gas emissions and that this effort is sustained through the year 2075. This scenario assumes that the efficiency-improving trends identified in the Partial Stabilization scenario continue to accelerate during the last 50 years of the period. It assumes that about 50 percent higher taxes are imposed on conventional and on synthetic fossil fuels and that solar and other renewable energy costs fall to the equivalent of $0.035/kWh -- a price approximately equal to the fuel cost of electricity generated by a conventional oil-fired steam plant burning oil at $19.50 per barrel oil in 2025 (in constant 1975 US$). By 2075 in this scenario, all coal is replaced as a boiler fuel with natural gas, biomass, hydropower, solar power or nuclear electricity. This reduces not only carbon dioxide emissions, but those of nitrous oxide, ozone and methane as well. This scenario further assumes that the Montreal Protocol is extended to require industrialized countries to reduce their use of CFCs to zero by 2020. By 2050, developing countries also reduce their CFC use to zero in this scenario. Deforestation is slowed, and forests and agricultural lands begin to be a net sink for carbon dioxide after 2025. By 2075, the biota is absorbing an additional 0.6 billion tons of carbon per year, approximately equal to the total CO_2 emissions from global fossil fuel usage. Regional population growth rates are held to the levels implied by the Low Population case, limiting global population to about 8 billion people in 2075.

The five scenarios reviewed here have substantially different ramifications for the pace of future global warming. Only the Stabilization Scenario ,however, produces a constant level of future greenhouse warming. All of the others imply an increasingly warmer world throughout the simulation period. Assuming that historical emissions will commit the planet to a warming of 2 to 5° F, with the exception of the Stabilization Scenario, all of the scenarios reviewed here will commit the planet before 2030 to a warming that exceeds the effect of a doubled-CO2 atmosphere.

The best recent estimates suggest that emissions occurring prior to 1985 have already committed the planet to a warming of 2 to 5 degrees F.[b] Assuming that this warming is already "in the pipeline" to the projected future warming commitment, the Base Case and Current Trends scenarios commit the planet to a total warming of 3 to 9 degrees F relative to the pre-industrial atmosphere (the doubled CO_2 effect) before 2020. By 2075, the planet is committed to an average surface warming of about 6 to 20 degrees F for the Base Case and about 8 to 24° F for the Current Trends case. (At the upper end of these temperature ranges, the parameterized, one-dimensional, radiative-convective model used in the MWC may no longer be highly accurate.)

A comparison of the remaining three scenarios is revealing. Halting population growth at 8 billion in the Low Population case postpones the doubled CO_2 effect by less than a decade. The warming commitment in 2075 is about 6 to 20 degrees F. The main reason that the doubling effect is not further delayed by the lowered population growth rate is that most of the decline in population occurs in developing countries whose per capita emissions of CO_2 and other greenhouse gases remains low during the period between 1985 and 2025.

In the Partial Stabilization scenario, the implementation of the Montreal Protocol and the introduction of technology with higher levels of energy efficiency reduce the rate of growth in greenhouse gas emissions. In this scenario the timing of the doubled CO_2 concentration in 2075 (about 450 ppmv in the Partial Stabilization scenario versus approximately 600 ppmv in the Base Case), lower concentrations of CFCs, and a slower rate of growth in the concentration of methane. Although methane emissions are not modelled directly, the rate of growth in methane concentrations is assumed to be lower (0.5 percent per year in the Partial Stabilization scenario versus 1.0 percent per year in the Base Case) in order to reflect the assumed impact of

---

[b] This estimate is significantly higher than the one incorporated in the WRI Research Report, A Matter of Degrees: The Potential for Controlling the Greenhouse Effect (Mintzer, 1987). The conventional wisdom (and the estimate in Mintzer, 1987) suggests that the warming effect of emissions occurring between 1880 and 1980 is approximately 0.5-1.5o C. The estimates cited here reflect the higher recent estimates suggested in testimony in the U.S. Senate given by V. Ramanathan, one of the world's leading analysts of greenhouse gas warming. (Ramanathan, 1988).

reduced rates of carbon monoxide (CO) emissions from fossil fuel combustion and of methane emissions from biomass burning and from the fossil fuel supply cycles. In this scenario, as in the Stabilization Case, the slower rate of warming is likely to reduce some of the feedback processes, including the rate of release of methane from beneath the tundra in high latitude areas. In the quantitative assessments included in this report, no "credit" is given for these lower feedback rates.

The Stabilization Scenario postpones the doubled CO_2 effect to well beyond 2075. In this scenario, the commitment to warming of 2.8 to 7.2 degrees F above the pre-industrial level by 2075 is entirely due to gases already in the atmosphere in 1988 and those that will be released during the transition period from now until 2050. The transition contribution is only 1.0-2.2 degrees F, after which no net future contributions are made through the end of the study period. · Preventing additional warming is accomplished in part by the rapid introduction of solar energy and other non-carbon dioxide releasing energy production technologies and in part by the full phase-out of CFCs.

The future warming commitment of each of the four scenarios are plotted in Figure 3.

IDENTIFYING THE MOST EFFECTIVE POLICIES FOR REDUCING EMISSIONS FOR REDUCING EMISSIONS

The conclusions from this analysis are straightforward: policy choices and technological change can substantially affect the timing and severity of future global warming. Over the long term, energy policy choices that improve the efficiency of energy use and shift the fuel mix to less carbon-intensive fuels can substantially slow the rate of growth in CO_2 emissions. Clearly, the largest near-term reductions in commitments to future global warming can be achieved by rapidly phasing-out the use of the most dangerous CFCs. The relative contribution of the smokeless production technologies such as solar, other renewables and nuclear energy will depend on their prices and the ability of the nuclear industry to solve its problems and regain public and investor confidence. Finally, growth in regional economies in all parts of the world is not found to be inconsistent with efforts to reduce the risks of future global warming. Achieving these combined goals requires, however, shaping industrial societies in efficient new ways, slow the growth of human populations, and become better stewards of our natural and managed ecosystems.

Achieving the dramatic reductions in future emissions of CO_2 and other greenhouse gases that are reflected in the Stabilization Scenario will be a prodigious challenge. Reducing emissions from the industrialized countries during the next 30 years will be key to achieving long-term success. To accomplish this goal, energy efficiency must be improved dramatically.

To achieve the necessary reductions in the residential sector in this stabilization scenario, for example, the average efficiency of electricity use must increase by 55 percent in the United States and by 70 percent in the Soviet Union between now and 2025. But the largest and most important improvements have to come in the industrial and transportation sectors. The efficiency of electricity use in production processes must also increase in order to achieve the reductions needed in national energy demand. But achieving these improvements in efficiency will not alone produce a sufficient reduction in industrial energy demand. Among other things, the per capita demand for such basic materials as cement, paper, chemicals, non-ferrous metals and steel must also be reduced. Recycling of many of these materials can significantly reduce demand for energy in the production of new products.

A low energy future does not require any undiscovered technologies between now and 2025. It will require, however, that a number of steps be taken to improve the economic efficiency of energy markets. In addition to improving the flow of information between various actors in the energy market, it will be absolutely critical to "get the prices right" and start giving people rational price signals. This may mean the introduction of various new taxes and fees and the elimination of subsidies that distort market prices and encourage greater use of energy generally. Whether price setting that internalizes all of the environmental and social costs will be sufficient to bring about the required energy-intensity reductions is a subject that deserves more attention than this paper can provide.

A carbon tax, combined with consumption taxes on energy use, may be needed to begin the process of internalizing the full cost of energy supply into the price of commercial fuels. In addition, the assumed rapid decline in the price of solar energy and of natural gas-derived energy in the Stabilization Scenario probably would require substantial additional funding of development projects or (preferably) a corresponding reduction in the subsidies to conventional technologies. As a consequence of these measures, smokeless energy-production technologies (including solar, wind, other renewable energy systems, and nuclear fission) compete for the marginal investment dollar.

Beyond pricing measures, performance standards will probably be required to encourage the rapid introduction of certain high-efficiency technologies. Encouraging manufacturers and consumers to choose high efficiency cars will not require any new laws of physics but will require strong political leadership in the United States and in other industrialized countries. Achieving the desired improvements in electricity efficiency may require that performance standards similar to those now imposed on household appliances be mandated both for electricity supply and all electricity-using equipment.

The only way to directly alter carbon absorption and release is to intervene in the carbon cycle by directly managing parts of the global biosphere. In the Stabilization Scenario net biological emissions of are reduced CO_2 to zero by 2025 and make the biota a net sink after that through a combination of slowing deforestation, replanting of trees, and building up organic matter in agricultural and forest soils.

The degradation of forest and agricultural ecosystems is in large part tied to the pressures of rapid population growth. Large population numbers can also compound even modest per capita levels of greenhouse gas emissions as the scenario clearly demonstrate. It is interesting to note that at the levels of energy consumption per person envisioned for newly industrializing countries, lower population figures reduce the rate of deforestation and play a critical role in lowering biological sources of the greenhouse gases carbon dioxide, methane and nitrous oxide.

Clearly, there is no single solution to the complex challenge of the climate change problem. Virtually all aspects of society are involved in creating the greenhouse heat trap, and all must play a role in the solution. Astute public policies can doubtless mitigate some of the negative economic impacts of the more forceful initiatives included in the Stabilization Scenario summarized above. At the same time, the possibility cannot be excluded that some of the required capital investments and other changes necessary to reshape energy supply along more efficient and environmentally benign lines could involve some economic penalty.

The desirability of policies designed to eliminate subsidies, pricing distortions, and other market imperfections which act as deterrents to efficient energy utilization is not in question. Rectifying these defects, by definition, enhances economic welfare. It is in the introduction of additional policy interventions where the issue remains sufficiently unsettled as to merit further study of the tradeoffs between energy and environmental goals, on the one hand, and broad economic objectives, on the other.

REFERENCES

1. Wigley, TMC, Jones, PD and Kelly, P.M., "Empirical Climate Studies" in Bolin et al., The Greenhouse Effect, Climate Change and Ecosystems, 1987.

2. Dickinson, Robert, "How Much Will the Climate Warm and When?" in Bolin et al., eds., The Greenhouse Effect, Climate Change, and Ecosystems, 1986, op cit.

   Lashof, D., "The Dynamic Greenhouse: Feedback Processes that may Affect Future Concentrations of Atmospheric Trace Gases and Climatic Change, Climatic Change, Vol. 14, 1989.

3. World Resources Institute, World Resources Report: 1988-89.

4. Crutzen, P.J., "Tropospheric Ozone: An Overview" in Isaksen, I.S.A., Ed. Tropospheric Ozone, Reidel, Dordrecht, 1988.

5. World Resources Institute, World Resources Report 88-89, Op. Cit.

6. Seinfeld, 1988. Some recent evidence suggests that clear-cutting forest areas increase the release N_2O from soils.

7. Crutzen, Op. Cit., 1988.

8. Mintzer, I., 1987. A Matter of Degrees: The Potential for Controlling the Greenhouse Effect, Research Report no. 5, World Resources Institute, Washington, D.C.

9. Mintzer, 1987. Op. cit.

10. Vu, My T., World Population to 2100, World Bank, Washington, D.C. 1987.

Table 1.

Science to Policy Translation Matrix
for Global Warming from Greenhouse Gases
PERCENT OF GLOBAL WARMING

Gas / Sector	Carbon Dioxide	Methane	Ozone	Nitrous Oxide	Chlorofluorocarbons (and others)	Percent Warming by Sector
Energy Direct	35	_3_		_4_	---	49
Indirect		_1_	_6_			
Deforestation	_10_	_4_	---	---	---	14
Agriculture	_3_	_8_	---	_2_	---	13
Industry	2	---	_2_	---	20°	24°
Natural Feedbacks	?	?	?	?	---	?
Percent Warming by Gas	50	16	8	6	20°	100

Values are estimates of the percent of global warming between 1980 and 2030 arising from each gas and by policy sector, assuming current trends. Numbers italicized and underlined are very uncertain.

This assumes current trends for CFCs. If the Montreal Protocol is fully implemented, CFC contributions to warming would drop to less than half the projected value by 2030.

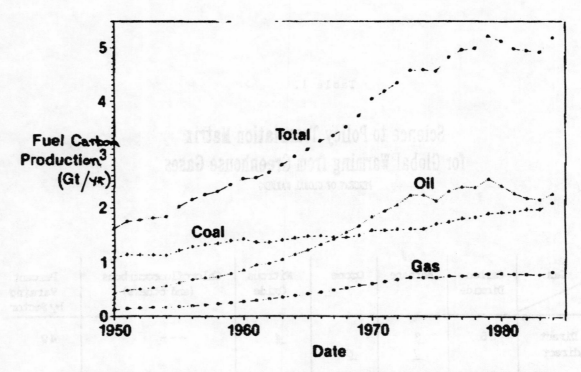

Global Fossil Fuel Emissions of Carbon Dioxide   Fig 1

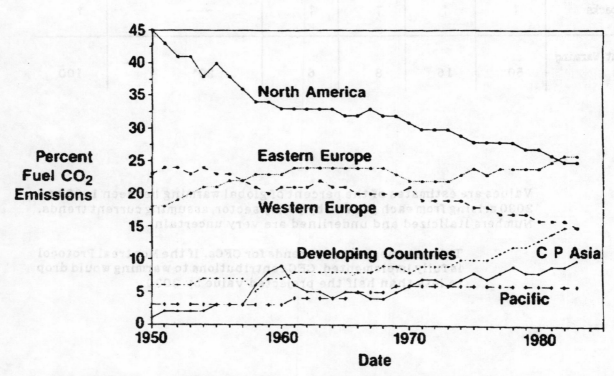

Regional Distribution of Carbon Dioxide Emissions   Fig 2

Figure 3.

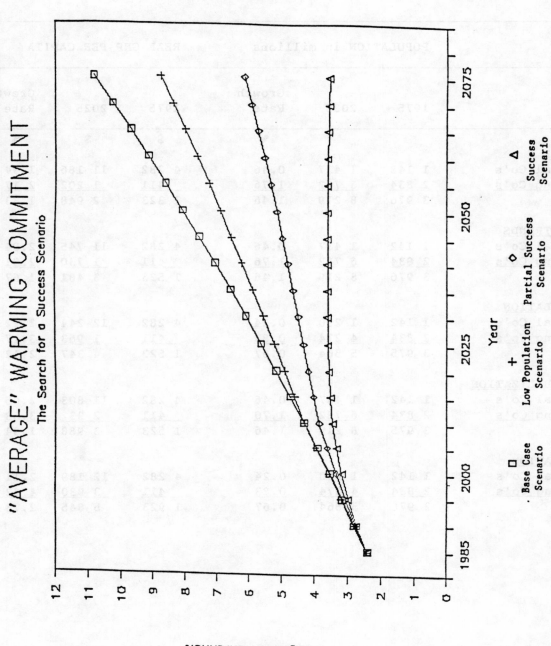

"AVERAGE" WARMING COMMITMENT
The Search for a Success Scenario

TABLE 2

COMPARISON OF POPULATION AND GNP IN THE NEW WRI SCENARIO

1975 – 2025

Region	POPULATION in millions			REAL GNP PER CAPITA		
	1975	2025	Growth Rate	1975	2025	Growth Rate
			%	$	$	%
BASE CASE						
Industrial Co's	1 142	1 437	0.46	4 282	11 186	1.94
Developing Co's	2 834	6 782	1.76	411	1 202	2.17
Total	3 976	8 219	1.46	1 323	2 948	1.33
CURRENT TRENDS						
Industrial Co's	1 142	1 437	0.46	4 282	11 745	2.04
Developing Co's	2 834	6 782	1.76	411	1 730	2.91
Total	3 976	8 219	1.46	1 523	3 481	1.67
LOW POPULATION						
Industrial Co's	1 142	1 290	0.24	4 282	12 244	2.12
Developing Co's	2 834	4 274	0.83	411	1 963	3.17
Total	3 975	5 564	0.67	1 523	4 347	2.12
PART STABILIZATION						
Industrial Co's	1 142	1 437	0.46	4 282	11 803	2.05
Developing Co's	2 834	6 782	1.70	411	2 332	3.53
Total	3 975	8 219	1.46	1 523	3 988	1.94
STABILIZATION						
Industrial Co's	1 142	1 290	0.24	4 282	12 189	2.11
Developing Co's	2 834	4 274	0.83	411	3 930	4.62
Total	3 976	5 564	0.67	1 923	5 845	2.73

THERMONUCLEAR FUSION :

A POTENTIAL ENERGY SOURCE FOR THE NEXT CENTURY

K. Steinmetz, Ch. Maisonnier, J. Darvas

Commission of the European Communities
200, rue de la Loi, B-1049 Brussels

INTRODUCTION

Whatever the scenarios on the growth of the world's energy consumption, new sources of energy will be needed in the next century to complement the present ones based essentially on fossil fuels and nuclear fission. Environmental considerations such as acid rain, greenhouse effect and nuclear safety introduce a substantial pressure to improve present energy systems and to develop new sources that offer the prospects of being more benign to the environment.

In the following a brief introduction on the principles and basic features of fusion will be presented, and the prospects on the expected moderate environmental impact of fusion, and the path towards fusion as a commercial energy source will briefly be outlined.

Nuclear Fusion

Fusion is the source of energy of the sun and of the stars, in which nuclear reactions take place at temperatures of about 15 million degrees Celsius. To make a fusion power source on earth, however, requires even more stringent conditions.

Nuclear energy is released when the nuclei of light elements fuse together to form heavier ones. The fusion reaction of greatest interest for

441

energy production, because it imposes the least stringent conditions, is that between the nuclei of the two heavy isotopes of hydrogen – deuterium and tritium (Table 1). Deuterium is very plentiful on earth. In contrast, tritium does not occur naturally and must be manufactured. This can be achieved by incorporating lithium into a (jacket called) blanket surrounding the reacting zone (see also Fig. 2). The neutrons absorbed in the blanket will react with lithium to form tritium and helium gas (breeding reaction).

Table 1 : Basic Nuclear Reactions

Nuclear Fusion :

$$D + T \longrightarrow {}^4He + n + 17.6 \text{ MeV}$$

Tritium Breeding :

$$^6Li + n \longrightarrow {}^4He + T + 4.8 \text{ MeV}$$
$$^7Li + n \longrightarrow {}^4He + T + n - 2.5 \text{ MeV}$$

Overall :

$$D + Li \longrightarrow 2\,{}^4He + \text{Energy}$$

Some of the neutrons will also find their way to react with the reactor structures producing unwanted radioactivity. However, the fusion reaction itself does not lead to radioactive products ; the induced radioactivity may be avoided, or at least reduced, by appropriate choice of the structural materials.

Other fusion reactions, such as D-^{3}He and D-D, would offer the advantage of producing much less energetic neutrons and could, therefore, increase substantially the capability to realise more fully the potential of fusion as a clean energy source. As the level of plasma performance required for a reactor is much more demanding, the realisation of D-^{3}He and D-D fusion at reactor level is one of the long-term goals of the fusion community.

442

Conditions for Fusion

Fusion reactions can only take place if the nuclei are brought in close proximity to one another, which is difficult to achieve because all nuclei carry a positive electric charge and therefore repel one another. By heating the gaseous fuels to very high temperatures (then called plasma), enough energy can be given to the nuclei to overcome the repulsive force and to fuse together. In the case of the deuterium-tritium reaction, temperatures in excess of 100 million degrees Celsius are required. In order to produce a net output of energy a minimum number of fuel particles must be heated to this high temperature and confined in thermal insulation for long enough periods from its material surroundings. Different approaches exist and are investigated to REALIZE nuclear FUSION : magnetic confinement, inertial confinement and muon-catalysed fusion.

The European Fusion Programme focusses its efforts to magnetic fusion in toroidal configurations, by far the most effective scheme today, where the plasma is confined and controlled by suitable magnetic fields. Figure 1 shows a schematic of a tokamak which is at present considered the most favourable system for the achievement of reactor conditions. Other toroidal configurations actively investigated are the Stellarator and the Reversed Field Pinch.

Fusion Reactors

A number of conceptual fusion reactor designs have been carried out. A schematic of a fusion reactor is shown in Fig. 2. The plasma is surrounded by a blanket which has two primary functions : heat extraction and tritium breeding. In slowing down the neutrons from the plasma, the blanket is heated. The heat is extracted from the blanket and can be conveyed, for example, to a conventional steam/turbine generator for electricity production. The tritium bred in the blanket will be extracted and be fed back into the plasma for refuelling. The fusion reactor can therefore be ideally represented as a closed system whose primary fuels (deuterium and lithium) and output (energy and helium) are non-radioactive. However, inside the reactor there will be two sources of radioactivity : the intermediate fuel (tritium) and the structure of the reactor which will be activated by neutron bombardment.

ENVIRONMENTAL IMPACT OF FUSION

An important aim of European fusion research and development is to lead the way towards a fusion power plant that is economically acceptable, technically reliable and benign to the environment. Fusion energy, when available, will not automatically fulfil these criteria. Studies in the European Communities /1/ and e.g. the United States /2/ indicate, however, that an environmentally benign fusion reactor may be achieved by proper engineering aimed at maintaining a low tritium inventory, and by developing special materials to keep the activation of the reactor structures at low level. This is not to say that a consistent design along these lines is in hand. Although great progress has been achieved on the way of fusion towards reactor conditions (as shown in the next chapter), it remains a formidable challenge to integrate all desirable environmental, safety and economic features into a coherent design.

Fuels, Output, Emissions to the Environment

As already mentioned neither the externally supplied primary fuels - deuterium and lithium - nor the ultimate fusion reaction product - helium gas - are radioactive or toxic. Fusion plants do not emit any of the Greenhouse gases : no CO_2, SO_2, NO_x or any other biotoxic chemicals.

The conversion of nuclear heat into electricity is presently foreseen to be provided by conventional techniques ; therefore the generation of waste heat will be the same as in any other type of steam raising plant. There is the ultimate potential of direct energy conversion in advanced fuel cycles.

Inherent Safety Features

Fusion reactors will be complex nuclear installations but will nevertheless have a number of intrinsic safety features. A fusion power plant will be designed such that the effects of all credible accidents on the

444

environment will be kept small by generic safety features : in particular, an uncontrolled nuclear runaway is impossible since a change of operation conditions will lead to plasma instabilities and termination of the burn process (the total fuel present in the reaction chamber at any time is sufficient for a few seconds burn only while in fission it is of the order of one year ; the potential for damage in a fusion plant thus comes mainly from the magnetic energy stored in the coils) ; when the fuel flow is interrupted, the reaction stops ; the power density in the first wall and blanket structure is relatively low ; afterheat after shutdown is moderate ; the bulk of radioactive material is non-volatile structural material.

Studies /3,4,5,6/ based on plausible extrapolations from todays physics and technology to reactor level indicate that the credible accident potential of fusion reactors excludes calamities disrupting normal life in the community outside the reactor site boundary. It is further concluded that releases of tritium and radioactive internal structural materials will cause no immediate harm to an individual outside the power station boundary even following a major accident or plant failure.

Radioactive Waste

The principle radioactive components of a fusion reactor will be the torus wall and the blanket structure, both of which become activated by the fusion neutrons.

Preliminary studies in that field indicate that the radioactive waste generated by fusion power plants will be quantitatively comparable to those generated in fission plants, but qualitatively it will be much less of a potential hazard. Work is under way to further reduce the already moderate environmental impact of fusion as derived from todays technology : a considerable development potential exists in the reduction of activation by choice of new structural low-activation materials and by limitation of waste quantities by improving life time of first wall and blanket components.

PATH TOWARDS FUSION AS A COMMERCIAL ENERGY SOURCE

Strategy of the European Fusion Programme

The path towards a commercial fusion reactor can be devided, albeit somewhat arbitrarily, into three stages : first, the demonstration of scientific feasibility ; then of technological feasibility, and finally, of commercial feasibility (Fig 3). These stages are, however, far from being independent of each other and certainly overlap in time. At present, with JET (Joint European Torus) and the specialised devices in the Associated Laboratories, the European Fusion Programme is primarily in the scientific stage. The Next Step, NET (Next European Torus) at the European level and ITER (International Thermonuclear Experimental Reactor) at the world level, is conceived as an engineering test reactor of the Tokamak type that should fully confirm the scientific feasibility of fusion in a first phase of operation and confront its technological feasibility in a second. The commercial feasibility will be proven in a subsequent step : the DEMOnstration reactor.

Within this overall three-stage strategy of the Community programme (JET and the specialised devices, the Next Step, and DEMO), the near-term objectives of the programme are :
- to establish the physics and technology basis necessary for the detailed engineering design of the Next Step;
- to embark on the detailed engineering design of the Next Step as soon as the necessary data base exists;
- to explore the reactor potential of certain toroidal magnetic configurations akin to the Tokamak.

Figure 3 also indicates the time horizons involved : the commercial feasibility is expected to be demonstrated around 2025, so that fusion could make a substantial contribution to the energy supply by the mid of the next century.

446

Status of Fusion Research

A figure of merit used to assess the progress of fusion on its way to reactor conditions is the so-called fusion product $n_i \tau_E T_i$ (n_i, T_i are respectively the central density and temperature of the fuel nuclei (ions), τ_E is the energy confinement time). Figure 4 visualizes the performances already achieved and the distance to the goal of reactor conditions.

So far the best performances have been achieved at the largest devices : JET (Joint European Torus), situated in Culham/UK, TFTR (Tokamak Fusion Test Reactor) at Princeton/USA, and JT-60 at Naka-Machi/Japan.

JET /7/, the largest tokamak of the world with a major plasma radius of about 3 m, minor radii of 1.2 to 2.1 m and plasma currents of up to 7 MA has achieved the highest fusion product so far and has routinely operated at reactor relevant temperatures. The evolution of device parameters and experimental values typically achieved, though not simultaneously, on devices operating when JET design started (1973), compared with such values presently obtained on JET and values foreseen for NET, are presented in Table 2. As indicated by Table 2 an enormous progress has been achieved. In terms of plasma parameters and performance the extrapolation from JET to NET/ITER is appreciably smaller than that made in 1973 when designing JET, but NET/ITER will incorporate new, reactor-relevant technologies which make it on the whole a big step towards a reactor.

THE NEXT STEP

Following initiatives taken at the highest political level, four parties (the European Community, Japan, USA and USSR) agreed in early 1988 to participate to an equal quadripartite basis in the joint development of a conceptual design for an engineering test reactor of the tokamak type and of supporting R&D activities. The ITER conceptual design activities, conducted under the auspices of the IAEA, are to be concluded by the end of 1990. The single conceptual design thus developed would then be available to each of the Parties to use either in their own national programme or as part of a

Table 2 : Evolution Of Device Parameters

	Before JET (1973)	JET (1989)	NET (foreseen)
Plasma volume (m^3)	≈1	≈140	≈1100
Plasma current (MA)	0.4	7	25
Pulse length (s)	≈1	≈20	700 – 6000
Energy confinement time (s)	≈0.02	1	3
Additional power (MW)	<1	35	≈85
Fusion product (m^{-3} keV s)	≈3x10^{18}	2.4x10^{20}	> 5x10^{21}

larger international collaborative venture. Canada ensured its participation in the ITER conceptual design work via a formal involvement in the Community contribution.

Both NET and ITER are currently in their conceptual design phase. NET /8/ and ITER /9,10/ have substantially similar objectives and therefore the configurations and solutions chosen for the two basic devices are also rather similar. The differences in their respective parameters are essentially due to the following :

- different safety margins are taken in the two designs : NET uses more conservative assumptions than ITER in particular as far as confinement and inductive plasma current drive capabilities are concerned : therefore, the major radius in NET is 6.3 m compared to 5.8 m in ITER;

448

-   the emphasis on steady-state operation, i.e. on non-inductive drive of
    the full plasma current during burn, is stronger for ITER than for NET.

A list of the main NET parameters is given in Table 3.

Table 3 : NET Parameters And Expected Performance

Major radius (m)	6.3
Minor radius (m)	2.05
Magnetic field on axis (T)	6.0
Elongation	2.2
Plasma current (MA)	25
Average electron density ($10^{20}m^{-3}$)	0.9
Neutron wall loading (MW/m^2)	1
Fusion power (MW)	1100
Pulse length (s)	up to 6000
Divertor exhaust power (MW)	140
Divertor peak loading (MW/m^2)	7

The basic R&D needs for ITER and NET are similar, as are their time
scales. The main R&D requirements for the Next Step can be summarized as
follows :

Physics :

*   plasma-wall interaction
*   helium (ash) removal
*   impact of plasma disruptions
*   operational limits
*   plasma heating and plasma current drive

Technology :

* magnets (superconductors)
* blanket
* tritium handling
* remote maintenance
* low activation materials
* plasma heating and current drive systems

For the time being, the European Fusion Programme is aiming at covering all R&D needs of the Next Step.

SUMMARY AND CONCLUSIONS

From the considerations mentioned above, thermonuclear fusion - a potential energy source for the next centuries - is attractive for the following reasons :

* it has the potential to become a major long-term source of energy ;
* its primary fuels (deuterium and lithium) are plentiful also in the European Community ;
* its primary fuels and output ashes are non-radioactive ;
* it does not produce any actinides ;
* there is no emission of any of the Greenhouse gases ;
* fusion has inherent safety features : a nuclear or thermal runaway can be excluded, and its credible accident potential excludes calamities disrupting normal life ;
* the development of an environmentally benign fusion reactor is a question of proper engineering and of successful materials development.

International collaboration will certainly play an important role in the detailed engineering design phase of the Next Step, either through the joint planning of the Next Step activities, or, more ambitiously, through the joint engineering of a single device.

The time horizon for fusion to become commercially available are : the demonstration of the commercial feasibility by around 2025 ; a significant contribution to the energy supply could be expected by mid of the next century.

REFERENCES

/1/ "Environmental Impact and Economic Prospects of Nuclear Fusion", Report EUR FU BRU/XII-828/86, Commission of the European Communities, Brussels (1986).

/2/ J.P. Holdren et al., "Summary of the Report of the Senior Committee on Environmental, Safety, and Economic Aspects of Magnetic Fusion Energy", Report UCRL-53766-Summary, LLNL Livermore (1987) ; and Proc. 12th International Conf. on Plasma Physics and Controlled Nuclear Fusion Research, Nice, 1988, IAEA-CN-50/G-1-5, Vienna (1989), to be published.

/3/ W.R. Spears, "DEMO and FCTR Parameters", NET-Report Nr. 41, EUR FU/XII - 361/85/41 (1985).

/4/ "STARFIRE - A Commercial Tokamak Fusion Power Plant Study", Argonne National Lab. Report, ANL/FPP-80-1 (1980).

/5/ INTOR, International Tokamak Reactor Workshop, Proc. Phase Two A, Part III, Vol. 1 and Vol. 2, IAEA, Vienna (1988).

/6/ P.I.H. Cooke, P. Reynolds et al., "A Demo Tokamak Reactor ; Aspects of a Conceptual Design", UKAEA, CLM-R254 (1985).

/7/ R.J. Bickerton et al., "Latest JET Results and Future Prospects", Proc. 12th Intern. Conf. on Plasma Physics and Contr. Nucl. Fusion Research, Nice, 1988, IAEA-CN-50, Vienna (1989), to be published ;

/8/ R. Toschi et al., "The NET Project : An Overview", ibid.

/9/ D. Post, "ITER : Physics Basis", ibid, IAEA-CN-50/F-II-1.

/10/ K. Tomabechi, "ITER : Concept Definition", ibid, IAEA-CN-50/F-I-4.

SCHEMATIC OF A TOKOMAK

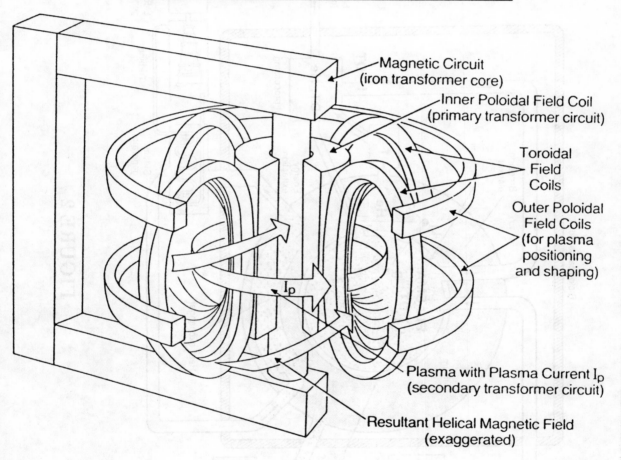

Magnetic Circuit
(iron transformer core)

Inner Poloidal Field Coil
(primary transformer circuit)

Toroidal
Field
Coils

Outer Poloidal
Field Coils
(for plasma
positioning
and shaping)

I_p

Plasma with Plasma Current I_p
(secondary transformer circuit)

Resultant Helical Magnetic Field
(exaggerated)

FIGURE 1

SCHEMATIC OF A FUSION REACTOR

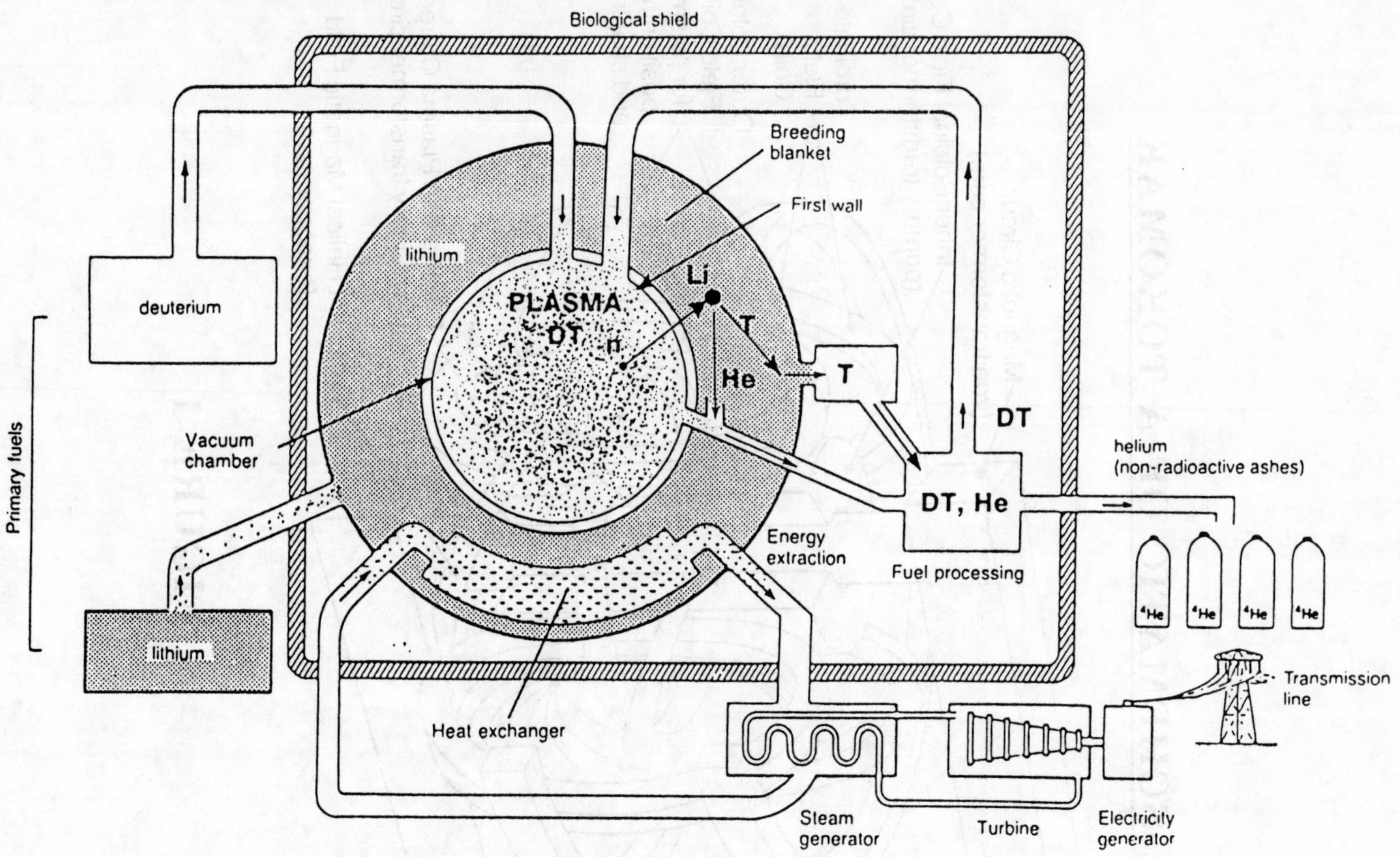

FIGURE 2

EUROPEAN FUSION PROGRAMME STRATEGY

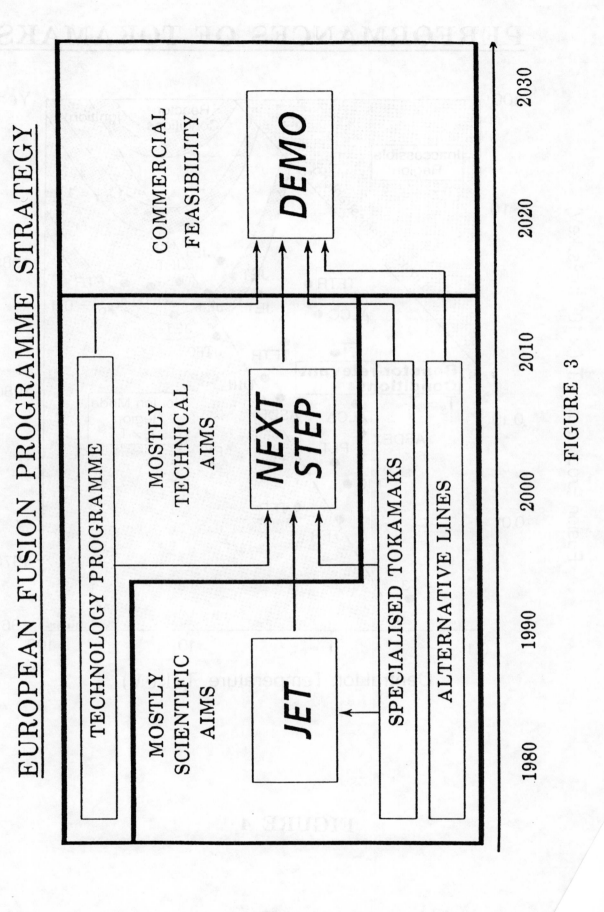

FIGURE 3

PERFORMANCES OF TOKAMAKS

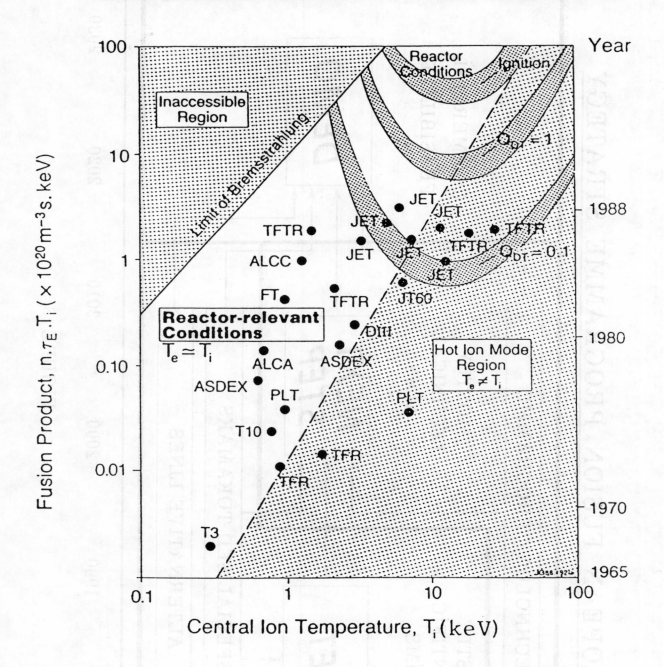

FIGURE 4

On Setting Targets for Reduction of Greenhouse Gas Induced Climatic Warming

J.A. Laurmann
Marine Science Institute
University of California, Santa Barbara
3372 Martin Rd.
Carmel, California 93923

1. Introduction

The last two years have seen a surge of public interest in a fundamental characteristic of planetary atmospheres and its implications for the future climate of the earth. The general realisation of the threat to global habitability from increasing concentrations of greenhouse gases is of very recent origin, even though the physics of the phenomenon have been known for some time. Actually, estimates of warming have changed little in the last fifteen years since first systematically studied by the scientific community. However, as we shall discuss below, their policy implications have only recently received attention. These can be viewed — perhaps should be viewed — from many perspectives. Thus, conclusions concerning the prospect range from doomsday forecasts of mass extinction to optimistic scenarios of a more benign planet. Reenforced in the United States by 1988 summer's presumably coincidental hot and dry summer, the last year has seen dominance of the former, pessimistic view, yet we have little assurance that next year will not see a shift away from the present influence of pro-active voices that are calling for immediate institution of greenhouse gas emission reductions. Such could either result from the rallying of latent powerfull interests reacting to the threat of economic loss if emissions controls are proposed, from a revision of our understanding of the threat, or from the ascendancy of politically more pressing and appealing issues.

In this paper we shall argue for a tempered view of the situation, in part because, once the threat of large climatic change is realised, or its imminence demonstrated, we see as essential the need for cooperation between countries, regions, political factions and interests who currently view the situation in quite different terms. Reasoned, non-provocative argument for consideration of this environmental problem, together with full display of the scientific facts as known and those that are not known, is important for achieving consensus on how, if and when to act. At the time when the need for action is generally perceived as a global commons necessity we would then hope that the reputation of scientific analysis would be intact and acceptable for important application.

In the following will cover various ways of assessing the importance of the portended climatic change and the degree of urgency for acting to mitigate it. Central to every argument on mitigation is the evaluation of the effects (costs) of the recommended actions and the costs of inaction if the problem is ignored. Recognition of the existence of two sides to the equation implies balance as the optimum solution, and we shall devote a considerable part of this paper to consideration of what constitutes "acceptable" climatic change. The fact that every inhabitant of the earth of this or the next generation will be subject in different ways to the effects of climate change if it occurs, means that a commonly held perception of its present day importance is unlikely. Consensus building and compromise will eventually dictate the choice of measures that should be taken. We intend here to indicate the range of differences that need to be spanned in such a future effort; at this time, exclusive dedication to achievment of a particular target for emissions reductions or for allowable climatic change that includes specification of means for attaining it, is likely only to prejudice a future negotation process. Thus the paper, although treating a variety of possible rationales for selecting such targets, will avoid recommending any one of them.

2. Physical Characteristics

The prime generic features of the greenhouse gas warming problem that influence all stages of argument on policy are

> its globally hetereogeneous nature in sources and effects,
>
> the high degree of uncertainty in magnitude and timing of its impacts,
>
> the decadal to centential time frame for its occurence.

We shall first briefly run through what is known concerning the physical sciences background of the greenhouse gas warming phenomenon as it relates to these important characteristics.

2.1 Sources

Figures 1 and 2 show the prime sources of greenhouse gas emissions by country, currently and projected to 2025. The latter date corresponds roughly to the current best guess of the time for reaching "effective CO_2 doubling" of atmospheric concentrations of infra-red absorbing trace gases. When globally averaged, the most probable value of the temperature rise is $3^\circ C$, a value generally, though not universally regarded as having serious negative impact on society. Observe from the figure the switch in roles of the developed and developing countries as prime producers of greenhouse gas emissions as time progresses. However, atmospheric levels are integrations of emission rates, and it can therefore be argued that, even in 2025 the presently industrialised world should be held largely responsible for climatic change effects that will be produced at that time. In fact, understanding of the rate of transfer of the gases from sources (mostly fossil fuels) to their eventual atmospheric, oceanic or terrestrial sinks is incomplete, so we cannot say definitively how much of those emissions released today will still be present in the atmosphere a given time in the future. For the largest single contributor to greenhouse warming - CO_2 - at current emission rates, about half of annual emissions remain airborne, and analysis give the relationships shown in figure 3 for future atmospheric CO_2 concentrations as functions of developed and developing country contributions (Kellogg and Schware, 1981). The results shown here are valid so long as CO_2 growth rates are high enough, perhaps greater than about 1% a year (Laurmann and Spreiter, 1983). At lower rates of growth, or for a falling emissions rate that would accompany an emissions reduction policy, the fraction of emissions atmospherically retained each year drops (c.f. figure 4), an important characteristic of carbon cycling in the consideration of climate change remediative actions (see section 4 on the Toronto target).

The characteristics just described are important when ascribing responsibility for climate change damages and hence in negotiations for institution of global climate change control measures in an equitable manner. Paying for insurance to cover costs of future potential climatic change induced damages involves similar considerations.

Finer definition of the sources of greenhouse gas emissions further compounds the political problems accompanying this set of facts. Thus, population growth and the associated increase in agriculture for food production is another source of increasing greenhouse gas emissions (primarily N_2O and CH_4), but more significant is industrial growth, and most burdensome of all, is increasing use of fossil fuels as the primary source of energy, and the associated production of CO_2, normally allowed to escape into the atmosphere. Of the three primary forms of fossil energy, oil, natural gas and coal, coal is the largest potential source of carbon dioxide. Distribution of coal resources, as currently known, is shown in Figure 5. Even if we accept that reduction in CO_2 emissions as part of an effort to reduce climatic warming is reasonably possible for the industrialized world via fuel switching and a large reduction in energy use growth rate, such a progam without a corresponding effort on the part of the developing world would not be sufficient, c.f. Figure 5 vis a vis the Chinese role. Moving China away from its present course of rapid industrialization using cheap indigeneous sources of coal is an obvious challenge that will require compromise and understanding on the part of all users of the global commons atmosphere.

2.2   Uncertainty

The first comprehensive model calculations of climate change due to increasing atmospheric greenhouse gases (then CO_2 only) were made some fifteen years ago with general circulation atmospheric models. Results then indicated a 1 - 5 $^\circ$C rise by 2030 for globally averaged

459

temperatures, with a mean value of about 2.5 $^{\circ}$ C, when atmospheric CO_2 concentrations were estimated to have doubled their 1860 value of about 280 ppmv (the so called pre-industrial level). The present value is 340 ppmv. Latest projections for this condition give a best guess value of 3° C with a large error range of from 1.5 to 4.5 $^{\circ}$C in 2025 (NAS 1983). (The exact meaning of the range estimate is not clear.) These figures are for warming due, not just to CO_2, but to other atmospheric greenhouse gases (N_2O, CH_4, O_3 and the CFCs). The fact that the timing of the predicted future warming has changed little in the last ten years results from the fact that projections of CO_2 emissions have dropped considerably since the '70s, due to reduction in energy use growth rates and a continued projection of this reduction into the future. Still, the latest (1988) figures for growth of global energy use indicate a return to the earlier pre-oil crisis growth rate, nearing 4% p.a. If continued, this will radically changes the timing of major climate change and the need for proximate action.

The uncertainty range just quoted is attributable to the error in understanding of the physics and chemistry needed for estimating the climatic effects of the greenhouse infra-red absorbing gases in the atmosphere. In addition, projections require knowledge of future emission rates of these gases. Here uncertainty is also very large and, moreover, is not by and large due to a lack of understanding that can be improved by research. Thus, large uncertainty is indigeneous to the problem and needs to be accepted as a permanent and unavoidable characteristic of this environmental issue that has to play a major role in decision making.

Figure 6 shows three projections developed by the 1987 Villach meeting of experts (Jaeger, 1988) for the "best guess" median increase in global temperature and two extremes. The latter encompass the 95% probability limits. The estimates allow for the transient response of climate warming due to the large thermal capacity of the oceans (Hansen et al., 1984). Thus the earlier cited date of 2025 for effective CO_2 doubling temperature rise has been delayed in this portrayal. We note that the so-called "ocean thermal inertia lag" effect means that temperatures will keep on rising even if atmospheric levels of the greenhouse gases stabilise, thereby introducing an irreversibility feature that is important for timing of responses to the threat.

If the uncertainty characteristics shown in figure 6 are introduced into climatic change impact costs, a series of depictions of costs of climatic change as influenced by the error in our knowledge can be produced. Figure 7 plots cumulative probability of climate change costs for a doubling of CO_2 levels (i.e. an equilibrium rise of 3°C). According to these cost estimates, a 3° global average temperature increase results in a 3% loss of world GNP (Laurmann, 1980). (The results shown in figure 7, in contrast to the plots of figure 6, do not allow for transient non-equilibrium effects.) The large uncertainty in costs has major implications concerning risk taking — figure 7 indicates a chance of 1 in 5 chance for a damage cost of 10% of world GNP rather than the median value of 3%. Under these circumstances risk aversity implies the need to act earlier in response to the threat. The examplar costs case of risk averse behavior illustrated in figure 8, shows that expected costs, calculated in a decision analytic framework (Laurmann,

1980) rise from 3% to 7% for a 2030 doubling date.

The most recent climate modeling calculations result in a median
estimate of between 2020 and 2040 for reaching CO_2 doubling temperatures
assuming thermal equilibrium, with a delay of perhaps 30 years for the
transient lag effect. If a $3^{\circ}C$ rise in global average temperature is
believed to be critical, this means we have to wait the order of 60 - 80
years to feel this level of effect. In evaluation of the need to take,
presumably expensive actions now to mitigate such long range future
potential damage, future discounting becomes the dominant consideration.
Figure 9 illustrates the effect of discounting the climate damage
function plotted in figure 7 for three discount rates, with very radical
effects on the present day perceived costs and hence the importance of a
2030 climatic change. Choice of discount rate applied over such a long
time period is critical for canonical decisions on the implications of
future climate change, and yet there is little agreement on appropriate
choice of discount rate. A zero value is advocated by a few, on the
grounds of intergenerational equity; others advocate a "standard"
business rate of return figure of 10% per annum. We have here an
instance of the need for attaining common understanding concerning a
technical feature of decision making in reaching global concensus on
emission reduction targets and shared costs.

2.3 Impacts

In reviews and policy oriented documentation it is common to emphasise
the deletarius consequences of greenhouse warming. For example, the
latest description of impacts of climate change coming from the U.S.
National Academy of Sciences (NAS et al., 1988) describes the expected
changes as follows:

> Climatic zones and storm tracks may be expected to shift
> Poleward. Crop Zones and natural ecosystems can be expected
> to migrate with the changing climate, although the extent of
> movement is uncertain. Major shifts could clearly have far-
> reaching economic, social and political consequences.

> Melting of land-borne glaciers and thermal expansion of sea
> water are expected to raise sea levels significantly over
> the next century. Projections of the amount of rise range
> from tens of centimeters to as much as 3 meters, with
> reasonable estimates centering on a 1 meter rise. Rising
> sea levels will increase the already troublesome rates of
> coastal erosion and loss of wetlands, while increasing
> saltwater intrusion would impair water supplies and
> agriculture in coastal areas.

> Warming is expected to be considerably greater in polar
> latitudes than in the tropics, and sea ice should diminish.
> A navigable Arctic Ocean would have major national security
> implications.

> Changes in rainfall patterns are likely, and some studies
> indicate greater summer dryness in mid latitude continental

regions. Regional changes in water supply and quality may
have significant economic and social consequences.

Adverse climate changes may be difficult to accommodate in
developing countries, where resources are not available to
adapt   to changing conditions. Growing pressues for
migration may result. Implications for human health may
result from changes in the range of disease vectors (e.g.,
insects and rodents) and in frequency and intensity of
extreme weather known to influence mortality rates.

The tone of this summary is decidely negative, and is suggestive of the
notion that life and society is optimally adapted to the present
climate, so that any change will be for the worse. Indeed the 3% of
world GNP loss for doubling of atmospheric CO_2 levels that we cited in
section 2.2. was in part derived with this principle in mind.

Although in agreement with another independently derived damage estimate
(Nordhaus, 1980), the 3% of GNP loss figure is uncertain in the extreme.
More economically secure estimates for local or sectoral, rather than
global damage costs have been made in the last year or two. Valid
aggregation of these small scale costs to the globe is not yet possible,
both because of insufficiency of micro-level calculated results and
because of the secondary interactive effects that arise when compounding
to the macro-scale. Additionally, local scale impact estimation is
fundamentally impaired since the primary tool for climate calculation –
the general circulation model – is unable to deliver climate information
at the sub-100 km scale needed for most climatic change impact
assessment. Even so, some generalised conclusions regarding climate
change impacts and their costs are possible. As we have indicated,
although very little hard data exist to substantiate it, there is indeed
general agreement that global CO_2 doubling temperature effects are
negative. For presently constituted natural ecosystems this is
certainly true; however, for human settlements, there can be argument.
For example, althogh the summary statement of the 1987 Villach meeting
of experts drew an overall negative conclusion (Jaeger, 1988), a more
studied view of the deliberations of that meeting gives a mixed picture
for a 2040 $3^{\circ}C$ rise in temperature, which we can encapsulate as follows:

At high latitudes   there is increased agricultural productivity,
                     better access to mineral and energy resources,
                     easier transportation,
                     loss of native habitat and ecosystems,
                     thawing of tundra.

At temperate latitudes   there are altered crop yields,
                          changes in agricultural practices,
                          destruction of ecosystems,
                          water resource problems,
                          altered energy demands.

At tropical latitudes   there are alterations in crop yields,
                         changed monsoonal patterns,
                         increases in drought and flood extremes.

A 1/2 m sea level rise   causes loss of valuable coastal zones.

In contrast to many summaries on the effects of climate change addressed to the policymaker, this description indicates the probability of important beneficial effects from climatic warning for some regions of the world, principally in northern latitudes, where access to raw material resources could improve, and major increases in agricultural productivity result. One Soviet scientist has so stated in no uncertain terms (Budyko, 1988). Under such circumstances we can no longer talk about a global commons - the greenhouse gas carrying capacity of the atmosphere - which it is in everyones interest to preserve, but rather we are confronted with an international scale environmental problem that is in some ways analogous to the trans-boundary transport of high altitude pollutants. Designation of a global figure for allowable greenhouse gas emission rates is then hardly an objective that can be easily agreed to by all, and benefits lost by avoidance of the climate change have to be included in consideration any overall global accords on control.

It is important to observe that the dilemma posed from such diversity of climatic effects in the task of obtaining global consensus is eliminated simply by considering larger climatic changes than obtain for the CO_2 doubled state. It is inevitable that the "winners/losers" problem will evaporate if times beyond CO_2 doubling are considered, or if conditions away from the most probable values of climate change are viewed seriously. Such a perspective on the problem makes clear the negative impact that would be felt by all, and immediately changes the set of decision making questions that have to be dealt with and the information base needed to answer them. (c.f. Laurmann, 1985a).

3. Target setting

The following sets out a number of fundamentally different perspectives on how targets for emission reduction levels can be established.

3.1 Sustainability

The projections cited above dealt mostly with warming attendant upon atmospheric CO_2 doubling - the most commonly considered greenhouse gas augmented state - and originally selected as a convenient if arbitrary standard condition for carrying out climate change modeling experiments. In fact, the $2-3^{o}C$ rise first predicted to result from this perturbation was initially regarded by many experts and non-experts as a minor change, and did not attract much attention outside the specialist group working on climate change. The realization that the expected warming at temperate and high latitudes would be much higher than the global mean, helped to change this impression. However, for obtaining a better appreciation of the dimensions of climatic change with which we are concerned it is more useful to compare predictions with the past long term history of the earth's climate.

Figure 10 reproduces the earth's temperature record extending a million

years into the past (Clark, 1982). Currently the earth is in a warm
interglacial era, at a high point in one of the seven glacial-
interglacial cycles that have occured over the last million years. Of
most note is the fact that during this time global temperatures have
rarely been as high as they are now, with extremes perhaps only about
$3^{o}C$ higher than at present. In fact the 4° rise given by the the
predictions of some of the current climate models for CO_2 doubling
(Hansen et al., 1988), was last experienced over a million years ago.
As far as human history is concerned we are thus contemplating
greenhouse gas induced warming of a magnitude never experienced by
socialized man. This fact alone provides sufficient grounds for some to
call for mitigative action, and it surely provides a prima facie case
for considering greenhouse gas warming as a potentially serious threat.
We note that the 3% loss of GNP estimated as the cost of climate change
damages for the CO_2 condition (section 2.3) can be considered to be on
the verge of sustainability, since it implies loss of all economic
growth according to historical experience and thus represents a
stationary socio-economic condition.

A rationale for avoiding at least the CO_2 doubling conditions therefore
exists now. Answering what the action should be, when it should be
taken and at what cost are not provided by this argument. In addition,
it has been supposed from the start of scientific investigation of
climate change that society (and indeed natural ecosystems as well)
would likely first be threatened by the speed of the climate change, and
failure to adapt to it. Spectral analysis of long term temperature
records indicates an intercentenial natural variability of climate of
$1^{o}C$, approximately matching the shorter term annual variability of
annual means. The target limit value of $0.01^{o}C$ maximum rise per year
suggested at the Villach conference mentioned above was based in part on
this information on natural climate variability, arguing that mankind
was able to survive at least this rate of change in the past, and could
therefore do so in the future. In fact this rate happens to coincide
with the condition for survival of boreal forests (Jaeger, 1988) in the
sense that, above this rate, natural migration of the species cannot
keep up with the movement of temperature isotherms, assuming the latter
progress steadily to higher lattitudes as global warming increases. For
comparison purposes, we plot in figure 11, two scenarios for climate
change — with fossil fuel growth rates of 1% p.a., and the other
greenhouse gases approximately doubling the rise due to CO_2 alone, but
with no ocean thermal inertia lag allowed for.

Information on the past variability of climate on all time scales is
available from the historical and paleo record (Berger, 1980). It thus
becomes possible to think in terms of estimates of specie survival rates
for increasing temperature that can be compared with the magnitude of
climate changes expected from greenhouse gas warming. We make note also
that mean rates of change are in general a function of averaging time
and hence the extremes of temperature variation. Sassin (Sassin and
Jaeger, 1988) has started to develop this theme, and we replicate one of
his results in figure 12. Most striking is the fact that the Villach
climate change projections (figure 6) mostly lie beyond the range of the
earth's past million years of climatic variation, when expressed in
terms of rates of climate change and its excursions.

These appeals to past climatic experience for setting both allowable rates of change and ceilings for climate change is appealing partly because it leads to the possibility of quantification of standards on a firm empirical basis. Of course we have no general way of knowing whether exceeding these limits will in net be detrimental to animal species, to complete ecosystems, or indeed to the functioning of entire human social and economic systems. On the other hand it may be that lower target figures would be needed for survivability of some species or be preferred so as to avoid excessive costs. Greater understanding of physiological responses amd human and animal ecosystem behavior is required to deal with these questions. This science is in a primitive stage of development, but we are already witnessing a blossoming of research effort on this broad general theme, stimulated by the global nature of the greenhouse gas warming problem (e.g. Sustainable Development of the Biosphere, Clark and Munn, 1987). Note that, as in all other aspects of greenhouse gas warming, we have ultimately to treat the problem at the global level, and although the micro-level, reductionist approach may be an essential part of answering all the questions we have in mind, aggregation to the macro-level is no mean task in itself, if only because of the interactive nature of the micro components and the non-linear feedbacks that we anticipate are involved. Setting of global targets for allowable temperature change and emission levels via such a perspective (the sustainable development theme) is a long way from being the logical and useful tool that we hope it one day will be.

Complementary arguments relating to rates of change occur when considering remediation measures for reducing warming, and when discussing maximum acheivable rates of greenhouse gas emission reduction. Most readily described is the macro-scale phenomenon first pointed out by Marchetti (Marchetti, 1975) that limits the rate of substitutability of one form of global primary energy production by another. In our case we are concerned with replacement – perhaps at a rapid rate – of fossil fuels with non-carbon based energy sources. Figure 13 is an example of the limitations imposed by assuming a 50 year market penetration time (the time taken for non-greenhouse gas emitting energy sources to increase their share of energy production fron 1 to 50%). These calculations were made originally to exemplify the role of the market penetration time constraint in hindering the ability to keep global levels of CO_2 from exceeding double the pre-industrial value (about 280 ppmv). The latter criterion for acceptable change, as we mentioned earlier, has been a favorite in arguing for control and reduction of carbon dioxide emissions (Laurmann, 1985b).

3.2 Benefit/cost analysis

Even if a standard for emission rate reduction could be derived using the ideas discussed above, its attainability involves a different set of issues. In the final analysis these require technical analysis. An extreme case is one in which overpopulation brings with it a concommitent and apparently unavoidable increase in greenhouse gas emissions that are associated solely with the need to satisfy basic food and habitat requirements. This variant of the Malthusian dilemma is not normally regarded as a likely consequence of anthropogenic production of greenhouse gases, but it is certainly true that reduction in the rate of population growth could significantly ease the greenhouse problem. A second technical feature that provides a limit to the means available to reducing emissions is intimately connected with time scales. Just as adaptabilty to changing climate change (section 3.1) is critical in evaluating the impacts of warming, the ability to effectuate emission reduction control in time to meet either rate of change targets or ceilings for global warming is critical in assessing possible responses to the threat of future warming. Thus, it is possible, even if all members of the global commons were to be in agreement and had authority to harness all available resources, that the latter would not suffice. It is not known whether or under which scenarios of future emission projections such a physical limitation would apply. But Marchetti, in referring to his well-known thesis of primary energy market penetration time limits, already discussed in section 3.1, certainly insists that such a fundamental road block exists. It is not clear if this restriction would apply to a mandated, rather than a free market transition into renewable, non-fossil energies, but it is clear that there are many barriers other than technological and economic to a rapid transformation of the world's energy base - institutional, bureacratic and political (See Laurmann, 1985c). All these delay implementation of greenhouse gas remediation efforts, and suggest the need for early initiation of target setting proceedures.

There is a middle ground between claims of impossibility of achieving useful emission reductions just mentioned, and the other extreme in which it is believed that emission reductions are both desirable, cost-effective and realisable, independent of their negative environmental effects (e.g. Lovins et al. 1981). This argues for balancing of the costs of emission reduction for reaching a pre-set target with the savings from the climate damage costs avoided. In considering means for reducing emissions we can look at:

> methods and costs for emission reduction to reach exogeneously set targets, for example, by means discussed in section 3.1. above, and

> ·calculated costs for emissions reduction and costs of induced climatic change, trading one off against the other.

3.3 Quantification; the climate damage function.

Cost/benfit analysis and its near cousin, optimum growth analysis, require specification of costs of climate change in terms of future greenhouse gas emissions and their sources. This operation is highly complex if complete, involving a multi-dimensional function of the action space describing the future world economic and political order. A large number of estimates of the impact of climate change on natural and man-made systems have been made, these usually for a new stationary climatic state, and most commonly, for a climate corresponding to a doubling of atmospheric CO_2 levels. Very few attempts to assign dollar cost figures to impacts at the global or national level have been made; we shall present some results from those the primitive cost calculations that have been attempted.

Global costs for CO_2 doubling have been surmised to amount to about 3% of world GNP, with very large, but unknown probable error (section 2.2). Both linear and quadratic dependence of costs on the size of climate change departure from the present have been assumed. Non-linear behavior seems the more probable, and it may be that a useful representation would employ a damage function that takes on catastrophic proportions (infinite cost) for some specified degree of warming of the planet, with ignorable effect before then. This might correspond to a situation in which a threshold for societal stability or sustainability is reached.

Comparison of costs for taking climate change abatement measures with the costs of the climate damage thereby avoided requires estimation of the effective present day costs of future climate damages, and this is a strong function of future discounting. Figure 14 illustrates the latter point for a linear climate change damage function model, assuming a 3% GNP loss at CO_2 doubling tempertures, and a range of discount rates for calculating cumulative present day costs of future climate change damages. The climate change future scenario assumed here is a "business as usual" projection, corresponding to the growth curves in figure 13. Levelised costs, i.e. the fixed annual payments (into perpetuity) that would cover borrowing the the total present day cumulative costs, would be 1/10 of the ordinate values at a 10% p.a. financing rate. We regard choice of the latter high discount rate as appropriate when considering expenditure of capital funds for emission control or for production of new, non-greenhouse gas producing energy plants, whereas, in calculating cumulative present day costs, the discount rate taken should represent a pure rate of time preference. This could be high or low, depending on the view taken on intergenerational equity. We should note that the calculations used in producing figure 14 assume economic growth at 3% p.a., with climate damage growing at the same rate - hence the infinite values of cumulative costs at a discount rate of 3% p.a.

Damage estimates such as shown in figure 14 can be used to argue about the current seriousness of the future climate change, and, when combined with cost estimates for reduction of climate change, to develop specific emission reduction policy recommendations. The only analysis done to date that uses such a comprehensive cost comparison technique is due to Nordhaus (1980). The optimal growth formulation that he applied implicitly produces future history of optimum CO_2 levels in the atmosphere, and hence emission rates. His conclusions are described in terms of a shadow price for CO_2 and tax rates that need to be applied to fossil fuels. Of course, an approach such as this that intimately connects the technical means and hence the costs for reducing emissions with their contemporaneous and future effects on climate in a global optimization scheme, will produce a future emissions reduction target path that is highly dependent on the specifics of the analysis and the methods employed to reduce emissions. Discussion on the ease or difficulty of reaching the target path is both difficult and irrelevant, since the method, if done correctly, automatically chooses the easiest (least total cost) emission reduction path. In the optimal growth method of analysis use of the emissions reduction target concept is therefore not particularly useful.

Damage estimates attendant upon a warmer climate are also available for the sea level rise sub-sector of the total climate damage cost function we have just discussed. Moreover, we believe these are considerably more reliable than other climate damage cost estimates, including the global aggregate. Costs for an assumed 1m rise in sea level (perhaps to be reached by the end of the next century) are shown in figure 15 for a number of countries. Figure 16 gives an estimate of cumulative damages from the sea level rise predicted to occur with a warming climate for the U.S.

3.4 Remediation

The last section covered essentially all the work done to date that has tried to quantify aggregated costs of climate change and relate them to the need for taking remedial steps. Numerous other analyses, especially for the energy sector, have been made that discuss emission abatement efforts to reduce atmospheric levels of greenhouse gases, especially CO_2, without assigning climate change damage levels. The most common criterion mentioned in these greenhouse gas remediation reports is avoidance of effective CO_2 doubling levels in the atmosphere, the possible effects of which we have discussed earlier. Qualitative reasons for avoiding exceeding a specific level of atmospheric CO_2 can be based on argument similar to that described earlier in section 3.1, concerning past climate variation. Since CO_2 and global temperature appear to have been remarkably well correlated in the paleo-record, such a CO_2 level based criterion might in fact act as a surrogate for climate damage, but a much better understanding of details of the connection between C-cycling and global climate is needed before such a relationship can be used with confidence.

Energy-economy models that incorporate the necessary range in transformability between energy sources and in emission control technologies and their costs can be used directly to assess costs of reducing emissions and achieving desired reductions in atmospheric CO_2 and other energy related gases. Again, results are often and conveniently put in terms of the taxes required on carbon containing fuels (these are the primary sources of CO_2 and also contribute to CH_4 and N_2O emissions). In these models, macro-parameters defined at the global or at national levels can be altered to reflect changing costs of certain energy forms induced by emission regulation or tax policies. The result in general will be a loss in GNP that will increase as the degree of CO_2 emission reduction increases. These losses could, but have not yet been compared with the type of climate damage cost estimates discussed above.

Quotation of tax rates of coal, oil or natural gas needed to achieve particular reductions in greenhouse gas emissions or atmospheric levels is the most common descriptive route. These are important in a practical and political sense since taxes are normally opposed by those interests that would suffer in their application, and so tax rate citation provides a common reference point useful in presentation to policy audiences. This mode of description is also a favorite in treating emission reduction strategy from the micro-level upwards. Much

468

discussed in several applications besides greenhouse gas warming is the potential for major energy, and hence greenhouse gas reduction through higher efficiency in production and use of energy, and through energy conservation measures. In many of these the cost savings from less energy use is claimed to exceed the cost of introduction of the energy efficiency enhancing techniques. We will not enter into this disputatious arena, other than making note of the primary authors advocating this route as a means for major reduction in climate warming (Goldenberg et al., 1985; Lovins et al., 1981; Chandler, 1985; Colombo & Bernadino, 1979). Counter arguments to this approach do exist (Bold, 1987; Cutler at al., 1984; Jorgenson, 1984; Schaefer, 1980). Here we only wish to suggest wariness in taking too simplistic an application of the numerous end-use calculations concerned with energy saving to derive macro-level (in our case, global) numbers. We are concerned with the neglect of major restructuring of basic industries as well as the secondary infra-structural consequences that massive introduction of energy efficiency technologies could have. We also not that in the macro-economic, free market reponse of the energy industry to reduction in net use or projected use of energy, and the associated excess capacity for energy production (and hence energy price reduction) could be discovery of outlets for new lucrative uses of energy. Additionally, if, and as we believe necessary, significant reduction in greenhouse gas associated warming will require extremely large and rapid assimilation of a new type of energy economy, with the near term introduction of non-fossil energy sources, the likelihood of major economic, technical, institutional and political problems exists if the transformation is to be achieved quickly. Overall, we claim that there are serious difficulties facing energy/economic modeling in defining the optimum path for emission reduction if the rate of reduction required is sufficiently large. Analysis to determine if the desirable limits to emission reduction fall within the bounds of serviceability of such models as we know them today has yet to be started.

4.   The Toronto Statement target

Until very recently it was not appreciated by energy analysts working on the greenhouse gas problem that reduction of CO_2 emission rates implied a fall below the commonly accepted value of approximately one half for the fraction of these emissions that were atmospherically retained (figure 4). Correcting this mis-application of a relatively well understood part of the complex climate system that governs climate change, notably eases the path to reduction of atmospheric concentrations of greenhouse gases. A recent paper by Harvey (1988) specifically addresses this feature. In fact Harvey has calculated the change in global mean temperatures that follow the Toronto prescription - a 20% reduction in CO_2 emissions by 2005, and eventual stabilization at 50% of current emissions. Figure 17 presents some of his results; here it is assumed that CO_2 doubling results in an equilibrium temperature rise of $4^{\circ}C$ - a value on the high side of most current modeling estimates. His calculations show that the Toronto target, if achieved, would yield a temperature rise only about $1^{\circ}C$ above today, and hence would likely not be cause for concern. Moreover, they also indicate that a 50% reduction in emissions (the long term Toronto goal) would in fact stabilize atmospheric greenhouse gas concentrations - a result not predicted by earlier studies.

There remains the question as to how the Toronto target could be achieved. A variety of recent calculations lead to differing conclusions. So far all treat only the first part of the target - a 20% cut of CO_2 emissions by 2005 - through massive efforts at energy use reduction, optimistically without loss of economic productivity, by increasing primary energy production efficiency and by improving end use energy efficiency and energy conservation practices. Specific modalities for achieving the target have been devised, at least for the developed, industrialized Western part of the world. Chandler (1988) has made the most comprehensive evaluation, concluding that a rate of increase in energy efficiency technology development at 3.5% per year, together with substantial taxes on coal and oil, would achieve the target for the United States, at no net loss in GNP. Analagous conclusions of a more general variety have been obtained earlier (Lovins et al. 1980; Krause, 1981; Bach, 1982). Most of these works have concentrated on detailed technological possibilities, and although questions concerning the transfer of conclusions on energy use reduction possibilities from the micro to the macro scale can be raised (as noted in section 3.3 above), even accepting the global applicability of such rosy conclusions, there remains the problem of acheiving the necessary reductions in the developing parts of the world. Here the outcome is more difficult to assess, and even though specific proposals for rapid economic growth in these countries without the normally associated increases in industrial and energy related emissions have been made (Goldenberg et al., 1988), major political problems issues abound. For example, the Toronto target would require China to drop its plans for economic development through large scale utilization of its coal resource base.

The analyses performed to date therefore imply that initially energy use reduction can accomodate the 2005 20% reduction limit provided political opposition in the industrialized countries can be overcome and provided the optimistic portrayals of economic development in the third world at much reduced rates of energy use can be fullfilled. Transition to a long term 50% reduction in emissions requires replacement of fossil fuels sources by renewable energy forms, and no analyses as to economic costs, if any, for such a deliberate move have been made.

If achieved, and accepting the global climate change damage costs values presented in section 2.2, the final global warming under the Toronto ceiling would result in about a 1% loss of global GNP, due to the 1° rise of temperture that Harvey gives. We should note that the 1% figure applies for a <u>linear</u> damage/temperature increase relationship. If, as seems likely, costs increase non-linearly with warming (Laurmann, 1980), GNP loss would be less than this for low ($1^{o}C$) temperature increases. Long term costs for acheiving the needed reduction in greenhouse gas emissions are not known, but in all likelihood they would be below this figure.

5. Summary

The limits to allowable change should be set by determining the most critical features of climate warming impacts. Perception of change and its significance could be as important as reality; extremes reflecting natural variability can be mis-read as indicating a trend, and failure to take note of a scientifically established trend because of the presence of natural variability is a likely sociological response that gives problems. A secondary, but extremely important consequence of warming will be increased frequency of extreme events, such as summer droughts, and changes in frequency and intensity of monsoons, floods and hurricanes. Growing seasons will lengthen at temperate and high latitudes. Rising seas will likely first be noticeable from rare high water and storm events. We need to decide whether targets for allowable greenhouse gas emissions should take such eventualities directly into account, or whether we can assume they are subsumed by targets that are based on the most obvious measure of warming, namely global mean temperature increase estimates. Target values will of course refer to a time ahead, and should be altered as time progresses and predictions change.

Choice of acceptable limits to climate change can be conveniently placed in the following generic classes:

1. Qualitatively based targets. Here predictions of climate change are compared with past experience or with long term past earth history, and judgement made as to unacceptability of projected anthropogenic warming by reference to empirical evidence of the past. The resultant target levels are either set as objectives, to be achieved by whatever means or costs, or judgement is again used as as to "acceptable" modifications of industial, agricultural and energy sector operations that are needed in an attempt to meet the target. Enforcement of use of "best available" emission reduction technology is a variant based on these non cost-based approaches.

2. Partially cost-based targets. The desired limits to climatic change are again set as in 1, but means for achieving the targets so defined are analysed quantitatively by calculating the costs of various means for emission reduction that can reach them. A search is made for the least cost and politically most attractive blend of steps that lead to a net reduction in all emissions that will result in attainment of the allowable temperature change limit, and thereby define a set of emission control targets for each greenhouse gas. In practice, it is likely that the temperature change target will be altered in a process of decision making that involves subjective judgement on what dollar sacrifice is worth the avoidance of the non-quantified cost of the climate change.

3. Full quantification of costs. Here future climate change damage costs, or the costs of adapting to climate change are estimated. The simplest approach is to then compare these figures with costs of the possible methods for avoiding climate change. Remediation cost ceilings are set at or below the estimated climate change costs, thereby establishing emission reduction targets.

4. Optimisation. Carrying 3 a step further, iteration between estimates of costs of ranges of emission reduction levels and costs of climate damage, so as to minimise the sum of the two, is the standard approach to cost/benefit analysis. The target figures thus obtained

defines the least cost solution. The optimal growth method of analysis carries this a degree of sophistication further by using an integrated total expected present day estimate for the sum of remediative and climate damage costs, usually in terms of (discounted) consumer surplus or utility measures. Maximization of the value of the integral with emission reduction as the variable parameter yields the time history of optimum emission levels.

All of the above methods present difficulties, either for acceptability of the rationale used, or in availability of a data base adequate for carrying through the analyses. At one extreme the criteria can be based on ecosystem or species survival as a means for obtaining consensus for limits to allowable change. Economics does not enter into this form of argument. At the opposite extreme, acceptability of a complete optimal growth analysis and its implications for emission reduction methods, depends on the perception of objectivity of the decision analytic tools used and on agreement on choice of values assigned to parameters in the model that involve ethical norms. The latter specifically concern the value of time and attitude toward risk, a well as the fundamental cannon that ~~that~~ utility and worth are expressible in economic terms.

The Toronto target for a 20% reduction in CO_2 emissions below todays rate by the year 2005, and an eventual steady state emission rate 50% below today's, if achieved will probably result in only moderate warming, not exceeding $1^{\circ}C$ above pre-industrial global temperatures. Cost for achieving these targets are unsure. The most likely optimal approach is early promulgation of total energy - and hence fossil fuel reduction - through energy efficiency enhancement, arguably at net profit or loss. Sufficiently rapid and large scale implementation of technologies for energy use reduction will surely be expensive. Associated with this step in energy policy modification will have to be others that reduce non fossil-fuel related greenhouse gas emissions. Apart from the CFC's, no analyses or policy recommendations have been made concerning ways to reduce agricultural and industry sources of emissions. However, reduction or elimination of deforestation has been actively promoted for reduction of CH_4 emissions as well as CO_2. We anticipate that global acceptance of the 20% CO_2 reduction target will be politically difficult to obtain. Even more difficult is the long term 50% reduction value, for which elimination of all fossil fuel sources is required. Costs and implementation issues for this stage of the Toronto global warming reduction strategy have yet to be defined.

References

Bach, W., 1982. Gefahr fur unser Klima. Wege aus der CO_2-Bedrohung durch Sinnvollen Energieeinsatz. Muller-Verlag, Karlsruhe.

Berger, A., 1980. Spectrum of Climate Variations and Possible Causes. In Climatic Variations and Variability: Facts and Theories. Ed: A. Berger. Reidel.

Bold, F.C. 1987. Responses to Energy Efficiency Regulations. Nature 323 286.

Budyko, M.I. & Yu. S. Sedunov, 1988. Anthropogenic Climatic Change. Paper presented at the Climate and Devlopment Conference, Hamburg FGR Nov. 1988.

Chandler, W.H., 1985. Energy Productivity: Key to Environmental Protection and Economic Progress. World Watch Paper 63. World Watch Institute, Washington D.C.

Colombo, U. & O Bernadini, 1979. A Low Growth Scenario and the Perspective of Western Europe. CEC Brussels.

Clark W. et al. Eds., 1982. The Carbon Dioxide Question: A Perspective for 1982. Oxford U. Press.

Clark, W. & R.E. Munn, Eds., 1987. Sustainable Development of the Biosphere. Cambridge U. Press.

Cutler, J., R. Constanza, C.A.S. Hall & R. Kaufmann, 1984. Energy and the U.S. Economy: A Biophysical Perspective. Science 225 890-877.

Goldenberg, J., T.B. Johansson, A.K.N. Reddy & R.H. Williams, 1988. Energy for a Sustainable World. Wiley Eastern, New Delhi.

Hansen, J.E., A. Lacis, D. Rind, & G. Russell, 1983. Climate Sensitivity: Analysis of Feedback Mechanisms. In Climate Processes and Climate Sensitivity. Eds: J. Hansen & T. Takahashi. Geophysical Monograph # 29, Am. Geophys. Union. Washington D.C.

Hansen, J., I. Fung, A. Lacis, D. Rind, G. Russel, S. Lebedeff, R. Ruedy & P. Stone, 1988. Global Climate Change as Forecast by the Goddard Institute for Space Studies Three-Dimensional Model. J. Geophys. Res. 73 9341-9364.

Harvey, L.D.D., 1988. Managing Atmospheric CO_2. Submitted for publication.

Kellogg, W.W. & Schware, 1981. Climate Change and Society. Westview Press, Boulder Colorado.

Krause, F., 1981. An Efficiency and Development Oriented Approach to World Energy Problems. International Project for Soft Energy Paths. San Francisco.

Jaeger, J. 1988. Developing Policies for Responding to Climatic Change. Summary of workshop held in Villach 28 Sept. - 20 Oct. 1987. Beijer Institute, Stockholm.

Jorgenson, D.W. 1982. The Role of Energy in Productivity Growth. The Energy J.5 (3) 11-26.

Laurmann, J.A. 1980. Assessing the Importance of CO_2 Induced Climatic Change Using Risk-Benefit Analysis. In Interactions of Energy and Climate. Eds: W.Bach, J. Pankrath & J. Williams. Reidel, Boston.

Laurmann, J.A. 1985a. The Global Greenhouse Warming Problem: Setting the Priorities.   Climatic Change <u>7</u> 261-265.

Laurmann, J.A. 1985b.   Market Penetration of Primary Energy and its Role in the Greenhouse Warming Problem.   Energy <u>10</u> 760-775.

Laurmann, J.A. 1985c.   Scientific Uncertainty and Decision Making: The case of Greenhouse Gases and Global Climate Change. <u>Science of the Total Environment</u>. <u>55</u> 177-186.

Laurmann, J.A. 1986.   Future Energy Use and Greenhouse Gas Induced Climatic Warming.   In <u>CEC Symposium on CO$_2$ and Other Greenhouse Gases: Climatic and Associated Impacts</u>. To be published.

Laurmann, J.A. & J.R. Spreiter, 1983.   The Effect of Carbon Cycle Model Error in Calculating Future Atmospheric Carbon Dioxide Levels. Climatic Change <u>5</u> 145-181.

Lave, L.B. 1984.   The Greenhouse Effect: the Socio-Economic Fallout. In <u>The Greenhouse Effect: Policy Implications of a Global Warming</u>. Eds. A. Abrahamson & P. Ciborowski. Center for Urban and Regional Affairs. Minneapolis.

Lovins, A.B., 1980.   Economically Efficient Energy Futures. In <u>Interaction of Energy and Climate</u>. Eds: W. Bach, J. Pankrath & J. Williams.

Lovins, A.B., L.H. Lovins F. Krause & W. Bach, 1981.   <u>Least Cost Energy Strategies: Solving the CO$_2$ Problem</u>.   Brick House, Cambridge, MA.

Marchetti, C. 1975.   Primary Energy Substitution Model.   On the Interaction between Energy and Society.   Chem. Econ. Eng. Rev. <u>7</u> 9.

National Academy of Sciences, 1983.   Changing Climate. National Academy Press, Washington D.C.

National Academy of Sciences, National Academy of Engineering & Institute of Medecine, 1988.   Global Environmental Change, Recommendations for President Elect Bush.   Washington D.C.

Nordhaus, W.D. 1980.   Thinking about Carbon Dioxide: Theoretical and Empirical Aspects of Optimum Control Strategies.   Cowles Foundation Discussion Paper #565.

Ramanthan, V., L. Collis, R. Cess, J. Hansen, I. Isakson, W. Kuhn, A. Lacis, F. Luther, J. Mahlman, R. Reck & M. Schlesinger, 1986.   Climate-Chemical Interactions and Effects of Changing Atmospheric Trace Gases. Rev. Geophys. <u>25</u> 1441-1482.

Rotty, R.M, & C. Masters, 1984.   Past and Future Releases of CO$_2$ from Fossil Fuels Combustion. IEA Oak Ridge TE.

Rotty, R.M. & G. Marland, 1980.   Constraints on Fossil Fuel Use.   In <u>Interactions of Energy and Climate</u>. Eds: W. Bach, J. Pankrath & J. Williams. Reidel.

Sassin W. & J.Jaeger, 1988. The Greenhouse Problem: A Step Towards Global Ecological Management. Draft report, June 1988.

Schaefer, H. 1980. Conservation Practices and Increased Efficiency of Energy Conversion and Usage. In Interactions of Energy and Climate. Eds: W. Bach, J. Pankrath & J. Williams. Reidel.

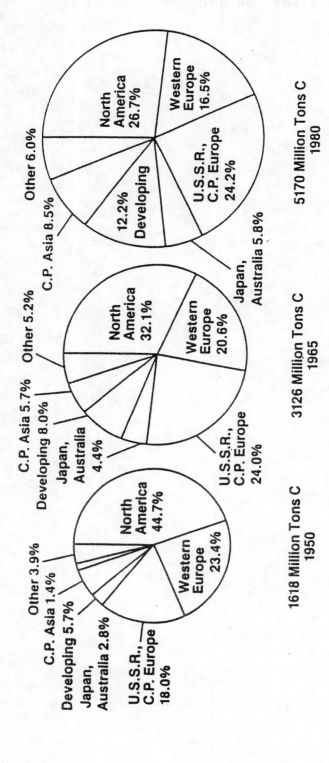

Figure 1. Historical Carbon Dioxide emission rates by region (Rotty & Marland, 1980)

476

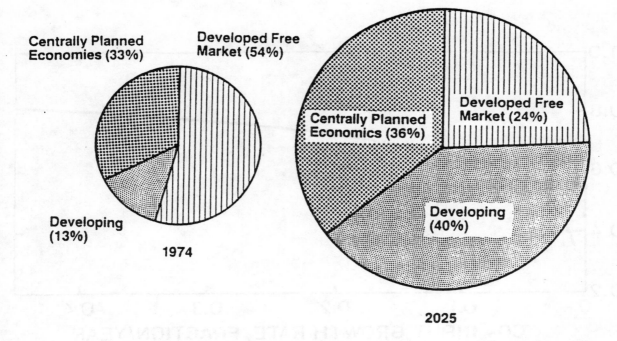

Figure 2.   Projected change in shares of carbon dioxide emissions
            for major world sectors. Taken from Rotty and Marland (1980)

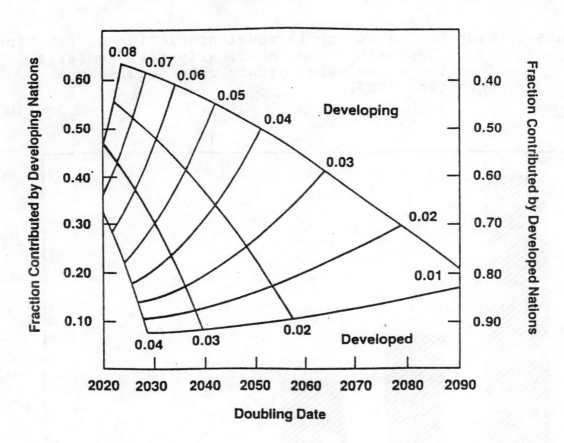

Figure 3.   Relative contribution of the developing nations to
            doubling of atmospheric CO_2 level as a function of
            energy growth rates in the developed and developing
            nations. Taken from Kellogg and Schware (1981)

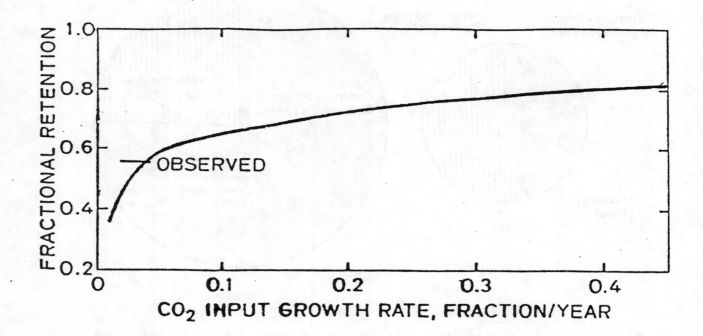

Figure 4. Four-box carbon cycle model prediction of fraction
of CO_2 emissions atmospherically retained as a
function of emission growth rate (Laurmann &
Spreiter, 1983).

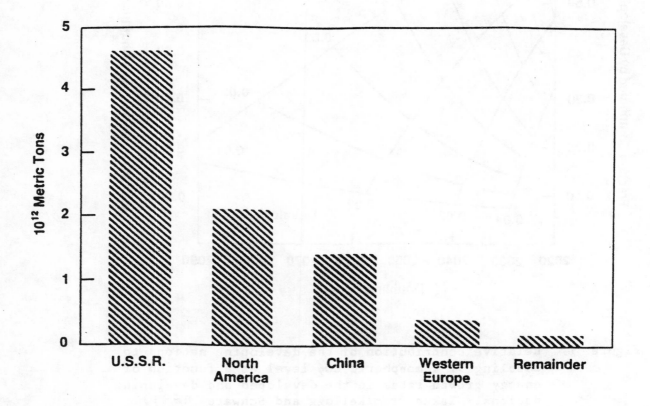

Figure 5. Global distribution of coal reserves.

478

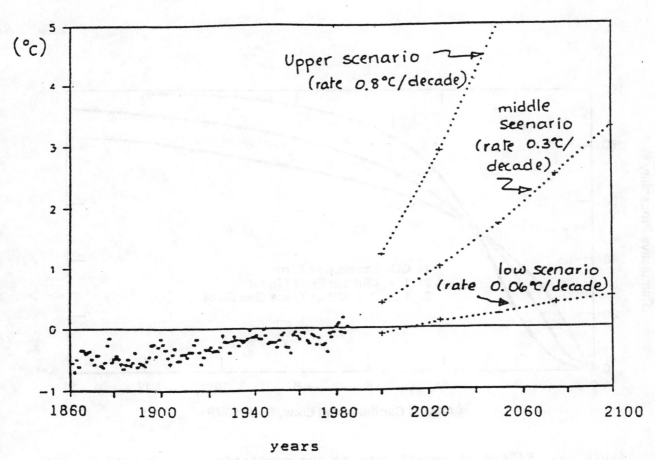

Figure 6.   Villach II warming projections (Jaeger 1988)

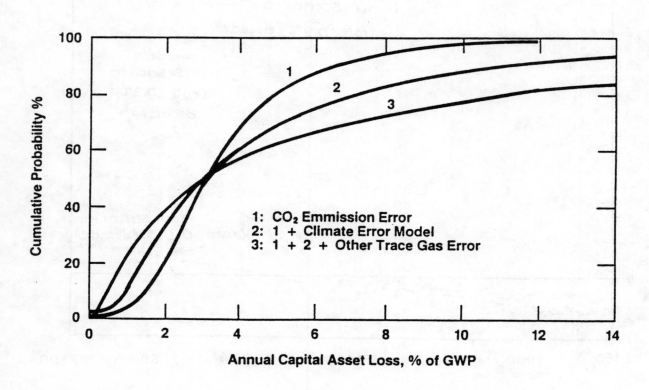

Figure 7. Effect of uncertainty on the cumulative probability
distribution of climate warming costs for the year 2025.

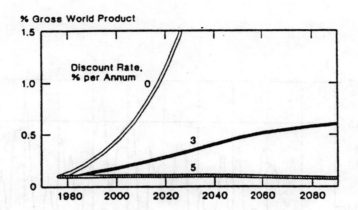

Figure 9. Effect of future discounting on present day expected costs of a year 2025 climatic change.

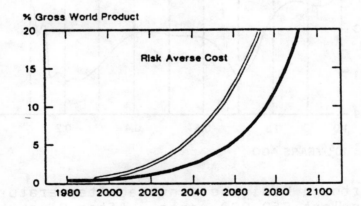

Figure 8. Effect of risk aversity on expected costs of climate change.

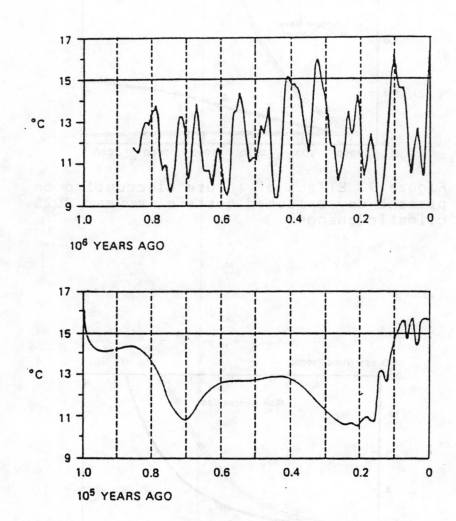

Figure 10. Northern hemisphere average temperature over the last 850,000 years. After Clark, 1981.

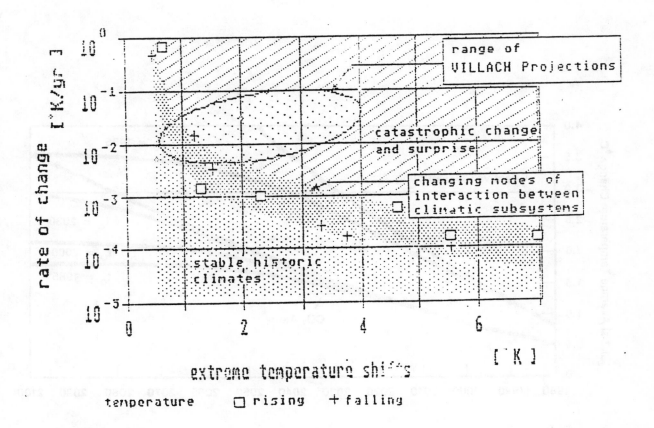

Figure 12. Domain of Villach (figure 6) climate change
projections in a rate-of-temperature-change/
magnitude-of-temperature-change space. The
heavy dotted area represents data extracted
from the climate record (figure 10), and
bounds the range of the past one million
years of climatic variations.

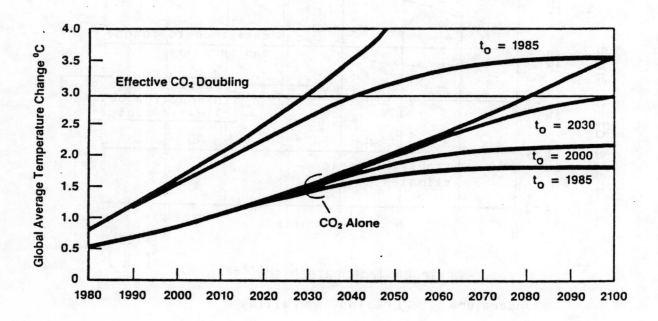

Figure 13. Effect of global temperature rise due to increasing
CO_2 level for logistic replacement of fossil fuels,
assuming various dates, t_o, for a 1% share of the
non-fossil replacement energy (lower curves). Upper
curves show the effect of including all trace
greenhouse gases, including CO_2.

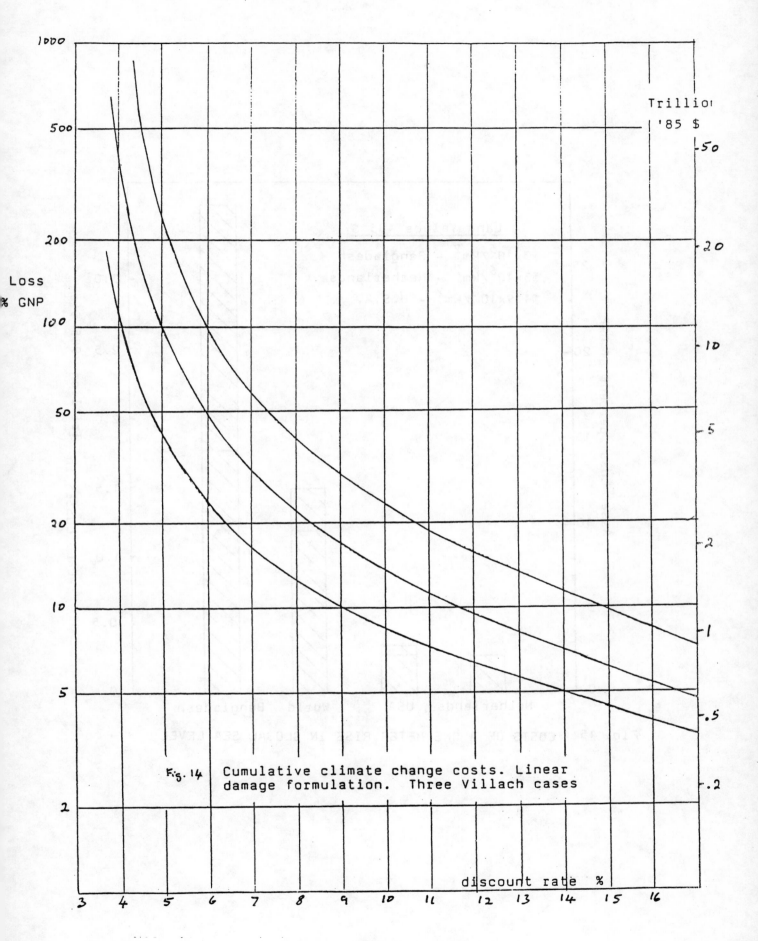

Fig. 14   Cumulative climate change costs. Linear damage formulation.   Three Villach cases

485

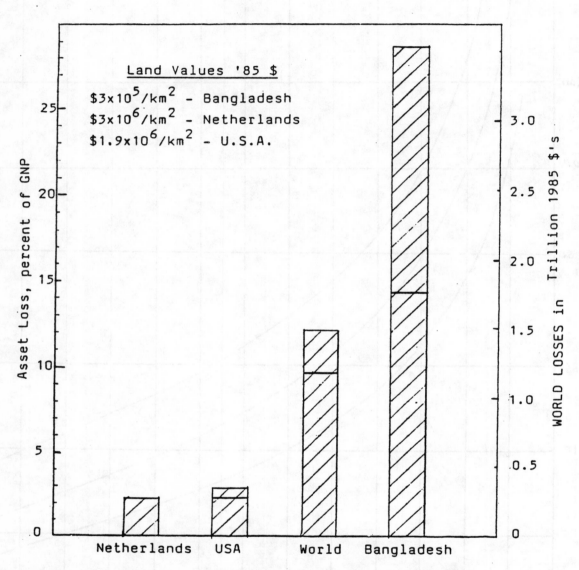

Fig. 15   COSTS OF A ONE METER RISE IN GLOBAL SEA LEVEL

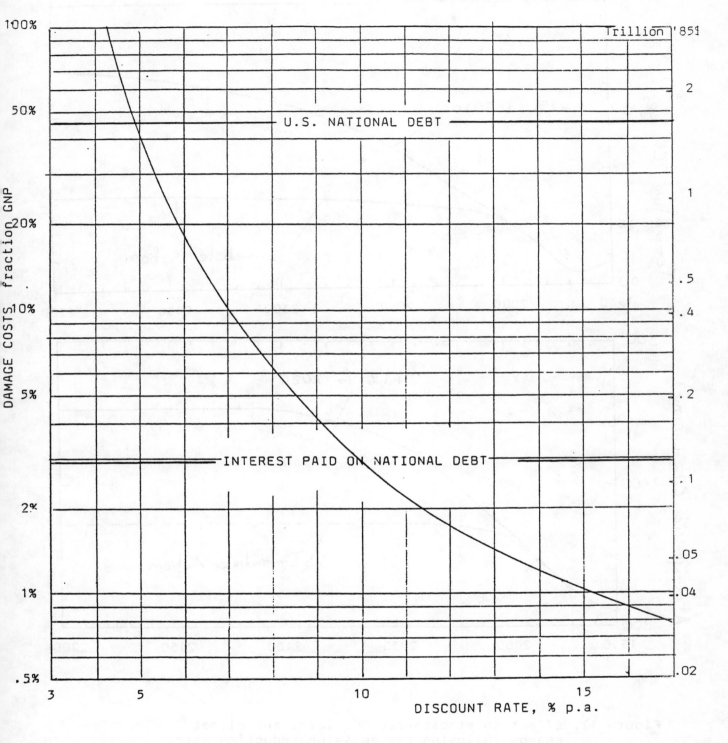

Figure 16.   Cumulative sea level rise asset loss for the U.S.,
discounted to 1985.

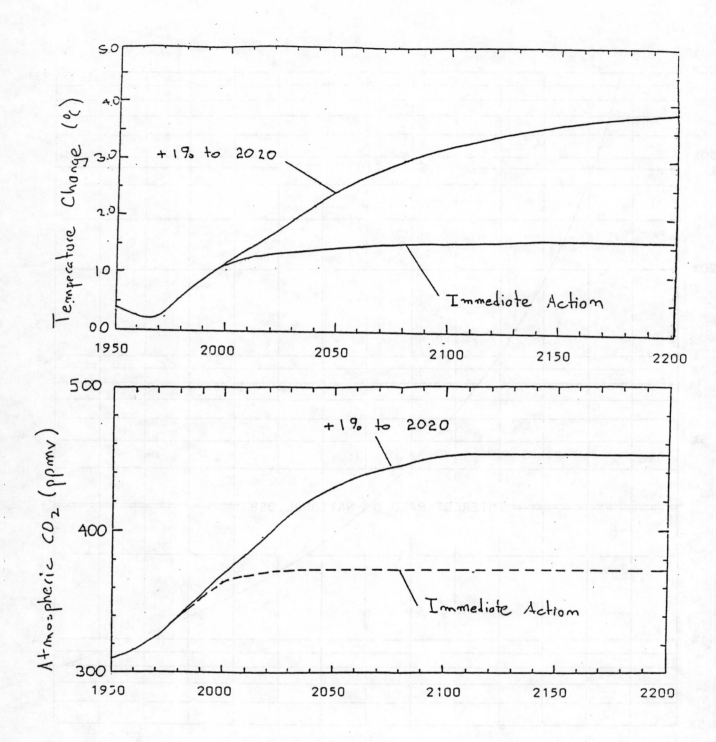

Figure 17. Effect on atmospheric CO₂ level and climate
change following two emission reduction paths:

"Immediate Action"	"1% to 2020"
(Toronto 2005 target)	(delayed action)
constant to 1991, decreased to -2%p.a. by 2001;	+1%p.a. to 2020, decreased to -1%p.a. by 2030;

- continued until emissions match uptake.

AN EFFECTIVE GREENHOUSE GAS EMISSION REDUCTION STRATEGY FOR THE PROTECTION OF THE GLOBAL CLIMATE

by

Wilfrid Bach

Center for Applied Climatology
and Environmental Studies
Department of Geography
Climate and Energy
Research Unit
University of Münster
Robert-Koch-Str. 26
4400 Münster
FR Germany

Conventional scenario analyses with their growth projections into the future have very little to do with climate protection. A new and effective climate protection strategy is needed. It consists of the following steps: Setting of a permissible ceiling for global warming through an International Convention - assessment of a combination of trace gas emission reduction scenarios which guarantee to stay below set ceiling using climate modeling for backcasting and not, as usual, for forecasting - allocation of the required emission reduction rates by nation through an International Convention - assessment of the existing emission reduction potential by gas, source, measure, and nation - development of optimized emission reduction scenarios for all major nations - agreement on emission reduction plans through International Convention and Protocol.

The strategy is applied to the FR Germany. For a set global warming ceiling of 1-2°C in 2100 relative to 1860 the required CO_2-emission reduction share of Germany in one specific scenario amounts to 36 Mio t/yr during 1990-2000. A variety of measures is available to meet the reduction quota in each year

1. INTRODUCTION

There is wide consensus that, at the continuation of the present emission trends of greenhouse gases, mankind is heading toward a climatic change whose impacts cannot be tolerated. The nations of the world have therefore no choice but to come together and, as a precautionary measure, work out jointly an agreement on how best to reduce the climatic risks.

To this end, much activity has been started at national and international levels. For example, in late 1987 the Parliament of the Federal Republic of Germany established an Enquete-Commission "Protection of the Earth's Atmosphere" consisting of 11 members each of the Parliament and the science community. The express mandate of the Commission is to

- probe into the destruction of the protective ozone layer in the stratosphere and the accumulation of deleterious ozone in the troposphere

- explore global climatic changes and their societal impacts

- investigate the destruction of forests and soils

- develop strategies for the reduction of the influencing factors threatening climate

- recommend specific countervailing measures and evaluate their political feasibility.

Another example, at the European level, is EUROSOLAR, an 'Association for the Solar Energy Age', which was founded in 1988 with the express purpose of "preserving or restoring the natural living conditions for Man and his environment by shedding the burden resulting from nuclear and fossil fuel energy use; working toward a solar energy age is seen as a centenarian task which will supply mankind with sustainable energy that is environmentally benign and socially acceptable."

International conferences are indispensible for further strengthening the scientific basis. Some examples are:

- The First and Second North American Conference on Preparing for Climatic Change: (Washington, 1987 and 1988)

- Energy and Climate Change: What Can Western Europe Do? (Brussels,1988)

- The Changing Atmosphere(Toronto, 1988)

- Climate and Development(Hamburg, 1988)

- The Second World Climate Conference(Geneva, 1990).

To prepare for the Second World Climate Conference and to work toward an International Convention, WMO and UNEP recently set up an 'Intergovernmental Panel on Climate Change' with three Working Groups to

-   assess the available scientific information on climate change

-   assess the environmental and socio-economic impacts of climate change

-   formulate response strategies.

Very important are also national policies, such as the "Global Warming Prevention Act of 1988" in the US, which was designed "to implement energy and natural resource conservation strategies appropriate to preventing the overheating of the Earth's atmosphere". Even more importantly, the Act sets national goals for the reduction of CO_2, NO_x, and CFCs, and encourages the adoption of a binding multilateral agreement no later than 1992.

The major task ahead is to come to a first agreement on the reduction of greenhouse gases by 1992. To be successful, such a reduction concept must be based on a strategy that takes equal account of the intricately related scientific, economic, social, and environmental aspects involved. The following reduction strategy is based, in part, on work done in the Enquete-Commission(1988).

2. REDUCTION STRATEGY

Man-made greenhouse gases (GHG) from fossil fuel burning are the main contributors to climatic change. To reduce the risk of impacts from such climatic change, it is prudent to reduce these GHG. A reduction strategy must take into account the following(Bach, 1989):

-   Myriads of individual sources all over the globe contribute to the global climate problem.

-   Past and present activities predetermine to a certain degree the magnitude of a future climatic change.

-   The sooner reduction measures are introduced, the smaller will be the GHG-induced climatic changes and impacts.

-   The key to buying time and decreasing climatic risk lies in a near-term reduction policy, when the rate of gas release is still very large, but the atmospheric concentration still relatively small, and not in a few decades hence, when the growth rates of the emissions have tailed off, while the concentrations have reached by then high atmospheric levels which influence the climate.

- It takes time to turn worldwide a wasteful and polluting energy/economic/land use system into one that is more efficient and environmentally benign.

- Beside the greenhouse gases, there are other contributing factors that are already identified (e.g. aerosols, land use changes), those already in the environment system but not yet identified as climate-relevant, and those that may be added in the future.

If society decides that the prospects outlined above pose an unacceptable risk, decision makers are called upon for coordinated action. The question is: what should be done, and, more so, how should it be done? The usual approach is an across the board percentage cut. While this often gives the impression of fairness, it makes neither economic sense nor does it necessarily lead to the required reduction of the climatic threat. The reason for this lies in the complexity of the economic and environment/climatic systems. The control measures for the individual gases and sources or uses are quite different and so is the ease or difficulty of their implementation. In view of the complex non-linear feedback processes between the environment and climate systems, one dimensional estimates of percentage changes must necessarily convey an erroneous picture.

A more rational approach is now presented. It starts from the premise that climate control is at the global level, and emission control of the greenhouse gases is at the national level, while agreement on all of these is at the International Convention Level. The individual steps of the reduction strategy are as follows:

- Setting a permissible ceiling for global warming through an International Convention

- Assessment of possible trace gas emission scenarios required to limit global warming to set ceiling using climate modeling

- Allocation of the required emission reduction rates by gas and nation through an International Convention

- Assessment of the existing emission reduction potential by gas, measure, and nation

- Development of optimized emission reduction scenarios by gas, source, measure, and nation

- Development of national emission reduction plans

- Agreement on emission reduction plans through International Conventions and implementation by Protocols.

2.1 Setting of a permissible ceiling for global warming through an International Convention

It is theoretically possible to set a ceiling for emission, for concentration, and for temperature. Based on the effects, a ceiling on emissions is, however, highly doubtful, because a certain emission reduction does not translate into the same reduction of concentration and temperature. Moreover, due to the inertia of the climate system, for a certain emission reduction, concentration and temperature may continue to rise for quite some time. Also, setting concentration limits for individual gases will not guarantee that a set temperature limit will not be exceeded. Consequently, for effective control, the ceiling must be set at the temperature level, because it is here that both the individual contributions of the climate-influencing factors and the inertia of the system are adequately taken into account.

Long term natural temperature changes between glacial and interglacial periods due to orbital changes have been shown to have a rate of the order of ca. $0.01^{\circ}C$/century (Enquete-Commission, 1988). The rate of change between the Industrial Revolution and the present has been about $0.6^{\circ}C$/century. If the current rate of ca. $0.1^{\circ}C$/decade could be maintained, temperature change could be limited to about $1^{\circ}C$ until 2100. Participants of the 1987 Villach Climate Conference were of the opinion that a global mean temperature rise of about $1^{\circ}C$/century was just about the limit to which most ecosystems could adapt. Nevertheless, there is cause for fear that magnitude and speed of such temperature change might overwhelm the adaptability of some ecosystems. Another criteria for setting a limit to temperature rise is the impact of climatic change on crops. Regression analyses for the US corn and wheat belts have shown that for a precipitation decrease or increase of 20% and a temperature increase of $1^{\circ}C$ and $2^{\circ}C$, corn yield was reduced between 8-14% and 20-26%, while wheat yield was reduced between 4-6% and 9-10%. In this connection it is well to realize that a global mean ceiling of "only" $1-2^{\circ}C$ entails a 2 to 3 fold temperature amplification toward the poles, and simultaneous changes in the circulation and precipitation patterns leading, as it were, to more extremes in terms of drought, floods, heat waves and violent storms, which will most likely become more frequent and hence receive most of the attention. On the basis of this argumentation, the Enquete-Commission set the ceiling of a mean global warming at $1-2^{\circ}C$ by year 2100 relative to 1860. In a joint statement, both the German Meteorological Society and the German Physical Society support this ceiling. Limiting global warming to the $1-2^{\circ}C$ ceiling requires a reversal of the present trends and a corresponding reduction of the climate-influencing anthropogenic factors. Due to the complexity of the climate system, the required emission reduction can only be assessed with sufficient accuracy with the help of climate modeling. Finally setting of a permissible ceiling for global warming can only be achieved through an International Convention.

493

2.2 Assessment of possible trace gas emission scenarios required to limit global warming to set ceiling using climate modeling

The usual approach is to develop scenarios which project trace gas emissions into the future and to calculate the resulting global temperature change for certain time targets. This may be of interest but has nothing to do with protecting the climate. If climate protection is the goal a different approach, called backcasting, is appropriate. A permissible ceiling for global warming is set, for example, at 1-2°C by 2100 relative to 1860 through an International Convention(see section 1.2), which determines the emission reductions required to meet set ceiling using climate modeling. The required emission reductions depend upon

- the permissible ceiling for global warming

- the number and type of greenhouse gases

- the time targets.

Therefore, within a set warming ceiling a number of reduction scenarios is possible. Fig.1 shows one example of an emission reduction scenario for the trace gases CO_2, CH_4, N_2O, CFC-11, and CFC-12. For demonstration purposes only the required CO_2-emission reduction changes and the resulting concentration changes are shown.

In this example the calculation is done with a parameterized form of a 1-d radiative convective model (RCM) of the atmosphere as developed by Hansen(1983) and a 1-d energy balance model (EBM) of the ocean of a type similar to that used by Cess and Goldenberg (1981). With this model combination the global mean transient surface temperature change is estimated from 1860 to 2100. The RCM takes into account the concentration changes of CO_2, CH_4, N_2O, CFC-11, and CFC-12 together with changes in surface temperature, volcanic activity and "solar constant". The latter two were ignored. Furthermore, neither the concentration changes in other CFCs and O_3 nor the rather complex chemical reactions of the ozone chemistry were considered. The CH_4 and N_2O expressions in the Hansen formulation were corrected.

The EBM computes the heat flux into the ocean which results from the trace gas concentration increases. For that purpose the ocean is subdivided into 42 vertical layers - analogous to the carbon cycle model. The ocean mixed layer (about 75 m) is assumed to be fully mixed. In the deep ocean (about 4000 m) energy transport is modeled as a vertical diffusion process. Of the many possible feedback processes only the ocean buffer factor is explicitly taken into account. Other feedbacks are included in

494

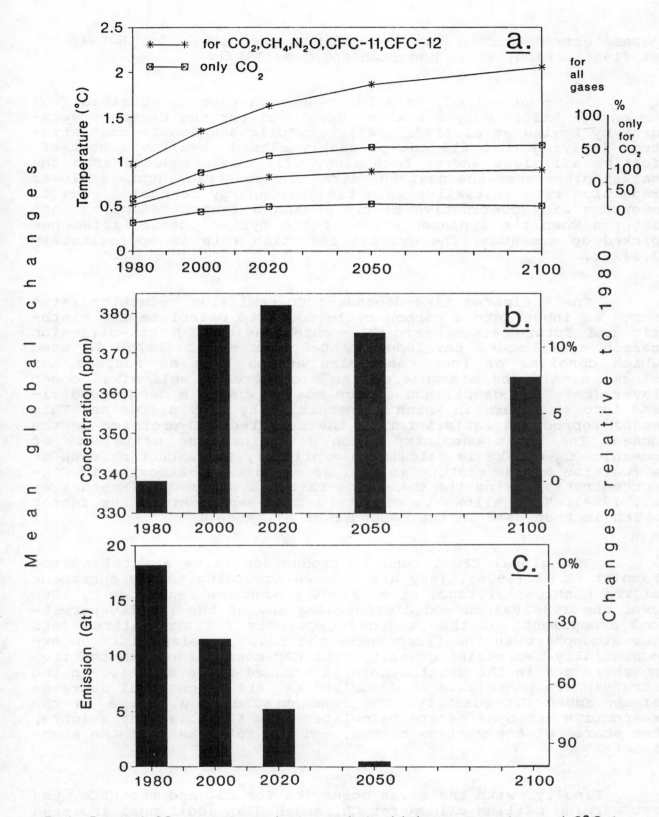

Fig.1 Required CO₂-emission reduction to limit global mean warming to 1-2°C in 2100 (c), resulting CO₂-concentration change (b), and temperature change due to CO₂ plus the other indicated trace gases (a).

parameterized form. Further details are given by Hansen et al.(1981), Bach(1985), and Bach and Jung(1987).

The required CO_2 emission reduction rate is obtained from an energy "Efficiency Scenario" developed for the German Government by Lovins et al.(1981/1983). It follows a least- cost strategy implying that all the presently already available cost-effective efficient energy technology will be introduced into the market place over the next 50 years. The scenario adopts a slower reduction rate initially, when rational energy technology has to overcome many obstructive barriers, and a faster reduction rate lateron when the implementation of the myriad possibilities has picked up momentum. The overall reduction rate is approximately 3.2%/yr.

The estimated time-dependent CO_2-emission reduction rates serve as input into a carbon cycle model to calculate the historic and future atmospheric CO_2-concentrations. A box-diffusion carbon cycle model developed by Oeschger et al.(1975) is used which consists of four reservoirs acting both as sources and sinks, namely the atmosphere, the biosphere, a well-mixed ocean layer (ca. 75 m deep), and a deep sea (ca. 4000 m deep) subdivided into 42 layers in which transport is by eddy diffusion. This model reproduces satisfactorily the measured ^{14}C-profiles in the ocean. The ocean chemistry which determines the efficiency of oceanic CO_2-uptake is calculated explicitly for each time step as a function of temperature as well as oceanic and atmospheric CO_2-concentration using the Oak Ridge carbon cycle model (Emanuel et al. 1984). This allows to compute a time-dependent buffer factor which is then used in the Oeschger et al. model.

The global CFC-11 and 12 production rates are taken from Hammitt et al.(1986). They are reduced according to the agreement of the Montreal Protocol, i.e. a 50% production reduction by 2000 over the 1986 values and disregarding any of the legally-permitted exemptions. Of the produced CFCs only 70% are emitted into the atmosphere in the first year, the rest is released in an exponentially decreasing quantity. The CFC-mass balance in the troposphere and in the stratosphere is assumed to be 90 : 10. In the stratosphere photolysis is described as the exponential decrease of an added CFC-quantity. The concentrations by volume in the respective atmospheres are calculated from the molecular weights, the shares of the various masses, and the total mass of the atmosphere.

Finally, with the given scenarios for CO_2 and the CFCs, the set warming ceiling allows for CH_4 and N_2O an additional increase of only 10% each from 1980 to 2100.

It is well to return again to Fig.1, because it gives a good demonstration of the fact that a certain emission rate does

496

not translate into equal concentration and temperature reduction rates. This gives a good indication of the enormous inertia of the climate system. Even with the drastic CO_2-emission reduction (Fig.1c) of the order of ca. 90% by 2050, the concentration (Fig.1b) continues to rise by more than 10% over the 1980 value, a trend, which is only reversed in the late 2020s. All of this translates into the indicated global mean surface temperature rise (Fig.1a) reaching $1-2^{O}C$ in 2100 over 1860, to which CO_2 contributes about one half. The temperature range of $1-2^{O}C$ takes into account the uncertainty in the current state-of-the-art climate modeling. The uncertainty in our presently best available climate models, namely the three dimensional general circulation models, is between 1.5 and $4.5^{O}C$ for a CO_2-doubling.

2.3 Allocation of the required emission reduction rates by gas and nation through an International Convention

Emission reduction occurs at the national level. Therefore, the emission reduction rates determined by the global climate model must be allotted to individual nations according to some fair scheme yet to be worked out by an International Convention. For simplicity, here we have applied the same reduction rates to all countries. The developing countries (DCs) might argue that the industrialized countries (ICs) have had in the past a disproportionately large share in fossil fuel use and that, therefore, the DCs should be required to reduce their emissions much less, if at all. This would automatically mean greater reduction rates for the ICs.

Specifically, the 1982 CO_2-emission rates by source have been obtained from U.N. statistics (Rotty, 1986). The 21 largest CO_2-emittors account for ca. 86% of the global CO_2-emission. To this, the ICs contribute 72.4% and the DCs 13.4%. Within the ICs, the USA and Canada as well as Eastern Europe contribute each ca. 26% which is about double that contributed by Western Europe. In Western Europe the FRG is the largest CO_2-contributor with ca. 3.8% of the world total. The USA, as the world's largest single CO_2-emittor, releases 6 times as much CO_2 as the FRG, while the population is only 4 times greater. The three largest CO_2 emittors, the USA, the USSR and China, account for more than half the world's CO_2-emissions. They have an obligation to set an example, and this would have a marked effect on the global CO_2-release.

Fig.2 shows the required CO_2-emission reduction rates for the world's largest emittors. For the initial year 1990, for the FRG this amounts to about 50 mill.t or ca. 7% of the 1982 CO_2-emission. For better visualization, this translates into a required fuel reduction(MtCE) of 8, 11, and 4 out of the 1982 consumption of 115, 160, and 55 for coal, oil and gas, respectively. The required CO_2 reduction rates in 1990 for the USA, USSR and China are about 310, 240 and 104 Mio t, respectively.

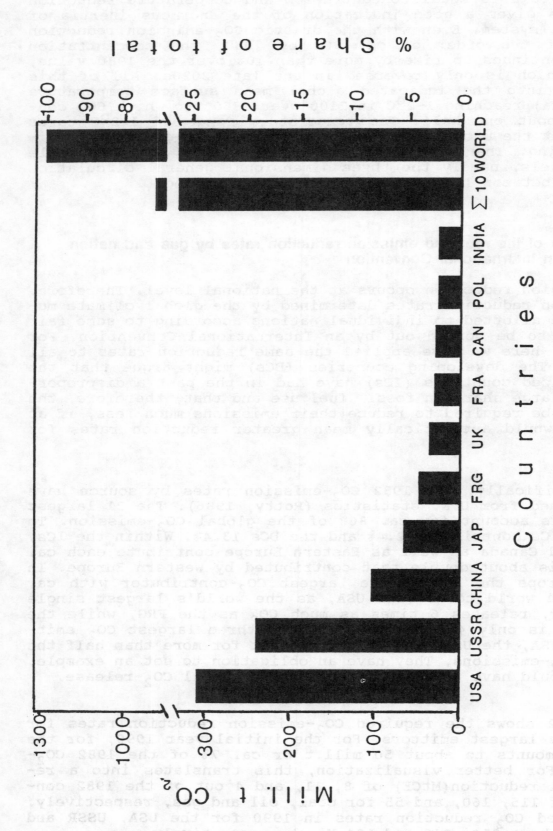

Fig.2 Required CO_2-emission reduction by major countries for initial year 1990 to limit global mean warming to 1-2°C in 2100

A large number of diverse measures is at our disposal to reduce the increasing trace gas trends. The next section highlights the main options available for reducing the risks.

2.4 Assessment of the existing emission reduction potential by gas, measure, and nation

At this stage the existing emission reduction potential is assessed for each measure with the purpose of deriving for each country an emission reduction menu which is at the same time

- technologically most advanced

- most cost-effective, and

- climatically and environmentally most benign.

The assessment of the reduction potential takes into account not only the technical and economical, but also the social and ecological aspects. The principle of minimizing both the costs and the environmental damages is followed, i.e. the reduction potential of the existing technologies and/or substitutes is considered only, if they are both more cost-effective as well as socially and environmentally more acceptable than those to be displaced. The cost comparison must not be based only on the operating costs, but also on the total social and environmental costs incurred on society, i.e. the external costs must be internalized. The influence of the reduction potential by economic and fiscal instruments(e.g. hidden subsidies, barriers, incentíves etc) must also be assessed.

The main question is, are the drastic reductions derived above attainable? For comparison reasons we convert the required percentage reductions to reduction quantities per decade as shown in Table 1. Also shown in this table is a selection of measures from a group at the Jülich Nuclear Research Center (Wagner and Walbeck, 1988) and another group at the German Federal Environmental Agency (Beck et al., 1988). Some of the measures in Table 1 can be introduced quickly, such as speed limit leading to a CO_2-cut of some 26 Mio t; others are longer-term, such as fossil fuel power plant efficiency improvement bringing a relief of 19 Mio t. Some measures will require the right political signals, others, such as the change in steel production, will come along with the ongoing structural changes resulting in a ca. 9 Mio t CO_2-reduction by year 2000.

Together, the existing reduction potential of ca. 430 Mio t could already avoid some 55% of the 1987 FRG CO_2-emission. Moreover, with the current potential, some 80% of the reduction requirement until 2010 can already be met. Equally important is the fact that the required reduction quantities decrease rapidly in subsequent decades as shown in Table 1. Furthermore, it is almost certain that all other countries have an even greater CO_2-emission reduction potential than the FRG. Even more significantly, also in the FRG there still exists a great untapped reduction potential. Latest estimates show that technical efficiency improvements in primary energy use of up to 90% are possible in industrialized countries (Enquete-Commission, 1988). More detailed studies are required to investigate to what extent the efficiency potential can be introduced into the energy-economy system. All of this is very good news, because the CO_2/climate problem appears to be manageable, despite the drastic reduction rates required. The Enquete-Commission has initiated the first analyses regarding the technological, economical and ecological aspects of the existing emission reduction potential of the major greenhouse gases.

2.5 Development of optimized emission reduction scenarios by gas, source, measure and nation

In this section we carry the analysis one step further by relating the required FRG emission reduction share to the emissions from each specific source and to the existing reduction potential based on specific measures (Table 2). So far this type of detailed analysis has only been done for CO_2. The required CO_2-emission reduction as obtained by climate modeling in section 2.2, and as allocated in section 2.3, for the FRG amounts to an average 36 Mio t/yr for the first decade 1990-2000. The existing reduction potential is taken from the diverse measures listed in Table 1. The current(1987) CO_2-emission amounts are calculated on the basis of the existing fossil fuel mix using as specific emission factors (t CO_2/t CE) 3.252, 2.683, 2.357, and 1.545 for lignite, hard coal, oil, and gas, respectively(Wagner and Walbeck,1988). In 1987 in the FRG, oil contributed almost one half to the total CO_2-emission of 790 Mio t, and almost as much as the other fossil fuels hard coal, lignite, and gas taken together. Hard coal, in turn, contributed as much as both lignite and gas. Interestingly, the CO_2-emission from gas was as large as that from lignite, despite the very low specific emission factor of gas. It is, however, not the emission factor, but rather the total amount brought into the atmosphere, that is climate-effective.

For demonstration purposes, reduction suggestions are made for the two years 1990 and 1991. The required average CO_2-emission reduction is 36 Mio t/yr. It is obvious that for Germany oil has to take the greatest cut. This can be effected most quickly by introducing a speed limit of 100/80/30 km/h for freeway/highway/inner city traffic reducing the CO_2-output by ca. 26 Mio t. The remaining 10 Mio t can come from improved building in-

Table 1   FRG CO_2-Emission Reduction Shares Required during the next Decades to Limit Global Warming to 1-2°C in 2100 and Selection of Existing FRG CO_2-Emission Reduction Potential

	Mill. t CO_2
CO_2-Emission in 1987[1]	790

Required Reduction[2]

during the decade 1990-2000	361
" 2000-2010	175
" 2010-2020	84
" 2020-2030	15

Existing reduction potential (a selection)[3]		cumulative
20 % increase in end-use energy efficiency for each electricity, oil, and gas	149	
Improved insulation in buildings	95	244
15 % substitution of hard coal and lignite used in electricity production by hydrogen	34	278
Speed limit (100/80/30 km/h for highway/freeway/inner city)	26	304
Substitution of industrial fuel oil S by natural gas	26	330
50 % substitution of fuel oil used in space heating by district heating	24	354
Improved efficiency in hard coal power plants	19	373
Improved efficiency in fuel oil appliances	18	391
33 % substitution of fuel oil in space heating by gas	16	407
Innovative technology in industry	12	419
Change in steel production	9	428
20 % addition of hydrogen to gas	6	434

Data extracted from: 1) Gesamtverband des deutschen Steinkohleberg-baus (1988); 2) Climate modeling; 3) Federal Environmental Agency (1988); Wagner und Walbeck (1988)

501

Table 2 FRG CO_2-Emission-Reduction-Scenario by Source and Measure for the Decade 1990-2000

Required emission reduction share of the FRG is ca. 36 Mio t/yr
to limit global warming to 1-2°C in 2100

Year	Measure	CO_2-Emission-Reduction-Scenario (Mio t)				
		Oil	Hard Coal	Lignite	Gas	Total
1987		385	203	101	101	790
1990		385	203	101	101	790
	- Speed limit (100/80/30 km/h)*	-26				
	- Improved insulation in buildings	-10				
1991		349	203	101	101	754
	- Improved efficiency in heating appliances	-2				
	- Improved efficiency in power plants and cogeneration		-2	-1		
	- Substitut. of heating oil by distr. heating	-2				
	- Improved insulation in buildings	-10				
	- Increase in end-use efficiency for oil, coal, gas	-10	-2	-2	-4	
	- Change in steel production		-1			
1992	dito	325	198	98	97	718
2000	"	133	158	74	65	430
1990	fossil fuel use	163.3	75.5	31.2	65.2	
2000	(Mio t SKE)	56.4	58.9	22.8	42.1	
1990-2000 Reduction (%)		66	22	27	36	46

* for freeway/highway/inner city traffic

sulation. Preliminary investigations show a reduction potential of about 100 Mio t, which, at an introduction rate of 10%/yr, could annnually contribute some 10 Mio t to the CO_2 reduction until the end of the century. While most reduction must continue to come from oil, also the other fossil fuels will have to make some contribution. Here the fact that Germany has practically no oil resources of her own and that coal is the major indigenous resource base must be taken into account. Improved end-use efficiency and cogeneration can bring substantial emission cuts in all fossil fuels. Gas pipelines will compete with the district heating pipelines from cogeneration. All of this has to be worked out with the goal of arriving at an optimized reduction mix that is best for the respective nation.

The numbers given here could form the basis for one CO_2-emission reduction scenario. Other scenarios are, of course, possible. This scenario would lead by year 2000 to reductions(%) of 66, 22, 27, 36, and 46 for oil, hard coal, lignite, gas, and total, respectively. Finally, it is this type of transparent listing in tabular form of the available information that is required by all involved in the decision-making process.

2.6 Development of national emission reduction plans

Effective climate control requires emission reduction plans for all major greenhouse gases. As a start it is, however, in order to focus on CO_2, with about 50% the main contributor to global warming. This has been done at the Toronto Climate Conference(1988) and, in the US, with the introduction of the "Global Warming Prevention Act of 1988". Table 3 shows the proposed reduction schemes. Unfortunately, these schemes are based on political numbers and have, therefore, no climatic relevance. With a complex non-linear system such as the climate system the emission reduction required to meet a set ceiling of global warming can only be derived with some confidence by using climate modeling. This has been done for the Hamburg Climate Conference(1988). But there was neither enough time allowed in the program to have a thorough discussion of the reduction strategy nor of the modeling results from which the reduction schemes were derived. Table 3 lists the required CO_2-emission reduction to meet the ceiling of a 1-2°C global warming in 2100 for the globe and the FR Germany, as derived from climate model calculations discussed in section 2.2. This is based on a scenario emphasizing efficient energy use. Other scenarios are possible depending on the specific goals, the type and number of climatic influencing factors as well as the time frame.

One of the major tasks ahead is the development of national reduction plans not only for CO_2 but also for the other greenhouse gases. Panels and working groups at international conferences have to work on the refinement of the concept and the details

503

Table 3 CO_2-Emission Reduction Plans

Reduction scheme

Toronto Climate Conference (1988)

by 2005 a reduction of 20 % of the 1988 level

US Global Warming Prevention Act of 1988

by 2005 a reduction of 20 % of the 1987 level

Own proposal

Global[1]

by 2000 a reduction of 35 % of the 1980 level
by 2020 "   "   " 70 % "   " 1980   "
by 2050 "   "   " 90 % "   " 1980   "
by 2100 a virtually closed carbon cycle

FR Germany[2]

by 2000 a reduction of 50 % of the 1980 level
by 2020 "   "   " 90 % "   " 1980   "
by 2050 "   "   " 95 % "   " 1980   "
by 2100 a transition to a closed carbon cycle

---

[1] CO_2-emission reduction share required to stay below a mean global temperature ceiling of 1-2°C in 2100 as deduced from climate modeling

[2] FRG reduction share proportional to her present contribution to CO_2-emission

of the reduction plans. Agreements have to be reached through International Conventions on the extent to which the world is prepared to protect the global climate.

3. NECESSARY MEASURES

The above strategies would lead us from the wasteful throughput economy to an environmentally benign sustainable economy, thereby ensuring a stable social order. The advantages of such a policy are many, from a decreased dependence on external resources, reduced deficits in the balance of payments, and job security, to a reduction of technological risks and associated acceptance problems as well as a decrease both in the destruction of the environment and the risks accompanied by climatic change.

We have no other choice but to act now. The pressing nature of the trace gases/climatic threat should give an impulse to the deployment of the above precautionary measures by which this danger can be reduced. The chances of success are by no means small, because what is most economical also causes the least climatic and environmental risks. This gives rise to cautious optimism, because the precautionary measures suggested above make sense also for other than mere climatic reasons and should therefore be taken anyway. The trace gas/climatic change problem need,therefore, not be fate because we have available to us the powerful tool of a low climatic risk strategy. The sooner we apply it the more gentle will be the transition to a sustainable future.

The necessary measures can be summarized as follows:

- The Montreal Protocol should be revised in 1990 with the goal of a complete phase-out by 2000 of all controlled halocarbons. All other halocarbons should also come under control of the Protocol and be phased-out by 2005.

- An International Convention for the Protection of the Global Climate should come to a consent no later than 1992 on a 70% CO_2-emission reduction by 2020 as compared to 1980.

- An intensive research program should be started immediately to investigate the emission reduction potential of all the major climate-influencing factors.

- There should be more emphasis on preventive than on curative measures supplemented by an adequate monitoring network.

- Transfer of best available technology from the industrialized countries(ICs) to the developing countries(DCs)
  should begin immediately in the interest of both saving
  our scarce resources and reducing the environmental and
  climatic risks.

- The best help would be the training of professionals in
  the DCs so that these countries can better help themselves.

- More stringent measures should immediately be taken for
  reducing both forest dieback in the ICs and the destruction of tropical forests in the DCs.

- Prompt steps must be taken to preserve and restore the
  fertility of topsoil so that food security of a growing
  world population is not further jeopardized.

- Emission reduction of climate-influencing greenhouse gases
  requires also a cut-down on feedlots, artificial fertilizer use, and monoculture.

- The investment in questionable space and military research
  must be stopped so that these funds can be made available
  for the development and use of climate-saving improvements in energy efficiency and renewable energy resources.

- An International Convention for the Protection of the Global Climate must think of even more far-reaching steps
  including a CO2-tax in the industrialized countries for
  alleviating the debt crisis and the environmental
  destruction in the developing countries.

REFERENCES

Bach, W.(1985): CO_2-Zunahme und Klima: Modellergebnisse, Geoöko-
dynamik 6 (3), 229-292.

Bach, W. and H.J. Jung (1987): Untersuchung der Beeinflussung des
Klimas durch Spurengase mit Hilfe von Modellrechnungen.
Münstersche Geogr. Arbeiten 26, 45-64.

Bach, W. (1989): The Endangered Climate. In: F. Krause and W.
Bach (eds.) Energy and Climate Change: What can Western
Europe do? Res. Rept. for Ministry of Housing, Phys.
Planning and Env., The Netherlands (in press).

Beck, P. et al. (1988): Stellungnahme zum Thema "Treibhaus-
effekt" der Enquete-Kommission, Federal Environmental
Agency, Berlin.

Cess, R. D. and S. D. Goldenberg (1981): The Effect of Ocean

Heat Capacity Upon Global Warming Due to Increasing Atmospheric Carbon Dioxide, J.Geophys.Res. 86, 498-502.

Emanuel, W.R., G.G. Killough, W.M. Post, H.H. Shugart and M.P. Stevenson (1984): Computer implementation of a globally averaged model of the world carbon cycle. Dept. of Energy, TR 010, Washington, D.C.

Enquete-Kommission (1988): Schutz der Erdatmosphäre, Eine Internationale Herausforderung, Zur Sache Bd. 5, Hrsg. Deutscher Bundestag, Bonn.

Gesamtverband des deutschen Steinkohlenbergbaus (o.J.) Steinkohle 1987/88, Essen.

Global Warming Prevention Act of 1988; HR 5460, 100th Congress, 2nd Session, Washington, D.C.

Hammitt, J.K. et al. (1987): Product Uses and Market Trends for Potential Ozone-Depleting Substances: 1985-2000. The Rand Corp., Santa Monica, CA.

Hansen, J.E. et al. (1981): Climatic impact of increasing atmospheric carbon dioxide, Science 213, 957-966.

Hansen, J.E. (1983) cit. Seidel, S. and D. Keyes: Can we Delay a Greenhouse Warming? US EPA; Washington, D.C.

Lovins, A. B. et al. (1981): Least-Cost Energy: Solving the CO_2 Problem, Brick House, Andover, Mass.

Oeschger, H., U. Siegenthaler, U. Schatterer and A. Gugelmann (1975): A box diffusion model to study the carbon dioxide exchange in nature, Tellus 27, 168-192.

Rotty, R. M. (1986): Estimates of Seasonal Variation in Fossil Fuel CO2-Emissions, Tellus 39 B, 184-202.

Walbeck, M und H. J. Wagner (1988): CO_2-Emission durch die Energieversorgung, Energiewirtschaftliche Tagesfragen, Heft 2, Februar.

THE NEAR- AND FAR-TERM TECHNOLOGIES, USES, AND FUTURE OF NATURAL GAS

Gordon J. MacDonald

Presented at the International Energy Agency/Organization for
Economic Co-operation and Development (IEA/OECD) Expert Seminar
on "Energy Technologies for Reducing Emissions of Greenhouse Gases"
Paris, France
April 12–14, 1989

THE NEAR- AND FAR-TERM TECHNOLOGIES, USES, AND FUTURE OF NATURAL GAS

Gordon J. MacDonald*

Abstract

The carbon dioxide emissions per unit of delivered thermal energy from burning natural gas are somewhat more than half (55–57 percent) those of coal. When used with efficient (43 percent) combined-cycle technologies, natural gas can produce a unit of electricity at a carbon dioxide cost that is 40 percent of the total CO_2 emissions for conventional coal-fired power plants. The greenhouse advantages of natural gas can be secured only by the continued development of technology to prevent leakages of unburned methane to the atmosphere, because in terms of radiative heat absorption, one methane molecule is equivalent to about twenty-five carbon dioxide molecules. The widespread assumption of a small resource base has slowed the development of appropriate energy conversion and production technologies for natural gas. However, unconventional sources of natural gas contain vast stores of energy. The energy content of methane trapped in clathrates in permafrost regions of the world and on the continental slopes is estimated at five times the energy content of the global coal resource base. The resource base of methane is large enough to serve as an eventual source of hydrogen, should climate change consideration force a movement toward a hydrogen-based economy. The energy required to obtain hydrogen from methane is one-sixth that required to obtain hydrogen from water. With current production technologies, the cost of developing unconventional sources of natural gas is prohibitive in today's energy market. Further development of these technologies and of the natural gas resource base will require a pricing mechanism, such as a carbon tax, that adequately values the environmental advantages of methane.

*Gordon J. MacDonald is Vice President and Chief Scientist of The MITRE Corporation, 7525 Colshire Drive, McLean, VA 22102 (USA).

Table of Contents

Introduction

Compared with the intense heat of Venus, or the stark day-night contrast of Mars and the moon, the earth possesses a relatively equitable climate that makes possible the existence of man. The favorable nature of the earth's climate is made possible by the beneficial interactions of oceans, atmosphere, and biosphere. A primary descriptor of climate, though by no means the only one, is global average temperature, which depends on the delicate balance between incoming, short-wavelength radiation from the hot sun and the outgoing radiation of the much cooler earth.

If the atmosphere were stripped away, the earth's average surface temperature would be an inhospitable 35° C colder than it is today. The atmosphere provides a warming blanket for the surface through the action of infrared-absorbing constituents, whose concentrations are much smaller than those of the dominant gases, oxygen and nitrogen. Water vapor, carbon dioxide (CO_2), and both stratospheric and tropospheric ozone are the most important of the absorbing species. A myriad of other constituents, including methane (CH_4), nitrous oxide (N_2O), and chlorofluorocarbons (CFCs), also contribute to atmospheric warming.

Today, man is changing the composition of the atmosphere in unprecedented ways and with a rapidity that far exceeds the rate of natural changes. Since the beginning of the Industrial Revolution, the level of atmospheric carbon dioxide has grown exponentially. In the preindustrial atmosphere, the CO_2 concentration was 280 parts-per-million by volume (ppmv); in 1958, after two centuries, it had increased by 35 ppmv to 315 ppmv, and over the next 30 years, it grew another 35 ppmv, to reach 350 ppmv in 1988. The rate of increase for atmospheric methane has been even more rapid; CH_4 has grown in abundance at a rate of one percent per year from preindustrial values of about 0.7 ppmv to the present concentration of 1.75 ppmv. The rate of increase of nitrous oxide has approximately equaled that of carbon dioxide. Data on tropospheric ozone are inadequate, but indicate that tropospheric ozone has increased in parts of the industrialized world at rates of 1 to 2 percent per year since 1950.

The exponential growth in the atmospheric concentrations of carbon dioxide and nitrous oxide correlates with the increase in the rate of burning of coal, oil, natural gas, and wood. The origin of increased atmospheric methane is a matter of dispute, but appears to be associated with developments in wetland agriculture, biomass burning, fossil fuel use, and livestock production, as well as increased bacterial production in soils caused by a warming climate. Ozone is primarily a product of photochemical reactions that require oxides of nitrogen, NO and NO_2, carbon monoxide, and hydrocarbons and is thus the result of energy use, particularly in the transportation sector. Among the rapidly increasing greenhouse gases, only the chlorofluorocarbons have an origin unrelated to the burning of fossil fuels.

The total amount of carbon dioxide that is added to the atmosphere by using fossil fuels depends on both the mix of fuels and the quantity of each fuel burned. The amount of carbon dioxide delivered to the atmosphere per unit of heat energy produced depends on the composition of the fuel. For natural gas, the low ratio of carbon to hydrogen leads to a lower carbon dioxide emission per unit of energy delivered than that emitted by other fossil fuels, as illustrated in table 1. When "synthetic" fuels are used, the amount of carbon dioxide released to the atmosphere is even higher than with direct use of fossil fuels (see table 2). Energy used in the production of synthetic liquids or gaseous fuels (methanol from coal, oil from shale, or gas from coal) is itself a source of CO_2, unless the production process is fueled by a non-fossil energy source such as nuclear. The total amount of carbon dioxide emitted through the use of synthetic fuels includes the quantity released during their production; this conversion process adds further to the atmospheric burden of carbon dioxide.

Table 1

Carbon Dioxide Emission from the Direct Combustion of Various Fuels

Fuel	CO_2 Emission Rate (kg C/10^9J)	Ratio Relative to Methane
Methane	13.5	1
Ethane	15.5	1.15
Propane	16.3	1.21
Butane	16.8	1.24
Gasoline	18.9	1.40
Diesel Oil	19.7	1.46
No. 6 Fuel Oil	20.0	1.48
Bituminous Coal	23.8	1.73
Subbituminous Coal	25.3	1.87

Table 2

Carbon Dioxide Emissions from the Production and Burning of Various Synthetic Fuels

Fuel	CO_2 Emission Rate (kg C/10^9J)	Ratio Relative to Methane
Shale Oil		
In situ 28 gal/ton shale	48	3.5
High temperature 25 gal/ton shale	66	4.9
High temperature 10 gal/ton shale	104	7.7
Liquids from Coal		
Sasol technology; Eastern coal	42	3.1
EXXON donor solvent; Eastern coal	39	2.9
Gasoline from methanol from coal	51	3.8
High-Btu Gas from Coal		
Lurgi	41	3.0
Hygas	40	3.0
Methanol		
From natural gas	21	1.5
From coal	36–44	2.7–3.2

Methane has long been recognized as a fuel with many advantages: it is environmentally clean relative to classical pollutants such as sulfur dioxide and carbon monoxide, it may be readily transported, and it is relatively simple to produce. To these advantages has been added the realization that methane is significantly more benign as a producer of greenhouse gases than either oil or coal. However, methane itself is an infrared absorber, and any leakage of natural gas into the atmosphere during the processes of production, transportation, and burning will enhance the greenhouse effect. In fact, the penalty for methane leakage is high, because one methane molecule has the equivalent greenhouse effect of about 25 carbon dioxide molecules (Donner and Ramanathan, 1980). Fortunately, accidental releases of methane can easily be controlled, provided the adverse effects of such leakages are recognized and the appropriate technology is applied. The lifetime of released methane in the atmosphere is short; it oxidizes to carbon dioxide in roughly ten years.

The major hurdle confronting methane as a fuel for the future, either as a power producer or a source of hydrogen, has been the widespread assumption that the resource base for methane is limited and that exhaustion of recoverable supplies is only a few decades away. The current glut in the worldwide supply of methane, accentuated by the persistent gas "bubble" in North America, has led to a failure to explore for sources of methane other than in conventional oil and gas fields. Unconventional sources of methane, such as tight sandstone formations, shales, geopressurized aquifers, coal deposits, deep reservoirs, and methane clathrates have not been explored or developed, even though their potential was identified following the oil price shock of 1973.

In this paper I explore the future role of methane as an energy source in a world where the use of fuels is constrained by considerations of the greenhouse effect. After a brief overview of the technologies required for the increased use of natural gas in the near term, the difficult question of the availability of gas resources is addressed. Of the future sources of methane reviewed here, methane clathrates are identified as the most promising, both because of the huge resource base and the high energy density contained within clathrate deposits. The new estimate I give for the clathrate resource base is founded on detailed geologic studies of the last decade. Even though energy balance considerations support the concept of producing natural gas from methane clathrates, data are not available to provide hard estimates of production costs. Another possible consequence of greenhouse constraints is an acceleration of the movement toward a hydrogen-based energy economy. I will briefly consider the possibility of using methane derived from clathrates as a source for hydrogen, because the energy required to separate hydrogen from methane is significantly less than that required to produce hydrogen from water.

Enhanced Use of Natural Gas in the Near Term

Natural gas is currently used throughout much of the industrialized world in several key end-use applications; particularly important are a wide variety of industrial energy applications and residential and commercial space heating. Natural gas also offers many efficiency and environmental advantages as a transportation fuel in the form of compressed natural gas (CNG). During the last decade, substantial CNG programs have been developed in New Zealand and the Netherlands, while CNG-powered vehicles have been in widespread use in Italy for over forty years. A recently published review of international programs noted that most Asian countries and many Latin American countries that possess domestic gas resources are conducting feasibility studies of CNG use, and many have pilot programs in place (Sathage et al., 1988). In addition to specialized end-use applications, natural gas can be employed as a fuel for electricity generation.

Use of Natural Gas as a Transportation Fuel

CNG is currently used in converted gasoline and diesel engines. The natural gas fuel system is a simple addition to the conventional engine and consists of a pressure regulator, gas/air mixer, fuel-switching mechanism, CNG refueling connector, and storage tanks. With natural gas, a standard-sized U.S. car requires 3.90 to 4.43 MJ/km. This amount includes the energy used in extraction, processing, generation, transmission, distribution, vehicle-filling, and propulsion, making a CNG vehicle more energy efficient than one powered by gasoline (4.58 MJ/km) or methanol (6.16 MJ/km) (Hay, 1985). A number of automotive companies have produced experimental vehicles designed to operate exclusively on natural gas. Because natural gas has an octane rating of 130, the compression ratio must be double that of conventional vehicles if full advantage is to be taken of its high octane content. In terms of carbon dioxide added to the atmosphere, a CNG vehicle would produce between 52.6 and 59.8 grams of carbon per kilometer, as contrasted with 86.6 and 129.4 g of C/km for gasoline and methanol-powered vehicles, respectively. However, this favorable comparison for CNG vehicles assumes that there are no unburnt methane emissions or leakages during vehicle operation.

The widespread use of CNG vehicles for alleviating the greenhouse problem would require the development of several technologies. The most important of these is the ability to fuel and operate the vehicle without leakages of methane into the atmosphere.

514

The leakage of only 1.4 to 1.8 grams of methane per kilometer traveled would negate the carbon dioxide advantage of CNG as compared with gasoline. In addition, conventional objections to CNG vehicles must be overcome: storage cylinders of sufficient capacity must be developed to provide an acceptable vehicle operating range (500 km). Storage capacity could possibly be increased by the use of absorbents that operate at pressures less than 2,000 kPa and through employment of strong, lightweight alloys for the tanks. There is also a need for the development of a safe, low-cost, reliable, maintenance- and leak-free residential compressor for tank refilling.

Use of Natural Gas at Existing Power Plants

Currently, several options are available for increasing the use of natural gas in the production of electricity. These options are attractive from the greenhouse viewpoint, both because of the lower carbon dioxide emissions and the relatively high efficiency of natural gas combustion technologies. One relatively inexpensive option would be to increase utilization of existing gas and oil-fired power plants (most oil-fired power plants can also burn natural gas). For example, in 1987, average capacity utilization rates for U.S. oil and gas power plants were about 40 percent, as contrasted to 58 percent for coal power plants. Gas and oil plants are utilized less for a variety of reasons including the greater speed with which they can be powered up and down in response to fluctuations in electrical demand. The variable cost of power at gas and oil plants is higher than at coal-fired plants.

Natural Gas Combustion Turbines—Simple and Combined-Cycle

In simple-cycle combustion turbines, as in jet engines, fuel is burned in compressed air, and the combustion gases turn a turbine to generate electricity. Afterward, the exhaust gases can be converted to steam, which can be used to generate additional power. The thermal efficiency of simple-cycle turbines is comparable to modern coal power plants (32 percent), but because they can be turned on and off rapidly they are employed to meet peak power demands. For combined-cycle turbines, capital costs are higher; they are also more efficient, and so are preferred for base power rather than peak demand.

A simple gas turbine would place into the atmosphere 151.9 grams of carbon for each kilowatt hour (kwh) of electricity generated. The comparable figure for a combined-cycle turbine operating at 43 percent efficiency is 113.0 g C/kwh. The carbon penalty for a 1,000 MW coal-fired power plant with scrubbers and operating at 32 percent efficiency is 278.1 g C/kwh. A conventional coal-fired power plant thus places about two and a half times more carbon dioxide into the atmosphere than does a natural gas combined-cycle power plant.

Recent advances in jet engine technology, including new materials and designs that allow higher combustion temperatures, will make both simple- and combined-cycle turbines more efficient, though at a cost of higher NO_x emissions. For example, one technology that has recently been commercialized is the steam-injected gas turbine (STIG). In this technology, steam not needed for process heat is injected back into the combustor for added power and efficiency. A single-cycle turbine with an electric output of 33 MW and an efficiency of 32 percent could, with full steam injection, operate at 40 percent efficiency with a 51 MW output.

Potential Sources of Natural Gas

In 1986, the Potential Gas Committee estimated a total of 20.9 Tm^3 for total natural gas reserves in the probable, possible, and speculative categories for the United States (Potential Gas Committee, 1987). In today's natural gas market, unconventional sources of natural gas, as well as ultra-deep reservoirs (below 25,000 ft or 7,620 m) are not economically attractive, because production technologies for these supplies are far more expensive than those for conventional sources, and because short-term supplies of conventional sources are plentiful. Indeed, significant amounts of gas are still being wasted—mostly in the Persian Gulf—and are needlessly adding to the CO_2 burden of the atmosphere. Approximately 30 Gm^3 of gas was flared during 1988, an amount equivalent in energy to 180 million bbl of oil. However, despite current wastage, the need for ultra-deep and unconventional sources will grow as conventional sources are depleted.

Ultra-deep gas sources exist in sedimentary deposits that can extend to depths greater than 15,000 m. Despite temperature on the order of 350° C at these depths, methane is still a stable phase, provided carbon in the form of graphite or kerogen is present in the host rock. Although the United States has many more wells and a much higher density of drilling than any country in the world, only a small fraction of deep sediments have been explored. However, very large resources have already been identified at depths in the 4,500- to 7,500-m range. For instance, the deep Anadarko Basin in Oklahoma may contain between 1.7 Tm^3 and 10 Tm^3 of methane. The success ratio for exploration in the Anadarko Basin is more than triple the U.S. average, with two out of three wildcats finding methane. The Tuscaloosa Trend in the southeastern United States has been estimated to contain 1.2 Tm^3; another 6 Tm^3 may be contained in the Rocky Mountain Overthrust Belt. These resources are generally located between 4,500 and 7,500 m, and even though they are not in the ultra-deep classification (below 7,620 m), the clear implication is that there are very large sources of deep natural gas.

The existence of conventional gas in ultra-deep sources depends on the presence of organic material in the deeply buried sediments. The future of ultra-deep gas depends on finding reservoir rocks with the necessary porosity and permeability to allow high rates of production for reasonable economic returns. Speculation that some gas may have an abiogenic origin has stimulated additional interest in deep gas, and exploration technology applicable to deep exploration has advanced rapidly with the availability of supercomputer technology. Abiogenic gas has been discovered in a number of localities, including the exploratory well currently being drilled in granite in Siljan, Sweden, but at quantities insufficient for commercial production. Also, new technology will be required to reduce the drilling costs, which is by far the largest cost factor. The experience of the Soviets with the Kola Peninsula well, at a depth greater than 11,500 m, and that of the Swedes in putting down the Siljan test well (now at a depth of 6,920 m) will be of great value in developing new deep drilling techniques.

During the 1970s, the United States government and industry together carried out an extensive effort to evaluate unconventional resources. Their findings describe a very large resource base of natural gas that includes methane in eastern shales, western tight sands, coal seams, and geopressurized aquifers. Table 3 lists the estimated resource base of these unconventional sources. The amounts of gas actually recoverable will depend on the types of technologies that become available. The estimated amount of gas recoverable at $7.50 per thousand cubic feet (equivalent to $43.50 per barrel of oil) with today's technology is also shown.

Table 3

Unconventional Sources of Natural Gas

Source	Estimated Resource Base (Tm³)	Estimated Recoverable Gas at Marginal Cost Up to $7.50/1,000 cu ft Using Current Technology (Tm³)
Eastern Shales	17	1
Western Tight Sands	17	5
Coal Seams	70	10
Geopressurized Aquifers	85–2,800	4.5

Source: Energy Resource and Development Administration (ERDA), Market Oriented Program Planning Study, June 1977, Washington, D.C.

Some western tight sands are natural gas formations of low permeability in which the resistance to gas flow through the tight rock is so high that conventional well production is uneconomic at current market prices. The deposits also tend to be lenticular in geometry, causing further production problems. Depending on how formations may be linked, stimulation of gas flow by massive hydraulic fractoring or explosives could result in producible reservoirs.

Large amounts of natural gas are trapped in the low-permeability, gas-bearing Devonian and Mississippian shale deposits in the eastern United States. A small amount of gas has been produced from these deposits for over a century. The U.S. Department of Energy sponsored research in the 1970s to devise means for increasing the rate of production of these shale deposits through fracturing. Substantial amounts of natural gas are also contained in the large coal deposits of the United States. Historically, coal-bed methane has been viewed as a safety problem during coal mining because it can accumulate in the mine and explode. Methane extraction from coal beds, now being conducted in a number of areas in the United States, differs from conventional natural gas production. The density of wells drilled is greater and the production profile is one in which the maximum rate of production is reached after two to three years rather than on well completion, as is usually the case. Coal-bed wells are shallow, less than 1,000 m. Again, the principal impediment to production is an inadequate means of draining the gas from the coal seams. Even with these production uncertainties, the U.S. Geological Survey added 2.5 Tm³ of coal gas to the reserve base early in 1989.

The most abundant of the unconventional deposits listed in table 3 is gas contained within the geopressurized aquifers in Texas and Louisiana, in which the gas is held in a hot brine. However, the technology required for development of this resource is the least advanced. Ideally, the gas would be separated from the brine at the surface, and the heat contained by the brine would be used to produce electricity. Although the production base is enormous, a massive research and development program is needed to solve the technological and environmental problems associated with this resource.

Methane Clathrates

Methane clathrates have to be added to the list of unconventional sources listed in table 3. Clathrates are ice-like compounds in which methane and other gases are caged by water molecules. On freezing, water ordinarily forms ice in a hexagonal crystal structure. In the presence of methane and other gases, water crystallizes in a cubic lattice that traps the gas molecules. Interest in methane clathrates as an energy resource has grown with the recognition that very large deposits of clathrate exist at shallow depths in permafrost regions and on the continental shelf. The energy density of methane clathrates is an order of magnitude greater than the energy density of other unconventional sources of gas, as is indicated in table 4.

Table 4

Energy Density of Various Sources of Natural Gas in Terms of Cubic Meters of Gas per Cubic Meter of Reservoir Rock

Natural Gas Resource	Energy Density m^3 of CH_4/m^3 of Rock
Coal Bed Methane	8–12
Western Tight Sands	5–10
Eastern Gas Shales	1–2
Geopressurized Aquifers	1–2
Methane Clathrates (30% porosity of reservoir)	45–50
Conventional Natural Gas	10–20

Properties of Methane Clathrates

In 1810, Sir Humphrey Davy of the Royal Institution in London discovered that at low temperatures, chlorine will enter into a crystal structure with water molecules to form an ice-like substance. Since the initial synthesis of this unusual compound, a variety of gases (CH_4, CO_2, H_2S, C_2H_6, etc.) have been found to form crystalline compounds in which the ice lattice expands to form cages that trap the gas molecules. The practical significance of these compounds, known collectively as clathrates, became apparent during the 1930s, when it was discovered that clathrate formation was a major problem in the pipeline transportation of natural gas under cold conditions.

Clathrates resemble wet snow or ice in physical appearance and can form two distinct structures, the smaller of which traps methane. The small clathrate unit structure contains 46 water molecules with up to 8 molecules of methane and leads to the formula $CH_4 \cdot 5.75 \, H_2O$, though not all the gas sites in the structure need be filled (Davidson, 1983). The laboratory density of a fully filled methane clathrate is 910 kg m^{-3}, so that one cubic meter of methane clathrate contains 170.7 m^3 of methane gas at standard conditions (STP). If only 90 percent of the sites are filled with methane, 156 m^3 of methane gas are contained within a cubic meter of clathrate. In terms of mass, a cubic meter of fully filled methane clathrate contains 122 kg of methane and 789 kg of water. When less than about 80 percent of the gas sites are filled, the methane clathrate is no longer a stable phase (Davidson, 1983).

The acoustic wave velocity in methane clathrate is about 3.3 km s^{-1}, or roughly 13 percent less than that of ice (Pandit and King, 1983). Sonic velocities of natural massive clathrates are significantly greater than the 2.0 to 2.5 km s^{-1} velocities of water-saturated sediments, with down-well log measurements indicating velocities of 3.3 to 3.8 km s^{-1} (Kvenvolden and McDonald, 1985). The higher acoustic velocity of clathrate provides a useful seismic indicator for clathrate zones, because it produces a large impedance mismatch between clathrates and water-saturated sediments. The heat of dissociation of methane clathrates to either ice or water and methane gas is 208.22 − 0.382 T MJ m^{-3} for T < 273.18° K and 430.84 − 0.127 T MJ m^{-3} for T > 273.18° K, where T is the temperature in degrees Kelvin (Kuuskraa et al., 1983).

Thermodynamic Stability of Methane Clathrate

The phase diagram shown in figure 1 outlines the region of stability for methane clathrate (Vysniauskas and Bishnoi, 1983). At low temperatures, high pressure, and in the presence of methane and water, methane clathrate is the stable phase. An important fact is that methane clathrate can exist at a range of temperatures and pressures at which pore ice would not be stable. Beginning at pressures greater than 2,600 kPa, methane clathrate is stable at temperatures greater than 0° C.

Impurities in the chemical system will shift the clathrate phase boundary. Sodium chloride, which reduces water vapor pressure, will move the phase boundary toward lower pressures. The addition of larger gas molecules—carbon dioxide, ethane, propane, and hydrogen sulfide—tends to stabilize clathrates and to shift the equilibrium curve to higher temperatures. In natural sediments, the effects of salinity in pore water approximately cancel the effects of the larger molecules (Kvenvolden and McMenamin, 1980), except in relatively rare instances of high ethane and propane contents. Because of these canceling effects, the curve for a pure methane-water system provides a reasonable estimate for the stability region of naturally occurring methane clathrates.

Energy Balance in Recovering Methane from Methane Clathrates

Two principal methods for recovering methane from clathrate deposits are thermal stimulation and depressurization. The thermal stimulation of a clathrate deposit requires the injection of thermal energy into the rock in order to raise the local temperature to the point at which clathrate dissociates. In depressurization, the pressure in the deposit is lowered until the phase boundary is crossed. The energy to dissociate the clathrate in depressurization is provided by heat flowing out of the interior of the earth. Clathrates become a potential energy resource only if the energy required to liberate the methane from the clathrate is a small fraction of the thermal energy contained within the released methane.

In order to estimate typical energy requirements for clathrate deposit development, I consider conditions representative of the Arctic Islands. The clathrate deposit is thus assumed to be located at a depth of 600 m and to have a temperature of −6° C. At 600-m depths, dissociation of clathrate to methane and liquid water takes place at 8° C. The porosity of the sediments is taken as 30 percent, with the pore space filled by clathrate; all the gas sites within the clathrate lattice are occupied by methane. The heat required to raise the temperature of the rock-clathrate mixture by 14° C is 31.3 MJ m^{-3}. The heat of dissociation of the clathrate at 8° C is 395.1 MJ m^{-3}. To dissociate the

519

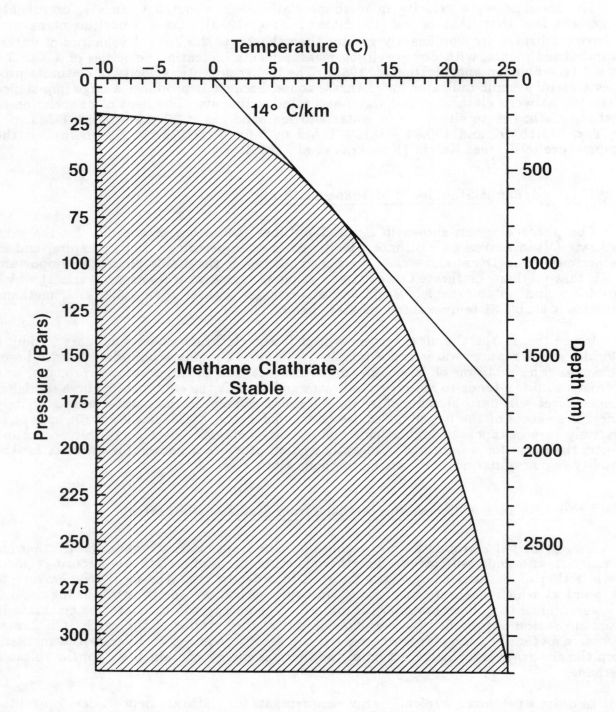

Figure 1. *Phase Diagram Showing the Boundary Between Free Methane Gas and Methane Clathrate for the H₂O-CH₄ System.* The ice-water phase boundary would be a vertical line through 0°C. The depth scale is drawn assuming hydrostatic equilibrium for pore water. The limiting geothermal gradient of 14°C/km for the stability of methane clathrate outside permafrost regions is also indicated.

clathrate, 118.5 MJ per cubic meter of rock plus clathrate are required, giving a total of 149.8 MJ of thermal energy needed to release 51.2 m^3 of methane. The heat content of the released methane is 1,960 MJ, so the ratio of the energy out to the energy required to produce the methane is 13. Total energy used in production is then 7.6 percent of energy developed. The energy balance, which is equal to the difference between output and input energies, is 1,810 MJ per cubic meter of reservoir. If the rock porosity is 40 percent, the ratio of energy out to energy in is 16.5 and the energy balance is 2,470 MJ m^{-3}; the corresponding figures for 20 percent porosity are 11.9 and 1,200 MJ m^{-3}. Because heat losses are neglected, the energy requirements are minimum figures, but they demonstrate the positive energy returns for developing a clathrate deposit.

In terms of total carbon emissions to the atmosphere, methane developed from clathrates at 30 percent porosity would place 14.5 kg C/GJ into the atmosphere, as compared with 13.5 kg C/GJ for methane from conventional deposits, and 23.8 kg C/GJ for bituminous coal (see table 1). The energy required to develop methane clathrates marginally lowers their attractiveness, but such methane would be far better in greenhouse terms than coal.

Natural Occurrences of Methane Clathrates

During the 1940s, difficulties with pipeline gas transmission in cold regions led to the widespread recognition in the exploration industry that methane clathrate would be stable below ground in permafrost regions (Katz et al., 1959). Drilling operators were initially concerned that penetration of a clathrate zone during drilling in cold regions might result in destabilization, causing rapid gas release and possible well loss. Development of the Russian gas fields in the Arctic during the 1960s proved such difficulties could be overcome and made clear the geologic significance of clathrate deposits (Makogon, 1978). Well-logging and formation tests revealed billions of cubic meters of natural gas frozen as hydrates in the Messoyakha gas field of Western Siberia. The field was partially developed by injecting methanol, which decomposed the clathrate and released large volumes of natural gas, though in most circumstances, use of methanol for production is prohibitively expensive.

During the 1960s, evidence of clathrates in ocean sediments appeared in seismic surveys as persistent reflectors that parallel the topography of the ocean bottom. Stoll et al. (1971) suggested that these reflectors were most probably the bottom of a zone of high-velocity methane clathrate. Since then, marine seismic surveys have found numerous examples of bottom-simulating reflectors in continental shelf areas.

The global distribution of clathrates was first summarized by Kvenvolden and McMenamin (1980). Their compilation was updated by MacDonald (1983) and Kvenvolden (1988). The present discussion focuses on those aspects of global clathrate distribution that are relevant to the production of natural gas.

Direct Evidence for Natural Clathrates

The first confirmed recovery of a natural methane clathrate occurred on March 15, 1972 in Prudhoe Bay. Samples in pressurized core barrels were recovered at several depths between 577 and 776 m from the Northwest Eileen Well No. 2, operated by EXXON and ARCO (Godbole et al., 1988). Well logs from the Eileen well confirmed the presence of several clathrate zones (Kvenvolden and McMenamin, 1980). Although strong gas shows in the drilling mud suggested free gas in the formation, anomalies in both the sonic and resistivity logs indicated ice or ice-like materials. The density log also indicated ice rather than free gas. Equivalent logging indications had been used by the Soviets in developing the Messoyakha field.

The first direct observations of gas hydrate in oceanic sediments were made by Yefremova and Zhizhchenko (1975) in shallow core samples from the Black Sea. At a water depth of 1950 m, microcrystalline aggregates of clathrates were recovered 6.5 to 8 m below the sea floor. The hydrate decomposed rapidly at the surface, releasing mainly methane and carbon dioxide.

An almost pure, 1.05-m-long core sample of clathrate was unexpectedly recovered from Site 570 (offshore Guatemala) of the Deep Sea Drilling Project in 1982 (Kvenvolden et al., 1984). The water depth was 1718 m, and the depth in the sediment was 218 m. Wireline logging showed that the sample came from a solid interval of clathrate 3 to 4 m thick. The 1982 sample was the first natural clathrate taken at sea that was preserved for shore-based studies. Gas pressure measurements made during decomposition indicated that the clathrate was very nearly filled to the formula $CH_4 \cdot 5.75\,H_2O$. The released gas was 99.4 percent methane, $\sim$ 0.2 percent ethane, and $\sim$ 0.4 percent CO_2. Dispersed clathrates have been observed in Deep Sea Drilling Cores in the Blake Outer Ridge of the Atlantic (DSDP Leg 76, Brooks et al., 1983). Clathrates were also found to be common in the slope sediments of the Middle America Trench off the shores of Mexico, Guatemala, and Costa Rica (Shipley and Didyk, 1982; Harrison and Curiale, 1982). Recently, clathrates have been found in sediments offshore Peru (Kvenvolden, 1988).

In the Gulf of Mexico, clathrates are found as 0.5- to 50-mm nodules, interspersed in layers 1 to 10 mm thick, or as solid masses greater than 15 cm thick. Eight sites on the Louisiana slope of the Gulf of Mexico have produced clathrate observations (Brooks et al., 1983). Of 800 shallow piston cores taken in water depths greater than 500 m, eight contained clathrates. Of the eight recovered samples, which were taken at depths ranging from 530 to 2,400 m, three differed from other natural clathrates in having a low ratio of methane to ethane plus butane, $C_1/(C_2 + C_3)$, of 1.9 to 4.4. All these samples are associated with oil-stained sediments containing up to 7 percent extractable oil. A further feature of the clathrates associated with oil sediments is the presence of a crystal structure large enough to enclose propane and other similarly sized molecules (Handa, 1988).

As indicated above, relatively few hydrate samples have been recovered. The scarcity of samples can be attributed to the difficulty in recovering and preserving material that rapidly decomposes when brought to the surface. Additionally, with the exceptions of Leg 70, Site 533 (Kvenvolden and Barnard, 1983) and Leg 84, Site 568 (Claypool et al., 1985), no specific attempts to recover clathrates have been made. Despite these limitations, clathrates have been found in widely differing geologic settings: Arctic permafrost, the tectonically active Middle America Trench, a sedimentary ridge off a continental passive margin (Blake Plateau), the very rapid depositional environment of the Gulf of Mexico, and finally, the stable environment of the Black Sea. The sediments enclosing clathrates vary from coarse sands on the North Slope to fine-grained hemipelagic muds and volcanic ash in the Middle America Trench. Clathrates are most commonly found as small crystals dispersed within the sediments or as larger nodules ($\sim$ 2 cm). Occasionally, clathrates occur in massive layers, with only a small admixture of contaminating sediments. Such a wide range of geologic environments suggests that where methane and water are available, clathrates may be ubiquitous, provided conditions in the sediments are within the methane clathrate stability region (see fig. 1).

Indirect Evidence for Natural Clathrates

Well data from Arctic regions has provided abundant indirect evidence for the presence of clathrates. Two exploratory wells drilled in the permafrost of the Mackenzie Delta showed sharp jumps in the methane content of their drilling muds after penetrating sand layers. Although the sands were very porous, their permeability was extremely low; the sonic log indicated a high velocity and also high resistivity. Bily and Dick (1974) interpreted the anomalous observations to result from clathrate-plugged interstices. Similar data indicate the existence of clathrates in the Arctic Archipelago of Canada (Hitchon, 1974) and the Viluy gas field in Yakutia, Soviet Union (Makogon et al., 1971).

The most commonly used indicator of clathrate deposits in marine sediments is the appearance of an anomalous seismic reflection that mimics the ocean bottom in its appearance (Shipley et al., 1979). Bottom-simulating reflectors (BSRs) are best recognized on seismic records where they cut across other reflectors produced by sedimentary structure, etc. BSRs arise from an impedance mismatch between overlying, high-velocity clathrates and underlying, lower-velocity sediments. The impedance contrast is heightened if the underlying sediments are water- or gas-saturated. Such reflectors are usually characterized by both a large reflection coefficient and a polarity reversal.

If the geothermal gradient is known, the depth to the bottom of a clathrate stability layer can be calculated. Calculated depths correlate well with many observed BSRs. Alternatively, the depth to the BSR can be used to estimate the geothermal gradient (Field and Kvenvolden, 1985).

The interpretation of BSRs as acoustic contrast due to the presence of clathrate has been strengthened by the discovery of clathrates in the Blake Outer Ridge, offshore Guatemala, offshore Mexico, and offshore Peru. Near-surface clathrates in the Gulf of Mexico and Black Sea have been found in regions without distinct BSRs. A number of areas with BSRs—offshore North Carolina to New Jersey and offshore Newfoundland— have been sampled without encountering clathrates. Not all BSRs need be associated with clathrates. Changes in lithology may also parallel bottom topography. For example, Hein et al. (1978) found that opal-A is transformed to opal-CT at the depth of the BSR in the Bering Sea. However, at depths consistent with the estimated limit of stability of clathrate, a BSR is a reliable indicator of the presence of clathrates in the sedimentary sequence.

Global Occurrences of Clathrate

Figure 2 summarizes the above discussion by showing the geographical location of all reported clathrate deposits, either inferred or actually sampled. Relatively few regions in the Arctic permafrost have been drilled and reported, so sampling in these areas is not as great as in the continental slopes sampled by the Deep Sea Drilling Project. There are no known occurrences of hydrates in Antarctica, but the sediments underlying the ice sheet should all be within the clathrate stability region (MacDonald, 1983).

Origin of Methane Clathrates

In order for clathrates to form, the temperature and pressure conditions existing at depth must overlap the stability region of clathrates, and methane must be present in sufficient abundance. The thickness of the clathrate stability zone depends on several factors: the geothermal gradient; the average temperature at the fluid-sediment interface; the depth of the ocean, for sea sediments; and the thickness of the ice, for regions covered by glaciers. The abundance of methane depends on the source of methane and the porosity and permeability of the rocks.

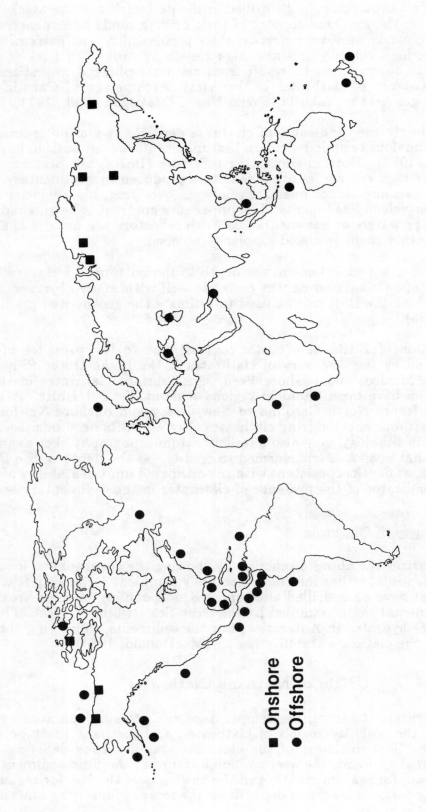

Figure 2. *Geographical Distribution of Methane Clathrate Occurrences.*
The map is updated from Kvenvolden and McMenamin (1980) and Kvenvolden (1988).

Onshore ■
Offshore ●

For regions not covered by ice or water, the phase diagram for clathrate (see fig. 1) strongly limits the subsurface stability zone. If the thermal gradient is greater than 14° C/km, the average surface temperature must be less than 0° C for clathrate to be stable at any depth. Because near-surface thermal gradients in cold regions range from 16 to 53° C/km (Lachenbruch et al., 1987), subsurface continental clathrates will be found only in regions of permafrost.

The thickness of the clathrate stability zone in ocean sediments is dependent on ocean depth and thermal gradient (as illustrated in figure 3) which gives an ocean-bottom temperature of 0° C. The bulk of ocean sediments are on continental slopes and shelves, as well as continental rises. Budyko et al. (1987) estimate the volume of sediments on continental margins at 250×10^6 km^3, as contrasted to 100×10^6 km^3 in the abyssal basins. If 40° C/km is taken as a mean thermal gradient on continental shelves (Lee and Clark, 1966), then a rough estimate of the average thickness of the clathrate stability zone in heavily sedimented areas of the ocean is 0.5 km (see fig. 3).

Because the appropriate pressure and temperature conditions for clathrate stability in ocean areas are widespread—62.6×10^6 km^2 of the ocean floor lies in the depth range of 0.2 to 3 km (Menard and Smith, 1966)—the availability of methane is the limiting factor for clathrate formation. Methane can form in place in sediments as a result of microbial methanogenesis or by thermochemical decomposition of organic matter at high temperatures (Tissot and Welte, 1978; Hunt, 1979). The shallow depths to which clathrates are limited favor a bacterial methane source. Alternatively, methane can originate at great depths and migrate upward to be trapped in the zone of stability.

A general picture of methane generation by bacteria in ocean sediments has evolved, largely as a result of the study of cores taken through the Deep Sea Drilling Program (Claypool and Kvenvolden, 1983). The decomposition of fermentable organic matter in marine sediments begins at deposition below the sediment-water contact. Organic matter is first attacked by aerobic bacteria that produce carbon dioxide, which dissolves in the water or escapes to the atmosphere. If the rate of sedimentation is low (< 1 cm/kyr), the organic content is small (≪ 1% C), and oxygen is abundant, aerobic bacteria can remove most or all of the organic material. When sedimentation is rapid—several centimeters per thousand years—the aerobic environment is replaced by anaerobic conditions at depths as shallow as a few centimeters. In the upper part of the aerobic region, bacterial action reduces sulfates and generates carbon dioxide and hydrogen sulfide. Deeper down, where sulfate has been removed, methane becomes the dominant product.

The anaerobic production of methane under sedimentary conditions is a complex biological process requiring synergism among a variety of microorganisms (Oremland, 1988). Bacterial action on the source organic matter leads to the production of carbon dioxide and hydrogen. The dissolved carbon dioxide forms bicarbonate, which is combined by methanogenic bacteria with hydrogen ions to form methane and water. Methanogenesis requires a highly reducing environment (Eh below −400 mV), and a pH in the range of 6 to 8. The temperature range for methanogenesis is broad; in lake sediments, methanogens function from 4 to 55° C, with an optimum between 35 and 42° C (Zeikus and Winfrey, 1976). Compounds other than carbon dioxide, such as acetate, can serve as substrates, but these are probably of less importance to the production of methane within sediments.

At a great enough depth, biological reactions cease, either from increased temperature and pressure, buildup of toxic products, or exhaustion of organic material. The depth at which biological activity ceases is not known, but is almost certainly in excess of a

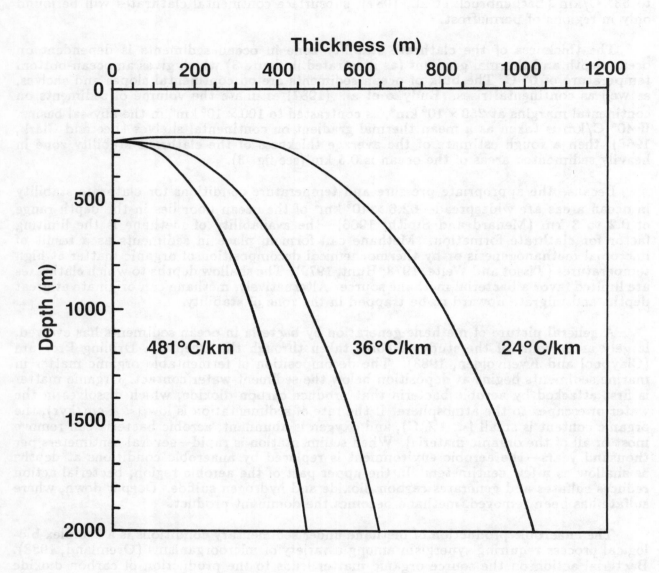

Figure 3. *Variation of Methane Clathrate Stability Zone as a Function of Ocean Depth and Thermal Gradient.*

kilometer and is dependent on the local geothermal gradient (Claypool and Kvenvolden, 1983). At higher temperatures ($> 80 - 120°$ C) and pressures (> 200 bars), thermogenic and catagenic reactions convert kerogen into petroleum and wet natural gases (methane plus higher alkanes).

If sedimentary clathrates are to form, the methane concentration must exceed the solubility of methane in the pore-space waters (Makogon, 1978). At sedimentary depths down to 3,000 m, and at water depths of 200 to 3,000 m, the solubility of methane in water having a salinity of 3.5 percent ranges from about 80 to 160 mmol/ℓ (Rice and Claypool, 1981), or about 1.8 to 3.6 volumes of gas at STP per volume of water. This amount of gas per volume of water contrasts with the 170 volumes of gas per volume of solids for clathrates. The amount of methane needed to form clathrate places limits on the organic carbon content of the sediment. Consider a sediment with a porosity of 50 percent and a bulk density of 1.7 g/cm^3, which corresponds to a grain density of 2.4 g/cm^{-3}. Given a carbon concentration by weight of 1 percent, the sediment would yield 1.5 moles of methane per liter of water, or a volume ratio of gas to fluid of 33. This ratio is well above saturation, if all the carbon were converted to methane. If all the methane were incorporated as clathrate, 33/170, or about 20 percent of the pore space would contain clathrate, and the remainder would be water. In order for methane to exceed saturation, the carbon content of the sediment must be in excess of 0.1 percent. Thus, in order for a significant fraction of the pore space to contain clathrate, the carbon content must be 1 percent or greater. Clathrate formation in carbon-depleted sediments is unlikely unless methane is introduced from below.

Estimates of Methane Stored in Clathrates

In order to estimate the amount of carbon currently stored in clathrates, both the volume of the clathrate zone and the fraction of that volume occupied by clathrate must be estimated. Determining the volume of the clathrate stability zone depends to first order on the temperature gradient and to second order on the composition of pore waters. Geothermal gradients are well enough known to determine the volume of the clathrate stability zone to a factor of two, but the remainder of the task, determining the fraction of the zone filled by clathrate, must remain speculative. Factors essential to making estimations, such as porosity, carbon abundance, etc., are poorly determined or simply not known.

Methane in Clathrates in Permafrost Regions

Cherskiy et al. (1985) have compiled geothermal data for the permafrost regions of Siberia. These data can be used to roughly estimate the areal extent, thickness, and volume of the clathrate stability zone, if one assumes that the effects of a deviation in the composition of pore waters compensates for the presence of gases other than methane in the clathrate. Table 5 lists estimated volumes of the clathrate stability zone. For those regions in which the hydrate zone contains Quaternary sediments (Timon-Pechora Province and the West Siberian Platform), a porosity of 0.4 is assumed. In regions where older sediments are found in the clathrate stability zone, the porosity is taken to be 0.2. The fraction of the available clathrate volume—porosity times total volume—that is actually occupied by clathrates is taken to be 1 percent. This percentage is based on the observation that in the logs of the Northwest Eileen State No. 2 and the Kuparuk D-8 and Kuparuk C-1 wells on the North Slope, about 10 percent of the clathrate stability zone is actually filled by clathrates (Collett and Ehlig-Economides, 1983). Makogon (1978) indicates that the percentage of filling by clathrates in the Messoyakha and Viluy fields is higher. However, because these two fields are in known regions of methane accumulation, a 10 percent occupancy estimate is deemed high. Instead, an arbitrary 1 percent figure has been used in constructing table 5. Based on a 1 percent filling, 654 m³ of methane are stored as clathrate in the permafrost regions of Siberia and 96 Tm³ of methane in the North American Arctic, giving a total of about 750 Tm³ of methane for the world's permafrost regions. As indicated, this estimate is highly speculative, but given the very large volume of near-surface sediments that lie within the clathrate stability zone, it is unlikely that as little as 75 Tm³ of methane are stored in permafrost clathrate. Indeed, an estimate higher than 750 Tm³ may actually be more appropriate, particularly if Antarctica were included in the estimate.

Table 5

Estimates of Methane Stored in Clathrates Permafrost Regions of Siberia

Region	Area of Hydrate Stability (km²)	Volume of Clathrate Zone of Stability (m³)	Estimated Methane Stored in Clathrates (Tm³)
Timon-Pechora Province	6.7×10^4	2.66×10^{13}	19
West Siberian Platform	1.1×10^6	3.3×10^{14}	225
East Siberian Craton	1.8×10^6	8.1×10^{14}	280
Northeast Siberia	6.1×10^5	3.7×10^{14}	130

Methane in Clathrates in Oceanic Sediments

Over the sediment-rich continental shelves, the great variation in geothermal gradient produces a zone of clathrate stability that ranges in thickness from 200 to 1,200 m (see fig. 3). An average value of 0.5 km is used to obtain an estimate of methane stored in oceanic clathrates. The total volume of sediments in the clathrate stability zone on slopes between 200 and 3,000 m is 31.3×10^6 km³. If sediment porosity is taken as 0.4, then 12.5×10^6 km³ are available for clathrate formation. Only a fraction of the total sediments will have a high enough carbon content to generate the amount of methane required to form clathrate. Premuzic et al. (1982) find that about 10 percent of the world's sea floor sediments contain concentrations of organic carbon greater than 1 percent, a value consistent with estimates made by Budyko et al. (1987). If 10 percent of the available pore space of carbon-rich sediments (1 percent of total sedimentary pore space) within the clathrate stability zone actually contain clathrate, the total volume of clathrate is 1.25×10^5 km³. The total volume corresponds to a 20,000 Tm³ of methane for 90 percent-filled methane clathrate. This estimate does not include clathrate in abyssal plains sediments at depths greater than 3,000 m.

Global Total for Methane Stored in Clathrates

The estimated total resource base of methane in clathrate (about 21,000 Tm³) is far greater than other unconventional sources (see table 3). Estimates for methane in geopressurized brines, which may represent former clathrate deposits buried in a rapidly evolving sedimentary column, are an order of magnitude less than that of clathrate deposits. The energy content of the clathrate resource base is 8×10^{23} J. If the coal resource base is taken as 5,000 Gt (MacDonald, 1982), the energy content of coal resources is 1.5×10^{23} J, or about 20 percent of the energy contained in the estimated clathrate base. The large clathrate resource base, the high energy density of clathrate deposits, the favorable energy balance, and low overall carbon dioxide emissions all argue that the possibility of developing the clathrate resource base should be pursued.

Despite unfavorable environmental conditions, it is likely that the commercial development of clathrate deposits will begin with the shallow clathrates zones in Arctic permafrost regions. Because of this, I will focus the following discussion of production on land-based, permafrost regions.

Production of Methane from Clathrate Deposits

As noted above, several methods for producing gas from clathrate zones have been proposed, including thermal recovery through steam injection, hot-water injection and fire flooding, depressurization, and injection of clathrate inhibitors such as methanol (Holder et al., 1984). Injection of methanol or glycols as inhibitors is expensive, owing to the large volumes of these chemicals that are required. In steam injection and fire flooding, the heat losses are large, and fire flooding dilutes methane with carbon dioxide. A combination of depressurization and hot-water injection appears most promising.

Two models of hot-water injection for methane production have been developed (McGuire, 1981). One of these is modeled on steam flooding of heavy-oil reservoirs and makes use of a central injection well surrounded by a number of production wells. The injected hot water pushes liberated gas toward the production wells. In a fracture flow model, water penetration is assisted by hydraulic fracturing of the reservoir between the injection and production wells. Parametric studies show that optimum water temperatures should be low enough to avoid excessive heat losses, but high enough to avoid high injection rates. The models also indicate that the clathrate zone should be at least 5 m thick and have a porosity of at least 15 percent, if thermal stimulation techniques are to work.

In the depressurization technique, clathrates dissociate upon pressure relief, absorbing energy and reducing the temperature of the remaining hydrates and rock matrix. The clathrates continue to dissociate until they generate enough gas to raise the pressure to the dissociation equilibrium value for the lowered temperature. As a temperature gradient between the clathrates and surrounding rock is established, heat flows into the reservoir. The temperature gradient and the heat flux into the clathrate zone are maintained by controlling the rate and amount of depressurization. Depressurization can be combined with hot water injection.

The detailed production mechanism will depend on a number of factors: if free gas is present along with clathrates, depressurization is favored; if free water or ice is present, a combination of techniques would be called for. In either case, the extent of the clathrate deposit will largely determine the economics of recovery. High-porosity sediments filled with hydrate and free gas present the most favorable conditions. The permeability and porosity of the surrounding rock largely determine the amount of heat loss. A hydrate deposit capped with an impermeable rock is one in which heat losses could be minimized.

Despite the lack of operational experience with wells in clathrate deposits, it is clear that such operations would face a myriad of technical difficulties. In the cold regions where clathrates are found, the injection of hot water may result in particularly severe heat loss. Also, because permeability in clathrates is reduced, gas flow and recovery might be prevented. In particular, as the gas flows to the production well it expands and cools itself by the Joule-Thompson effect. Such cooling could lead to the reformation of clathrate, which would then block the path of the released gas to the production well. All of these problems need to be investigated, though none appear to be fatal to the concept of producing gas from clathrates.

The great bulk of estimated clathrate deposits are in the ocean. The minimum water depth for clathrate stability is about 300 m, and observed deposits are often in sediments below water depths of 1,000 m or more. Gas production wells are currently not drilled in water of these depths, but the technologies for drilling in deep water are advancing rapidly.

Economics of Producing Methane from Clathrates

In the absence of hard data on recovery factors and heat losses for methane production from clathrate deposits, any discussion of economics must be viewed with great caution. However, McGuire (1981) and Holder et al. (1984) have prepared useful analyses indicating the general dimensions of the problem. Table 6 lists the computed costs for production from a 7.5 m-thick methane clathrate zone that has a porosity of 40 percent, a permeability of 600 millidarcies, and is located on the North Slope of Alaska. The recovery rate is determined by an injection rate of 30,000 bbl per day of water at 150° C.

Table 6

Comparison of the Economics of Production of Natural Gas from Clathrates and from Conventional Deposits on the North Slope of Alaska (in 1988 dollars)

	Thermal Injection	Depressurization	Conventional Free Gas
Total Investment ($)	5,084,000	3,320,000	3,150,000
Total Annual Equivalent Cost ($/year)	3,200,000	2,510,000	2,000,000
Total Production (10^6 sc ft/year)	900	1,100	1,100
Transportation Cost ($/$10^3$ sc ft)	4.25	4.25	4.25
Break-Even Price including royalties & fees ($/$10^3$ sc ft)	8.75	7.10	6.50

Costs for transportation of this gas are high; they involve transport first by pipeline from the North Slope to Cook Inlet, and then by LNG tanker to markets in Japan and the Pacific West Coast. The total cost, including the transportation cost of $7 to $9 per thousand cubic feet, is about equal to the cost of gas produced from other unconventional sources. The cost of production by depressurization alone—about $2.85 per thousand cubic feet—corresponds to an equivalent oil price of $16.50 per barrel. These lower costs suggest that clathrate deposits near market, for example, offshore northern California, may become economical sooner than those in the Arctic.

Production of Hydrogen from Methane Clathrates

The use of natural gas instead of coal as an energy source lessens overall carbon dioxide emissions and greenhouse potential, provided adequate care is taken to prevent the release of methane to the atmosphere. Nevertheless, replacement of oil and coal by methane would still result in increased atmospheric carbon dioxide. If the climatic consequences of such an increase are severe enough, pressure will arise for alternative fuels that do not emit any greenhouse gases. In the past, there has been speculation that hydrogen produced from water using solar or nuclear energy might be such a fuel (Bockris, 1980). The advantage of using water as the source of hydrogen is its very great abundance; a disadvantage is the large amount of energy required to dissociate the hydrogen from it. At standard conditions, the heat of formation of water per mole of hydrogen is −241.8 kJ, while the heat of formation of methane per mole of hydrogen is −37.4 kJ. Thus, breaking down water to form a mole of hydrogen requires about 6.5 times as much energy as deriving the hydrogen from methane. Previously, the assumption of a vanishing supply of natural gas led to the dismissal of methane as a source of hydrogen. Now, the large estimated resource base for methane in the form of clathrate and other unconventional sources of natural gas opens the possibility of using methane as a source of hydrogen.

At one bar pressure and in the absence of oxygen, methane breaks down to hydrogen and graphite at 545° C. At 1,000° K, the equilibrium ratio of the partial pressure of hydrogen to methane is 11.5; at 1,100° K, the ratio is 32. The energy balance for using hydrogen derived from methane as a fuel is easily calculated for the ideal case of zero heat losses during conversion. The amount of energy needed to dissociate methane to molecular hydrogen is the sum of the heat required to raise the temperature of methane from a producing temperature of 25° C to 545° C, which is 12.9 kJ per mole of hydrogen, plus the heat of dissociation, which is 43.7 kJ per mole of hydrogen, for a total energy requirement of 56.6 kJ/mole H_2. If the heat of combustion of hydrogen is taken as 241.8 kJ/mole, the resulting energy balance is positive and yields a net energy of 370.4 kJ for each mole of dissociated methane yielding two moles of hydrogen. This amount contrasts with the 890.4 kJ per mole of methane released by burning methane directly to carbon dioxide plus water, so that a penalty of 520 kJ/mole must be paid to prevent carbon dioxide from reaching the atmosphere. When water is used as the source of hydrogen, there is no net energy return. However, the use of methane as a source of hydrogen involves a positive energy return, even though the conversion process presents a host of engineering challenges.

The energy balance of methane-to-hydrogen production can be improved by using the sensible heat in the hydrogen to heat the incoming methane stream by means of a counter-flow heat exchanger, which reduces the 12.9 kJ/mole H_2. Efficient counter-flow heat exchangers might capture as much as 80 percent of the sensible heat, thereby improving the energy balance to 391 kJ per mole CH_4 and increasing the energy ratio of burning hydrogen to burning methane to 0.44. On dissociation, the thermal breakdown of methane produces fine-grained graphite; this process is used in the current manufacture of carbon black. Techniques would have to be developed to remove the fine particles efficiently or to promote formation of large graphite particles. The quantities of solid carbon produced would be massive, 0.53 Gt of solids for every Tm^3 of natural gas processed to hydrogen. Because the carbon is inert, burial would appear to be a safe means of disposal, though such disposal would pose large-scale material handling problems.

Summary

Observed and anticipated changes in atmospheric composition alter the earth's radiative equilibrium temperature. The resulting warming is not disputed. What is uncertain is the potential impact of the feedback processes that are capable of enhancing the radiative warming. Of even greater uncertainty are the specifics of the imminent changes in climate and weather over different parts of the globe: What will be the intensities and frequencies of extreme events such as droughts, floods, hurricanes, and other severe events that have great economic consequences? In the absence of definite information regarding the probability of such happenings, it is prudent to develop alternative energy paths that can be taken in the future if the anticipated consequences of climate change are sufficiently serious.

Nuclear, solar, and renewable biomass can provide energy without greenhouse penalties. Of the fossil fuels, natural gas is most benign with respect to climate change. For some time, it has been fashionable to dismiss natural gas as a potential future fuel, because the resource base was thought to be highly limited. However, it is gradually being recognized that methane in the form of an icy compound—methane clathrate—is present in very large quantities. Estimates presented in this paper show the energy content of the clathrate resource base to be five times that of the coal resource base. Thus far, development of technologies for the production of clathrate deposits has been hindered by prevailing low energy prices.

Methane clathrates, because of their limited range of thermodynamic stability, are found in shallow deposits less than 2,000 m deep, both in permafrost regions and on the deep continental shelf and slope. The energy density within clathrates is high compared to either conventional or unconventional gas sources. Moreover, energy balance considerations show that clathrate development is favored at a minimum penalty in terms of carbon dioxide emissions.

Clathrates also form a potential source of hydrogen, should climate considerations force a move to a hydrogen economy. Although clathrates are less abundant and less readily available than water, the energy requirements for securing hydrogen from clathrates are far more favorable than for obtaining hydrogen from water.

The ultimate development of large resources of unconventional natural gas depends on the construction of a pricing mechanism that adequately values the environmental qualities of methane. Currently, the environmental costs of energy use, particularly the greenhouse costs of coal, are not reflected in fuel prices. As a result, prevailing economics drive the continued development of coal resources, which have much lower recovery costs than does unconventional natural gas. Further, the immensity of the coal resource base and the low cost of producing coal together have retarded development of production technologies for another energy resource of vast proportion, methane clathrates, as well as other unconventional sources of natural gas. This unsatisfactory situation will persist unless actions are taken, first among industrialized states, to value more adequately the environmental qualities of natural gas. The carbon tax, under discussion both in Europe and in the United States, would provide a mechanism for correcting the current inadequacies in the pricing of these two fuels.

REFERENCES

Bily, C. and J. Dick (1974), Naturally occurring gas hydrates in the Mackenzie Delta, N.W.T., *Bull. Canadian Petrol. Geol.*, **22**, 340–352.

Bockris, J. (1980), *Energy Options*, Halsted Press, New York.

Brooks, J., L. Barnard, D. Wiesenburg, M. Kennicutt, and K. Kvenvolden (1983), Molecular and isotopic compositions of hydrocarbons at Site 533, Deep Sea Drilling Project Leg 76, *Initial Reports of the Deep Sea Drilling Project*, **76**, 377–390, Washington, DC: U.S. Government Printing Office.

Budyko, M., A. Ronov, and A. Yanshin (1987), *History of the Earth's Atmosphere*, New York: Springer.

Cherskiy, N., V. Tsarev, and S. Nikitin (1985), Investigation and prediction of conditions of accumulation of gas resources in gas-hydrate pools, *Pet. Geol.*, **21**, 65–89.

Claypool, G. and K. Kvenvolden (1983), Methane and other hydrocarbon gases in marine sediment, *Ann. Rev. Earth Planet. Sci.*, **11**, 299–327.

Claypool, G., C. Threlkeld, P. Mankiewicz, M. Arthur, and T. Anderson (1985), Isotopic composition of interstitial fluids and origin of methane in slope sediment of the Middle America Trench, Deep Sea Drilling Project Leg 84, *Initial Reports of the Deep Sea Drilling Project*, **84**, 683–691, Washington, DC: U.S. Government Printing Office.

Collett, T. and C. Ehlig-Economides (1983), Detection and evaluation of the in-situ natural gas hydrates in the North Slope Region, Alaska, paper SPE 11673 presented at March 23-25, 1983 meeting of the Society of Petroleum Engineers, California Regional Meeting.

Davidson, D. (1983), "Gas hydrates as clathrate ices" in *Natural Gas Hydrates: Properties, Occurrence and Recovery*, ed. J. Cox, Woburn, MA: Butterworth, 1-16.

Donner, L. and V. Ramanathan (1980), Methane and nitrous oxide: Their effects on the terrestrial climate, *J. Atmos. Sci.*, **37**, 119–124.

Field, M. and K. Kvenvolden (1985), Gas hydrates on the northern California continental margin, *Geology*, **13**, 517–520.

Godbole, S., V. Kamath, and C. Ehlig-Economides (1988), "Natural gas hydrates in the Alaskan Arctic," *SPE Formation Evaluation*, **3**, 263–266.

Handa, Y. (1988), A calorimetric study of naturally occurring gas hydrates, *Ind. Eng. Chem. Res.*, **27**, 872–874.

Harrison, W. and J. Curiale (1982), "Gas hydrates in sediments of holes 497 and 498A, Deep Sea Drilling Project Leg 67," *Initial Reports of the Deep Sea Drilling Project*, **67**, 591–594, Washington, DC: U.S. Government Printing Office.

Hay, N. (ed.) (1985), *Guide to New Natural Gas Utilization Technologies*, Atlanta, Ga.: Fairmont Press.

Hein, J., D. Scholl, J. Barran, M. Jones, and S. Miller (1978), Diagenesis of late Cenozoic diatomaceous deposits and formation of the bottom simulating reflector in the Southern Bering Sea, *Sedimentology*, **25**, 155–181.

Hitchon, B. (1974), "Occurrence of natural gas hydrates in sedimentary basins," in *Natural Gases in Marine Sediments*, ed. I. Kaplan, New York: Plenum Press.

Holder, G., V. Kamath, and S. Godbole (1984), The potential of natural gas hydrates as an energy resource, *Ann. Rev. Energy*, **9**, 427–445.

Hunt, J. (1979), *Petroleum Geochemistry and Geology*, San Francisco: W. H. Freeman.

Katz, D., D. Cornell, R. Kobayashi, F. Poettmann, J. Vary, J. Elenbass, and C. Weinaug (1959), *Handbook of Natural Gas Engineering*, New York: McGraw-Hill.

Kuuskraa, V., E. Hammershaimb, G. Holder, and E. Sloan (1983), *Handbook of Gas Hydrate: Properties and Occurrence*, U.S. Department of Energy, DOE/MC/1923a-1546, Washington, D.C.

Kvenvolden, K. and M. McMenamin (1980), "Hydrates of natural gas: A review of their geologic occurrence," *U.S. Geological Survey, Circular 825*, Washington, D.C.

Kvenvolden, K. and Barnard, L. (1983), "Hydrates of Natural Gas in Continental Margins," in J. Watkins and C. Drake (eds.), *Studies in Continental Margin Geology, Mem. 34*, Tulsa, OK: Amer. Assoc. Petrol. Geol., 631–640.

Kvenvolden, K., G. Claypool, C. Threlkeld, and E. Sloan (1984), Geochemistry of a naturally occurring massive marine gas hydrate, *Org. Geochem.*, **6**, 703–713.

Kvenvolden, K. and T. McDonald (1985), Gas hydrates of the Middle America Trench - Deep Sea Drilling Project Leg 84," *Initial Reports of the Deep Sea Drilling Project*, **84**, 667–682, Washington, DC: U.S. Government Printing Office.

Kvenvolden, K. (1988), Methane hydrates and global climate, *Global Biogeochemical Cycles*, **2**, 221–229.

Lachenbruch, A., J. Sass, L. Lawver, M. Brewer, B. Marshall, R. Munroe, J. Kennelly, S. Galanis, and T. Moses (1987), Temperature and depth of permafrost on the Alaskan Arctic Slope, in *Alaskan North Slope Geology*, ed. I. Tailleur and P. Weimer, Pacific Section Soc. Ec. Paliort Mineral. and Alaskan Geol. Soc., **50**.

Lee, W. and S. Clark (1966), Heat flow and volcanic temperatures, in *Handbook of Physical Constants*, ed. S. Clark, *Geol. Soc. Am. Memoir*, **97**, 483–511.

MacDonald, G. (1982), *The Long-Term Impacts of Increasing Atmospheric Carbon Dioxide Levels*, Cambridge, Mass.: Ballinger.

MacDonald, G. (1983), The many origins of natural gas, *J. Petrol. Geol.*. **5**, 341–362.

Makogon, Y., F. Trebin, A. Trofimuk, V. Tsarev, and N. Cherskiy (1971), Detection of a pool of natural gas in a solid (hydrated gas) state, *Doklady Acad. Sci. USSR, Earth Science*, **196**, 197–200.

Makogon, Y. (1978), *Hydrates of Natural Gas* (Transl. from Russia by W. Cieslewicz), Denver: Geoexplorer Associates, Inc.

McGuire, P. (1981), Methane hydrate gas production: An assessment of conventional production technology as applied to hydrate gas recovery, *Los Alamos Sci. Lab. Rep.* LA–91–MS, Los Alamos, N.M.

Menard, H. and S. Smith (1966), Hypsometry of ocean basin provinces, *J. Geophys. Res.*, **71**, 4305–4325.

Oremland, R. (1988), Biogeochemistry of methanogenic bacteria, in *Biology of Anaerobic Microorganisms*, ed. A. Zehnder, New York: J. Wiley, 641–705.

Pandit, B. and M. King (1983), "Elastic wave velocities of propane gas hydrates," in *Natural Gas Hydrates: Properties, Occurrence and Recovery*, ed. J. Cox, Woburn, MA: Butterworth, 49-62.

Potential Gas Committee (1987), *Potential Supply of Natural Gas in the United States (December 31, 1986)*, Golden, Colorado: Potential Gas Agency, Colorado School of Mines.

Premuzic, E., C. Benkovitz, J. Gaffney, and J. Walsh (1982), The nature and distribution of organic matter in the surface sediments of world oceans and seas, *Organic Geochemistry*, **4**, 63–77.

Rice, D. and G. Claypool (1981), Generation, accumulation, and resource potential of biogenic gas, *Am. Assoc. Pet. Geol. Bull.*, **65**, 5–25.

Sathage, J., B. Atkinson, and S. Meyers (1988), Alternative fuels assessment: The international experience, Lawrence Berkeley Laboratory Mimeo, March 1988, Berkeley, California.

Shipley, T., M. Houston, R. Buffler, F. Shaub, K. McMillen, J. Ladd, and J. Worzel (1979), Seismic evidence for widespread possible gas hydrate horizons on continental slopes and rises, *Am. Assoc. Petrol. Geol. Bull.*, **63**, 2204–2213.

Shipley, T. and B. Didyk (1982), "Occurrence of methane hydrates offshore southern Mexico," *Initial Reports of the Deep Sea Drilling Project*, **66**, 547-555, Washington, DC: U.S. Government Printing Office.

Stoll, R., J. Ewing, and G. Bryan (1971), Anomalous wave velocities in sediments containing gas hydrates, *J. Geophys. Res.*, **76**, 2090–2094.

Tissot, B. and D. Welte (1978), *Petroleum Formation and Occurrence*, New York: Springer-Verlag.

Vysniauskas, A. and P. Bishnoi (1983), "Thermodynamics and kinetics of gas hydrate formation," in *Natural Gas Hydrates: Properties, Occurrence and Recovery*, ed. J. Cox, Woburn, MA: Butterworth, 35-48.

Yefremova, A. and Zhizhchenko, B. (1975), Occurrence of Crystal Hydrates of Gases in the Sediments of Modern Marine Basins, *Doklady-Earth Science*, **214**, 219–220.

Zeikus, J. and M. Winfrey (1976), Temperature limitation of methanogenesis in aquatic sediments, *Appl. Env. Microbiol.*, **31**, 99–107.

ON COEFFICIENTS FOR DETERMINING GREENHOUSE GAS EMISSIONS FROM FOSSIL FUEL PRODUCTION AND CONSUMPTION

M.J.Grubb

Energy and Environmental Programme

Royal Institute of International Affairs

10 St James's Square

London SW1Y 4LE

Paper prepared for the IEA/OECD expert seminar on energy technologies for reducing emissions of greenhouse gases, Paris, 12-14th April 1989

SUMMARY

This paper examines the emissions of greenhouse gases from the energy sector, to determine the coefficients which need to be used in estimating greenhouse gas emissions on the basis of fuel production and consumption data, and the impact of altering the demand for fuel products.

Emissions are examined with respect to the carbon content of the fuels, non-oxidised components, associated CO_2 emissions, parasitic energy use in fuel production and delivery, and the associated production of non-CO_2 gases. The uncertainty in each component is estimated.

The carbon content of the fuels is the dominant factor in nearly all cases, but other important factors include:

    non-energy uses of oil
    parasitic energy use in the oil and gas industries
    methane emissions in the gas and oil industries
    hydrocarbon emissions from the transport sector

There are major uncertainties and/or variations between countries in each of these, and further work is required to reduce these uncertainties. The results do serve to emphasise, however, that there are some steps which can be taken to reduce greenhouse gas emissions from energy sector beyond simply modifying the demand for fuels.

CONTENTS

1. Introduction

The emissions of carbon dioxide associated with the production of fossil fuels has been examined in detail by Marland and Rotty (1984) for the US Department of Energy, and others. This paper revisits that work briefly, drawing upon detailed studies of the carbon content of different fuels, and extends it by including the parasitic use of energy within the energy industries (which need to be allowed for in estimating the impact of modifying the final demand for fuels) and the emission of non-CO_2 greenhouse gases associated with energy production and delivery.

The appropriate emission coefficient depends upon the issue/data under consideration, and hence in principle different coefficients are required according to the nature of the application. I have therefore sought to break down the discussion into the following issues:

*   Fuel carbon content
*   Carbon release coefficients, primary fuel basis
*   Carbon release coefficients, delivered fuel basis
*   Other Greenhouse Gas addition coefficients

This paper takes these in turn, starting with a summary of the work carried out on the carbon coefficient of fossil fuels by various contributors to the Energy and Environmental Programme study on the "Implications of the Greenhouse Effect." These valuable contributions, by I. Hughes (British Coal), Andrew Gordon (BP International), John Eyre (Shell International), D. Fortune (Shell Coal), and Peter Brackley, are gratefully acknowledged, along with earlier studies by Richard Beresford (British Gas). I am indebted to Peter Brackley for further discussion of the issues.

2. Carbon-content of primary fuels

The first issue to be resolved is the actual carbon content of fuels (and hence CO_2 production upon complete combustion) per unit energy production. Each is discussed in turn, and the conclusions tabulated at the end of the section. <u>All heating values cited are gross (higher) heating values.</u>

2.1 Natural Gas

Natural gas is readily amenable to analysis based upon chemical composition, using the following molecular data:

Molecule	CO_2 coefficient, tC/TJ
Methane	13.5
Ethane	15.4
Propane	16.2
Butane	16.7

<u>Source</u>: BP

Natural gas composition varies somewhat, but the methane fraction is nearly always between 90% and 96% (adjusted dry gas basis: see Marland & Rotty, 1984). This leads to a C-content coefficient of about 13.8tC/TJ for the dry gas with a range of ±1%.

2.2 Oil

As a highly complex mix of hydrocarbons, a crude oil coefficient cannot be derived from a molecular analysis. From extensive samples, Marland and Rotty (1984) for the US DoE, estimate 19.0tC/TJ ±2%; Gordon cites 18.5 for a "light sweet crude" and suggests 19.0 for an average crude coefficient (suggesting a slightly wider range).

Most other data offered relates to various oil fractions. Gordon cites products spanning 18.0tC/TJ for Gasoline, 18.5 for Kerosene, 19.0 for Gas oil (with Diesel oil 19.2), to 20.0 for fuel oils and heavy oil. Eyre estimates 18.3-19.4 for mogas, 18.8-19.7 for diesel fuel, and 20.2-20.8 for fuel oil, figures which are around 2-5% higher for reasons which are not

541

apparent but probably within the range of uncertainties. It is likely that the various published figures of around 20.0 are for fuel oil rather than crude.

2.3 Coal

Preliminary calculations for the Energy and Environmental Programme's study by one source suggested that coefficients could vary over a range 21-34tC/TJ according to the type of coal, and I suggested[1] that on chemical grounds a fairly broad range seemed plausible if the volatile matter was mostly hydrocarbons, and anthracites approximated to graphite, though even this sets an upper limit of 30.5tC/TJ. Marland also confirmed that he had not looked into the matter in detail in his work for the US DoE (private communication).

However, the extensive work carried out by Ian Hughes of British Coal (using two approaches) shows clearly that this wide range is erroneous, and other surveys confirm this, with the results shown in table 1 demonstrating a range of no more than ±5% about a central value of around 24.5tC/TJ.

Table 1. Results of coal composition studies		
Author	Source	Reported C-content, tC/TJ (dmf, gcv)[1]
Fortune (Shell Coal)	Wide sample	23.5-26.4
Gordon (BP)	Wide sample	23.2-25.4
Hughes (British Coal)	Seyler's Chart (exc. anthrac)	23.2-25.6
Hughes (British Coal)	Coal rank classific[n]	24.0-26.4
(1) dmf = Dry, mineral-free basis gcv = Gross (higher) calorific value		

------

[1] Grubb M.J., _Carbon coefficients for fossil fuels_, paper prepared for the Energy and Environmental Programme study group on implications of the greenhouse effect, December 1988

Hughes carried out an extensive analysis on the cause of variation, including subgroups. His work demonstrates that volatile matter content is by far the most important determinant of variation. Anthracites form the higher end of the range; excluding them gives a range for Bituminous coals of 23.2-25.2, which can be further narrowed if the calorific value of the coal is known by noting that the variation corresponds to roughly:

Carbon content, tC/TJ = 32.15 - 0.234 x GCV

for GCV (Gross calorific value) in the range 31-37 GJ/t on dmf basis. The range of uncertainty is then of the order of 1%. Anthracites fall outside this scheme; a value of 25.5 tC/TJ ±3% seems suitable for these.

On this basis Hughes notes that typical UK power station coals have a value of 24.5 tC/TJ, with industrial and domestic bituminous at 24.2.

Fortune also cites values for other solid fuels, based upon 'Technical Data on Fuels':

Wood	27.3	tC/TJ
Peat	27.5	"
Lignite	26.2	"

2.4 Summary: carbon content of fossil fuels

The conclusions are summarised in table 2.

3. Carbon release coefficients for primary fuels

The C-release coefficient for primary fuels is the coefficient for deriving the total carbon emitted as (or rapidly converted to) CO_2 on the basis of fuel production statistics. This is the goal on which the studies by Marland and Rotty (1983) for the US DoE concentrated. It differs from the

```
┌─────────────────────────────────────────────────────────────────────┐
│ Table 2. Fuel carbon content coefficients                           │
│                                                                      │
│          Natural Gas          13.8 ±1%                              │
│          Crude oil                 19.0 ±2%                         │
│          Bit. Coals           24.5 ±4%                              │
│          Anthracites          25.5 ±3%                              │
│                                                                      │
│ Oil products (all ±2% at max):                                      │
│                                                                      │
│          Gasoline             18.0                                  │
│          Kerosene             18.5                                  │
│          Diesel/Gas oil       19.0                                  │
│          Fuel oils            20.0                                  │
│                                                                      │
│ Note: Bituminous coal may be placed more precisely using calorific  │
│ value by using the equation given in the text.                      │
└─────────────────────────────────────────────────────────────────────┘
```

carbon content in that it needs to take account of non-oxidation of the primary source, and any additional CO_2 release in the fuel production and delivery.

3.1 Natural gas

Marland and Rotty cite non-combustion uses for natural gas as accounting for 3.4% of production in the US, but most of this is for Ammonia production and is soon oxidised: they suggest a maximum of 1% for non-oxidising applications, plus a maximum of 1% for non-oxidisation in combustion. This appears to neglect losses in production and in the network. Figures for methane leaks (see section 5) suggest losses of 1-5%. Some CO_2 is released in the flaring of gas which cannot be tapped in the production wells; say 1±1%.

3.2 Oil

Oil is applied to a number of non-energy uses and statistics vary, as shown in table 3.

Brackley's figures for allocation are derived from IP petroleum statistics with an assumption that a fair proportion of lubes etc are burned, and some wax consumed in candles, etc; bitumen, with no oxidation, is the major

Table 3. Estimates of non-oxidised oil fraction					
	Lubes/bitm/etc		Chemical feedstock		Result.
	Frc. oil	Frc.ox	Frc. oil	Frc.ox.	non-ox %
BP	7%	0	4%	20%	10%
M&R	See text				6.7%
Brackley	4%	20%	5%	80%	4.5%

factor however. Marland & Rotty also assume 50% oxidation of lubes, but derive 20% oxidation for chemical feedstocks as does Gordon. The figure is probably 5-10% non-oxidation but cannot be pinned down further on this data; say 7±3%.

M&R further estimate that 1.5±1.0% of fuel in burners is deposited as non-oxidised. Eyre states that although some burners can achieve nearly 100% conversion, a "ball park figure" for average gasoline conversion is 92%, implying 1-2% non-oxidation of all oils input to combustion. Much of this is emitted as CO, however, which is soon converted to CO2, and technical developments (in particular lean-burn engines and catalytic convertors) will result in much higher conversion: a figure of 0.5±0.5% may apply to projections, especially in industrialised nations.

Finally gas flaring needs to be added. The final figures given in M&R suggest that this accounts for about 2% of total CO_2 emissions from fossil fuels. There is no indication of how this is broken down between oil & gas production: if it is mostly from oil production it implies a penalty of 4-5%. UK statistics show that 40,000bcm of gas was flared at oil fields in 1986, corresponding to about 2% of oil production. A figure of 3±2% is suggested.

3.3 Coal

Marland & Rotty estimate that about 4.5% of coal sent to coking plants remains unoxidised, mostly as tar, and give overall non-oxidation by this route as 0.8% of coal production (US data). The unburnt fraction in burners is given as at most 1% but this would also be declining, and must be quite negligible in power stations. SO2 scrubbing however is essentially exchange of C for S, so that 2%-Sulphur coal by weight adds (12/32)x2% of CO_2. Additional small amounts arise from combustion in deposits and waste tips. The net adjustment must be very close to zero.

More significant are the questions raised by synthetic fuels and gas from coal. To a good approximation the effective carbon contribution will be very close to that of the source material unless the by-product is a solid form of carbon which would be very costly. Synfuels and gasification are discussed by Marland, 1983, and I shall not discuss them further here.

3.4 Summary: carbon release coefficients, primary fuel basis

The conclusion is that fuel production coefficients are very close to the fuel carbon content (though a little lower for gas and oil). In the case of oil particularly this is due to a partial cancelling of non-oxidised applications against additional CO_2 emissions (flaring), with substantial uncertainties in both. For completeness the implied coefficients are summarised in table 4.

Table 4. Summary of carbon release coefficients, primary fuel basis

Fuel	Carbon release coefficient, primary fuel basis (tC/TJ)
Natural Gas	13.6 ±2%
Crude oil	18.4 ±5%
Bit. Coals	24.5 ±4%
Anthracites	25.5 ±3%

The figure for bituminous coal may again be placed more precisely on the basis of calorific value by using the equation given in the text.

---

4. Carbon release coefficients, delivered fuel basis

4.1 Introduction

An assessment based upon the consumption of processed fuels differs from gross production assessment because (a) non-oxidising applications are not relevant, (b) statistics may be disaggregated according to fuel products, and (c) parasitic energy consumption, such as oil use in refineries, needs to be accounted for as a multiplier on the non-parasitic product use. It is this fuel product consumption coefficient which is relevant to most assessments of policy responses, whether the focus is upon end-use measures or upon fuel switching.

The inclusion of parasitic energy raises a fundamental problem in that the associated CO_2 release depends upon the fuel used in the process. However for fossil fuels it is reasonable to assume that the bulk of parasitic consumption is of the same type as the fuel itself, in which case we simply need to know the parasitic energy as a fraction of the fuel throughput.

Data on parasitic energy for all fuels were derived by Hannon and Casler for the US economy in 1974, and quoted in Marland (1983). These are described as showing "only direct and indirect energy consumption during processing". Some figures can be derived from the "own use" entries in standard energy statistics tables, though this may be a subset of the total which needs to be supplemented by use of other fuels.

4.2 Natural Gas

Gas is combusted almost entirely in its original state (after drying & removal of contaminants). The figures of Hannon and Cassler suggest a very large parasitic use of 19%. UK statistics (HMSO, 1987) show "own use" in gas at 7.5% for 1986. Own use in the Soviet gas industry is about 10%. The figures are sufficiently large and disparate that further analysis would be useful. In principle the correct figure is likely to vary according to country and applications - in Holland, with a relatively small dense network, it is said to be no more than 1%, while uses of Liquified Natural Gas (LNG) in particular must have much higher than average parasitic use. A broad figure of 10±5% may be suitable for many purposes, but country-specific data is really required.

4.3 Oil

Oil is transformed into a wide range of products, and the carbon content of some of most important have been listed in table 2. The Hannon & Cassler figures suggest a 12% correction for total parasitic use, which is not inconsistent with the IP petroleum statistics figure of 7.3% refineries consumption for UK 1986 (this fraction appears to have remained of similar magnitude since 1973).

The real complication with oil is that the parasitic energy should apply differently to different products. Increasing fuel-oil use, for example, would reduce the extent of cracking and could even reduce overall refineries consumption. Reforming is a major energy user, as is purification and drying of the product for some markets. **The uncertainty in how parasitic energy for oil products should be allocated is of at least the same order as the range in carbon content of the main products.**

Furthermore the penalties are likely to be higher for the lighter products near the top of the barrel, which have the lower C-content. **In lieu of better data I therefore suggest that an across-the board consumption coefficient is applied for all oil products:** a figure of about 21 tC/TJ before other factors are considered would roughly account for the total parasitic use. To this, gas flaring again adds around 4±2%.

4.4 Coal

The only significant transformation of coal is into coke. Even this is a very minor use, for which I have not found data on parasitic consumption, so the simple carbon-content is suggested. The impact on the general coal coefficient is negligible. The only figure of overall parasitic consumption is the 2.2% of Hannon & Cassler. In principle a further correction should be applied for transport (as it should for oil) but care would be required to avoid double counting and again figures would vary according to the transport involved. Transport of coal from Australia to Europe is said to involved a parasitic use of about 5%. Say 3±2% overall for domestic consumption, with a slightly higher figures for internationally traded coal.

4.5 Summary: carbon release coefficients for fossil fuels, delivered fuel basis

The consequences for fuel-product consumption coefficients are shown in table 5. Note that the ranking of fuels is still clear but the range between them is significantly reduced.

Table 5. Carbon release coefficients for fossil fuels, delivered fuels basis

Fuel consumed	Carbon release coefficient tC/TJ
Natural Gas	16.1 ±5%
Oil products	22.1 ±8%
Bit. Coals	25.3 ±5%
Anthracites	26.3 ±4%

Note also that in this I am excluding questions of combustion efficiency which obviously depends on the particular technology. This is likely to be of particular importance for substitution in the electricity sector, where the efficiency difference between coal steam with FGD and the gas combined cycle may increase the CO_2 advantage of the gas technology by perhaps 50% above that indicated by the coefficients shown in table 5.

4.6 Primary electricity sources

With primary electricity sources the assumptions that all parasitic energy use is of the same fuel type as the product breaks down irretrievably for two reasons. First, a major part of parasitic use in technologies such as hydro and nuclear is in plant construction and so comes before any energy is produced, so consumption could only be of the same energy type in a steady state situation, which is irrelevant. Secondly, some of the energy required is thermal processing and will inevitably be supplied from fossil sources.

In my view it is essential to separate the issue of construction energy from that of parasitic energy in the fuel cycle. The former may be very important but because of the general problems of energy accounting compounded by the complications introduced by time lag between consumption and first production, I do not propose to consider it further. A thoughtful paper on the subject would be valuable, however.

Parasitic thermal energy in the fuel cycle (in practice only relevant for nuclear) is more amenable to consistent analysis, in principle at least. For this it is not unreasonable to assume that electricity input is "own use" and simply serves to derate the useful output from the plant. Nigel Mortimer[2] has provided some preliminary calculations for the thermal part of the fuel cycle, dominated by ore separation, for which he states the energy requirement as:

Energy requirement = (24900G+700)/(611300G-360) MJ(t)/MJ(e)

where G is the ore concentration in per cent U_3O_8. On this basis he estimates the effective CO_2 emission to be about 3.4 tCO_2/TJ(e) for an ore grade of 0.2% uranium oxide, compared with a figure of 200-400 for fossil sources. At current ore grades thermal parasitic use in the fuel cycle can thus be readily ignored. However, Mortimer does note that (if the equation above holds) it does set a limit to the ore concentration useful for CO_2 abatement. The equation implies that exploiting an ore concentration of about 10ppm uranium oxide would emit as much CO_2 as a coal station. This still leaves very large reserves, but it could be a limiting factor in the long-term exploitation of thermal nuclear power as a response if this energy requirement is a fundamental one.

As with parasitic energy use in nuclear and hydro construction, it is an area in which further study would be useful.

5. Including other Greenhouse Gases

The final step in estimating the Greenhouse impact of fossil fuel consumption is to consider other energy-related Greenhouse Gas emissions,

---

[2] N. Mortimer, CO_2 release from non-fossil fuels (preliminary draft analysis), Sheffield City Polytechnic, UK

namely methane, nitrous oxide, and tropospheric ozone. By weight, these are all negligible in comparison with CO_2, but their radiative impact in the atmosphere is much greater per molecule, with factors relative to CO_2 of about 27 for methane and 200 for nitrous oxide.

Nitrous oxide is projected to account for perhaps 5% of the total greenhouse effect and figures cited in Smith, 1988, suggest that fossil fuels account for under 20% of the total emissions, so we can reasonably ignore it.

For methane, Smith cites a global estimate of 35mt/yr release from coal mines and under 35mt/yr from the natural gas industry. Cicerone (1989) cites the same coal figure; his figure for natural gas is 45mt/yr, but this includes venting and other gas sources, not all of which are directly attributable to the industry. Cicerone emphasises the paucity of data and approximate nature of the estimates. Total "unaccounted for" in the US gas supply industry is said to be 2-3% of the primary input, but much of this could be unaccounted burning; on the other hand it does not include losses in production fields, which is probably the major factor, particularly as pipelines improve.

If we take a modest 60mt total methane from coal+gas and multiply by the relative forcing of 27, we obtain the notable result that methane release apparently adds around 25% to the greenhouse impact of fossil fuel consumption, and over 35% to the effect of gas and coal.

This is not consistent with figures cited in Smith suggesting that fossil fuels account for 10-20% of the global methane emitted, which in total contributes about 20% to the current greenhouse forcing. By contrast, these two numbers taken together suggest that fossil fuel-related methane emissions account for at most 10% and probably closer to 5% of the impact of fossil fuel CO_2 emissions. There is still substantial discrepancy if one uses a higher estimate (from isotopic measurements) that 30% of methane is from fossil sources (this figure probably includes some from non-anthropogenic sources, including releases from peat bogs and from clathrates).

The answer to this apparent inconsistency lies in the nature of the atmospheric cycles. About 50% of emitted CO_2 is estimated to remain in the atmosphere but it has a long atmospheric lifetime. Methane has no such sink, but has a much shorter lifetime - about 10% of atmospheric methane is thought to be oxidised each year. The time-average impact of methane emissions is thus much reduced. Simple accounting of total matter emitted is thus not the relevant factor, and the latter approach to estimating the importance of methane (i.e. based on modelling projections of warming) is the correct one, though necessarily approximate since one is really comparing two incompatible quantities.

The methane breaks down to water+CO_2. The figure of 60mt correspond to a little over 1% of the total fossil carbon emissions, suggesting that methane emissions account for a long-term addition of 1-2% to the total carbon emissions of coal and gas, formed as the methane itself is destroyed.

A final factor to note is that the marginal impact of changing emissions is not in proportion to the total contribution for either CO_2 or methane (i.e. radiative forcing does not increase linearly with atmospheric concentration: a logarithmic relationship is usually used for CO_2, and a square root formulation for methane). This further complicates the comparison, but probably gives further relative weight to methane reductions.

Richard Warwick of the Climatic Research Unit has used a 5% methane addition to fossil fuel CO_2 emissions to obtain the CO_2 equivalent. On the basis of the figures above and the observation that methane emissions from the oil industry are small, I estimate a methane-related addition of 10±5% for coal and gas, with 3±2% for oil. This addition needs to be applied to any of the figures produced earlier to derive total forcing coefficients in tCeq/TJ. The crudity of both the data and the assumptions on which methane & CO2 are compared in this manner need to be emphasised, however.

Another greenhouse gas which needs to be considered is tropopheric ozone. Its total contribution to greenhouse forcing is thought to be about 10%, but this is quite uncertain as its impact depends upon height, and concentrations are not well know.

The formation of tropospheric ozone is very complex, but emissions of unburnt hydrocarbons, particularly the aromatics, appear to be the dominant factor (NO_x also plays a role, but its impact depends upon conditions and can be positive or negative). These emissions arise mostly from transport (catalytic converters will greatly reduce emissions) and the solvent industry. Statistics for the UK suggest that transport accounts for at least a third and possibly half of total aromatic HC emissions, the majority coming from petrol exhaust but with substantial contributions from petrol evaporation (based on data from UK photochemical oxidants review group, 1987). Assuming transport to account for around 40% of hydrocarbon emissions, and noting the other wide uncertainties, the coefficient for this sector needs to be augmented by perhaps 40±35% - which swamps all other uncertainties for the transport sector.

In principle, nitrogen oxides and carbon monoxide (from incomplete combustion in many applications) are also relevant because of their indirect involvement in reactions destroying ozone and methane respectively. No figures were obtained indicating the significance or otherwise of these emissions.

6. Conclusions

This paper has summarised the work carried out by several contributors, primarily on the carbon content of fuels, and reviewed the broader issues surrounding CO_2 emission coefficients. The results are summarised in table 6. The following observations can be drawn concerning these results:

*   The carbon content of fuels dominates their impact on radiative forcing in most cases, but other factors are significant
*   Production carbon release coefficients are not the same as consumption coefficients: the latter are 15-20% larger for oil and gas use
*   Uncertainties/variations in carbon release coefficients are around 6-8%. Dominant uncertainties are:
        coal - carbon content, + parasitic use if shipped
        gas   - parasitic energy consumption

553

Table 6. Summary of results for estimating total greenhouse gas emissions from fossil fuels[1]

Fuel	C content tC/TJ	Non-oxid^n correct^{n,x}	Addition[1] CO_2, %	Parasitic Use, %	Methane corrct^n,%
Gas[2]	13.8±0.1	−3±2%	1±1%	10±5%	10±6%
Oil[3]	19.0±0.5	−7±3%	4±2%	8±5%	3±2%
Coal[4]	24.5±1.0	−1±1%	1±1%	3±2%	10±6%

Notes

1. Carbon release coefficients on primary fuel basis are obtained by summing columns 1+2+3. Coefficients on delivered fuel basis are obtained by summing 1+3+4. The methane addition coefficient must be added to either to obtain the C-equivalent release for radiative forcing.

2. Natural gas by pipeline. Parasitic use, in particular, varies greatly between countries and applications; figure for CNG were not obtained.

3. Transport use incurs an additional penalty due to ozone-forming hydrocarbon release. This may swamp other factors, with an addition of 40±35% for cars without catalytic convertors.

4. Bituminous and brown coal. Anthracite has a C-content of 25.5tC/TJ; other factors are as for bitumous coal. Internationally traded coal would have probably have a larger parasitic use associated with the transport.

            oil - several factors (flaring, parasitic,etc)

* Methane emissions add 10±6% to the CO_2 forcing impact from the coal and gas industries

* Ozone-forming hydrocarbon emissions may add 40% to the CO_2 impact of the transport sector, but this is very uncertain. There are additional interractions of ozone with nitrogen oxides, and methane with carbon monoxide, which were not estimated in this study.

* CO_2 release from uranium ore processing is negligible, but could become significant if very much lower-grade ores were exploited. The CO_2 use in construction of non-fossil generating facilities was not evaluated.

A number of policy-related conclusions follow from these results:

* The radiative benefits of coal to gas switching are more like 2/3 than the commonly cited 1/2 in many cases, depending somewhat on local circumstances

*   More data is required on:
    - methane emissions from the coal and gas industries
    - parasitic energy use in all energy industries
    - hydrocarbon emissions from the transport sector
*   Much more work is required to quantify:
    - the radiative importance of tropospheric ozone
    - the indirect impact of CO on methane, and NOx on ozone
    - the theoretical basis for comparing the impact of ground level emissions of CO_2, methane, and hydrocarbons
*   Reducing gas flaring, parasitic energy use, methane releases and hydrocarbon emissions are options to be considered alongside fuel switching and demand measures.

Finally, it should be emphasised that using methane from sources where it would otherwise leak into the atmosphere, such as coal mines and landfill waste sites, reduces overall radiative forcing (and may have safety benefits).

<u>References to published papers:</u>

Cicerone R.J, <u>Biogeochemical aspects of atmospheric methane,</u> <u>Global biogeochemical cycles,</u> Vol.2 no.4, Dec 1988

HMSO, <u>Digest of United Kingdom Energy Statistics 1987</u>

Institute of Petroleum Information Service, <u>Petroleum statistics</u> 1987

Marland G and R. Rotty, <u>Carbon dioxide emissions from fossil fuels: a procedure for estimation and results for 1950-1982.</u> <u>Tellus 36B,</u> 1984

Marland G., <u>Carbon dioxide emission rates for conventional and synthetic fuels,</u> <u>Energy</u> vol.8 no.12, 1983

Smith G., <u>CO₂ and climatic change,</u> IEA Coal Research, 1988.

UK photochemical oxidants review group (1987), <u>Ozone in the UK,</u> Interim report prepared at the request of the Deparment of the Environment, pub DoE. Cited in Greenpeace evidence to the House of Commons Energy Committee study on Energy Implications of the Greennhouse Effect.

Other works referred to are communications sent to me by the authors listed in the introduction.

NEW PROSPECTS FOR SOLAR HYDROGEN ENERGY:
IMPLICATIONS OF ADVANCES IN THIN-FILM SOLAR CELL TECHNOLOGY

Joan M. Ogden
Robert H. Williams

Center for Energy and Environmental Studies
Princeton University
Princeton, NJ 08544, USA

presented at the IEA/OECD Expert Seminar on
Energy Technologies to Reduce Emissions of Greenhouse Gases

OECD Headquarters
Paris, FRANCE
April 12-14, 1989

NEW PROSPECTS FOR SOLAR HYDROGEN ENERGY:
IMPLICATIONS OF ADVANCES IN THIN-FILM SOLAR CELL TECHNOLOGY (*)

Joan M. Ogden
Robert H. Williams

Center for Energy and Environmental Studies
Princeton University
Princeton, NJ 08544 USA

ABSTRACT

In recent years, there have been rapid advances in thin-film amorphous silicon solar cell technology. Industry projections indicate that 12-18% efficient solar photovoltaic (PV) modules costing $0.2-0.4 per peak Watt could become available around the year 2000. If these goals are realized, hydrogen produced via PV-powered electrolysis in a sunny area such as the Southwestern US would cost $9-14 per Gigajoule, which would be competitive with other synthetic liquid fuels from coal or biomass (projected to cost $8-16 per Gigajoule), and probably less expensive than electrolytic hydrogen from nuclear power (projected to cost $14-24 per Gigajoule). Moreover, the lifecycle costs of using PV hydrogen for applications such as automotive transport and residential heating would be comparable to these costs for other synfuels or electricity, if efficient use of energy is stressed. If concerns about the greenhouse effect mitigated against the continued use of fossil fuels, it appears that PV hydrogen could offer an economically acceptable and environmentally benign alternative beginning in the early part of the next century. A possible path toward a large-scale PV hydrogen energy system in the United States is described, beginning with local transportation systems in the Southwest.

---

* The findings presented in this paper are given in more detail in Joan M. Ogden and Robert H. Williams, "Hydrogen and the Revolution in Amorphous Silicon Solar Cell Technology," PU/CEES Report No. 231, Center for Energy and Environmental Studies, Princeton University, Princeton, New Jersey 08544, USA, February 15, 1989. A less technical version of this report will be published as Solar Hydrogen: Moving Beyond Fossil Fuels by the World Resources Institute, Washington, DC in 1989.

NEW PROSPECTS FOR SOLAR HYDROGEN ENERGY:
IMPLICATIONS OF ADVANCES IN THIN-FILM SOLAR CELL TECHNOLOGY

Joan M. Ogden
Robert H. Williams

Center for Energy and Environmental Studies
Princeton University
Princeton, NJ 08544, USA

A.   Introduction

It has long been recognized that electrolytic hydrogen produced from photovoltaic (PV) electricity would have strong environmental advantages over fossil fuel-based energy supply options [1]. Hydrogen is a high quality fuel, which could replace oil and natural gas for transportation, heating and power applications. When hydrogen is burned in air the primary combustion product is water vapor with traces of NOx. And if hydrogen is produced via electrolysis from PV electricity no carbon dioxide would be released in its production or combustion. A PV hydrogen energy system is one of the few long-term energy supply options which could supply the world's energy needs without contributing to the "greenhouse warming" of the planet.   Moreover, the absence of carbon monoxide, volatile organic compounds, oxides of sulfur and other noxious pollutants in the production and combustion of PV hydrogen would help solve serious local and regional air pollution problems such as the deterioration of urban air quality and acid deposition on lakes and forests.

PV hydrogen systems were studied intensively in the 1970s [2]. Research indicated that PV hydrogen production and use for transportation, heating and power would be technically feasible. However, economic assessments published the early 1980s [3-5] concluded that PV electricity would probably always be too expensive for PV hydrogen to be economically competitive with other synthetic fuels. This conclusion, based in large part on expectations at that time of the future prospects for solar cell technology, has remained the "conventional wisdom" regarding PV hydrogen -- at least in the United States. But several factors motivated us to to reexamine the prospects for PV hydrogen.

First, since the early 1980s there has been rapid progress in the development of new solar cell technologies that require much less raw material and energy to manufacture and are amenable to inexpensive mass production techniques. In particular, advances in the technology of thin-film amorphous silicon (a-Si) solar cells suggest that DC electricity from PV systems (and therefore PV hydrogen) could become much less costly than was previously thought feasible.

Second, there is a growing awareness of the high environmental costs of continued dependence on carbon-based fossil fuels. Rising concerns about global climate change from the greenhouse effect [6], the long-term effects of acid deposition [7] and poor urban air quality [8] are providing strong motivation for developing low polluting, non-fossil energy sources.

In this paper, we discuss the implications of advances in a-Si solar cell technology for the prospects for PV hydrogen production. First, we review recent progress and projections for a-Si solar cells. Next we present a conceptual design of a PV hydrogen energy system based on a-Si solar cells and estimate the cost of PV hydrogen production in the US Southwest.

We then compare our estimated cost of PV hydrogen with estimates of the costs of electrolytic hydrogen produced from other electricity sources (wind, hydropower and nuclear) and other synthetic fuels from coal and biomass, for the timeframe near the turn of the century. To facilitate a comparison, all energy supply options are treated on the same economic basis. While economic costs of externalities are not explicitly included, we compare the various supply options with regard to emissions of CO_2 and other pollutants, land and water use constraints and security issues.

To better understand the economic competitiveness of PV hydrogen, we have also estimated the lifecycle costs to the consumer for energy services provided by PV and other energy sources. We have compared the lifecycle cost for automotive transportation with PV hydrogen fuel to the cost of using other alternative liquid and gaseous synthetic fuels and electricity from coal or nuclear power. We have also estimated the cost of residential heating with hydrogen, synthetic natural gas from coal and electricity. As part of this comparative economic analysis, we explore the synergism between energy efficient end-use systems and the lifecycle cost.

Finally, we sketch a possible scenario for the development of a PV hydrogen energy system in the United States and suggest policies that could facilitate such development.

B. Progess in Amorphous Silicon Solar Cell Technology

1. Background

Before about 1980, the only commercially available solar cells were made of high-grade, single-crystal silicon, using a materials- and energy-intensive crystal-growing process. In the early 1980s the industry began marketing solar cells made of polycrystalline silicon, a less expensive (but somewhat less efficient) material composed of many small crystallites. Also around 1980, the first thin-film a-Si solar cells were introduced [9].

The cost of solar cells has dropped steadily, as new solar cell materials and improved manufacturing methods were developed. In the early 1970s solar cells were priced at over $100 per peak Watt. At present, single crystalline cells cost perhaps $5 per peak Watt to manufacture, polycrystalline cells $3 per peak Watt and a-Si about $1 6 per peak Watt [10]. While improvements in efficiency and manufacturing techniques may eventually reduce the production cost of crystalline and polycrystalline solar cells to $1-2 per peak Watt [11], a-Si technology appears to have the greatest potential for reaching the low solar cell costs needed to make PV hydrogen economically competitive.

2. The Amorphous Silicon Solar Cell

In 1974 it was discovered that thin films of amorphous silicon could be used to convert sunlight directly into electricity. Amorphous silicon thin film cells are about 1 micron thick, using much less raw material than crystalline solar cells, which are typically 100-200 microns thick. In contrast to the crystal growing and cutting process, a-Si solar cells are produced by vapor depositing silicon on an inexpensive substrate such as glass, plastic or stainless steel. Because of the speed with which vapor deposition can be done, the ease with which electrical connections can be made, and lower energy and material requirements, amorphous modules can be mass produced much more quickly and cheaply than crystalline modules [12].

Progress in amorphous silicon technology has been rapid. Efficiencies have increased from 1% for the first cells in 1976 to over 13% for small-area laboratory cells, to over 11% for larger area laboratory modules and to 5-7% for large-area, commercially available mass-produced modules in 1987 (Figure 1a). One of the leading scientists involved in the development of a-Si solar cells projects that an efficiency of 18% will be achieved in the laboratory in the early 1990s [13]. Since the first commercial a-Si solar cells were introduced in 1980, production has steadily increased, reaching 11.9 MW (or 41 percent of the total worldwide PV market [14])in 1987 (Figure 1b).

Present indications are that this rapid growth will continue. Chronar Corporation of Princeton, NJ, started construction on a 10 MW per year a-Si solar cell factory toward the end of 1988 [15] and has recently announced plans to build a 50 MW a-Si power plant in California [16]. Solarex, of Newtown, PA, is planning to build a 10 MW per year production facility in the early 1990s [17]. Arco Solar, Inc., of Chatsworth, CA, is building a manufacturing facility in Camarillo, California, capable of producing 5 MW per year and is designing a separate plant capable of 70 MW of yearly production [18].

Although the future costs of a-Si solar cells cannot be predicted with certainty, it seems likely that commercial modules with efficiencies of 10-12% and costs of $0.50 per peak Watt will be available by the mid-1990s. Moreover, manufacturers project that by the turn of the century or shortly thereafter 12-18% efficient PV modules costing $0.20-0.40/Wp could be available (Table 1).

C. Key Technical Assumptions Relating to Amorphous Silicon Solar Cells

Our analysis of the prospects for PV hydrogen in the period near the turn of the century or shortly thereafter is based on four assumptions relating to a-Si solar cell technology in that time frame:

o   that stable efficiencies of 12-18% will be realized in commercial modules;

o   that production costs of a-Si modules will be in the range $0.20-$0.40 per peak Watt;

o   that a-Si modules will have lifetimes of 30 years;

o   that area-related balance of system costs will be $33 per square meter for large, fixed, flat-plate arrays.

In this section we review the technical basis for each of these assumptions.

1.   Assumption 1: a-Si PV Modules Will Have Stable Efficiencies of 12-18%

While today's commercially available single-layered a-Si modules have modest conversion efficiencies of about 5-7%, efficiencies achieved in the laboratory with small-area, multi-layered cells (*) have exceeded 13%, and may reach 18% by the early 1990s. To date, efficiencies achieved in the laboratory have been realized in commercial modules some 5-6 years later (Figure 1a). However, this time lag could shorten in the future, as manufacturers move toward computer-integrated manufacturing of a-Si cells, resulting in a generally higher level of quality control in production than in the laboratory--a phenomenon that has occurred for a variety of products in the semiconductor industry [19]. By the early to mid-1990s, efficiencies for commercial modules are expected to be 10% (for single-layered modules) and 13% (for multi-layered modules) [20]. By the year 2000, commercial modules approaching the "practical limit" values of 12-14% for single-layered solar cells and 18-20% for multi-layered cells may well become available [21,22,23].

One problem that plagued the early development of a-Si solar cells is that the cells experience an initial loss of efficiency (known as the Staebler-Wronski effect [24]) when exposed to light. While the Staebler-Wronski effect is not yet fully understood theoretically, the problem has been largely solved in practice. The initial efficiency degradation can be completely reversed by reheating the cells to their annealing temperature (about 200°C) for a few minutes. At typical outdoor solar cell operating temperatures of 50-60 $^{\circ}$C some annealing takes place, which tends to counterbalance the Staebler-Wronski effect [25]. Also, making single-layered cells thinner and using multiple layers tends to retard the initial loss of efficiency. Single-layered and multi-layered modules can now be made for which the efficiencies stabilize, after a few months exposure to sunlight, at about 80% and 90% of their initial values, respectively [26].

On the basis of these results, we assume that stabilized solar module efficiencies of 12-18% will be reached around the turn of the century. We further assume that the efficiency of a large PV system is 85% of the individual module efficiency, due to electrical losses in wiring and perhaps to wind-blown dirt or dust on the modules [27]. Thus, a PV system constructed of 12-18% efficient modules would have an overall efficiency of 10.2-15.3%.

2.   Assumption 2: The Production Cost of Amorphous Silicon PV Modules Will Be in the Range $0.2-$0.4 per peak Watt

Today most commercial a-Si solar cells are produced in batch operations in small facilities with production capacities of the order of 1 Megawatt peak per year or less. The current production cost is estimated at $1.5-1.6 per peak Watt for 6% efficient cells [28]. In larger plants, considerable economies of

---

* Multi-layered cells, made by depositing several thin-film layers, each tuned to absorb a different part of the solar spectrum, are more efficient than single-layered cells, because they utilize more of the solar spectrum. The different layers can be tuned to different parts of the solar spectrum by alloying a-Si with different materials in different ways.

scale could be realized.   In 1988, according to Dr. E. S. Sabisky, then manager
of the Amorphous Silicon Research Project, at the Solar Energy Research
Institute, in Golden, Colorado [29]:

> "...if today's thin-film amorphous silicon modules of 6-8% efficiency are
> combined with a 10 megawatt annual production plant, the module cost
> target of $1 per peak Watt can be reached..."

Sabisky's prognosis is reflected in the announcement in September 1988 that
Chronar, of Trenton, New Jersey, will build a 50 MW plant at a site 60 miles
north of Los Angeles for $125 million [30]. This plant, which will sell the
electricity to the Southern California Edison Company, represents an enormous
scale-up from the largest amorphous silicon facility built to date, a 100 kW
generating field operated by Alabama Power Company. The PV modules in this
plant are expected to have initial efficiencies of 7 1/2% and guaranteed
stabilized efficiencies of 5% and are expected to cost $1.25 per peak Watt
[31].   Similar cost estimates have been projected by other manufacturers and
are shown in Figure 2.

In Table 2, we show how costs might evolve from current levels to the
range of $0.2-0.4/Wp. These estimates are based on a detailed cost evaluation
by Solarex researchers for a 10 MW per year plant (Table 3), which would
produce 6% efficient modules costing $1.16 per peak Watt [32].

Increasing the scale of production from 10 MW to 100 MW per year would
lead to savings mainly in labor costs, and the cost of 6% efficient modules
would be reduced from $1.16 to $0.94 per peak Watt (Table 2).

Increasing the efficiency of thin-film a-Si cells generally involves fine
tuning the deposition process by adding slightly different amounts of dopant, or
varying slightly the layer thickness, or adding additional layers, or fine
tuning the alloy composition of various layers. Because so little material is
associated with the active layers of PV cells, because this material makes such
a small contribution to the cost of the module (Table 3), and because such
modifications will probably entail at most minor minor changes in labor costs,
it is reasonable to expect that the unit cost of an a-Si module will vary
inversely with efficiency [33]. Thus, a doubling or a tripling of cell
efficiency would probably lead to a two-fold or three-fold reduction in the
cost per peak Watt. The production cost of 12% (18%) modules from the 200
MW/yr (300 MW/yr) Solarex plant would be $0.47/Wp ($0.31/Wp).

Materials costs could also be reduced. For the Solarex design of a
production plant producing 6% efficient cells, glass accounts for $0.23 per
peak Watt or half of the materials cost. The cost of glass could be reduced
$0.06 per peak Watt, if chemical strengthening were not required. A further
reduction in the cost of glass could be realized in large-scale (> 60 MW)
facilities, where it would be possible to integrate a float-glass manufacturing
plant with an a-Si plant. Recovering silane (SiH_4) gas (the primary feedstock
for amorphous silicon deposition) during processing, and reducing module
framing costs could further reduce material costs. Researchers at Solarex
estimate that the overall cost of materials could potentially be reduced to
about $0.11 per peak Watt through innovations such as these, for 12% efficient

cells produced in a 200 MW per year production facility.

Finally, as the technology matures, the rate of equipment obsolescence will slow, making it possible to increase the equipment depreciation period. The effect of increasing the depreciation period from 5 to 10 years plus the effects of the materials innovations mentioned above are summarized for a 1.67 million square meters per year production facility in the last three columns of Table 2. Note that total production costs with such innovations would be in the range $0.16 to $0.25 per peak Watt for 12-18% cells, lower than the range we assume for our PV hydrogen analysis.

3. **Assumption 3: Amorphous Silicon PV Modules Will Have Lifetimes of 30 Years**

Because amorphous silicon solar cells are a new technology, field tests of more than a few years have not yet been completed. However, present-day commercial modules pass a battery of accelerated environmental tests. These tests are designed to simulate many years of use in a short time by subjecting the solar modules to rapidly varying extremes of light, temperature, humidity, hail impacts, etc. A preliminary judgment (which must be verified by further field testing), based on the results of such tests and expected processing improvements, is that a 30-year lifetime is a reasonable expectation [34].

4. **Assumption 4: Area-Related Balance of Systems Costs Will Be $33 per Square Meter for Large, Fixed, Flat-plate, A-Si Based PV Arrays**

Area-related balance of systems (BOS) costs include the support structure holding the PV modules, the array wiring and electrical equipment, land, site preparation and other construction costs. Previous conceptual design studies and analysis of data from experimental PV arrays and demonstration projects indicate that area-related BOS costs of $50 per square meter could be readily achieved with present technology [35,36]. If low-cost support structures using pre-fabricated PV panels were employed, this cost could be reduced to perhaps $37 per square meter. With a low-current, high-voltage electrical design, which is especially well suited to a-Si cells, wiring costs could be reduced to give a total area-related BOS cost of $33 per square meter, as assumed in this study [37].

D. **Economic Assumptions**

In comparing PV hydrogen to other long-term energy supply options, it is important to use the same accounting rules for the various alternatives. In our study we made several economic assumptions, which were used in all cases:

o   In estimating the production cost of energy (for PV hydrogen, synfuels, electricity, etc.), we assume that the production facility is a utility-owned plant. We assume a discount rate of 6.1% and annual insurance costs equal to 0.5% of the capital cost. These values are suggested by the Electric Power Research Institute, for evaluating alternative electric power technologies for US utilities.

o   We neglect federal and state corporate income taxes and property taxes in our analysis of energy production costs. As the tax structure varies in different countries, this allows a more general comparison, one which is not specific to the US. More importantly, these taxes discriminate against capital-intensive technologies such as PV hydrogen. [In our policy discussion (Section J), we recommend that energy supplies be taxed in ways that do not discriminate against such technologies.]

o   Environmental and security externalities associated with energy production and use are not explicitly taken into account in our economic analysis. Instead we calculate only direct production costs for energy and direct lifecycle costs for delivered energy services.

o   For calculations of the lifecycle costs of particular energy services delivered to consumers (for example, the lifecycle cost per kilometer of owning and operating an automobile), a discount rate of 10% is assumed.

o   All costs are given in 1986 US dollars. Fuel costs are given in $ per gigajoule, calculated on a higher heating value basis.

E.   An Amorphous Silicon Based PV Electrolyzer System

1.   Design of a Solar Photovoltaic Electrolytic Hydrogen System Based on Amorphous Silicon Solar Cells

Figure 3 is a sketch of a PV hydrogen system based on a-Si solar cells. DC electricity produced in a PV array would power an electrolyzer, which would split water into hydrogen and oxygen. Hydrogen would be compressed, if neccessary, for onsite use, storage or pipeline transmission. The technical characteristics of the PV hydrogen system and economic assumptions used in our study are summarized in Table 4.

Because at least 50% of the cost of PV hydrogen is due to the cost of DC electricity, it would generally be less expensive to locate a large PV hydrogen energy system in a sunny location and transport the produced hydrogen to distant users via high pressure pipelines than to produce the hydrogen locally. In our study for the US, we assumed that the PV system is located in the Southwest, and that the average annual insolation incident on a tilted, fixed, flat-plate array is 271 Watts/m^2, the measured value in El Paso, Texas.

2.   The Cost of PV Electricity

The production cost of DC electricity from an amorphous silicon solar array located in El Paso is plotted as a function of solar module and balance of systems costs for 12% and 18% efficient PV modules in Figure 4. These graphs can be used to estimate the cost of DC electricity given the solar module and BOS costs. Alternatively, if a certain DC electricity cost is desired, the required PV module and BOS costs can be inferred. For 12% (18%) efficient PV modules costing $0.4/Wp ($0.2/Wp), and BOS costs of $33/m^2, DC electricity would cost $0.035/kWh ($0.020/kWh).

3.   The Cost of PV Hydrogen

For DC electricity costing $0.020-0.035/kWh, PV hydrogen would cost $9.1-

14.0/GJ to produce. For a large scale hydrogen energy system, the total cost of PV hydrogen at the end of the pipeline (including the costs of hydrogen production, storage, compression, and pipeline transport for 1600 km) is calculated as a function of the PV DC electricity cost in Figure 5. Storage in depleted gas wells would add perhaps $0.18/GJ, compression to 6.9 MPa (1000 psia) $1.5/GJ and pipeline transmission 1600 km (1000 miles), $0.35/GJ [38].

F. Comparison of PV Hydrogen and Other Synthetic Fuels

1. Production Cost

(a) PV Hydrogen vs. Electrolytic Hydrogen from Wind, Hydro and Nuclear

In Table 5 we show estimated costs for electrolytic hydrogen from PV, nuclear, hydropower, and wind sources. We find that PV hydrogen costing $9.1 to $14 per GJ would be cheaper than nuclear-based electrolytic hydrogen ($14 to $24 per GJ), hydrogen from new hydroelectric sources ($21 to $26 per GJ) (*), and hydrogen from wind sources ($17 to $20 per GJ).

(b) PV Hydrogen vs. Synthetic Fuels from Coal and Biomass

In Table 6 we show estimated production costs for synthetic liquids and gases derived from coal and biomass and compared to projected production costs for PV hydrogen [39]. We find that PV hydrogen would be approximately competitive with synthetic liquids from coal or biomass (projected to cost $8-16/GJ), but probably more expensive than synthetic gases from coal (projected to cost $4-8/GJ).

(c) Economies of Scale

Because PV arrays and electrolyzers are modular technologies, it would be possible to achieve low production costs in facilities of modest scale. There are essentially no economies of scale for electrolyzers larger than about 2 MW [40] and for PV systems larger than 5-10 MW [41]. Thus, PV hydrogen systems could be highly modularized, with typical module capacity in the range 5-10 MW and characteristic capital costs of 5-15 million dollars. This is contrasts with coal-based synfuels, where billion dollar plants would be needed to exploit economies of scale (Table 6). The fact that the transition to a PV hydrogen economy can be initiated by "starting small" should greatly facilitate a societal decision to choose this course.

(d) Sensitivity of PV Hydrogen Cost to Assumptions

The estimated production cost of PV hydrogen depends on the technical and economic parameters assumed in Table 4. We have illustrated in Table 7 how the cost of PV hydrogen production changes, if the assumptions are changed.

----

* For large-scale hydrogen production at new sites. Hydrogen could be produced at much lower cost at many existing hydroelectric sites and some new sites. However, most sites for producing hydroelectricity at low cost are already being exploited.

For example, if the area-related balance of systems cost were $50 per square meter (the US Department of Energy target for fixed, flat-plate arrays by the year 2000) instead of $33, the cost of PV hydrogen would be $10.6-16.3/GJ instead of $9.1 to 14.0/GJ. Similarly, if the lifetime of a-Si modules turned out to be 15 years instead of 30 years, the cost of PV hydrogen would be $10.4-18.2/GJ.

2. Emissions of CO_2 and Other Pollutants

PV hydrogen is one of the few energy supply options which would release no carbon dioxide into atmosphere. In contrast, synthetic fuels from coal would release almost twice as much carbon dioxide per unit of energy consumed as oil and almost three times as much as natural gas (Figure 6). No sulfur oxides, volatile organic compounds, carbon monoxide or particulates would be emitted in hydrogen combustion. While hydrogen-powered engines can produce significant levels of nitrogen oxides, this could be controlled to low levels with various techniques. With hydrogen-powered fuel cells, even NOx emissions would be negligible.

3. Water and Land Use

PV hydrogen production capacity could be concentrated in relatively small areas. Producing hydrogen equivalent to total US oil use would require a collector field in the Southwest of some 64 thousand square kilometers (24 thousand square miles). This is equivalent to 0.5% of total US land use or about 7% of the desert area in the US. Figure 7 shows the areas required to produce hydrogen equivalent to all oil, all oil and gas, and all fossil fuel use in the US in a single circular collector field located in the Southwest.

For comparison, the amount of land required to produce synthetic liquid fuels equivalent to current US oil use from strip-minable coal would average about 22 thousand square kilometers, and from biomass about 1790 thousand square kilometers. Thus photovoltaic hydrogen production would require 3 times as much land use as production of coal-based synthetic fuels, but only about 1/30th as much land use as would be required for the production of synthetic fuels from biomass. The land requirements for PV hydrogen, biomass synfuels, and coal synfuels are shown in Table 8, in relation to land use for other purposes.

PV hydrogen could be produced even in very arid regions. In El Paso, Texas, for example, where the insolation is high but precipitation very light [with annual rainfall amounting to only 20 centimeters (8 inches)], the feed-water required for electrolysis would be only 12 to 17 percent of the average precipitation falling on an area equivalent to the solar collector area (Table 9). This is much less than the water required for biomass energy production.

4. A Comparison of PV Hydrogen and Nuclear Hydrogen

A large-scale energy system based on electrolytic hydrogen from nuclear power would have many of the same advantages as a PV hydrogen energy system--e.g. no CO_2 would be produced. Despite the problems associated with nuclear power, widespread concern about the greenhouse problem is leading to calls to reconsider the nuclear power option. For example, the Conference Statement of the 1988 World Conference on the The Changing Atmosphere: Implications for Global Security recommended [42]:

> "There is a need to revisit the nuclear power option. If the problems of safety, waste, and nuclear arms proliferation can be solved, nuclear power could have a role to play in lowering CO_2 emissions."

Two of the problems mentioned here, though unsolved, are in principle amenable to technical fixes that could greatly reduce the risks involved: nuclear safety and nuclear waste disposal. There is already considerable interest in some quarters of the nuclear community in fundamentally redesigning nuclear power systems in ways that would make them inherently safe, thus addressing the major public anxiety about nuclear power at present, its safety [43].

But the nuclear weapons connection to nuclear power is more troubling. Inherent in nuclear technology is the fact that the fissile plutonium (*) produced as a byproduct of reactor operation in uranium-fueled nuclear power plants is also the stuff from which nuclear weapons are made. A present-day 1000 MW nuclear power plant discharges some 141 kg of fissile plutonium annually in its spent fuel [44]. For comparison, it takes less than 10 kg of fissile plutonium to make a nuclear explosive.

For the present generation of nuclear power plants the risks of proliferation and criminal diversion of nuclear weapons-usable materials are limited by the fact that the produced plutonium is locked in the spent fuel of these power plants, where it is partially protected against diversion to nuclear weapons use by the intensely radioactive products of nuclear fission. However, the nuclear weapons connection to nuclear power would come into sharp focus if the nuclear power industry were to be reborn, and nuclear power were to come to play a major role in the world energy economy. If nuclear power were developed on a large scale, concerns about limited world supplies of uranium would force a shift from the present generation of nuclear power plants based on "once-through" nuclear fuel cycles (for which no efforts are made to recover the plutonium and unused uranium in the spent fuel elements after they are discharged from the reactors) to fuel cycles that involve the reprocessing of spent nuclear fuel and the recycling of the recovered plutonium in fresh fuel for both present reactor types and a new generation of plutonium breeder reactors.

Under current plans for this second-generation nuclear power technology some 30,000 kg of fissile plutonium will be circulating in worldwide commerce in hundreds of shipments annually by the year 2000--in trucks, trains, ships, and planes [45]. This scenario poses formidable institutional challenges for safeguarding this material against occasional diversion to nuclear weapons purposes.

------

* Fissile plutonium is plutonium made up of those isotopes that are fissionable by slow neutrons (e.g. those that mediate the controlled nuclear chain reactions in present-day nuclear power plants).

For a "born-again" nuclear industry the institutional challenges would be much larger than those implicit in present plans. Suppose that concern about the greenhouse problem were to lead to the resurrection of the nuclear power to the extent that nuclear energy were to replace 1/4 of coal use worldwide (by substituting nuclear power for coal-based power generation) plus 1/4 of oil and gas (by providing hydrogen derived from nuclear power plants). Such an emphasis on nuclear power would require a global installed nuclear capacity of more than 3000 GW (up from about 300 GW in 1986), equivalent to about twice the present level of electricity generation from all sources. About 7/8 of the nuclear energy required for this scenario would be for hydrogen production production [46]. At such a high level of nuclear power development, uranium resource constraints would probably create large incentives for plutonium recycling. With plutonium recycle technologies each 1 GW of installed capacity would discharge in its spent fuel and have reprocessed and recycled nearly 1000 kg of fissile plutonium per year, so that the amount of plutonium circulating in global nuclear commerce would be some 3 million kg per year. It is difficult to imagine human institutions capable of safeguarding these plutonium flows against occasional diversions of tiny but significant fractions to nuclear weapons purposes.

Fortunately, the economics of a large-scale shift to nuclear hydrogen are not compelling. At present nuclear power costs in the United States, electrolytic hydrogen would cost almost twice as much as the high end of the costs targeted here for PV hydrogen in the year 2000 (Table 5). But even if the nuclear industry's cost reduction goals for a reborn nuclear industry were met--involving nearly 50% reductions in both capital and O&M costs--the cost of electrolytic hydrogen based on these cost targets would still just reach the upper end of the costs targeted for PV hydrogen in the period near the year 2000 (Table 5).

G. PV Hydrogen as a Transportation Fuel

If the cost and performance targets for a-Si solar cells near the turn of the century are realized, PV hydrogen could become comparable in cost to other liquid synthetic fuels (Table 6). This suggests that the first large market for PV hydrogen might be as a transportation fuel. Moreover, automotive manufacturers in Germany and Japan have ongoing programs to develop hydrogen-powered cars, making it likely that hydrogen-based transport technology could be commercialized by the early 21st century. In this section, we compare the lifecycle costs of automotive transport with PV hydrogen and other options (gasoline, methanol from coal, synthetic natural gas (SNG) from coal, and electricity from coal or nuclear energy).

1. Hydrogen-Powered Cars: The Storage Challenge and Energy-Efficient Cars

Because of gaseous hydrogen's low volumetric energy density, storing enough fuel onboard for a long travelling range is a major challenge. This suggests limiting the use of hydrogen to short-range vehicles or using liquid hydrogen as fuel. However, liquid hydrogen fuel would be more considerably more expensive than gaseous hydrogen, because of the high cost of liquefaction, which could add $6-10/GJ to the fuel cost [47].

An alternative approach to the short-range problem is to increase the fuel economy of the vehicle (so that less fuel is needed to attain a reasonable travelling range) and use compressed hydrogen gas or metal hydride storage. It is now technically feasible to improve automotive fuel economy to the range 2.9 to 2.3 liters per hundred kilometers (1/100 km) [80 to 100 miles per gallon (mpg)] of gasoline-equivalent fuel [48]. For example, Toyota recently demonstrated the AXV, a prototype 4-5 passenger car which gets 2.4 1/100 km (98 mpg) on diesel fuel [49]. High fuel economy was achieved by using a direct-injection diesel engine, reducing the weight of the car to 650 kg through extensive use of plastics and aluminum, streamlining to reduce aerodynamic drag, and using a continuously variable transmission. If these innovations were applied to hydrogen-powered cars, limited range would not be a serious constraint on the use of compressed hydrogen gas or metal hydride storage. In our study, we assumed that a hydride storage tank is used, containing 1.5% hydrogen by weight (*).

2.   The Economics of Cars Fueled with PV Hydrogen and Other Fuels

Figure 8 shows the lifecycle cost per kilometer of owning and operating a car powered by gasoline (at the present US price), methanol from coal, synthetic natural gas from coal, electric batteries, and PV hydrogen for a range of delivered prices for the alternatives to gasoline and for three levels of fuel economy [corresponding approximately to 7.8, 4.7, and 2.6 1/100 km (30, 50 and 90 mpg of gasoline equivalent, respectively)]. We find that fuel costs are a relatively small part of the total cost [less than 25% at 7.8 1/100 km (30 mpg), and less than 10% at 2.6 1/100 km (90 mpg)]. Moreover, there is little difference among fuels in the total lifecycle cost of owning and operating a car. The consumer would pay about the same per km for a low-polluting, PV hydrogen-powered car, as for one fuelled with synthetic fuels from coal or electricity from coal or nuclear. And, for highly efficient cars, a travelling range of 400 km could be achieved with all fuels except electricity, where battery weight and bulk would limit range to about 170 km, even with advanced batteries and efficiency improvements.

H.   Hydrogen for Residential Heating

Although hydrogen sounds like an exotic choice to most people in the US today, hydrogen-rich gases have been used for home heating for over a hundred years (**).

---

* The cost for the compressed hydrogen gas [at 16.5 MPa (2400 psia)] alternative would be $1.5/GJ higher than for hydrides, because of the extra compression cost at the filling station [50].

** "Town gas" (a mixture of approximately half hydrogen and half carbon monoxide with traces of methane, which can be derived from coal, wastes or wood), was piped into millions of urban homes in the US earlier in this century, and is still used in parts of Europe, Asia, and South America. In some regions of the US, natural gas did not supplant town gas as a residential fuel until after World War II.

Hydrogen appliances using open flame burners would be similar in cost and efficiency to today's natural gas heating systems and appliances. In fact, existing natural gas appliances could perhaps be converted to hydrogen use, if the burners and some metering devices were replaced [51]. Because of NOx production, appliances using an open hydrogen flame would have to be vented, as with natural gas flames.

Because the estimated cost of producing synthetic natural gas (SNG) is so much less than that for PV hydrogen ($5.7 to 7.9 per GJ for SNG vs. $9.1 to $14.0 per GJ for PV hydrogen--see Table 6), it would seem that the residential market, where these fuels could compete, would be an unpromising one for PV hydrogen. However, the competition between alternative energy sources should be judged on the basis of a comparison of lifecycle costs for the energy services delivered, not the relative fuel prices. The prospects for PV hydrogen look much better when energy services are considered instead and if emphasis is given to the efficient use of energy in providing these services.

To show how the economic prospects for PV hydrogen improve with emphasis on energy efficiency, we have estimated the annual cost to consumers of providing space heating and water heating with hydrogen, SNG derived from coal, and electricity, and for three alternative sets of energy end-use technologies:

1. Low First-Cost Technologies for Conventional New Houses: conventional furnaces for natural gas and hydrogen; conventional tank-type water heaters for natural gas and hydrogen; electric resistance space and water heating for electricity.

2. Energy-Efficient Technologies for Conventional New Houses: condensing furnaces for natural gas and hydrogen; tankless water heater for natural gas; catalytic water heater for hydrogen; and heat pumps for space and water heating with electricity.

3. Energy-Efficient Technologies for Superinsulated Houses: catalytic space heaters for both natural gas and hydrogen; tank-type water heater for natural gas (*), catalytic water heater for hydrogen, resistance space heating (**) and heat pump water heater for electricity.

The levelized annual cost (in dollars per year) of space and water heating with each energy carrier and with each set of end-use technologies is shown in Figure 9 [54], for a New Jersey climate. The energy prices assumed in each case are equal to the total cost of delivering the energy carrier to the household. [In both the PV hydrogen and the SNG cases, the cost of transporting the gas 1600 km (1000 miles) from the production site is included.]

---

* A tank-type water heater is used here because tankless units are usually installed in conjunction with a central furnace.

** In a superinsulated house the heating load is so small and the heating season so short that it would not be cost-effective to install a heat pump. It would be more economical to install a low capital cost resistive heating unit.

The relative costs for space and water heating with the different energy carriers are roughly what one might expect in the "base case" of low first-cost end-use technologies used in conventional energy houses: SNG systems are the least costly, followed by electricity, with hydrogen coming in last.

When energy-efficient end-use technologies are deployed in conventional houses overall costs are reduced for hydrogen and electric systems, though not for SNG systems; and hydrogen fares somewhat better than in the base case in relation to the other energy carriers: when priced at the low end of its estimated cost range, hydrogen-fueled systems would be able provide heat at about the same cost as for electrical systems and SNG systems fueled with SNG priced at the high end of its estimated cost range.

But when energy-efficient end-use technologies are deployed in superinsulated houses overall costs drop sharply in all cases--even though superinsulated houses are much more costly to build; and, suprisingly, hydrogen becomes more competitive with the other energy carriers. Using hydrogen is clearly less expensive than using electricity for space and water heating and hydrogen would be competitive with SNG at the high end of the price ranges for both fuels and slightly (about 5%) less expensive at the low end of these ranges. This shift arises because of the capital savings and efficiency improvements that are possible in end-use equipment used with superinsulated housing designs.

When a superinsulated housing design is used, it is no longer necessary to install a costly central heating system (furnace or heat pump plus ductwork) in a New Jersey climate. Instead houses can be heated and maintained comfortably with a few low-cost space heaters--resistive heaters for electricity or gas-fired heaters for hydrogen and SNG. Moreover, high-efficiency catalytic combustors can be used with the gas-fired heaters.

In a catalytic heater, fuel gas combines with the oxygen in air at a relatively low temperature in the presence of a catalyst, such as platinum or stainless steel [52]. Instead of a flame, the catalytic reaction produces a radiant glow. Thus, catalytic space heaters are generally used in the room to be warmed, to take advantage of radiant heating.

With hydrogen fuel, catalytic combustion could be carried out at a low enough temperature that there would be negligible NOx production [52], and the combustion products (mainly water vapor) can be discharged directly into the heated space. Thus hydrogen-fired catalytic heaters (unlike SNG units) require no vent for the exhaust gases. And whereas SNG catalytic heaters would be about 85% efficient, the hydrogen unit would be close to 100% efficient. Moreover, the catalytic hydrogen heater would also improve comfort by acting as a humidifier, as well as a heater. With a superinsulated house so little fuel is required for space heating that the humidity level does not become excessive, and a comfortable relative humidity of 40-50% can be provided on cold days [53].

I.   A Possible Path for Commercialization of PV Hydrogen in the US

Table 10 shows a list of potential markets for PV electricity and PV hydrogen as a function of the solar module price required to "break even" with current sources. The first markets for PV hydrogen (merchant hydrogen, small chemical users), could open at about the same time as utility peaking markets for PV electricity, sometime in the early to mid 1990s. While merchant hydrogen and small chemical markets would be small compared to energy markets, they could play a role in the near-term development of PV hydrogen systems.

If performance and cost goals for a-Si solar cells are met, PV hydrogen produced in the Southwestern US could become cost competitive with other transport fuels from coal or biomass around the year 2000. While economics alone would not compel a switch from gasoline to PV hydrogen-powered transport, environmental considerations might provide an early impetus for developing PV hydrogen as a transport fuel.

As an example of how a local PV hydrogen transport system might develop, we developed a scenario for the city of Phoenix, Arizona. Located in a sunny area, Phoenix is ideal for PV hydrogen production and has a severe and growing air pollution problem. In 1987, levels of ozone, carbon monoxide and particulates in Phoenix exceeded federal health standards 33 days out of 365 [55]. The majority of this pollution is due to automotive emissions. Over the next 30 years the population is expected to increase from 2 to 5 million people, and without controls on automotive emissions, air pollution from cars could triple [56].

Much tougher automotive emissions standards are needed to prevent further deterioration of and to improve urban air quality. In the near term, implementing proposed new emissions standards [57] could cut pollution in half by 2000. But the introduction of hydrogen-powered vehicles could also have a small, but significant, near-term impact.

Since PV hydrogen is not likely to be sufficiently low in cost to introduce before the turn of the century, one way to initiate a transition to a hydrogen economy in the 1990s would be to use offpeak power from conventional power plants to provide hydrogen for fleet vehicles. In our scenario half of all fleet cars (4% of all cars but accounting for 10% of all car miles travelled) in Phoenix are operated on hydrogen derived from this offpeak power by the year 2000. After the year 2000 PV hydrogen would start to be introduced, so that by 2015 half of all cars in Phoenix would be operating on hydrogen. In this scenario automotive emissions would fall to half the present level by 2000 and would still be below present levels by 2015, despite a projected three-fold increase in the number of vehicle-miles traveled.

Fleets offer the oft-mentioned advantage that they could be centrally fuelled and maintained, which would minimize the initial investment in a fuel distribution system. The first hydrogen fleets might belong to utilities. which would probably have some experience with PV systems, or to city or state governments. Because there are no significant economies of scale for PV hydrogen systems larger than perhaps 5-10 MW, the initial investments in PV hydrogen systems could be small. More capacity could be added in modest increments, as the demand increased.

Once the technology is established for fleets, local utilities could begin to offer hydrogen fuel for private automobiles. In order to interest consumers, it would be important to have multi-fuel capable vehicles, which could use other fuels in areas where hydrogen was not yet available.

If local hydrogen transport systems were successful in reducing air pollution levels in the Southwest, cities in the Northeast and the Midwest might decide to convert to hydrogen as well. If just 10% of fleet vehicles in the US converted to hydrogen, there would be enough demand to justify building a pipeline to bring hydrogen from the Southwest to the Midwest or Northeast. Hydrogen-powered transport might then follow a similar pattern in northern cities, first as a fuel for fleets and then for private vehicles. If hydrogen became readily available in the North, it might eventually find uses as a fuel for residential space and water heating as well as for urban transportation.

While a transition to a hydrogen economy could begin as early as the 1990s, it would take many decades to complete, largely because of capital constraints on the evolution of the energy system [58]. One of the most important interim strategies needed to facilitate this transition is an emphasis on the improved efficiency of energy use--to improve both consumer acceptability of hydrogen as an energy carrier (e.g. to give hydrogen cars a range per fueling that would allow them to be general use vehicles) and to improve the economics of hydrogen energy (e.g. to help hydrogen be competitive in space and water heating applications). Of course, the more efficient use of energy derived from present sources also makes sense for economic, environmental, and security reasons as well [59].

J. Public Policy Issues

What should the role of the public sector be in promoting a transition to a hydrogen economy? Policies aimed at coping with externalities would be helpful--tough local air pollution standards, an oil tariff [60] or an oil products tax [61] to help reduce the risks of importing oil, and a carbon tax or alternative measure to help cope with the greenhouse problem.

Also a tough policy promoting the more efficient use of energy would greatly facilitate a transition to hydrogen in the period beyond 2000. As urban transportation appears to be the most promising initial market for PV hydrogen, the promotion of improved automotive fuel economy is crucial. While an oil tax or tariff would help promote improved fuel economy, it is not likely that such a measure by itself would be be adequate to generate the high fuel economy levels needed to overcome the range constraint on hydrogen-fueled vehicles, because, at high fuel economy levels, the cost of fuel contributes so little to the total cost of owning and operating a car, and this total cost varies little with fuel economy [62]. Thus it would be desirable to complement an oil tariff or tax with increases in the federally mandated fuel economy standards or the levy of "gas guzzler" taxes or similar measures.

Another important issue is public policy relating to the taxation of energy. The costs of producing PV electricity and PV hydrogen are mostly capital costs. Consideration should be given to eliminating taxes that discriminate against capital-intensive energy sources in favor of alternatives (*). If governments need tax revenues from the energy industries in excess of what can be provided by oil and carbon taxes, the additional taxes should be designed so as not to discriminate against capital-intensive energy systems--e.g., a "gigajoule" tax, proportional to the energy content of the energy carrier.

Increased R&D is warranted on thin-film PV technologies to hasten the achievement of the PV performance and cost goals. Increased R&D should also be focussed on low cost PV balance of systems designs, on PV hydrogen system design, and on hydrogen end-use systems.

Since a hydrogen energy economy in the US will probably be initiated through the application of hydrogen to urban fleet vehicles in the Southwestern states, state or city government procurement of hydrogen-powered vehicles could help the early development of hydrogen facilities in sunny areas, with the hydrogen initially produced from offpeak power at conventional power plants.

Some fraction of US federal government auto, bus, and truck fleets might also be converted to hydrogen. Not only would so doing help create initial markets for hydrogen vehicles but also it would provide impetus for investments in the first hydrogen pipelines.

K.   Conclusion

The prospects for dramatic improvement in the performance and cost parameters for a-Si solar cell technology indicate that in the period near the turn of the century PV hydrogen could be an economically acceptable alternative to synthetic liquid and gaseous fossil fuels. Moreover, a shift to a PV hydrogen system would be easier to inititiate than a shift to fossil synfuels, because it is possible to start small and evolve in modest increments.

The primary impetus for a shift to PV hydrogen would probably be growing environmental concerns about fossil fuels--urban and regional air pollution and the global greenhouse problem. While these problems are also forcing a reconsideration of nuclear power, nuclear hydrogen would probably be more costly, and the PV hydrogen alternative would avoid the risks of diversions of nuclear weapons-usable materials from the nuclear fuel cycle that are inherent in nuclear power technology and that would become acute at the high levels of nuclear power development required to have a significant impact on the greenhouse problem.

The benefits of a PV hydrogen economy appear to be so great that serious consideration should be given to policies that would hasten the transition.

———

* In this study corporate income tax and property taxes were neglected, as the analysis was aimed at comparing the costs to society of alternative energy systems. Has these taxes been included as they are presently assessed, the outlook for capital intensive-energy technologies (PV hydrogen and electricity, hydropower, wind power and nuclear) would have been less favorable than indicated relative to fossil fuel-based technologies.

REFERENCES

1. J.B.S. Haldane was the first to suggest electrolytic production of hydrogen fuel from wind power in 1923, and in 1927, A.J. Stuart suggested using hydropower as a source of electricity for hydrogen production. The use of solar power to produce hydrogen was first suggested by J. O'M. Bockris in 1962. (see: J. O'M. Bockris, Energy Options, Halsted Press, New York, 1980.)

2. For detailed information about the uses of hydrogen see, for example:

   K.E. Cox and K.D. Williamson, Hydrogen Its Technology and Implications, five volumes, CRC Press, Boca Raton, Florida, (1979).

   T.N. Veziroglu, Hydrogen Energy: Parts A and B, 2 volumes, Plenum Press, New York, (1975).

   The International Journal of Hydrogen Energy, started in 1976, is the primary publication of hydrogen energy research community.

   See also reports from the Brookhaven National Laboratory hydrogen program (mid-1970s to 1987).

   T.N. Veziroglu, ed., Hydrogen Energy Progress, Proceedings of the World Hydrogen Energy Conferences, Pergamon Press, 1974-1988.

3. W.J.D. Escher, R.W. Foster, R.R. Tison and J.A.Hanson, "Solar/Hydrogen Systems Assessment," DOE/JPL-9559492, 1980.

4. E. Fein and T. Munson, "An Assessment of Non-Fossil Hydrogen", Gas Research Institute Report GRI 70/0108, 1980.

5. P.D. Metz, Chapter 3.0, "Technoeconomic Analysis of PV Hydrogen Systems," BNL Report B199SPE(3), 1985.

6. Philip Shabecoff, "Global Warming Has Begun, Expert Tells Senate," New York Times, June 24, 1988, p.1.

7. Committee on the Monitoring and Assessment of Trends in Acid Deposition, National Research Council, Acid Deposition: Long-Term Trends, National Academy Press, Washington, DC, 1986.

8. "Urban Ozone and The Clean Air Act : Problems and Proposals for Change," Staff Paper, Oceans and Environments Program, Office of Technology Assessment, April 1988.

9. P.D. Maycock and E.N. Stirewalt, The Photovoltaic Revolution, Rodale Press, NY, NY, 1985.

10. S. Kaplan, Chronar Corp., private communications, 1988.

11. R.A. Whisnant, P.T. Champagne, S.R. Wright, K.C. Brookshire, and G.J.

Zuckerman, "Comparison of the Required Price for Amorphous Silicon, Dendritic Web and Czochralski Flat Plate Modules and Concentrators," RTI Associates Report, Research Triangle, North Carolina, 1985.

12. D.E. Carlson, Solarex Thin Films Division, private communications, 1988; D.E. Carlson, "Low-Cost Power from Thin-Film Photovoltaics," in Electricity, T.B. Johansson, B. Boglund, and R.H. Williams, eds., University of Lund Press, Lund, Sweden, 1989.

13. Ibid.

14. Photovoltaic Insiders' Report, February 1988.

15. S. Kaplan, Chronar Corp., private communications, 1988.

16. M. L. Wald, "Solar Power Plant Planned for California," Business Day Section of The New York Times, September 6, 1988.

17. D.E. Carlson, General Manager, Thin Film Division, Solarex, private communications, 1987.

18. Photovoltaic Insiders' Report, January 1987.

19. Private communications from Sigurd Wagner, Electrical Engineering Department, Princeton University, July 1988.

20. Four US manufacturers (Chronar, Solarex, Arco Solar, and ECD) are in a cost shared program with the US Department of Energy to produce modules of these efficiencies by 1990, PVIR, February 1987.

21. D.E. Carlson, Solarex Thin Films Division, private communications, 1988; D.E. Carlson, "Low-Cost Power from Thin-Film Photovoltaics," in Electricity, T.B. Johansson, B. Boglund, and R.H. Williams, eds., University of Lund Press, Lund, Sweden, 1989.

22. E.A. DeMeo and R.W. Taylor, "Solar Photovoltaic Power Systems: An Electric Utility Perspective," Science, v. 224, April 20, 1984.

23. Private communications from Sigurd Wagner, Electrical Engineering Department, Princeton University, July 1988.

24. D.L. Staebler and C.R. Wronski, Applied Physics Letters, v. 31, p. 292, 1977.

25. D.E. Carlson, "Solar Cells," in Semiconductors and Semimetals, v.21, Part D, ed. J.I. Pankove, Academic Press, NY, p.7, 1984

26. D.E. Carlson, 8th European Photovoltaic Solar Energy Conference, Florence, Italy, May 9-13, 1988.

27. US Department of Energy, DOE/CH10093-19, January 1988. Also EPRI, Sandia reports.

28. S. Kaplan, Chronar Corp., private communications, 1988.

29. "$1 per Wp Module Cost Target Seen Obtainable by Early 1990s Without Efficiency Gains," <u>PVIR</u>, p. 4, May 1988.

30. M. L. Wald, op. cit.

31. "Chronar Negotiating to Install 60 MW, $150 Million PV System in Southern California," <u>PVIR</u>, p. 1, September 1988.

32. D.E. Carlson, "Low-Cost Power from Thin-Film Photovoltaics," in <u>Electricity</u>, T.B. Johansson, B. Boglund, and R.H. Williams, eds., University of Lund Press, Lund, Sweden, 1989.

33. Private communications from Sigurd Wagner, Electrical Engineering Department, Princeton University, July 1988.

34. D.E. Carlson, General Manager, Thin Film Division, Solarex, private communications, 1988.

35. S.L. Levy and L.E. Stoddard, "Integrated Photovoltaic Central Station Conceptual Designs," EPRI Report AP-3264, June 1984.

36. G.T. Noel, D.C. Carmichael, R.W. Smith, and J.H. Broehl, "Optimization and Modularity Studies for Large-Size, Flat-Panel Array Fields," Battelle-Columbus, 18th IEEE PV Specialists' Conference, Las Vegas, Nevada, October 1985.

37. J.M. Ogden and R.H. Williams, "Solar Hydrogen and the Revolution in Amorphous Silicon Solar Cell Technology," Princeton University CEES Report No. 231, February 1989.

38. Ibid.

39. Ibid.

40. E. Fein and K. Edwards, "Market Potential of Electrolytic Hydrogen Production in Three Northeastern Utilities' Service Territories," EPRI Report EM-3561, May 1984; R.F. Craft, Electrolyser Corp., private communications, 1985.

41. G.T. Noel, D.C. Carmichael, R.W. Smith, and J.H. Broehl, op. cit.

42. Conference Statement of the 1988 World Conference on the The Changing Atmosphere: Implications for Global Security, op. cit.

43. R.K. Lester, "Rethinking Nuclear Power," <u>Scientific American</u>, vol. 254, no. 1, pp. 31-39, March 1986.

44. Present day nuclear power plants fueled with low-enriched uranium and operated on once-through nuclear fuel cycles at 65% average capacity

discharge 141 kg of fissile plutonium per year in their spent fuel, some 24.8 kg per billion kWh (H.A. Feiveson, F. von Hippel, and R.H. Williams, "Fission Power: an Evolutionary Strategy," Science, vol. 203, pp. 330-337, January 26, 1979).

45. D. Albright and H. Feiveson, "Why Plutonium Recycle?" Science, vol. 235, pp. 1555-1556, March 27, 1987; D. Albright and H. Feiveson, "Plutonium Recycling and the Problem of Nuclear Weapons Proliferation," Annual Review of Energy, vol. 13, pp. 239-265, 1988.

46. In a full-blown plutonium economy, involving plutonium breeder reactors and light water reactors operated on closed fuel cycles, spent fuel would be reprocessed to recover the plutonium and recycle it in fresh fuel. The fissile plutonium discharge rate would be about 181 kg per billion kWh for a liquid metal fast breeder reactor (H.A. Feiveson, F. von Hippel, and R.H. Williams, "Fission Power: an Evolutionary Strategy," Science, vol. 203, pp. 330-337, January 26, 1979) and 142 kg per billion kWh for a light water reactor operated on natural uranium and plutonium (C.E. Till, "Fuel Cycle Options and Fueling Modes," Argonne National Laboratory, 1978, unpublished). In equilibrium each 1 GW of breeder capacity could produce enough fuel to meet its own recurring needs and support 0.42 GW of light water reactor capacity (H.A. Feiveson, F. von Hippel, and R.H. Williams, "Fission Power: an Evolutionary Strategy," Science, vol. 203, pp. 330-337, January 26, 1979). Thus in a full-blown plutonium economy the fissile plutonium discharge rate would average about 170 kg per billion kWh--or 966 kg per installed GW(e), for nuclear plants operating at a 65% average capacity factor.

The global consumption of fossil fuels in 1986 was as follows (British Petroleum Company, "BP Statistical Review of World Energy," London, June, 1987):

oil	2881.0	MTOE	-	120.6	EJ
natural gas	1507.1	MTOE	-	63.1	EJ
coal	2309.1	MTOE	-	96.7	EJ
Total	6697.2	MTOE	-	280.4	EJ

Replacing 1/4 of coal with nuclear electricity would require 2.22 trillion kWh per year of nuclear electricity (assuming that nuclear power displaces coal-fired power plants that are 33% efficient). Replacing 1/4 of oil and gas with hydrogen derived via electrolysis would require some 15.84 trillion kWh of nuclear electricity (assuming a rectifier efficiency of 96% and an electrolyser efficiency of 84%). Thus displacing 1/4 of fossil fuels would require the annual production of some 18.06 trillion kWh of nuclear electricity (compared to total world electricity production of 9.27 trillion kWh in 1984). At an average capacity factor of 65%, this would require 3200 GW of installed nuclear capacity and the annual generaton of some 3.1 million kg of fissile plutonium.

47. J.M. Ogden and R.H. Williams, op.cit.

48. D. Bleviss, Preparing for the 1990's: The World Automotive Industry and Prospects for Future Fuel Economy Innovation in Light Vehicles, Federation of American Scientists, Washington, DC, January 1987.

49. Toyota Press Release, October 23, 1985.

50. J.M. Ogden and R.H. Williams, op.cit.

51. J. Pangborn and M.I. Scott, "Domestic Uses of Hydrogen," in Hydrogen: Its Technology and Implications, David A. Mathis, ed., Energy Technology Review No. 9, Noyes Data Corporation, (1976).

52. J.M. Ogden and R.H. Williams, op.cit.

53. Ibid.

54. Ibid.

55. "Urban Ozone and The Clean Air Act : Problems and Proposals for Change," Staff Paper, Oceans and Environments Program, Office of Technology Assessment, April 1988.

56. Maricopa County Regional Public Transit Authority, "Building Mobility : Transit 2020," draft report, Phoenix, Arizona, 1988.

57. M. Walsh, "Pollution on Wheels," Report to the American Lung Association, February 11, 1988.

58. J.M. Ogden and R.H. Williams, op.cit.

59. J. Goldemberg, T. B. Johansson, A. K. N. Reddy, and R. H. Williams, Energy for a Sustainable World, World Resources Institute, Washington DC, 1987.

60. H.G. Broadman and W.W. Hogan, "Oil Import Policy in an Uncertain Market," Energy and Environmental Policy Discusssion Paper No. E-86-11, Johan F. Kennedy School of Government, Harvard University, November 1986.

61. W.U. Chandler, H.S. Geller, and M.R. Ledbetter, Energy Efficiency: A New Agenda, American Council for an Energy Efficient Economy, Washinton, DC, July 1988.

62. F. von Hippel and B.G. Levi, "Automotive Fuel Efficiency: the Opportunity and Weakness of Market Incentives," Resources and Conservation, vol. 10, pp. 103-124, 1983.

Table 2: Production Cost of Amorphous Silicon Solar Cells ($ per W_p)

Annual Production	5-Year Depreciation						10-Year Depreciation w/Reductions in Materials' Costs[c]		
Million sq. m.	0.167			1.67			1.67		
MWp	10	20	30	100	200	300	100	200	300
Efficiency	6%[a]	12%[a]	18%[b]	6%[a]	12%[a]	18%[b]	6%	12%	18%
Variable Costs									
Direct Labor	0.16	0.08	0.05	0.06	0.03	0.02	0.06	0.03	0.02
Materials	0.46	0.23	0.15	0.46	0.23	0.15	0.22	0.11	0.07
Fixed Costs									
Indirect Labor	0.07	0.035	0.023	0.01	0.005	0.003	0.01	0.005	0.003
Indirect Expenses	0.06	0.03	0.02	0.01	0.005	0.003	0.01	0.005	0.003
Depreciation	0.41	0.205	0.14	0.40	0.20	0.133	0.20	0.10	0.065
Total Cost	1.16	0.58	0.39	0.94	0.47	0.31	0.50	0.25	0.16

[a] These estimates were made by Carlson (D.E. Carlson, "Low-Cost Power from Thin-Film Photovoltaics," in Electricity, T.B. Johansson, B. Boglund, and R.H. Williams, eds., University of Lund Press, Lund, Sweden, 1989), based on computer integrated manufacturing technology under development at Solarex, in Newtown, Pennsylvania. In this analysis detailed cost estimates were made for present-day production technology using 6% efficient cells, and these cost estimates were extrapolated to 12% efficient cells, assuming that the costs per unit area will not change as the efficiency is increased.

[b] These cost estimates are obtained by extrapolating Carlson's estimates to 18% efficient cells, with the assumption that the cost per unit area will not change as the efficiency is increased.

[c] Two possibilities for cost reduction are taken into account here: (i) reduced materials costs and (ii) an extended equipment depreciation period. Carlson estimates that the collective effect of the materials cost reduction efforts discussed in the text could be to reduce the cost of materials from $0.46 per peak Watt for 6% efficient cells manufactured at a rate of 10 MW per year to perhaps $0.11 per peak Watt for 12% efficient cells produced at a rate of 200 MW per year (D.E. Carlson, "Low-Cost Power from Thin-Film Photovoltaics"). Since the pace of innovation will probably slow as the technology matures, it is assumed here that the depreciation period is doubled from five to ten years.

Table 3: Estimated Amorphous Silicon Solar Module Production Cost for a Factory
Producing 10 MW$_p$ per Year of 6 Percent Efficient Solar Modules (a).

	Dollars per W$_p$
Equipment Depreciation (b)	0.41
Direct Materials (c)	0.46
Direct Labor and Fringe Benefits (d)	0.16
Indirect Labor (e)	0.07
Indirect Expenses	0.06
Total	1.16

a   For a factory planned by Solarex.

b   For a five-year depreciation period and a capital cost for equipment
(computer integrated manufacturing) in a 20,000 square foot facility
estimated to be $16,500,000.

c   For 1 foot x 4 foot modules produced with an overall yield of 84%.   The
following is a breakdown of the materials cost:

Material	Cost ($ per Wp)
Glass (chemically strengthened)	0.23
Silane	0.07
Encapsulant	0.04
Frame	0.03
Diborane, phosphine	0.03
Stannic chloride	0.02
Wire, other process gases	0.02
Aluminum	0.02
Total	0.46

d   83 direct employees, 5 day work week, 2.5 shifts per day.

e   17 indirect employees.

Table 4: Solar Photovoltaic Electrolytic Hydrogen System Parameters[a]

PV ARRAY: Amorphous silicon solar cells
                                   Tilted, fixed, flat-plate array
                                   PV module efficiency - 12-18%
                                   PV module cost - $0.2-0.4/Wp
                                   Balance of systems efficiency - 85%
                                   Balance of systems cost - $33/m^2
                                   Annual operation and maintenance cost[b]
                                        - $0.45/m^2
                                   30-Year PV system lifetime
                                   Indirect costs add 25% to capital costs

PV ARRAY/ELECTROLYZER COUPLING: Direct connection, 93% coupling efficiency[c]

ELECTROLYZER: Atmospheric pressure, unipolar electrolyzer[d]
                                   Rated voltage - 1.74 Volts
                                   Rated current density - 134 mA/cm^2
                                   Operating current density - 268 mA/cm^2
                                   Efficiency at operating voltage - 84%
                                   Installed capital cost of system (including
                                        indirect costs) - $170/kWDC in
                                   Annual O&M - 2% of capital cost
                                   20-Year electrolyzer lifetime

ECONOMIC ASSUMPTIONS: All costs expressed in 1986 US$
                                   Discount rate - 6.1%[e]
                                   Annual insurance - 0.5% of capital cost[e]
                                   All income and property taxes are neglected

---

[a]   For PV hydrogen system > 10 MW in size.

[b]   Suggested by field data from large PV arrays. G.J. Shusnar, J.H.
       Caldwell, R.F. Reinoehl and J.H. Wilson, "ARCO Solar Field Data for Flat
       Plate PV Arrays," 18th IEEE PV Specialists' Conference, Las Vegas, October
       1985.

[c]   C. Carpetis, International Journal of Hydrogen Energy, v. 7, p. 287, 1982;
       C. Carpetis, IJHE, v. 9, p. 969, 1984; R.W. Leigh, P.D. Metz and K.
       Michalek, Brookhaven National Laboratory Report BNL-34081, December 1983;
       P.D. Metz and M. Piraino, BNL-51940, July 1985.

[d]   Electrolyzer operating characteristics and costs are based on currently
       available unipolar technology. We have assumed that no rectifier is
       needed. R.L. Leroy and A.K. Stuart, "Advanced Unipolar Electrolysis," and
       M. Hammerli, "When Will Electrolytic Hydrogen Become Competitive?" IJHE,
       v. 9, pp.25-51, 1984.

[e]   These values are used by the Electric Power Research Institute in
       evaluating utility scale power production facilities (EPRI, Technical
       Assessment Guide, Vol.1: Electricity Supply, EPRI P-4463-SR, 1986).

Table 5: Estimated Costs for Alternative Sources of Electrolytic Hydrogen

| | PV[a] | | NUCLEAR[b] | | HYDROPOWER[c] (GLOBAL AVERAGES) | | WIND[d] | |
	n-12% $0.4/Wp	n-18% $0.2/Wp	Current	Target	2000L	2000H	LOW	HIGH
ELECTRICITY GENERATION								
System Size (MW)	10	10	1100	1100	-	-	(40 x 2.5)	
Capital Cost ($/kW)	992	564	2970	1620	3260	4000	1340	1580
Plant Life (years)	30	30	30	30	50	50	30	30
Capacity Factor	0.271	0.271	0.566	0.65	0.47	0.47	0.35	0.35
O & M (mills/kWh)	1.9	1.3	12.0	6.5	2.9	2.9	8.7	10.3
Fuel (mills/kwh)	-	-	7.5	7.5	-	-	-	-
Electricity Production Cost (mills/kwh)	34.7	19.8	66.5	36.3	55.4	70.2	43.0	50.8
HYDROGEN PRODUCTION COST[e] ($/GJ)								
Electricity	11.47	6.55	22.91	12.51	19.01	24.18	14.83	17.48
Electrolyzer	2.52	2.52	1.47	1.27	1.76	1.76	2.36	2.36
Total	14.0	9.10	24.4	13.8	20.8	25.9	17.2	19.8

----

[a] See Tables 4 and 6.

[b] The capital costs indicated for nuclear power are estimates made by the Electric Power Research Institute (EPRI, Technical Assessment Guide 1: Electricity Supply, 1986). The higher value is EPRI's estimate of the cost of a plant that would be ordered in the US at present. The lower value is EPRI's target for "improved conditions" in the United States-- resulting from higher construction labor productivity, a shorter construction period, a streamlined licencing process, etc. The "current" capacity factor (56.6%) is the actual average for US nuclear plants in the period 1983-1987; for "target" conditions it is assumed that this increases to 65%. The fuel cost of 0.75 cents per kilowatthour and the "current" O&M cost are actual average values for nuclear plants in 1986 [Energy Information Administration, "Historical Plant Costs and Annual Production Expenses for Selected Electric Plants, 1986", DOE/EIA-0455(86), May 27, 1988]. The "target" O&M cost is a an value set by EPRI (EPRI, 1986).

[c] The indicated hydro capital costs and capacity factors are global average estimates for the year 2000 (H. K. Schneider and W. Schulz, Investment Requirements of the World Energy Industries 1980-2000, World Energy Conference, 1987). The hydro O&M costs are EPRI estimates for the indicated capacity factor (EPRI, 1986).

[d] For mass-produced 2.5 MW wind turbines configured to produce 100 MW (J. I. Lerner, "A Status Report on Wind Farm Energy Commercialization in the United States, with Emphasis on California," paper presented at the 4th International Solar Forum, Berlin, FRG, October 6-9, 1982). The O&M cost is assumed to be 2% of the initial capital cost per year.

[e] For non-PV electricity sources, unit electrolyzer capital costs are assumed to be 25% higher because of the rectifier, which is assumed to be 96% efficient.

Table 6: Estimated Production Costs for Synthetic Liquid and Gaseous Fuels[a]

| FUEL | SYNFUEL PLANT | | | Fuel Production Cost[b] |
	Plant Size (1000 GJ/day)	Installed Cost ($10^6)	($/Wp)	($/GJ)
Gasoline Derived from Coal	176	5107	2.26	16.2
Methanol Derived from Coal	322	3860	0.93	7.9
	32.2	567	1.37	10.0
Ethanol Derived from				
Sugar cane (Brazil)	2.68	7.7	0.22	8.0
Corn (US)	13	95.2	0.57	14.4
Synthetic Gas Derived from Coal (c)				
High heating value gas	264	1820	0.54	5.7
	88	756	0.67	6.4
	26.4	341	1.00	7.9
Intermediate heating value gas	264	822	0.24	4.0
PV Hydrogen (Southwestern US)				
n-18%, $0.2/Wp	0.183[d]	7.8	0.92	9.1
n-12%, $0.2/Wp		9.2	1.10	10.7
n-18%, $0.4/Wp		10.9	1.30	12.3
n-12%, $0.4/Wp		12.3	1.47	14.0

[a]   The cost estimates presented here were derived using a self-consistent set of assumptions to ensure that the cost comparisons for alternative technologies are meaningful. For details see Reference [37].

[b]   The production cost includes plant capital, feedstock and operation and maintenance costs, but not the costs of transmission or storage.

[c]   For Lurgi Dry Ash process with Western US coal.

[d]   This corresponds to a 10.8 MWp PV array coupled to an 84% efficient unipolar electrolyzer, where the coupling efficiency is 93%. The peak PV hydrogen output is 8.4 MW, and the average output is 183 GJ/day for a system located in the Southwestern US, with average insolation of 271 Watts per square meter. There should be no significant economies of scale for PV powered electrolyzers above about 5-10 MW.

Table 7: Sensitivity of PV Hydrogen Cost to Assumptions

	Plant Size (1000 GJ/day)	Installed Cost ($10^6)	Installed Cost ($/Wp)	H2 Production Cost ($/GJ)
BOS – $33/m², PV System Lifetime – 30 years				
n–18%, $0.2/Wp	0.183	7.8	0.92	9.1
n–12%, $0.2/Wp		9.2	1.10	10.7
n–18%, $0.4/Wp		10.9	1.30	12.3
n–12%, $0.4/Wp		12.3	1.47	14.0
BOS – $50/m², PV System Lifetime – 30 years				
n–18%, $0.2/Wp	0.183	7.8	0.92	10.6
n–12%, $0.4/Wp		12.3	1.47	16.3
BOS – $33/m², PV System Lifetime – 15 years				
n–18%, $0.2/Wp	0.183	7.8	0.92	10.4
n–12%, $0.4/Wp		12.3	1.47	18.2

For PV hydrogen produced in the Southwestern US, as in Table 6.

Table 8: Land Use Requirements for PV Hydrogen, Biomass and Coal Synfuels

Fossil Fuel Displaced at the Present US Average Per Capita Consumption Rate	Fossil Fuel Displaced/Year (GJ/person)	LAND REQUIRED (hectares/person)		
		w/PV H2(a)	w/Biomass(b)	w/Coal Synfuels(c)
	287	0.053	0.72	0.018

US Land Area Per Capita for (d):

Forests and Woodland	1.1
Cropland	0.8
Permanent Pasture	1.0

Average Land Area per Capita (d) in:

World	2.62
Africa	5.02
Rwanda	0.38
North America	5.18
El Salvador	0.35
Asia	0.92
Bangladesh	0.13
Europe	0.96
Netherlands	0.23
South America	6.29
Equador	2.79

Fossil Fuel Displaced for the World at the Present Consumption Rate	Fossil Fuel Displaced/Year (Exajoules)	LAND REQUIRED (million hectares)		
		w/PV H2(a)	w/Biomass(b)	w/Coal Synfuels(c)
	283	53	670	18

Total World Land in:

Forests and Woodland (d)	4090
Cropland (d)	1470
Permanent Pasture (d)	1350
Deserts (e)	3140

---

a   For PV electricity produced in 15% efficient solar modules and converted to hydrogen at 84% efficiency on tilted collectors requiring a ground area twice as large. For average insolation on tilted collectors of 271 Watts/square meter (250 Watts/square meter on the ground).

b   For biomass grown @ 40 tonnes of dry biomass (@ 18 GJ/tonne) per hectare per year and converted to fluid fuels at 65% efficiency, on average.

c   For surface-mined coal.

d   World Resources Institute, World Resources 1987, Basic Books, Washington, DC, 1987.

e   M. P. Petrov, Deserts of the World, John Wiley & Sons, 1976.

Table 9:   Water Requirements for PV-Hydrogen and Biomass Energy Systems

	Useful Energy Production Rate (GJ/sq.m/year)	Water Requirements (cm of rainfall/year)
Efficiency of PV modules (%) (a)		
10	0.35	2.3
15	0.53	3.4
Biomass Productivity (b) (tonnes/hectare/year)		
10	0.012	30- 70
20	0.023	60-140
30	0.035	90-210
40	0.046	120-280

---

a   See Ref. [37].

b   Biomass production requires for photosynthesis and transpiration some 300 to 700 tonnes of water for each tonne of dry biomass produced. Assuming that fluid fuels are produced from biomass at an average efficiency of 65% (and that the heating value of dry biomass is 18 GJ per tonne), the water requirements become 25,000 to 60,000 liters per GJ of produced biofuel.

Table 10: Potential Markets for PV Electricity and PV Hydrogen

Electricity Market	Breakeven Solar Module Price ($ per peak Watt)	Potential Market Size (MWp)	Hydrogen Market
Corrosion protection	20-100		
Buoys	60		
Consumer products: calculators, etc.	10	100	
Remote water pumping	4-7	2,000	
Diesel generator replacement, remote power	5	10,000	
US Utility electric peaking (Total US peaking)	2-3	50,000	
	1-2	400	Merchant hydrogen
Daytime power for grid connected residences in US	0.7-1.5	100,000	
	0.5-1.0	3,000	Intermediate chemical
	0.20-0.40	7,000	10% of US automotive fleets
		4,000	Heating in 2 million super-insulated houses
		60,000	Large chemical
		600,000	All US home heating
US baseload electric power (with storage)		600,000	
		1,100,000	All US Automobiles

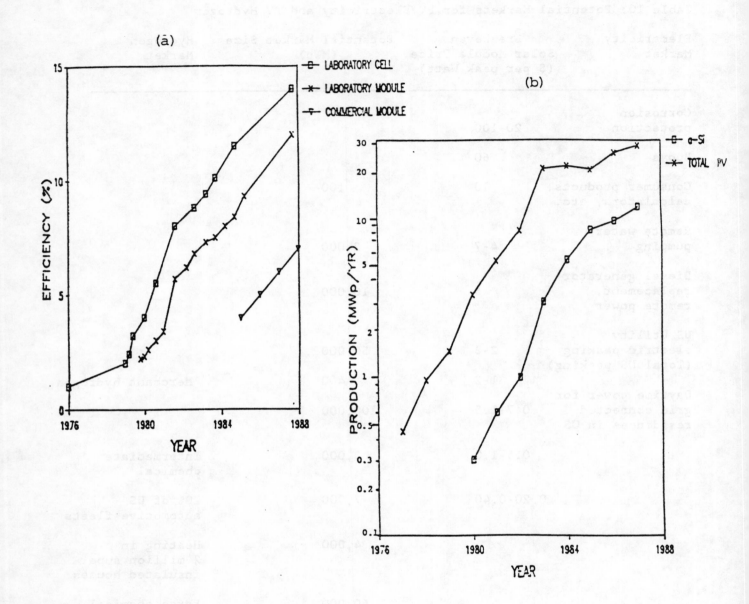

Figure 1. Progress in amorphous silicon solar cell technology. The trends in the efficiency of amorphous silicon (a-Si) small-area laboratory cells (typically 1 cm x 1 cm in size), larger area laboratory modules (≥ 100 square centimeters), and commercially available modules are shown vs. calendar year in the top graph. The bottom graph shows the total annual PV production volume and the annual production of amorphous silicon solar modules in megawatts of peak power manufactured per year.

Sources: Y. Kuwano, SANYO, private communications, 1985; Z Erol Smith, private communications, 1987; D. Carlson, Solarex, private communications, 1987; Photovoltaic Insider's Report, February 1987 and January 1988.

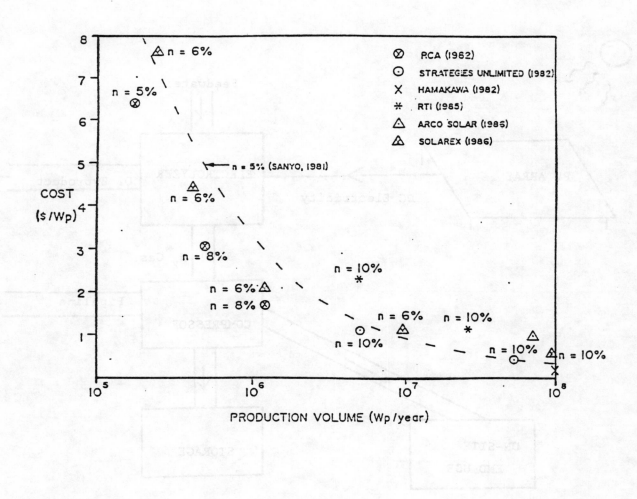

Fig. 2. Estimated cost versus annual production volume for amorphous silicon solar cell modules, based on recent projections by various manufacturers and researchers. The annual production volume is for an individual factory.

Source: Private communication from David Carlson, General Manager, Thin Film Division, Solarex, November 1987.

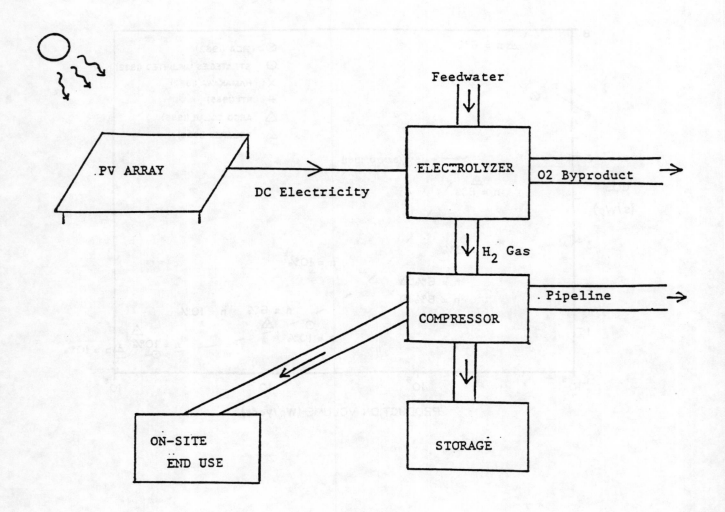

Fig. 3.   A solar photovoltaic electrolytic hydrogen system.

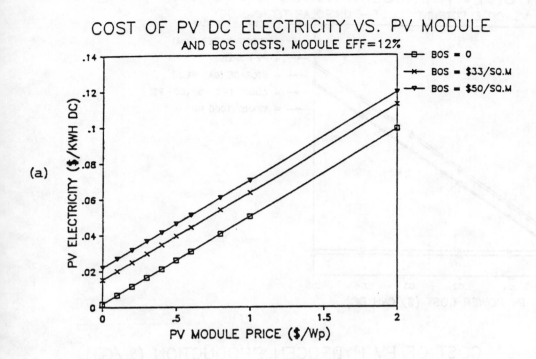

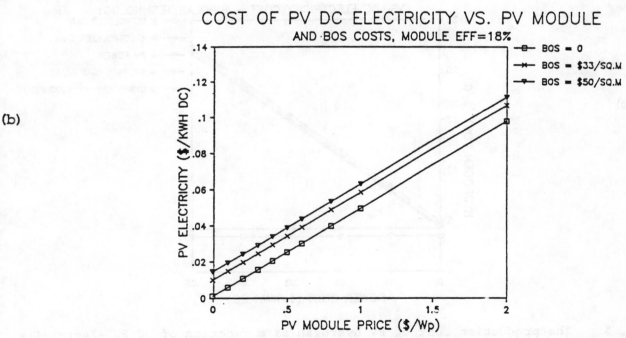

Fig. 4.   The cost of DC PV electricity vs. solar module and balance of systems costs for 12% efficient PV modules (top) and 18% efficient PV modules (bottom) located in the Southwestern US with average insolation of 271 Watts per square meter, assuming a system lifetime of 30 years and an annual operation and maintenance cost of $0.45 per square meter.

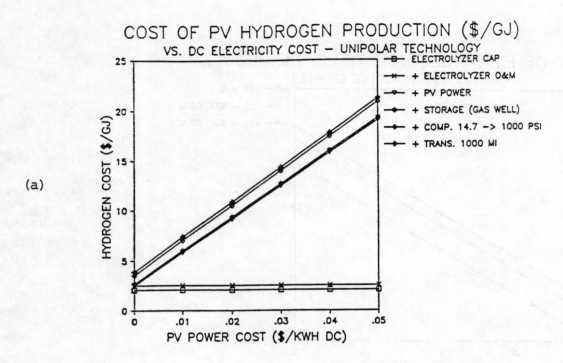

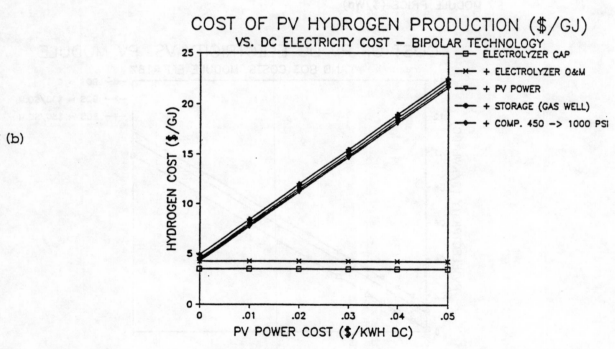

Fig. 5. The production cost of PV hydrogen as a function of DC PV electricity cost for unipolar (top) and bipolar (bottom) electrolyzer technologies. The hydrogen production cost includes electrolyzer capital costs, operation and maintenance costs, the cost of DC PV electricity, the cost of storage, the cost of compression from electrolyzer pressure [1 atmosphere for unipolar technology and 30 atmospheres (450 psia) for bipolar technology] to a pipeline pressure of 68 atmospheres (1000 psia), and the cost for 1600 km (1000 miles) of pipeline transmission.

594

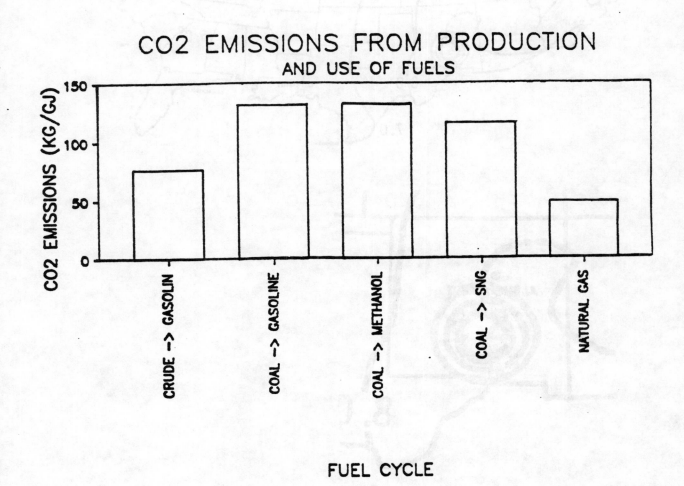

Figure 6. Carbon dioxide emissions from production and use of fuels.

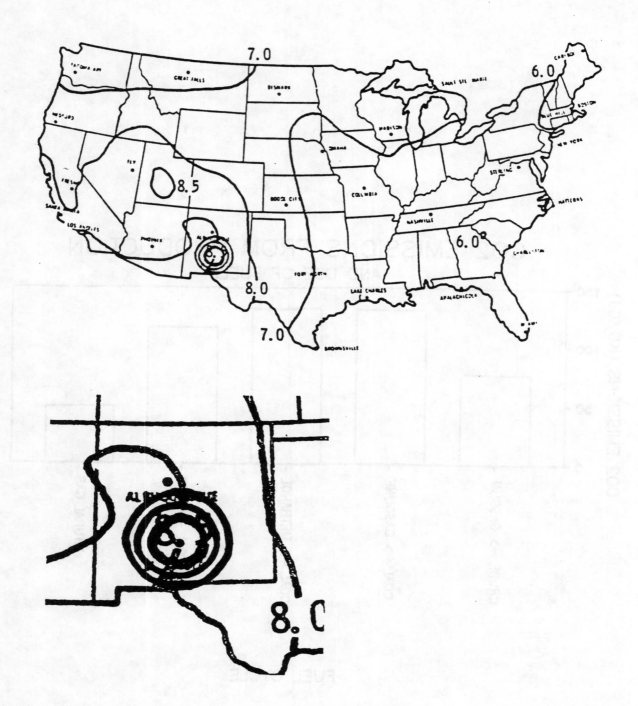

Fig. 7. The land area required to produce an amount of PV hydrogen equivalent in energy to 1986 US oil consumption (33.6 EJ), oil + natural gas consumption (51.4 EJ), and oil + natural gas + coal consumption (65.7 EJ).

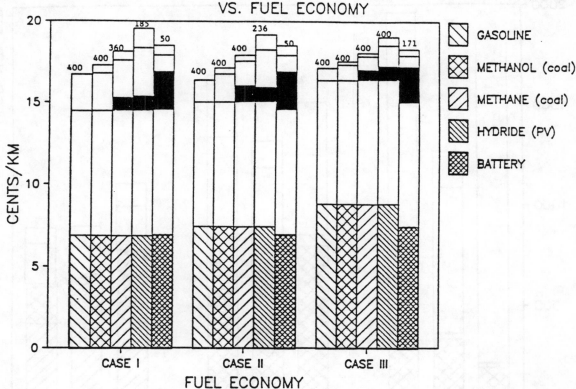

Fig. 8.   A comparison of levelized costs for owning and operating cars on
various fuels for three levels of automotive fuel economy. The levelized cost
in cents per kilometer is shown for automobiles fueled with gasoline, methanol
from coal, synthetic natural gas from coal, electricity stored in batteries,
and PV hydrogen. Case I is roughly equivalent to a present day sub-compact car
with a fuel economy of 7.8 1/100 km (30 mpg) of gasoline equivalent. Case II
is roughly equivalent to a subcompact car with a more efficient engine (Diesel
or stratified charge), with a fuel economy of 4.7 1/100 km (50 mpg). Case III,
for which the gasoline-equivalent fuel economy is 2.6 1/100 km (90 mpg), is
what could be achieved in a car with a more efficient engine, aerodynamic
styling and a continuously variable transmission. For each case, the levelized
cost has four components: initial capital cost of the vehicle (which is shown
in hatched patterns and includes the purchase price of the car, exclusive of
extra storage system costs), miscellaneous expenses (shown in white above the
initial capital cost of the vehicle, this includes tolls, registration fees,
insurance, parking, repairs and maintenance), the storage cost (shown in black
and includes any extra cost for a special fuel storage system such as
batteries, compressed gas cylinders or hydride tanks), and the fuel cost (shown
in white at the tops of the bars). The delivered costs of fuels are: $8.19/GJ
($1/gallon) for gasoline, $9.9-$12.1/GJ for methanol from coal, $8.9-$11.0/GJ
for synthetic natural gas from coal, $0.06-0.10/kwh for electricity, and $12.9-
$18.0/GJ for PV hydrogen. The numbers at the tops of the bars indicate the
travelling range (in km).

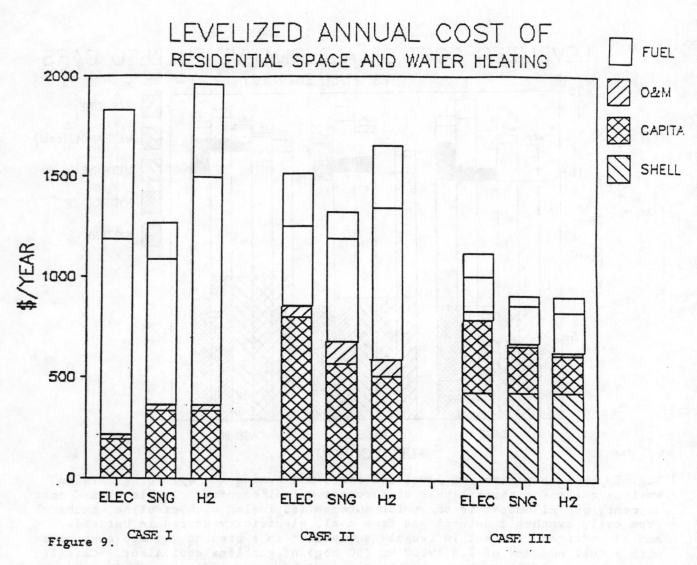

Figure 9.   CASE I   CASE II   CASE III

Levelized costs to consumers for residential space and water heating.

The costs are for houses located in the state of New Jersey in the US and
heated with electricity, synthetic natural gas from coal, and PV hydrogen, for
three levels of energy end-use technology. Level I involves low first-cost
equipment in conventionally-constructed houses. Level II involves the use of
energy-efficient equipment in conventionally-constructed houses. Level III
involves the use of energy-efficient equipment in super-insulated houses. The
levelized cost has four components: building shell retrofit costs (which apply
to the super-insulated house), heating system capital costs, operation and
maintenance costs and fuel costs. A range of delivered fuel costs is shown,
corresponding to the high and low energy cost estimates made in this study: PV
hydrogen ranges from $12.5-17.6/GJ, SNG from $8.4-10.5/GJ, and electricity from
$0.06-0.10/kwh. A ten percent discount rate is assumed for the lifecycle cost
calculations [37].

Carbon Dioxide Recovery and Utilisation in the Synthesis of Fine Chemicals and Fuels: a Strategy for Controlling the Greenhouse Effect.

by **Michele Aresta**
Dipartimento di Chimica,Università, 70126 BARI, Italy

Summary

The control of the emission of the so called "greenhouse gases", especially carbon dioxide, is a must for avoiding a climate change of unknown effects. Several options are possible for reaching the goal of the 20% reduction of the emission of carbon dioxide from 1986 level (or 40% on the 6.8 Gt y^{-1} forecast) by 2010 (Montreal Protocol, 1987).
A strategy that might contribute to mitigate the effects caused by the greenhouse gases emission is the recovery and utilisation of carbon dioxide. In this paper the state of the art and perspectives are presented and discussed.

Paper presented at the IEA-OECD Expert Seminar on Energy Technologies for Reducing Emission of Greenhouse Gases, Paris 12-14 April, 1989.

Carbon Dioxide Recovery and Utilisation in the Synthesis of Fine Chemicals and Fuels : a Strategy for Controlling the Greenhouse Effect.

by **Michele Aresta**
Dipartimento di Chimica,Università, 70126 BARI, Italy

1.Introduction

There is a general feeling nowadays that the steady increase of the atmospheric carbon dioxide concentration will cause unexpected effects on climate and primary agricultural productivity. Today there are two effective tools for fighting the accumulation of carbon dioxide in the atmosphere: limiting the use of carbon based fuels enhancing the efficiency of energy production, and calling a halt to mass deforestation. However, a new strategy is currently being developed in order to deal with this problem:
the recovery and utilisation of carbon dioxide.
Different tactics can be envisaged that might produce benefits in the short-, medium-, long-term:

  i) **Utilisation of large masses of carbon dioxide in synthetic chemistry.**
  ii) **Utilisation of biotechnology for carbon dioxide conversion into other C1 molecules (essentially methane).**
  iii) **Amelioration of the yield of photosynthetic processes in both C3 and C4 plants using the genetic engineering tool.**

In this paper the possible routes to the utilisation of carbon dioxide are discussed with particular emphasis on "carbon dioxide based industrial processes".

2. Anthropogenic Emission of Carbon Dioxide.

Fuel (coal,gas,oil) burning and cement manufacture are the main industrial sources of carbon dioxide. The total anthropogenic CO_2 emission in the atmosphere is evaluated as ranging around $5 \cdot 10^9$ t of carbon per year.(Fig 1)
This throws the natural carbon dioxide cycle off-balance and is causing the present accumulation of carbon dioxide into the atmosphere.

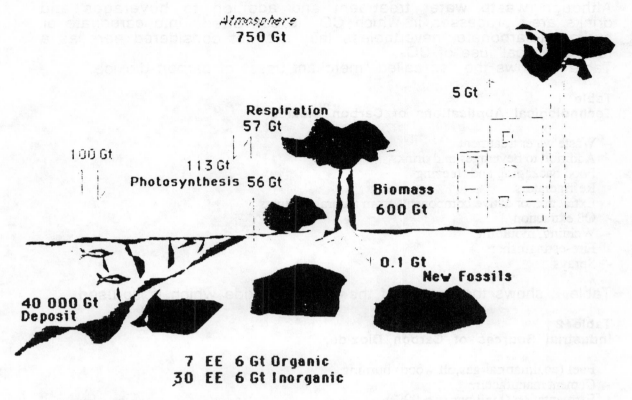

Carbon Accumulation and Fluxes

Atmosphere
750 Gt

5 Gt

**Respiration
57 Gt**

100 Gt

113 Gt
Photosynthesis 56 Gt

**Biomass
600 Gt**

0.1 Gt
New Fossils

**40 000 Gt
Deposit**

7 EE 6 Gt Organic
30 EE 6 Gt Inorganic

Bold characters give accumulation; plain characters indicate fluxes.

Fig. 1

Natural photosynthesis (113 Gt estimated per year) and the oceans (the adsorption-release balance approximates 100 Gt per year), for a long time considered as infinite CO_2-buffers, are not able to fix all the carbon dioxide generated by human activities. It is a fact that the rate at which carbon based fuels are burned is approximately 50-100 times greater than the rate of carbon fixation in new fossils by natural processes. This gives an idea of the net rate of consumption of the carbon-fuels available in Nature.

However CO_2 is a cheap and ubiquitous source of carbon. The amount present in the atmosphere ranges around 750 Gt and the amount fixed in carbonates approximates $30 \cdot 10^6$ Gt. It is of great potential intellectual significance to convert it into molecules of commercial utility or to fix it in disposable species.

3. Industrial Uses of Carbon Dioxide.

At present, the carbon dioxide uses in industrial activities can be categorized as follows :

- Chemical uses
- Merchant uses

according to whether carbon dioxide is or is not used as a building block for other chemicals.

Although waste water treatment and addition to beverages and drinks are processes in which CO_2 is converted into carbonate or hydrogen-carbonate, nevertheless they are not considered here as a real "chemical" use of CO_2.

Table 1 shows the so called "merchant uses" of carbon dioxide.

Table 1
Technological Applications of Carbon Dioxide.

- Waste water treatment
- Addition to beverages and drinks
- Food packaging and freezing
- Refrigerators
- Extraction of active components from natural products
- Oil extraction
- Welding, moulding
- Fire-estinguishers
- Sprays

Table 2 shows the source of the carbon dioxide which is so-used.

Table 2
Industrial Sources of Carbon Dioxide.

- Fuel (coal,natural gas,oil,wood) burning
- Cement manufacture
- Fermentation (high purity > 99%)
- Other industrial processes
- Natural wells (high purity >99%)

The estimated world wide amount of carbon dioxide used in the processes listed above ranges around 8 Mt per year and the carbon dioxide recovered from flue gases represents only a few per cent of the total amount.The recovery from industrial processes and from fermentation plants and the extraction from natural wells accounting for most of the CO_2 used. As the cost of the extraction is often lower than that of the recovery, the former practice is quite extensively pursued in those Countries where pure carbon dioxide is available in wells.

4. Recovery of Carbon Dioxide from Flue Gases.

The recovery of carbon dioxide generated in the combustion of fuels or in an industrial process is performed according to the procedures listed in Table 3.

Table 3
Processes for Carbon Dioxide Recovery.

Trapping by Condensed Phases.
- Amines
- Alcohols and glycols
- Other solvents (Organic carbonates, ethers)
- Basic oxides
- Molecular sieves

Separation Using Selective Membranes.

When condensed phases (liquid or solids) are used, the adsorption of carbon dioxide is carried out under pressure and at low temperature. Then it is recovered upon increasing the temperature and lowering the pressure. The energy of the process varies according to the nature of the interaction carbon dioxide-substrate.

The simply physical adsorption is characterized by a low energy content, while the chemical interaction (see for example the case of the reaction with basic oxides) can sometimes imply quite a high energy content. Most commonly used solvents are amines, namely ethanolamine.

The direct recovery of carbon dioxide from flue gases, prior to their emission into the atmosphere, has not been intensively pursued so far, because of the cost of the process, that requires also the NOx and SOx elimination, and because of the limited utilisation of CO_2. However, if we consider the *per se* cost of the recovery process and the benefit that derives from cutting down the carbon dioxide emission into the atmosphere, we can conclude that the recovery will soon become an economically viable practice.

This is particularly true if we consider that new uses of carbon dioxide are currently being developed.

It is obvious that if we wish to recover most of the carbon dioxide generated by a power station fired with carbon based fuels, then we must think about a recovery process able to treat in due time a huge amount of gas, of the order of several kt per day.

In this respect the selective membrane process is very promising as it would avoid the use of large volumes of liquids or large masses of solids, the treatment plants being, thus, more easy to build and to drive. Technological questions are still open at the moment.

The storage and transportation of carbon dioxide can be considered as quite easy and safe. In fact, the critical properties of the cumulene (Θ_C = 304.0 K, P_C = 7.38 MPa) make gaseous carbon dioxide liquifiable without great technological problems and its "inertness" guarantees for its safe storage and transport.

5. Utilisation of Carbon Dioxide in Synthetic Chemistry.

Thus, to recover and recycle carbon dioxide may prove to be a promising way to control the CO_2 concentration in the atmosphere. Obviously, the **utilisation** of carbon dioxide as a source of carbon in **synthetic chemistry** appears very attractive as this molecule is cheap, abundant and ubiquitous.

Indeed, several Research Groups are engaged in developing new processes based on carbon dioxide for the synthesis of:

   i) **Bulk and fine chemicals.**
   ii) **Energy rich C1 molecules.**
 iii) **Cn hydrocarbons and their derivatives *via* the Fischer-Tropsch process.**

The development of a "carbon dioxide based industry" for the synthesis of **bulk and fine chemicals** will allow the following major objectives to be fulfilled:

a)   The recovery of carbon.
b)   The replacement of multistep processes by direct synthetic procedures.

c) The finding an alternative to processes based on toxic and more expensive starting materials (e.g. phosgene, isocyanates, chloroformates, carbon monoxide, etc.).

It is evident that, for this measures to be really efficient in controlling the carbon dioxide level in the atmosphere, we must give preference to the development of processes for the synthesis of chemicals used in very large amounts (several hundreds of Mt per year), rather than those for the synthesis of specialities, that would use most probably only a few kt per year of carbon dioxide.
In my opinion, the utilisation of CO_2 in both the synthesis of fuels and of specialities is of extreme economical interest and ecological usefulness. In fact, as far as the synthesis of specialities is concerned, the use of carbon dioxide can help to solve other ecological problems, as we shall see later.
It is worth noting that only a very few industrial processes based on carbon dioxide are available today.
The amount of CO_2 that is currently used for the synthesis of chemicals is around several millions of tons per year, the major applications being in the areas reported in Table 4.

Table 4
Uses of Carbon Dioxide in Synthetic Chemistry.

- **Urea**   30 Mt
  (Fertilizers,food additives,resins)
- **Organic carbonates**
  (Polycarbonates,carbamates,other chemicals)
- **Inorganic carbonates**
  (Sodium,potassium,barium,other salts)
- **Pharmaceuticals**   20 kt
  (Aspirin,others)
- **Synthesis of methanol**   ~30 Mt

The prospects for a wider use of carbon dioxide are quite interesting.
The reactions in which CO_2 can be used can be divided into three main categories, as shown in Table 5.

Table 5
Utilisation in Synthetic Chemistry.

Building Block for:
- Bulk or fine chemicals
- Energy-rich C1 molecules
- Cn Hydrocarbons and their derivatives
  (Fischer-Tropsch process)

From a formal point of view all the processes that utilise carbon dioxide have to be considered as reduction reactions and thus require energy. Indeed, all those processes that use the entire CO_2 molecule (the carboxylation reactions or, in general, the reactions affording the formation of C-C,C-N, C-O bonds) present (if any) a low energy demand (of the order of a few kJ mol^{-1}), while those processes in which a deoxygenation reaction takes place require a higher amount of energy (often greater than 100 kJ mol^{-1}). As an example,the $\Delta G°(298 \ K)$ for the conversion of CO_2 into CO and oxygen is 257.15

604

kJ mol^{-1}.

For this reason the synthesis of chemicals in which the entire CO_2 molecule is incorporated will be discussed separately from the reactions bearing to C_1 or C_n energy richer molecules.

6. Low Energy Processes.

In this section we shall consider the utilisation of carbon dioxide in the synthesis of molecules containing functionalities such as

- C(O)O -   acids, esters, lactones.
- O-C(O)O -   organic carbonates and polycarbonates.
- N-C(O)O -   carbamates.
- N-C(O) -   ureas, amides, polyureas, polyamides.

The synthesis of these chemicals is at present based on many-step procedures (as, for example, in the case of acids, lactones, esters, amides) or on the utilisation of toxic species such as phosgene (this is the case of the synthesis of organic carbonates and carbamates). Possible routes to organic carbonates and to carbamates based on CO_2 rather than on $COCl_2$ are represented in Fig. 2.

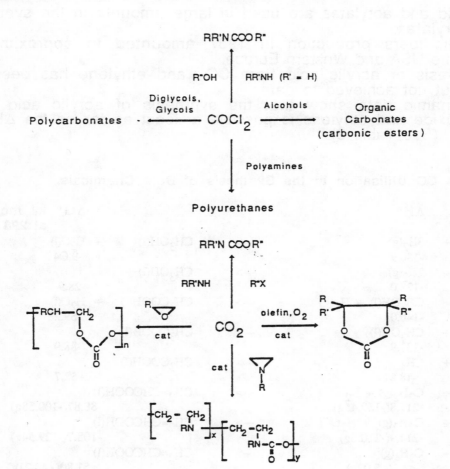

Fig 2. CO_2 vs CO routes to chemicals containing the carbonate and carbamate functionalities.

It is worth noting that phosgene (LC 3 mg L^{-1}) has a current market of a few Mt per year all over the world.

As we shall see the global amount of carbon dioxide that could be used, in future, in the synthesis of this type of chemicals might be of the order of several tens of Mt per year.

Let us consider separately the different classes of products.

6.1 Carboxylates.

The utilisation of carbon dioxide in the one-step synthesis of carboxylic acids and their derivatives (esters, lactones) is of great importance. Several studies have shown that alkenes (monoenes, dienes) and alkynes react with carbon dioxide in mild conditions to afford carboxylated products whose nature varies according to the reaction conditions and to the catalyst used.

It would be of interest to industry to make the following syntheses operative:

6.1.1 Acrylic Acid (CH$_2$=CHCOOH) from Ethylene and Carbon Dioxide.

Acrylic acid and acrylates are used in large amounts in the synthesis of polyacrylates.

The acrylic-fibers production in 1987 amounted to approximately 1.5 Mt in the USA and Western Europe.

The synthesis of acrylic acid from CO$_2$ and ethylene has been attempted but not achieved to date.

Thermodynamic data show that the synthesis of acrylic acid from carbon dioxide and ethylene is possible at least as far as the ΔH_R is concerned. (See Table 6)

Table 6
CO$_2$ versus CO Utilisation in the Synthesis of Bulk Chemicals.

		ΔH°				ΔG° kJ mol^{-1} at 298 K
1) CO$_2$(g)	+	3H$_2$(g) -130.9	--->	CH$_3$OH(l)	+ H$_2$O(l)	-9.04
1') CO(g)	+	2H$_2$(g) -128.0	--->	CH$_3$OH(l)		-28.9
2) CO$_2$(g)	+	CH$_3$OH(l) + H$_2$(g) -140.0	--->	CH$_3$COOH(l)	+ H$_2$O(l)	-69.28
2') CO(g)	+	CH$_3$OH(l) -137.9	--->	CH$_3$COOH(l)		-88.9
2") CO$_2$(g)	+	CH$_4$ -18.61	--->	CH$_3$COOH(l)		+52.7
3) CO$_2$(g)	+	C$_2$H$_2$(g) + H$_2$ -217.8(-169.45g)	--->	CH$_2$=CHCOOH(l)		-88.87(-100.88g)
3') CO(g)	+	C$_2$H$_2$(g) + H$_2$O(l) -214.9(-210.6g)	--->	CH$_2$=CHCOOH(l)		-108.7(-129.54g)
4) CO(g)	+	C$_2$H$_4$(g) -43.3(+5.02g)	--->	CH$_2$=CHCOOH(l)		+51.80(+40.21g)
4') CO(g)	+	C$_2$H$_4$(g) +1/2 O$_2$ -326.3(-277.9g)	--->	CH$_2$=CHCOOH(l)		-205.0(-216.90g)

Conversely, the free energy value $\Delta G°$ for this reaction does not seem to be too favourable. It seems likely that both thermodynamic and kinetic factors hinder such an achievement which considering the uses of acrylic acid, would be of real industrial interest.

Synthesis from carbon dioxide, acetylene and dihydrogen seems more favourable.

6.1.2 Adipic Acid from Butadiene ($CH_2=CH-CH=CH_2$) , Carbon Dioxide and Dihydrogen.

Adipic acid, $HOOC(CH_2)_4COOH$, is used in the synthesis of Nylon and other amidic fibers whose market ranges around several Mt per year. It would be of great interest to synthetise this acid directly from butadiene and carbon dioxide.

Today, the synthesis of dicarboxylic acids from olefins and carbon dioxide has been achieved. However, new catalysts (transition metal systems) characterized by a high selectivity and a high turnover number have yet to be developed.

6.1.3 Derivatives of Saturated Lactons (Five and Six Membered).

Flavorings and perfumes can have a lactone structure. A few kt of CO_2 per year could be used in this case.

The synthesis of this kind of products has been attempted and in some cases encouraging results have been obtained.

6.1.4 α or, ß-Butenolides from Olefins.

The butenolide structure occurs in many natural products.

For example the angelica lactons are γ-methyl Δ^{α} - and Δ^{β} - butenolides.

Several antibiotics such as

PENICILLIL ACID   PROTOANEMONIN   PATULIN

and natural pigments

VULPINIC ACID

contain the butenolide structure.
A market of the order of a few kt per year can be envisaged for chemicals of this type.

6.1.5. Formic Acid , HCOOH.

This species is of great interest both as an intermediate for the synthesis of chemicals and as a vehycle of hydrogen.
Formic acid,in fact,is easily converted to CO_2 and H_2 in the presence of palladium at room temperature. This aspect can be relevant to the project aiming to use hydrogen as a energy vector.
In fact, while hydrogen itself poses the problem of the storage and transport,formic acid has the correct requisites for an easy handling.

6.1.6 Oxalic Acid , (COOH)$_2$.

The synthesis of oxalates from CO_2 can be accomplished easily with an acceptable selectivity using the electrochemical method.
In this case sacrificial anodes are used. Despite the consumption of the electrodes, a cost estimate reveals that this process is competitive with those actually operating, which are based on chemical oxidation of hydrocarbons or on methanol and CO.
A remarkable amount of carbon dioxide could be used in this process in view of the many applications of oxalates.

6.1.7 Citric Acid,HOOC-CH$_2$CH(COOH)CH$_2$COOH,from Acetone or other Acids from Molecules Containing Active Hydrogens.

Citric acid can be prepared from acetone and CO_2. This carboxylation reaction can be extended to other active-hydrogen-containing hydrocarbons.The coupling of the C-H activation with the C-C bond formation is one of the most interesting processes to be developed.

608

6.1.8 Phenylpropanoic Acid.

The synthesis of this product from styrene and carbon dioxide is quite interesting. Phenylpropanoic acid and its derivatives are used as antiflammatory agents. The amount of carbon dioxide used in this way may range around several kt per year in the future.

6.2 Ureas, Carbamates and Polyurethanes.

The utilisation of carbon dioxide in the synthesis of these chemicals is of noticeable interest. Several Mt of carbon dioxide per year could be used.

6.2.1 Ureas.

The synthesis of urea (Eq.1) is the process in which the largest amount of CO_2 is used.

$$2\,NH_3 \quad + \quad CO_2 \qquad ---> \qquad (H_2N)_2CO + H_2O \qquad (Eq.1)$$

An interesting use of CO_2 would be in the synthesis of substituted ureas:

$$\begin{array}{ll} (RNH)_2CO & RNHC(O)NHR' \\ \text{Symetrical} & \text{Asymetrical} \end{array}$$

The synthesis of polyureas (Eq.2) is also of great industrial importance

$$n\,CO_2 \quad + \quad 2n\,H_2NRNH_2 \ ---> \qquad [-HNRNHC(O)NHRNH-]_n + H_2O \quad (Eq.2)$$

in view of the use of these chemicals.
Tens of Mt of carbon dioxide would be used in these processes.

6.2.2 Carbamates.

Carbamates, used as agrochemicals (herbicides, pesticides) and as pharmaceuticals can be obtained from carbon dioxide (Eq.3), replacing, thus, the more costly process based on toxic chemicals such as phosgene (Eq. 4-5) and isocyanates (Eq.6).

$$CO_2 \quad + \quad R'NH_2 + RX \ ---> \qquad R'NHCOOR \qquad (Eq.3)$$

$$COCl_2 \quad + \quad ROH \qquad ---> \qquad ROCOCl \qquad (Eq.4)$$

$$ROCOCl \quad + \quad R'NH_2 \qquad ---> \qquad ROC(O)NHR' \qquad (Eq.5)$$

$$R'NCO \quad + \quad ROH \qquad ---> \qquad R'NHCOOR \qquad (Eq.6)$$

The reaction represented in Eq.3 affords quite interesting results when both R and R' are alkyl groups. The development of appropriate catalysts to promote the reaction when R,R' = phenyl and the use of alkylating agents other than halides are the key points to make this process really operative.

A potential use of several hundreds kt of carbon dioxide per year can be estimated.

6.2.3 Polyurethanes.

These plastics are used virtually all over the world and something like 20 Countries produce the raw materials. Originally obtained by reaction of toluene diisocyanate with alcohols (Bayer Patent, 1927), nowadays they can be also prepared by reaction of carbon dioxide with aziridine.
By the 90's their market is expected to reach 5 Mt per year.

6.3 Carbonates and Polycarbonates.

These species can find a market both as specialities and as bulk chemicals,according to their molecular structure and properties. Some monomers can be used as intermediates in the synthesis of other chemicals. As an example, the case of dimethylcarbonate (DMC) is discussed in detail.
The total consumption of carbon dioxide in this industry would be of the order of tens of Mt.

6.3.1 Carbonates.

The reaction of CO_2 with alcohols represents a convenient route to organic carbonates of formula $ROC(O)OR$. In partcular, dimethyl carbonate,DMC,which has several uses (Table 7)

Table 7
Uses of Dimethylcarbonate (DMC) , $(CH_3O)_2C=O$.

As a Phosgene Substitute in the Synthesis of :
- Carbamates (herbicides,pesticides,pharmaceuticals)
- Long chain alkyl carbonates (lubricants)
- Oligocarbonates (intermediates)
- Allylglycol carbonates
- Special carbonates
- Polyurethanes

As a Dimethylsulphate Substitute in the Synthesis of:
- Methylated phenols
- Methylated amines
- Other methylated products

in industry (as a solvent, as an alkylating agent replacing the more toxic dimethyl sulphate, or as an acylating agent) could be synthetized from methanol and CO_2 (Eq. 7a),

$$2\,CH_3OH + CO_2 \quad ---> \quad CH_3OC(O)OCH_3 + H_2O \qquad (Eq.7a)$$

or from dimethyl ether and CO_2 (Eq. 7b).

$$(CH_3)_2O + CO_2 \quad ---> \quad CH_3OC(O)OCH_3 \qquad (Eq.7b)$$

Most probably this synthesis could be accomplished advantageously

610

using heterogeneous catalysts.
Industrially operating processes for the synthesis of DMC are based on carbon monoxide and methanol (Eq. 8). The reaction is catalyzed by copper salts (Cu(I) Enichem Synthesis, Cu(II) Dow Chemicals).

$$2\,CH_3OH \;+\; CO \;+1/2O_2 \;\text{---}> \quad (CH_3O)_2CO + H_2O \qquad \text{(Eq.8)}$$

Another reaction to be exploited is the "oxidative carboxylation of olefins" (Eq.9).

$$RCH{=}CH_2 + CO_2 \;+1/2O_2 \quad\text{---}>\quad RCH\!-\!CH_2\!-\!OC(O) \overset{\overline{\qquad O \qquad}}{} \qquad \text{(Eq.9)}$$

This reaction is of great interest as "natural gas" or "flue gases" containing CO_2 and variable amounts of dioxygen could be used as the reaction mixture.
The interest in developing this process lays in the potentially large utilisation of cyclic carbonates in industrial processes (see Tab. 8).

Table 8
Cyclic Carbonates
can be Used in the Synthesis of:

$$RCH\!\!-\!\!CH_2\!-\!OC(O) \overset{\overline{\qquad O \qquad}}{}$$

- Diols and derivatives (RCHCH-CH$_2$; R'OCH$_2$CHOH)
  $\qquad\qquad\qquad\qquad\quad$ OH OH $\qquad\qquad\quad$ R

- Esters R'C(O)OCH$_2$CHOH
  $\qquad\qquad\qquad\qquad\qquad$ R

- Aminoalcohols (ArNHCH$_2$CHOH)
  $\qquad\qquad\qquad\qquad\qquad\quad$ R

- Polyureas,polycarbonates and polyesters
- N-Heterocycles of the oxazolydone type
- Thiiranes
- Other unsaturated carbonates

6.3.2 Polycarbonates.

Polycarbonates with molecular weight ranging from 100 to 150 kDa are easily obtained using Zn-Al systems as catalysts. These are able to polymerize oxyranes with CO_2 effectively.
Quite recently Air Products Chemicals, Arco Chemicals and Mitsui Petrochemical started a joint venture programme for the synthesis of special "air-made ecological polymers" obtained from epoxides and carbon dioxide.
The polycarbonates investigated decompose thermally above 450 K and are used as blinders in ceramics and as adhesives in the electronic Industry.
Quite a specialistic use (a market of the order of a few kt per year) can be envisaged for these chemicals. But other polycarbonates are used for making more commonly used items and in this case the market could easily reach a several tens of Mt per year size.

7. High Energy Processes.

The reduction of carbon dioxide has been attempted successfully in several ways. The key points in this process are the "source of energy" and the "oxygen sink".

611

7.1 Chemical reduction.

The **reduction** of CO_2 to other **C1 molecules** richer in energy (Eq. 10) is a process of enormous interest

$$CO_2 \longrightarrow HCOOH \longrightarrow CO \longrightarrow H_2CO \longrightarrow CH_3OH \longrightarrow CH_4 \qquad (Eq.10)$$

requiring an energy input that varies with the end product.
At present the largest use of carbon dioxide is in the methanol industry,but other processes that can reduce CO_2 to CO and CH_4 selectively have been investigated and are still the target of a great deal of research interest.
The reaction of CO_2 with high molecular weight hydrocarbons in the presence of catalysts is industrially operated to produce CO (Calcor 1985) or short-chain carboxylic acids (Chevron 1977).
The possibility of a selective chemical reduction to other C1 molecules has been also investigated
Up to now, carbon monoxide is the product most commonly obtained in mild conditions.
The reduction of carbon dioxide to C1 molecules other than formic acid (HCOOH) poses the problem of finding an "oxygen sink", as one or two atoms of oxygen per molecule of carbon dioxide must be released. The processes developed to date are "metal-assisted" but frequently an oxidation of the metal system is observed that causes the inactivation of the catalyst. Other oxygen fixing agents are phosphines,silanes,protons.
The dream is to transfer the oxygen atom onto unsaturated hydrocarbons in order to use CO_2 as an oxidant agent (Eq. 11).

$$CO_2 \quad + \quad RHC = CH_2 \quad \longrightarrow \quad CO + RHC - CH_2 \qquad (Eq. 11)$$
$$\overset{}{\underset{O}{\diagdown \diagup}}$$

The energetics of this reaction, when the olefin is ethylene, seem to be favourable ($\Delta H_R = -3.1$ kcal mol^{-1}).
The possibility of generating CO catalitically from CO_2 in mild conditions would allow a more economical route to those industrial processes that use carbon monoxide as the feeding stock to be found. Alternatively, CO_2 might be directly used.
Table 6 compares the thermodynamic data ($\Delta H°$ and $\Delta G°$ at 298 K) for the synthesis, from CO or CO_2, of some C_1,C_2 and C_3 species that can be considered as bulk starting materials commonly used in the chemical industry.
It is evident that the $\Delta H°$ values are often quite similar for the two processes, but in some cases the $\Delta G°$ values can be quite different.
Values given in parentheses and marked as (g) correspond to the reactions in which all products are gaseous.
A strong entropic contribution is evident in some cases. This is emphasized for the synthesis of methanol from CO and CO_2 in Table 6 and 9.

7.2 Electro- and Photo-chemical Reduction of Carbon Dioxide.

The direct reduction of CO_2 to other C_1 molecules has been also attempted, with varying success, using electrochemical techniques.
The yield, the rate of conversion,and the selectivity are crucial points in this process. Attempts have been made to get good results

Table 9
Carbon Dioxide Reduction to Methanol:
Role of the Entropic Contribution.
(evaluated at 298 K)

$$\Delta H° \qquad\qquad \Delta G° \ kJ \ mol^{-1}$$

$$CO_2(g) \ + \ 3H_2(g) \ \longrightarrow \ CH_3OH(g) \ + \ H_2O(g)$$
$$\qquad\qquad -49.50 \qquad\qquad +3.89$$

$$CO_2(g) \ + \ 3H_2(g) \ \longrightarrow \ CH_3OH(l) \ + \ H_2O(l)$$
$$\qquad\qquad -130.8 \qquad\qquad -9.04$$

by using catalysts (transition metal complexes, light, or both) or by modifying the surface of the electrodes.

Efforts have been made to use **solar energy** for CO_2 reduction. Two different approaches have been attempted.

Semiconductors have been used as direct energy-transfer agents with water as the electron source for CO_2 reduction.
This process mimics the natural one in which CO_2 and water are converted into sugars and dioxygen according to the simplified overall reaction

$$CO_2 \ + \ H_2O \ \xrightarrow{h\nu} \ 1/n \ (CH_2O)_n + O_2$$

The best quantum yield actually observed is still very low, just higher than 0.01 per cent. Moreover, the process is not selective. These open questions must find an answer if this technique is to be really practical for carbon dioxide reduction.
On the other hand, the utilisation of solar cells as energy source for the indirect reduction of carbon dioxide by solar energy in the presence of water is also of great potential interest.
The human dream is to reverse the combustion reaction

$$C_nH_{2n+2} \ + \ (3n+1)/2 \ O_2 \ \begin{smallmatrix} ---> \\ <--- \end{smallmatrix} \ n \ CO_2 \ + \ (n+1) \ H_2O + h\nu$$

using solar energy.

The coupling of the **water-splitting reaction** (Fig 3) with the **carbon dioxide reduction** might allow **fuels** to be obtained from CO_2 and water.

7.3 Utilisation in Fischer-Tropsch Processes.

The direct utilisation of carbon dioxide in **Fischer-Tropsch processes** is also investigated with the aim of replacing CO.
Although more hydrogen is required for the reduction to Cn hydrocarbons (and their derivatives) of CO_2 than of CO, a higher selectivity and less drastic operative conditions (lower temperature and pressure) might make the process based on carbon dioxide competitive
A full development of this process would mean that huge amounts of carbon dioxide per year could be used. It is worth recalling that at Sasol (South Africa) the potentiality of FT plants using a CO/H_2 mixture is of the order of Mt of a few hydrocarbons per year.

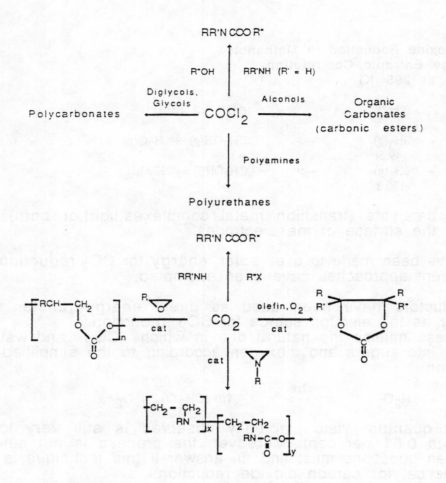

Fig 3.

8. Utilisation of Methanogens.

Anaerobic bacteria are able to reduce CO_2 to methane.
The methanation of carbon dioxide is of considerable potential
interest as this reduction process could be coupled with any process
generating pure carbon dioxide (fermentation,pure gas recovery from
flue gases,other sources).
The conversion rate (78.4 mL $min^{-1}g^{-1}$ biocatalyst) seems very
attractive for the development of industrial applications. The
extraction of the pure enzyme from bacteria does not seem to be
strictly required. Moreover,these bacteria should not produce
environmental contamination, owing to their properties.
In this area a great effort is currently being made to understand the
mode of action of the enzymes and their structure. Mimicking
systems are also synthetized and used as in vitro catalysts for
carbon dioxide reduction to CO or methane. This biotechnological
application to fuel generation from carbon dioxide is of great
potential interest.
If necessary, nutrients for bacteria and heat could be recovered
from wastes.
The utilisation of thermophilic methanogens is of practical interest
as these can assure a higher rate of methane production. In the case
of biogas production from biomasses, an increment of up to seven

614

times the amount of methane has been found when the temperature is increased from average room temperature (293 K) to 313 K.
The crucial point in this case is that the bacteria are very sensitive to oxygen and the methanation process must be performed out of all contact with air.

9. Amelioration of the RuBisCo Selectivity towards Carboxylation.

This is one of the most attractive intellectual approaches to the control of CO_2 concentration in the atmosphere.It is known that RuBisCO (Ribulose-Bisphosphate-Carboxylase-Oxygenase),an enzyme present in green leaves of both C3 and C4 plants, is able to catalyze both the carboxylation and the oxidation of Ribulose with a 50% efficiency for each process.
The amount of biomass produced by photosynthesis seems to depend on the amount of RuBisCO and on its ability to fix carbon dioxide. Therefore, this enzyme is at present the target of intensive research all over the world by several Research Groups who are investigating the site-directed mutagenesis of RuBisCO.
The final purpose of this long-term research is to get a better RuBisCO that would fix CO_2 with a greater efficiency with respect to the natural one. The result would be highly efficient crop plants obtained through technological manipulation of the enzyme.
This would allow plants to reduce the atmospheric CO_2 concentration more efficiently than observed at present.

10. Utilisation of CO_2 : Dream or Reality?

It is time we started thinking positively about the recovery of carbon dioxide (and of nitrogen and sulphur oxides) from flue gases. Figure 4 gives a representation of an integrated scheme of activities that might operate around a Power Station fuelled with carbon based fuels. The recovery of flue gases,their separation and recycling are represented. The conversion of carbon dioxide into energy richer species by means of different techniques (chemical,electro- and photo-chemical,biotechnological) is considered. Moreover carbon dioxide can be used in other industrial applications or green houses.
Figure 4 seems to represent an ideal cycle. This might be true a few years ago. Today it is not pure imagination : a few steps are reality.
The recovery of heat and CO_2 from flue gases, the CO_2 conversion using different approaches are feasible. More effective technologies must be developed, new catalysts must be found so that the yield, the turnover number, the selectivity of the CO_2 conversion can be improved. An interdisciplinary approach is necessary.
The key question is: when it will be possible to operate processes for converting amounts of CO_2 of the order of Gt per year? At the moment we can use amounts of the order of tens of Mt per year.
A forecast may be possible if we refer to the development of the FT process. This was applied in Germany at an industrial level (15 plants, global production of the order of Mt) in the '40s for fuel production from coal. Nowadays one single plant produces a few hundreds kt per year.
The CO_2 chemistry is really very young. Reference dates are 1973 (Oil crisis) and 1975 (structural characterisation by M.Aresta and

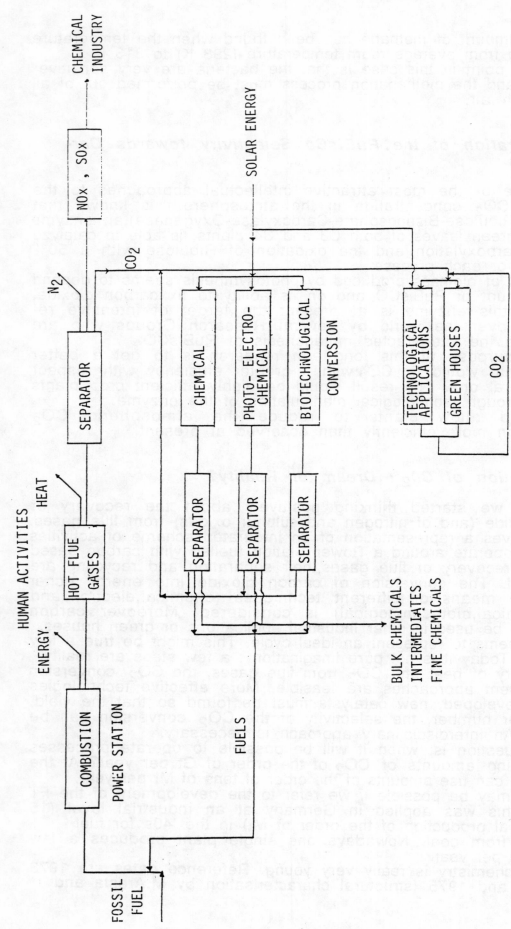

Figure 4

coll. of the first transition metal-carbon dioxide complex and evidence of a facile CO_2 to CO reduction at room temperature that gave a new approach to the studies on CO_2). If we take into account the actual rate of evolution of Chemistry then we can say that if we operate an integrated long-term research project we can be optimistic about the future. May be that by the 2010s the Scientific Community developed method for using CO_2 as a source of carbon and massive amounts of CO_2 could be used either as a "hydrogen vehycle" ($HCOOH, CH_3OH$) or for the synthesis of fuels (CH_4, Cn species) and disposable items.

Still, it comes out that the hydrogen and CO_2 conversion are problems that must be investigated in a parallel way. The utilisation of solar energy seems to be the connection bridge.

We will need carbon dioxide fuels in the 2000s as these will allow to keep the present organization of the private transportation means. The nuclear- and the hydrogen-solution to the energy needs of the society both pose the problem of a technological solution for the private sector of the transport.

11. Conclusion.

There are a few "must" for the utilisation of carbon dioxide could be a strategy for reducing the carbon dioxide emission in the atmosphere .

We must think to use amounts of carbon dioxide of the order of Gt per year.

The synthesis of specialities (kt per year),bulk chemicals (Mt per year) and fuels (Gt per year) requires a different energy input.

The cost of the energy used for the recovery and the utilisation process must be kept as low as possible. Solar energy should be thus used,mimicking Nature.

The life and the utilisisation of the species obtained from CO_2 must be considered.

In fact,as I have already pointed out (M. Aresta and G.Forti Eds., "Carbon Dioxide as a Source of Carbon", pag. 410, Reidel Publ.1987) when we think to use CO_2 as a source of carbon for the synthesis of chemicals we must consider that any molecule produced from CO_2 when used will give back CO_2. So the problem of the "life" of the chemicals originated from CO_2 is a fundamental one.

Going back to Scheme 4, it contains an important message.

It does not seem likely that only one class of Scientists will have the right solution at hand. We need to work with a great exchange of knowledge and technologies at an international level.

My personal opinion is that a Panel must be created in order to work out ecologically acceptable and economically convenient alternative routes for the recovery of carbon dioxide from flue gases and for its utilisation in the synthesis of chemicals and fuels as a means to control its level in the atmosphere.

The experiences of Laboratories working in different areas must be assembled and compared and a report on the costs and benefits for each practical application should be made available to National Governments.

Acknowledgements: The financial support from NATO (Grants 543/85, 44/88, 611/88) and the precious collaboration of Dr. Immacolata Tommasi are gratefully acknowledged.

OECD/IEA EXPERT SEMINAR ON

ENERGY TECHNOLOGIES FOR REDUCING EMISSIONS OF GREENHOUSE GASES

THE POTENTIAL LONGER TERM CONTRIBUTION OF NUCLEAR ENERGY IN REDUCING CO$_2$ EMISSIONS IN OECD COUNTRIES

by

W. Gehrisch, T. Haapalainen, Y.M. Park, G.H. Stevens, K. Todani
Nuclear Development Division
OECD Nuclear Energy Agency

THE POTENTIAL LONGER TERM CONTRIBUTION OF NUCLEAR ENERGY IN REDUCING CO$_2$ EMISSIONS IN OECD COUNTRIES

INTRODUCTION

1.   This paper sets out to consider the potential impact of rapid and widespread introduction of nuclear power in the OECD on the production of CO$_2$ from electricity generation. A basic and, at the time of writing, implausible assumption is that current constraints of public acceptability are totally removed. It is further assumed that high safety standards are maintained but that some of the complexity of regulatory difficulties in some countries will have been set aside.

2.   With the major current constraints on nuclear development suspended there would be the opportunity for greatly increased penetration of the electricity generation sector by nuclear plant. The factors which would then become potentially important constraints on the rate of penetration are identified and discussed in this paper. As an aid to this discussion the paper first presents short term forecasts for nuclear capacity and then elaborates on a simple model recently used in the NEA to develop ideas on the demand for uranium up to 2030. This results in a rough gauging of the contribution of nuclear power which could reasonably be expected by 2030 in OECD countries together with an estimate of CO$_2$ emissions avoided. The paper concludes with a discussion of some of the uncertainties in the analysis and suggestions for refining them.

CURRENT AND SHORT-TERM NUCLEAR CONTRIBUTIONS

3.   In 1987, 13 of the 24 OECD countries (Belgium, Canada, Finland, France, F.R. Germany, Italy, Japan, Netherlands, Spain, Sweden, Switzerland, United Kingdom and United States) used nuclear power for electricity generation. The electricity generation in OECD countries was 5838.2 TWh including 1312.6 TWh from nuclear generation (22.5%) (Ref. 1). The as yet incomplete figure for 1988 show that there has been a further increase in the nuclear share of growing electricity demand.

4.   According to NEA's latest published statistics (the 1988 Brown Book) the electricity generation of the total OECD countries will be 7840.2 TWh including 1828.6 TWh from nuclear generation in 2000. Also according to the Brown Book the electricity generation of the total OECD countries will be 8685 MWh including 1990 MWh from nuclear generation in 2005.

LONGER TERM SCENARIO OF NUCLEAR ELECTRICITY PRODUCTION

5.   The future beyond the turn of the century is obscured by large uncertainties which can only be treated qualitatively and any projections reaching into the first decades after 2000 can only show our appreciation of what is likely to happen under certain assumptions.

6.   We have nevertheless elaborated scenarios for nuclear development until 2030 for the purpose of analysis of the long term uranium supply and demand situation which will be prepared later this year, the 1989 version of the "Red Book".

7.   The scenarios adopted for that work took into account the present and future trend of electricity demand in OECD countries in relation to GDP growth using a single demand curve, and two different penetration rates for nuclear power into the electricity market. The scenarios were not intended to respond to concerns about the Greenhouse Effect.

8.   Up to the year 2000, the data for electricity generation and nuclear share were taken from the Brown Book. Beyond 2000, electricity demand was linked to GDP growth which was assumed to be 2.6% during the entire period from 2000 to 2030. The GDP-Electricity demand elasticity factor was assumed to be 0.7 in the early years, tending towards 0.5 at the end of the period.

9.   On the basis of this single electricity demand projection, two paths for nuclear power development were established. The lower path is not of great interest in the context of this paper.

10.   In seeking a plausible description for the higher path it was assumed that currently heightened concerns over acid rain and other pollution effects of fossil fuels would persist. However it was also assumed that while research over the next 5-10 years showed that although the Greenhouse Effect was real its consequences would not be as harmful or unavoidable as some present analyses indicate. By assumption there would be no further major incidents of the Three Mile Island type and a fortiori none with off-site consequences remotely resembling the Chernobyl accident. By the turn of the century there would also have been progress in setting up waste repositories with good experience of their early use and of the associated transport arrangements.

11.   Consequently it was thought plausible that there could be a significant shift in public attitudes to nuclear power reflected in a wave of renewed ordering of nuclear power plant in those OECD countries which currently have some. The vigour of this wave would vary between countries. The more complex the established regulatory mechanism the slower would be the wave in gathering momentum. However in several European countries with present nuclear shares at around 40%, such as Spain, Switzerland and the Federal Republic of Germany, significant new orders would be placed before 2000, leading to new nuclear capacity being commissioned before 2010. The strong nuclear countries in OECD, such as Canada, France and Japan, would be able to meet high targets for nuclear capacity. By about 2010 several (but not all) non-nuclear countries would begin to develop nuclear infrastructures and bring nuclear capacity into use in growing amounts after 2010. By 2030 the share of nuclear power in electricity production would approach an asymptotic level consistent with trends first established in the 1970s and early 1980s.

12.   No explicit assumption was made as to fossil fuel prices but implicitly these were expected to be at levels which gave nuclear a slight cost advantage over competing forms of electricity generation.

13.   In calculating the path with which this "plausible scenario" was
consistent the main parameters adopted were that penetration of the
electricity generation market by nuclear power would follow a logistic curve
after 2000 with an asymptotic nuclear share in electricity production and
halving time of 25% and 17 years for OECD America and 60% and 20 years for
OECD Europe + Pacific respectively. The results of the calculations are
displayed in Figure 1. This scenario requires relatively modest growth rates
of nuclear capacity and would pose no serious problem to the nuclear industry
in terms of industrial capability to build the required reactors.

"UNCONSTRAINED" NUCLEAR GROWTH

14.   It is clear that use of nuclear power cannot be a panacea but if the
Greenhouse Effect is as real and severe in its effects as the worst
forebodings of the pessimists it may become necessary to rely to the utmost on
nuclear power. What would that "utmost" be? In the longer term it might be a
contribution to all forms of use of energy given appropriate development of
technology but for the moment we will concentrate on the possibilities of
nuclear replacing fossil-fuel in electricity generation. How far can nuclear
penetration go? What will be the constraints? The remainder of this paper
offers some broad evaluation subject to considerable uncertainties and
discusses some of the factors which might become constraints.

15.   First as an aid to calculating some figures we introduce an assumption
that the asymptotic level of penetration for the OECD as a whole will be 70%.
Since France and Belgium currently derive 70% of their electricity from
nuclear this seems a reasonable starting point although there is no intrinsic
reason why this penetration level should not be exceeded. We also impose the
assumption that the halving time used in the logistic curve should be
20 years. We maintain the same trend line for growth of electricity demand.
The result of this calculation is displayed in Figures 2 and 3. An even more
extreme scenario with the same asymptotic penetration rate and a halving time
of 10 years is shown in Figures 4 and 5. This calculation is not considered
further in the remainder of this paper. We stress that these calculations are
for purposes of illustrating problems and possibilities and take no account of
current policies of OECD Member governments.

16.   We note, in passing, that no attention is paid here to electricity
generation in non-OECD countries, although in the long run, given improvement
in their economic circumstances and predictable population growth,
particularly in the developing countries, they will become predominant in
electricity production. Whether the long run falls within the time frame
considered could be argued, but it is difficult, for a number of reasons, to
envisage that developing countries as a whole would be able to make extensive
use of nuclear power in this period.

POTENTIAL CONSTRAINTS ON NUCLEAR CONTRIBUTION

17.   The calculation above of 1120 GWe being installed by 2030 ignores many
effects which might combine to prevent installed capacity reaching that level.
It is therefore worth identifying what might be the principal obstacles, if
only to indicate points on which greater certainty would be worth seeking.

Figure 1
NUCLEAR CAPACITY 2000-2030 IN OECD

Halving time: North America = 17 years
                  Europe+Pacific = 20 years

Asymptotic share: North America = 25%
                     Europe+Pacific = 60%

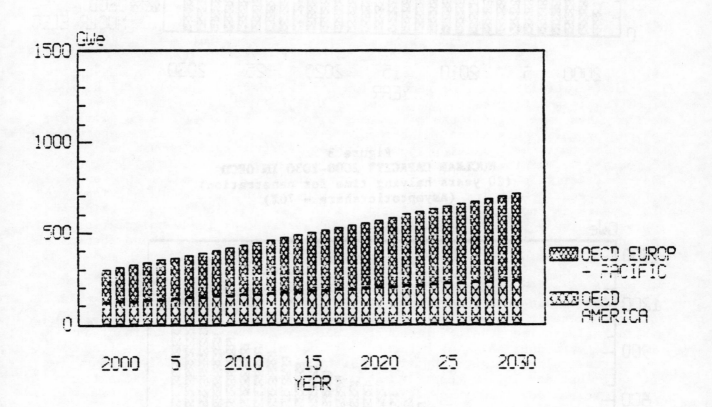

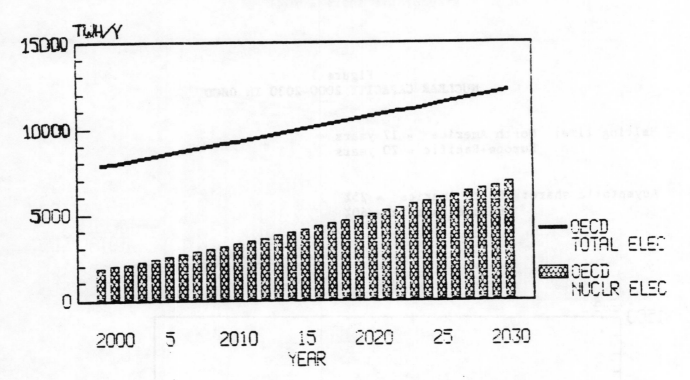

Figure 2
TOTAL AND NUCLEAR ELECTRICITY GENERATION 2000-2030 IN OECD
(20 years halving time for penetration)
(Asymptotic share = 70%)

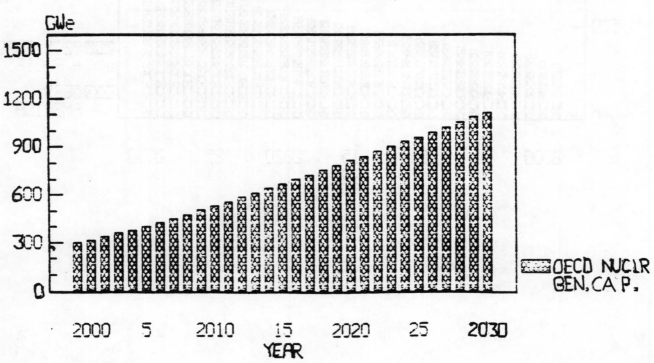

Figure 3
NUCLEAR CAPACITY 2000-2030 IN OECD
(20 years halving time for penetration)
(Asymptotic share = 70%)

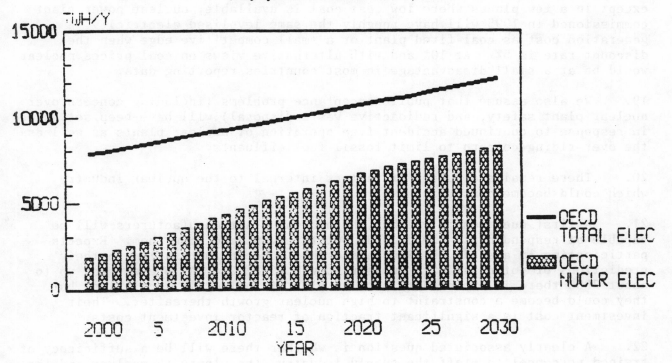

Figure 4
TOTAL AND NUCLEAR ELECTRICITY GENERATION 2000-2030 IN OECD
(10 years halving time for penetration)
(Asymptotic share = 70%)

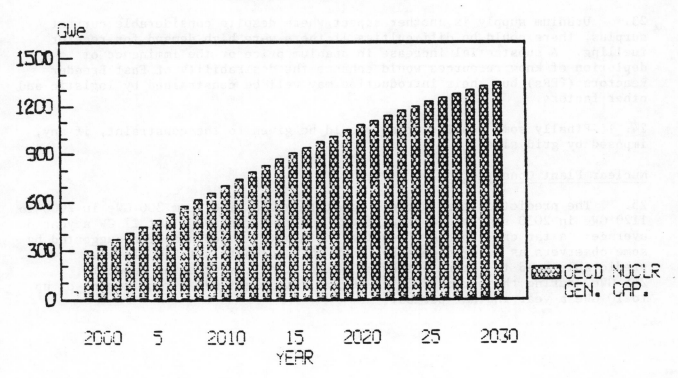

Figure 5
NUCLEAR CAPACITY 2000-2030 IN OECD
(10 years halving time for penetration)
(Asymptotic share = 70%)

18.   For the purposes of the exercise we are assuming that the relative economics of use of nuclear power are at least as good as indicated in recent work jointly by the IEA and NEA. Using the reference parameters of the study, except in a few places where low cost coal is available, nuclear power plant commissioned in 1995 will have roughly the same levelised electricity generation cost as coal-fired plant or a small competitive edge when the discount rate is 5%. At 10% and with alternative views on coal prices nuclear would be at a small disadvantage in most countries reporting data.

19.   We also assume that public acceptance problems (including concern over nuclear plant safety, and radioactive waste disposal) will have been set aside in response to continued accident free operation of nuclear plants as well as the over-riding concern to limit fossil-fuel effluents.

20.   There remains a number of factors internal to the nuclear industry which could become constraints.

21.   A first question is whether the nuclear plant manufacturers will be capable of responding to the demands putatively placed upon them. Experts participating in a study in 1986 raised this concern in relation to their predictions of potential high nuclear growth (Ref. 2). At present and up to about 2000 there is expected to be a surplus of fuel cycle facilities, but they could become a constraint to high nuclear growth thereafter. Their investment cost is a significant fraction of reactor investment costs.

22.   A clearly associated question is whether there will be a sufficiency of trained personnel to staff the supply industry, the plants themselves and the regulatory authorities, in addition to undertaking the decommissioning, at least partially, of some of the first generation nuclear plants. Recruitment of safeguards inspectors for the international system operated by the IAEA also needs to be taken into account.

23.   Uranium supply is another aspect where despite considerable current surplus, there could be difficulties if there were high demand for reactor fuelling. A substantial increase in uranium price or the imminence of depletion of know resources would enhance the desirability of Fast Breeder Reactors (FBRs) but their introduction may well be constrained by logistic and other factors.

24.   Finally some consideration should be given to the constraint, if any, imposed by grid sizes.

Nuclear Plant Construction

25.   The prediction that nuclear capacity would rise from 300 GWe in 2000 to 1120 GWe in 2030 suggests a requirement for constructing a net 27 GW a year on average; a far cry from the "reactor per minute" which has been suggested by some observers as being necessary, even when account is taken of the predictable need to retire a substantial part of the existing capacity, 245 GWe, during the period 2005 to 2025. This might add a requirement of up to 10 GWe a year in that period.

26.   Comparing the calculated requirement with historic and current performance of the industry it seems that the target while not impossible could be a fairly demanding one. Between 1975 and 1980 installed nuclear capacity in OECD countries increased by 16.7 GW a year. From 1980 to 1985 the average, again in terms of connection to the grid, was 17.6 GW a year.

27.   A current problem for the industry is, however, that the expected rate for the period 1985 to 2000 is only 6.5 GW a year, with the U.S. situation looking particular grim for the industry with only 3 GW expected to be connected between 1990 and 2000. The nuclear plant manufacturing industry is currently in a state of upheaval as firms merge, formulate joint ventures and seek other markets to exploit. The effect is intended to be to preserve the capability to respond to nuclear power plant orders. No doubt there will be some success in this endeavour but the longer the gap in orders the more difficult it will be quickly to relaunch a large construction programme. If orders were to be forthcoming in the immediate future then there would appear to be no great difficulty in achieving the rate of net additions to capacity shown in Figure 3, about 18 GW a year initially. If, on the other hand, as is implicit in the latest Brown Book figures, the rate of ordering of nuclear power plant does not pick-up until after 2000, then there might well be considerable problems in achieving a construction rate greater than about 10 GW a year in the early years of the next century.

28.   The re-direction of capacity to build fossil-fuel plants could obviously ease certain bottlenecks but some delay in commissioning or re-commissioning investment in the manufacturing capacity for the specifically nuclear plant components could be expected.

29.   The nuclear reactors which would be built after the turn of the century may differ significantly from those constructed recently. While France, Japan and the United Kingdom remain committed to large reactors i.e. more than 1000 MWe, there is increasing interest elsewhere in reactors of 600 MWe or smaller. The design aims for these reactors include making them simpler and quicker to construct. We presume that these features would off-set the additional work required to build a given amount of capacity in a larger number of smaller plants. However there is additional uncertainty introduced by this point.

30.   In summary, while it is not possible to provide firm conclusions on the achievable rate of nuclear plant construction it would seem very difficult for the industry to adjust quickly from the low construction rates expected in the 1990s (3.6 GW a year) to the 18 GW a year required to match the calculated growth in nuclear share. On the other hand there seems to be no intrinsic reason why the building rate should not increase rapidly later on so that the calculated capacity could be in place by 2030.

Training of Manpower

31.   One of the main determinants of how quickly the nuclear manufacturers could re-deploy would be the availability of educated and qualified manpower. Looking to the French experience of the 1970s, during which decade the construction of 41 PWRs was started, might give the impression that there would be little difficulty. However there are several reasons why this analysis may not be instructive. At that time there was a large pool of

qualified and experienced personnel available in the CEA who could be re-deployed to the manufacturing, utility and regulatory bodies. There was also a vigorous flow of newly qualified nuclear engineers from the universities and specialized schools. This appears to contrast with the current situation where, with the possible exception of Japan, there has been reduction of government expenditure on national nuclear research organisations, diversification of the tasks of those organisations, and, at least according to hearsay, at the same time there has been a considerable reduction in the numbers of students and places for students in nuclear engineering.

32. Certainly lack of timely training of reactor operators in utility companies and licensing engineers in regulatory agencies would be a serious limiting constraint for an orderly introduction of nuclear power into a country where previously there had been no major nuclear power development programme. The long lead time that it takes to develop such essential engineering disciplines, the need to establish a regulatory framework including that of licensing of reactor operators, and the feasibility of utilizing the existing engineers through re-training would be some of the critical factors that must be addressed at a very early stage.

33. The NEA is currently undertaking a study to assess, in a much broader sense, the availability of qualified manpower in all major sectors of the nuclear industry in supporting varying growth scenarios of nuclear power. This study, when completed sometime in 1990, should provide some quantitative indications in terms of, for example, the number of reactor operators or licensing engineers required as a function of a given growth scenario of nuclear power.

Fuel Cycle Capabilities

34. Unlike fossil fuels, the nuclear fuel cycle consists of many different processing steps as illustrated below:

° Before nuclear fuel is loaded into the reactor	– Uranium mining and milling, conversion to UF_6, enrichment, fabrication, and related transportation in between these processing steps.
° After nuclear fuel is discharged from the reactor	– Interim storage of spent nuclear fuel, reprocessing or permanent disposal of spent nuclear fuel, and related transportation in between these steps.

35. Availability of these fuel cycle capabilities on a global or regional basis could also become a limiting factor, particularly to those countries either "switching" to nuclear power or expanding their existing nuclear power development programmes.

36. The lead time requirements for construction of fuel cycle facilities will, of course, depend on individual fuel cycle steps. However, given sufficient resources and barring any technological or environmental

628

difficulties, the lead time for construction of most of these fuel cycle facilities will, with the probable exception of mines and reprocessing plants, be shorter than that of a typical nuclear power plant which takes 6 to 10 years. If current design aims for shortening the construction time of power reactors are realised then the lead times may become comparable for most of the front-end facilities other than mining. That might impose a requirement for closer coordination of plans for reactor and infrastructure investment but it seems unlikely that shortage of fuel cycle capacity would constrain the rate of growth of nuclear capacity.

37.   Another factor to be considered is the capital intensiveness for some of the fuel cycle facilities. While the useful data in this regard are scarce, partly due to commercial confidentiality, the capital outlays for enrichment plants, reprocessing plants and the geological repository for permanent disposal of spent nuclear fuel are quite significant, for example, the cost of the new THORP reprocessing plant being built at Sellafield is of the same order of capital cost as a PWR.

38.   In the absence of reprocessing, in addition to electricity a significant amount of spent fuel is produced. The unit of measurement is the tonnage of heavy metal contained; by 2000 it is expected that some 150 000 HM will have been produced in the OECD. According to the nuclear capacity curve in Figure 3 a further 650 000 tonnes HM would have been added by 2030 (in the absence of improvements in fuel efficiency). This would require a storage volume of the order of 260 000 to 130 000 cubic metres and there is no reason to suppose that this amount of storage could not be possible. However this level of unreprocessed spent fuel storage appears to be incompatible with the basic scenario, as the spent fuel will be wanted for reprocessing as discussed below (paragraph 46).

Network Sizes

39.   In order to maintain stability of the supply system it is desirable that no one electricity generating source should have more than about 10% of the system. In some of the smaller OECD countries, or in isolated parts of larger ones, it would be undesirable to introduce LWRs of the sizes currently being built. Even the advanced LWRs now under consideration in the United States and Japan could be too large at 600 MW for such applications. However there is a variety of designs in hand by manufacturers of both LWRs and HWRs which are sized between 200 MW and 600 MW. None of these has been built as yet, but the technology is essentially that already in use in other plants. We would anticipate that they would be in commercial use well before 2020, so that essentially network size is not going to be a significant constraint on nuclear penetration in OECD countries.

Uranium Resources

40.   While, from an industry and other viewpoints the high nuclear scenario seems to be feasible, it would run into problems when it comes to uranium supply. Reasonably Assured and Estimated Additional Resources (RAR + EAR), i.e. known resources, recoverable at a cost of up to $130/kg U are about 3.6 mmillion tonnes U. This uranium would almost all be consumed by 2030 if all reactors were present technology LWRs even considering some increase in efficiency of uranium use by improved fuel design and operating practices and

if no additional resources were explored. Up to 30% uranium could be saved by recycling of plutonium and reprocessed uranium but this would only extend the resource base slightly.

41.   These known resources do not represent a final view of the potential constraints in as much as future exploration will make more uranium available for production than is known today. These uncertain resources have been tentatively quantified under the term "Speculative Resources". Again the number itself (around 10 million tonnes of Speculative Resources are thought to be potentially discoverable) should not be taken as a reliable indicator of the eventual constraint. Even if the speculations were proved correct the further resources might not be available in practice for a variety of reasons e.g. geographical (remote areas), political (moratorium on uranium mining) as well as economic (poor grades). However, it is expected that several million tonnes of uranium from these extra deposits will probably be available in future years.

42.   A major increase of uranium demand in accordance with the scenario would cause prices to rise fairly rapidly. Immediately available supplies would attract premium prices even while low cost known resources remained to be exploited. The price rise would be rapid and quite long sustained because large uranium mines need long lead times before going into production, especially as the producers have to meet ever stiffer regulatory requirements and to present their plans at long public hearings. However, there is probably of the order of 5000 tonnes of production capacity which could be re-started fairly quickly and perhaps another 3000-5000 tonnes potential at small deposits which would also be brought into production fairly quickly. Uranium supply is therefore unlikely to be a direct constraint on the growth of nuclear capacity.

Constraints on FBR Introduction

43.   In the longer term the need to exploit higher cost resources would maintain higher price levels. Increased uranium prices would certainly favour development and deployment of FBRs earlier than now considered and render more certain their envisaged deployment by 2030. Their current state of development is such that there should be no technical impediment to this. Current problems lie rather in the economics of FBRs.

44.   Current designs of FBR appear to have higher investment costs than the LWRs in use. This implies that they would perhaps take more manufacturing and construction resources than for an LWR of equal capacity. However design efforts are underway to reduce the investment costs of FBRs and some engineers already believe that, in series production, an FBR need cost no more than an LWR. In the long run therefore it could be expected that there would be little effect on the ability of OECD countries to install nuclear plant if there were a significant switch to FBRs. Initially, however, it should be expected that there would be a reduction in the building rate as the manufacturers and constructors followed the learning curve on a new task.

45.   The benefit of FBRs can only be obtained by closing the fuel cycle with reprocessing plants. At present only small pilot plants exist for reprocessing spent FBR fuel. On the basis of experience with these plants France and the United Kingdom already have designs for plants to service the

needs of a few FBRs although we understand that these are not yet at a stage where commitments to build them could be undertaken. Nevertheless, the intention at one time was to have plants capable of servicing, say, 5 FBRs available within 10 years and, over the time scale we are considering, there seems no reason to doubt that adequate technology would be available. That remains the case, even considering the very long lead times to be expected for constructing plants to service a reasonably sized pool of FBRs. For comparison it should be noted that the uranium oxide fuel reprocessing plants at Cap La Hague (UP3) and the THORP plant at Sellafield are expected to take about 12 years to bring into operation. Thus introduction of a full FBR/fuel reprocessing capacity will have to be decided well in advance of need.

46.   A major factor in deciding on the timing and rate of introduction of FBRs will be the availability of plutonium to fuel them. By, say, 2020, there will be of the order of 650 000 tonnes of spent fuel containing about 6500 tonnes of plutonium in various forms of storage or repository. Depending on how much of the spent fuel is in retrievable storage (and as yet there are no operating final repositories for spent fuel) the majority of this plutonium would be available, after reprocessing, for FBR fuel fabrication. At present a fuel for LWRs but such usage seems unlikely to make significant inroads into the physically existing amount of plutonium.

47.   The main impediment to FBR introduction would be the availability of freshly reprocessed plutonium. That depends on decisions having been taken well in advance on setting up the necessary plants, including facilities for handling the highly radioactive wastes that are produced. Experts in the relevant subjects have confidence that appropriate technology is already available so from that point of view, and consistent with the broad assumptions of this paper, there should be no inherent technical constraint on a fairly widespread introduction of FBRs as from 2020. It is recognised, however, that this development would not be undertaken without some hard thinking on matters related to policy on non-proliferation of nuclear weapons, such as trade in and frequent movements of plutonium between fuel cycle facilities.

UPPER LIMIT OF NUCLEAR PENETRATION BY 2030

48.   As is evident from the discussion above there are many uncertainties in the analysis of nuclear penetration even if we take the electricity demand forecast as given. We have, however, found no cogent technical or industrial reason for not believing that a nuclear capacity of 1120 GWe could be installed by 2030 if governments were to pursue policies favouring the use of nuclear power. However, unless those policies came into effect very soon, we doubt that the calculated rate of nuclear power plant construction would be achieved early in the 21st century -- a sudden jump in the construction rate by a factor of about 5 does not seem plausible. For consistency we assume that FBRs would be introduced into service in about 2020 and that would tend to reduce the construction rate in the preceding decade. At that time there would be a need also to replace a large proportion of the currently existing nuclear capacity.

49.   The arguments set out above would lead to a projected path for growth of OECD nuclear capacity which would be less steep to begin with than illustrated in Figure 3, but would be considerably steeper after 2020.

50.   The reasoning adopted also suggests that a higher nuclear capacity could be achieved in 2030 if that were required for policy reasons, but that to achieve or exceed the levels shown up to about 2025 in Figure 3 would require early action by governments directed to that end.

CO_2 EMISSIONS AVOIDED

51.   To complete the picture we assume that the fuel replaced by the new nuclear plants would be coal. However, we believe it would be inappropriate to suggest that the replaced coal would not have been burnt more efficiently than today. For this calculation we reduce the avoided CO_2 emission by one fifth (cf. current expectation that efficiencies of, for example, integrated gasification combined cycle plants could be up to 50% instead of about 38% for current best practice). Accordingly by the end of the period the scenario we have produced suggests that annually some 1.2 G tonnes of carbon in CO_2 will be avoided.

POTENTIAL REFINEMENT OF ANALYSIS

52.   In this section brief indications are provided of actions which will lead to reduction of some of the uncertainties bearing on the analysis of constraints.

53.   The uncertainties inherent in the introduction of new reactor types and their effect on rate of construction are unlikely to be susceptible to significant reductions before there is commitment to building a given design of small LWR, HTR or an FBR.

54.   It would, however, be useful to obtain a clearer view of the order rate which could currently be satisfied by nuclear plant constructors in the OECD, and how that is likely to evolve if the current low ordering rates persist. It would also be worth identifying what specific additions to design and manufacturing capacity would be required to meet nominated higher construction rates. Complete precision would not be expected in the output of such studies, but an overall view on a reasonably quantitative basis would be helpful in considering potential responses to the Greenhouse Effect.

55.   The study on qualified manpower already mentioned (paragraph 33) is expected to illuminate part of the problem of capacity expansion.

56.   In the preparation of the Red Book 1989 revised estimates of uranium production capacities will be obtained. There would be some merit in seeking closer quantification of expected delays in achieving given higher rates of production and of the expenditure and delay associated with converting Speculative Uranium Resources into know resources.

REFERENCES

1.   Electricity, Nuclear Power and Fuel Cycle in OECD Countries, Main Data, Paris, OECD, 1988

2.   Nuclear Energy and its Fuel Cycle – Prospects to 2025, Paris, OECD, 1987

OECD PUBLICATIONS, 2, rue André-Pascal, 75775 PARIS CEDEX 16 - No. 44893 1989
PRINTED IN FRANCE
(61 89 09 1) ISBN 92-64-13267-8

DESIGNING THE GOOD HOME

DESIGNS OF

HUGH NEWELL JACOBSEN • BOHLIN CYWINSKI JACKSON • OBIE G. BOWMAN

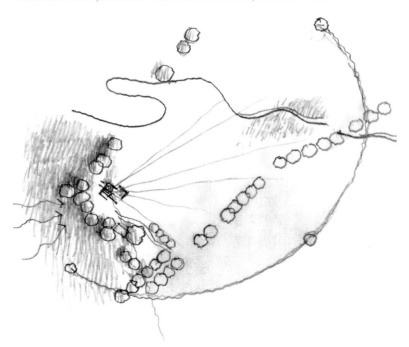

DESIGNING THE GOOD HOME

DESIGNS OF

HUGH NEWELL JACOBSEN · BOHLIN CYWINSKI JACKSON · OBIE G. BOWMAN

BY DENNIS WEDLICK
Written with Philip Langdon

HDi

HARPER
DESIGN
international

an imprint of HarperCollinsPublishers

Copyright © 2003 by Dennis Wedlick

First published in 2003 by
Harper Design International, an imprint of HarperCollins*Publishers*
10 East 53rd Street
New York, NY 10022-5299

Distributed throughout the world by
HarperCollins International
10 East 53rd Street
New York, NY 10022-5299
Fax: (212) 207-7654

ISBN: 0-06-008943-1
Library of Congress Control Number: 2003113949

Packaged by:
Grayson Publishing, LLC
James G. Trulove, Publisher
1250 28th Street NW
Washington, D.C. 20007
Tel: 202-337-1380
Fax: 202-337-1381
jtrulove@aol.com

Manufactured in Canada

First Printing, 2003

1 2 3 4 5 6 7 8 9 / 10 09 08 07 06 05 04 03

ACKNOWLEDGMENTS

Phil Landon and James Pittman collaborated with me on my first book The
Good Home *that featured the work of my own office. I am so pleased and
grateful that they were able to work with me again on* Designing The Good
Home; *without them I could not have realized the premise of this book.*

*I also would like to thank all those at the offices of Bohlin Cywinski Jackson,
Hugh Newell Jacobsen, Architect and Obie G. Bowman, Architect. Their contributions and cooperation in producing this book made it possible to present the
finest work of today's residential architects. In my own office, I would like to
give a special acknowledgement to Chris Hunt as the liaison to all participants
of the book; his talent is greatly appreciated by all.*

HALF-TITLE PAGE Peter Bohlin's Gaffney Residence

TITLE PAGE Peter Bohlin's Goosewing Farm

FOREWORD

In his previous book, *The Good Home*, Dennis Wedlick set out to establish a common language we can all use to explain how and why successful houses work. Nearly all of us perceive when a house is flawed in some substantial way, but most of us lack the vision and voice to identify what's wrong. Dennis' first book helped us see and express what troubles us and what delights us about the houses we live in and visit. This book continues those admirable goals and adds a new purpose: to introduce us to a kinder, gentler Modernism we can all live with.

Today's user-friendly Modernism is the style we've been waiting for—where form doesn't just follow function, it embraces feeling, too. Looking to the past for inspiration is no longer verboten, so a house can look fresh and still feel familiar. Applied by a gifted architect, the style is original and evocative. It combines honest materials, simple forms, open floor plans, and an intimate connection to the land. It differs from previous incarnations of Modernism because the center is the natural world and the human realm rather than the machine age. Devoid of dogma, it allows for great flexibility of artistic expression and personal fulfillment for both the client and the architect.

There are few more fluent interpreters of this approach than Peter Bohlin, Hugh Newell Jacobsen, and Obie Bowman. Each has an aesthetic—a different sensibility through which he filters the world—but none has a specific agenda he's committed to applying. These architects aren't interested in simply imposing their vision on a client's lot. Instead, they're open to inspiration from any source, whether houses of the past, industrial buildings, elements of nature, or the half-remembered experience a client wishes to conjure. Answering the client's needs and doing justice to the site are their paramount concerns.

Existing houses and new production houses are all about decisions made with someone else in mind. Although a production builder may allow selection of some finishes and fixtures, the important floor planning and house siting have already been done. How the house lives is predetermined by a laundry list of generic demographic assumptions. The result is usually a strange amalgam of a semi-updated floor plan wrapped in a bygone architectural style. It's all half-measures, trapped between how we live now and how we used to live years ago. At first, the house may look pleasingly familiar, but it grows disturbing as the distortions and compromises sink in. This is often true of the remodeled older house as well, especially when an addition tacks onto an existing building with little or no change to the original house. Unfortunately, buyers may not detect these flaws until they've moved in and lived with the place awhile.

Our best opportunity for a truly satisfying house is to plan one from the ground up, using the site, our needs and desires, and the architect's technical and artistic expertise to steer the design. Rooted in genuine specifics and freed from stylistic constraints, the house can't help but feel personal—full of warmth and character.

Today's Modern house doesn't turn its back on the past, nor does it copy the pattern book verbatim. The result is dramatic and comfortable, timeless and fresh. Indeed, it makes a very good home.

—S. Claire Conroy is editor of *Residential Architect*.

TABLE OF CONTENTS

TABLE OF CONTENTS

DESIGNING THE GOOD HOME
THE MODERN THINKING THAT GUIDES TODAY'S PICTURESQUE HOUSES

A good home is not just a neutral vessel, blandly waiting for human activity to occur within its walls. A good home possesses character, or what might more precisely be called emotional resonance—the capacity to inspire thoughts and feelings in its occupants. Because I've been designing houses since the 1980s and because I've seen so many houses fail to connect emotionally with their owners, I make it my habit to try to figure out how houses stir the imagination. This book is part of my continuing search for guideposts that designers, builders, and prospective homeowners can use in their own effort to create satisfying homes.

Seen from a certain angle, the quest for emotional qualities in a dwelling verges on the mystical. Surely something magical happens when a house—which is an inanimate amalgam of wood, glass, metal, and an untold number of other substances—touches an individual's spirit. But identifying the sources of that magic and dissecting its workings are, to a large extent, a matter of intuition and experience, at least for me.

I think I've begun to make sense of what goes on, and in these pages I lay out a series of techniques you can use to design a house that will arouse feelings. I concentrate on houses that register on people emotionally and that are rooted in modern aesthetics and technology. These modern picturesque or modern romantic houses have the ability to affect us emotionally and at the same time suit contemporary ways of living. This dual focus—on feeling and on modern methods—is for me a critical element in how to design a good house.

What the past teaches us

Historically, feeling has been an important ingredient in American domestic architecture. In the seventeenth century, when the first permanent settlers from the British Isles and northern Europe arrived on this continent, they encountered a terrain that was unfamiliar and often inhospitable. The resources available in the new land were largely unknown. What they found in North America was so arduous and alien, so different from Europe, that the newcomers might as well have been colonizing the moon. In such a setting, shelter, even a rude, primitive one, was the first necessity. The settlers fashioned dwellings from whatever materials could be gathered within walking distance.

Like children using their imagination to build forts in back yards or in the woods, they assembled rustic dwellings to protect themselves from all that seemed strange or threatening—the extreme weather, the wild animals, and of course the natives already living in the vicinity. The colonists' first rudimentary creations were shoebox homes of stacked logs with pitched roofs, sometimes of earth—gathered together in a village pattern, their boundaries fortified when possible. As soon as possible, however, the primitive homes were added onto, or entirely new houses were built. The colonists wanted their homes to resemble the houses they had known in the Old World. The settlers had no interest whatsoever in emulating the Native Americans' building customs. Nor did they intend to invent new forms for the New World. Rather, the progress of North American house design throughout the next three hundred years consisted mainly in translating and adapting European antecedents to the houses that would be built here.

Although American builders looked to European models, their impulse was more romantic than historicist. They didn't so much copy Old World designs as invoke imagery that recalled certain places—such as villages in England, Germany, and the Netherlands. Early Americans built with the materials they found readily available, even if those materials had not commonly been used on the houses they were remembering.

ABOVE   Some early American houses, like stone-walled Fort Klock, which Johannes Klock built in 1750 near St. Johnsville in New York's Mohawk Valley, were designed to ward off attacks. Because of their defensive intent and because glass was costly, houses often shut out the landscape, no matter how lush its beauty.

OPPOSITE   A picturesque house and a barn that I designed in Kinderhook, in New York's Hudson Valley. Odd proportions, like those of a steep-pitched roof punctuated by a single dormer, often rouse people's feelings. The barn is modern in its bold shape, which makes a surprising contrast against the diminutive, almost quaint character of the house.

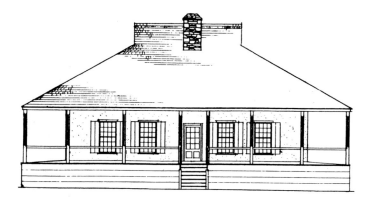

ABOVE The J.B. Valle house from about 1800 in Ste. Genevieve, Missouri, suggests an attempt to relate house to landscape. The raised porch beneath a high, hipped roof provides a sheltered outlook in the Mississippi River Valley, ruled at the time by France.

Americans erected houses predominantly of wood, even if the inspirations for the designs were European dwellings of masonry. Some wooden eighteenth-century structures became known as Dutch colonials despite the fact that they conveyed impressions of Dutch houses that had been built of stone.

Even when American homes used the same materials as those on the other side of the Atlantic, such as stone or brick, they were hardly ever identical to the European originals. Americans adapted European imagery and styles to new conditions, different landscapes. In the nineteenth century, Italianate and Queen Anne cottages were built on beaches or on oceanfront cliffs in styles that recalled homes in the Italian countryside or in English suburbs.

For Americans, literal reproductions of houses from the old country could rarely suit the lifestyles of a new industrious and democratic society. The mansions of the upper class overseas were generally considered too ostentatious for capitalist tycoons in the United States, even if the American versions did turn out to be quite elaborate. The cottages of Europe's lower class also did not lend themselves to copying—for an opposite reason: they were too inferior to house the New World's working class. Old homes in Europe's countryside had been designed without comfort-giving technologies such as central heating and indoor plumbing, and they had used glass sparingly—partly because it was terribly costly and partly because historic European architecture had sought to keep the world out. That defensive posture continued in some American houses, such as dwellings in the 1700s in New York's Mohawk Valley that were called "forts" because they were designed to ward off attacks by French and Indians. But as Hugh Newell Jacobsen pointed out to me, American architecture quickly shifted to a different path, and houses in the New World stopped being built for defense. American architects displayed an eagerness to let in the sun, the landscape, and the neighbors.

I believe that this embrace of the charm of European homes—even while avoiding the negatives associated with European life—is what makes American homes, and other colonial architecture throughout the world, so romantic and alluring. Today what's fascinating about American house design is that it draws imagery and sentiment from one world and mixes it with technology and methods from another. A number of contemporary architects, myself included, are inspired by the dynamics of this modern romantic sensibility. This approach is well suited to the many people who vow to build their own homes because their vision of the ideal home is not shared by "production home-builders" (the building industry's term for tract-home builders). Like the Americans of long ago, many people are torn between two worlds: one that arouses the feelings associated with a good home and one dictated by basic needs, location, and finances. In my experience, many clients crave a home that will satisfy complex aspirations—that will offer surprises and gratifying sensations while also being modern and functional.

Modern picturesque architects

Homeowners-to-be often seek out architects whose work resonates with those aspirations. I too have sought out such architects for inspiration. I have chosen three of them to be the subject of this book: Hugh Newell Jacobsen, Peter Bohlin, and Obie Bowman. The three have much in common, yet they produce work that is remarkably divergent. That should not be surprising, since these are architects who dare to test new recipes for designing a good home and who take a free-spirited approach to their work. What they share is a love of life and landscape, a passion about modern construction technologies, and a penchant for the familiar. Their work is not historicist, but it is picturesque. It is evocative—suggestive of ingrained associations—even as it is new. These designers are modern romantics—architects adept at producing contemporary houses that arouse emotion, that have soul.

The three I've selected invent new forms, but not for the sake of inventing new forms. They devise interesting shapes to engage the viewers' and inhabitants' feelings, just as romantic designers have done for generations. Their houses are wondrously unique, yet not alien. They

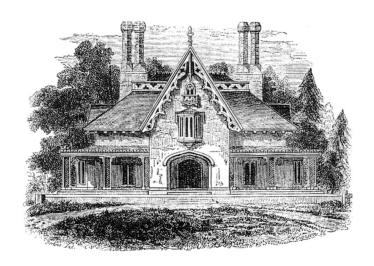

TOP When Andrew Jackson Downing published this "cottage-villa in the rural Gothic style" in 1850, he drew attention to its picturesque qualities: "The high pointed gable of the central and highest part of this design has a bold and spirited effect, which would be out of keeping with the cottage-like modesty of the drooping, hipped roof, were it not for the equally bold manner in which the chimney-tops spring upwards."

BOTTOM A modern house by Hugh Newell Jacobsen looks much simpler, yet its upward-thrusting shapes and a few well-chosen details, such as vertical battens, evoke the feeling of nineteenth-century Gothic Revival.

remind visitors of something familiar, whether it's a saltbox house that's common in their neighborhood or an old farm shed that they saw on a trip across the Midwest, or something they cannot quite put their finger on. Jacobsen builds homes, mainly in the eastern U.S. and Canada, whose profiles in the bright sunlight suggest the skyline of a village. Bohlin has used barn-like shapes for homes in various parts of America, and has designed provocative houses whose shapes bring to mind assemblies of garden walls and lean-tos. Bowman has created houses on the Pacific coast whose profiles look as if they might first have been seen by pioneers heading west in a wagon train. Some houses by Bowman are bold in design, but are wedged into the landscape so as not to intrude upon the context or the neighbors.

These three architects never tire of the art and craft of home-building. They push humble, affordable materials into mimicking the much more costly substances of which castles and monuments are made. Every detail from the nosing on a stair tread to the flashing of a chimney is for Jacobsen, Bohlin, and Bowman another opportunity to pursue a better way of building. Like old cobblers, they fret about whether their houses are comfortable to live in and whether they are built well enough to last. Their concern with how people respond is what makes these designers romantic and sets them apart from architects who detail with only the sculptural quality of construction, rather than feeling, in mind. I had the pleasure of seeing first-hand that their obsession with seeking and applying the newest technologies is not aimed at being avant-garde but at being sure they are producing good homes.

Each of the three adapts to the site, whether on the shore of an ocean, at the edge of a precipice, or in the middle of a farmer's field. All of them manipulate the

ABOVE LEFT The intricate yet powerful structure of an old barn that was framed in a German tradition that employs complex interior bents, at Springton Manor in Chester County, Pennsylvania.

LEFT A modern house by Peter Bohlin in Ontario, Canada, captures the flavor of the barn, only better, because ample sunlight plays across the timbers.

landscape's attributes in conjunction with the home's layout, hiding a view from the visitor until just the right moment. They meticulously fit the home to the site and wrench out every possible attribute the location can offer. These are architects who know that a single tree can be a wonderful landscape. They can make the most of exceedingly humble properties as well as magnificent ones.

Their buildings are not shelters whose sole purpose is to contain life. They are homes that let life in, let nature in. The three do not build "glass houses"; they frame nature with windows. The patterns of the glass transform views into portraits that are capable of revealing nature's complexity. They harness the sun so that it brightens corners, casts deep shadows, remains neutral when it ought to, and makes the homes richer and more complex. There is no doubt, though, that each of the architects in this book is his own person. Each of these architects has a flair for the picturesque.

The meaning of "picturesque" design

I use the word "picturesque" differently than most people do. Today picturesque is often taken to mean "pretty"—or pretty with an overlay of quaintness. I instead use the term the way art historians do—to refer to an aesthetic approach that emphasizes irregular and unexpected features that catch people's attention and engage their interest. That was the sense of picturesque that prevailed among many Americans in the middle of the nineteenth century, when architect Alexander Jackson Davis and landscape designer Andrew Jackson Downing ranked among the nation's most prominent tastemakers. Davis and Downing disliked the strictness of the Greek Revival; they rejected houses modeled on temples from the classical world, and they insisted that American houses should aim for a more informal and expressive character. In *The Architecture of Country Houses*, Downing wrote that domestic architecture "should exhibit more of the freedom and play of feeling of every-day life." In my first book, *The Good Home: Interiors and Exteriors*, I made the case that houses today should be designed in a picturesque manner, and presented several

TOP A house by Obie Bowman on the Oregon coast allows its occupants to feel as if they're outside, even while fully sheltered from the elements. This quest to bring the outdoors in is a strong theme of modern residential design.

BOTTOM Richard Neutra's VDL Research House, built in California in 1932, is a modern house that achieved the same aim at a time when great sheets of glass were uncommon.

TOP *The ground extends onto the roof of a Bowman house, folding it into the landscape. In effect, the house burrows into the earth—one way of relating a home to its setting.*

BOTTOM *Long ago, as Eric Sloane sketched in* American Barns and Covered Bridges, *structures in farm country were sometimes built with their long slope oriented to the north, obtaining a measure of protection against cold winter winds. A design of this kind imparts a strong sense of shelter.*

homes of my own design as examples.

By the 1840s and 1850s, thanks to Davis and Downing, the Picturesque Movement had started to influence American houses. The movement encompassed a number of styles, including Stick Style, Queen Anne, and—the mode that architectural historian Vincent Scully identified as a high point of American domestic design—the relaxed and relaxing Shingle Style. A picturesque house is not shy about incorporating irregularity and expressiveness or about arousing sentiment. The designer may exaggerate the shape and prominence of the roof, give windows unusual dimensions, or prolong the experience of entering the house. A room may rise surprisingly high, to generate a sense of expansiveness, or its walls may slope down, to create a cozy, enveloping refuge. Picturesque houses allow experimentation and departures from the norm. It is these departures, exaggerations, surprises, and mysteries that invest a house with feeling and give it soul. The picturesque is a form of romanticism, an approach that prizes emotion and imagination. So in this book I sometimes call these three "picturesque architects" and at other times call them "modern romantics." They brilliantly combine romantic traits and modern methods.

When I first proposed presenting the work of these three architects, the reaction was always the same: Why Hugh Newell Jacobsen, and who is Obie Bowman? Peter Bohlin was the only identifiable modern romantic of the bunch. When I met Peter at his home in Waverly, Pennsylvania, he was as confident as I that he would belong in any book on picturesque homes. Architects have a great fear of being misinterpreted (because it happens so often), and Peter Bohlin's initial concern was that I would show only his projects that had cottage-like profiles. Peter is a renowned designer, but many of his most published projects have traditional attributes: gable roofs, wood-frame construction, and common house parts such as columns, double-hung windows, and porches. The traditional projects have the virtue of being the most affordable of his designs; the cost is naturally higher for nontraditional buildings made entirely of custom-made components. Despite the quality of the work he's done in a relatively traditional vein, Peter didn't want readers to get a skewed view, one that would ignore his more nontraditional designs.

This book therefore features diverse projects of Peter's firm, Bohlin Cywinski Jackson. Some of the houses shown are relatively simple, with fairly traditional profiles, whereas others are more complex, exhibiting modernist elements such as flat roofs. Regardless of the budget or the character, the genesis of a Bohlin house does not vary. Peter starts, as each of these architects does, by intensely studying the property's potential to take advantage of natural features: sunlight, vegetation, and views, to name a few. The next step is to work out a simple geometrical organization of the house on its land. What that generally means is that Peter takes the plan of the house—which is based on the size and functions required by the homeowner—and incorporates it into a diagram of lines and rectangles showing where the views are, how people approach the house, how large the parking area or the patios should be, and where any special landscape features should be situated. The lines and rectangles are laid down according to an academic set of proportions. The goal, though, is not academic. Peter wants the house to please the homeowner with its beauty, and he wants the house to be well suited to its landscape.

Inside and out, Peter's houses exude warmth that transcends the nature of their materials. Why some of his houses feel warm is obvious—he's been known to line a home's interior with wood on nearly every surface. A preferred wood for Peter is clear Douglas fir finished with satin polyurethane that glows, like a flame, with the littlest bit of light. Other homes are constructed with rough concrete, metal connections, and sheetmetal sheathing, industrial materials all, yet as Peter assembles and details them, they take on the same warmth as their wood counterparts. The best way to describe how he accomplishes this is to refer to agricultural buildings. Farmers have rarely been shy about using crude or industrial materials to build the structures they need, such as grain silos, milking barns, and storage sheds. As these structures have aged, they have blended into the

TOP A house by Jacobsen practically floats above its landscape, creating a setting where people can feel immersed in nature.

BOTTOM Jacobsen's house strikingly recalls the tent platforms that vacationers often used in the late nineteenth century, when summer accommodations like these on Upper St. Regis Lake in the Adirondacks were rudimentary. A house can respect nature by making a purposefully light imprint on the land.

TOP *A California design by Bowman shows how a private, protected outdoor space can add to a house's allure.*

BOTTOM *Philip Johnson's Ash Street house in Cambridge, Massachusetts, in 1942, also featured a protected outdoor space—on view from inside the home.*

agricultural landscape, coming to look as natural as the crops and the livestock. Peter uses these same materials in ways that recall their practical and unpretentious application by farmers. They initially appear to have been chosen mostly just to get the job done, but the final result is enchanting.

No one could imagine Hugh Newell Jacobsen employing the warm palette of Peter Bohlin, which is why bringing these two architects together into one book surprised many of the people I spoke with. The fact is, both Jacobsen and Bohlin restrict themselves to a very limited palette in order to create beautiful houses; the palettes are simply different. Where Bohlin covers floors, walls, and ceilings with Douglas fir, Jacobsen makes all the surfaces white. That difference has made an impact on how the two architects are regarded. Being known for all-wood homes has made Bohlin touchy about not being thought of as a modernist. Being known as a master of all-white homes has made Jacobsen sensitive to being seen *only* as a modernist. When I spoke with Jacobsen, he complained about potential clients who avoid him because they think he is too much of a modernist.

In truth, few designers have a more romantic orientation than Jacobsen. His homes may be all-American in shape—perfect profiles of gable-roofed New England houses are part of his repertoire—yet he straddles two worlds, finishing those traditionally shaped homes with white-painted brick and sheets of glass: materials and details more often associated with modern office buildings. Whereas photographs of Bohlin's interiors easily exude warmth, presentations of Jacobsen's struggle to avoid appearing cold. "Warm and cozy" is a maternal attribute possessing undeniable appeal, but we should recognize that bright, airy, and clean are equally sensual attributes. Jacobsen designs homes that are meant to be loved for a lifetime, and he makes a point of telling his clients so. Many people feel at peace in a comfortable mess—a fact sometimes used in arguments against the purity of Modern design—but just as many, if not more, find peace in tidiness, orderliness, and Puritan principles, essential parts of Jacobsen's lasting appeal.

Jacobsen designs homes to be successful in their efficient use of space, in their spareness and avoidance of frivolous details, and in their no-nonsense relationship to the landscape. The effect is nevertheless romantic because efficiency, in Jacobsen's hands, achieves the character of poetry.

When Jacobsen chooses the site for a house, he finds the sweet spot that will make a striking impression on visitors and that will allow the landscape to change constantly as the visitors take each step toward it or around it. Jacobsen's houses do not need the assistance of a screen of foliage or the curve of a hill to make them vigorous. The shape and the layout of the house are dynamic in themselves. The house's profile may consist not of a single, simple gable form, or of the even simpler flat roof form, but of multiple volumes assembled like cards—one lapped over another, partly concealing and partly revealing the next. These are then arranged into V-configurations or into courtyards or rambling compounds, depending on the light, the topography, or the views. There is no dogmatic reason for laying out a house this way. Jacobsen simply knows that a design of this sort will be satisfying to live in and that with this layout, the house will make the best of the land it sits on. Jacobsen's compound layouts serve more than one purpose; they enliven the approach to the house, dramatize movement through the interior, and enhance ordinary experiences such as sitting at the dining room table—all good picturesque reasons for designing in this fashion.

Then there is Obie Bowman. Visiting Bowman at his office, you are not surprised that few people have heard of him. Nestled into beautiful countryside in northern California (where wealthy families start vineyards for the fun of it) lies a aluminum Airstream trailer small enough to be pulled by a Volkswagen Beetle. This is the main office of Obie Bowman Architects. All the bravado of Peter Bohlin and Hugh Newell Jacobsen combined would not be enough antimatter to negate the modesty of Obie Bowman. Yet just like those two, Bowman has the skills and the power to bring nature to its knees. The Pacific coast, Bowman territory, is far

TOP A roof trellis creates a lacy shadow pattern on the natural-toned wall of a Bowman house. Materials—and their appearance in natural and artificial lighting and under different atmospheric conditions—give a house part of its allure.

BOTTOM A similar sense of extending outward and attention to the intrinsic character of materials distinguish the Adelaide M. Tichenor house that the influential Pasadena firm Greene & Greene designed in 1905.

TOP   Restraint can be a powerful aesthetic, linking a new house to the character of dwellings built two or more centuries ago, as demonstrated by Jacobsen's Greene residence on Chesapeake Bay.

BOTTOM   At a Shaker settlement near Lexington, Kentucky, simplicity and restraint announce themselves through straightforward shapes, more wall than window, and a near-absence of decoration. Some houses achieve expressiveness through purity and discipline.

more fierce than places like the shores of Virginia where Jacobsen's clients have settled down in his troupe of colonial shapes. In settings where the wind refuses to let trees grow taller than flag poles, Bowman's structures wedge themselves underneath the landscape's skin. Permanently embedded, they are impossible to move.

The forms Bowman uses are not entirely unfamiliar, and they blend remarkably into the scenery. From the outside, they appear reclusive. A Bowman design may be a courtyard house on the coast, or a three-story dwelling with a tiny footprint squeezed between the trees in a forest, or an earth-bermed house with ground cover growing up and over the roof. Visitors must find their way to the front door, yet they are subtly led by what seems to be some mystical force. This is an aspect of good picturesque design. Like the view to the house itself, the view to the ocean, the distant hills, or just to the evening sky is held back, then revealed, then taken away again, to give the lucky homeowners more pleasure than any ordinary home would render. The exterior is visually forceful yet it seems less rigid or organized than Bohlin's or Jacobsen's designs, giving Bowman houses a deceptively hodge-podge composition.

All three of these architects inject a bit of levity into the mix. Jacobsen's prim and proper clapboard houses have big holes cut into them. Bohlin's houses have giant window boxes globbed on like big warts. Bowman leans absurdly large posts against a house; the house would cave in from the weight if it weren't for the hidden structure. The textures, colors, and scale of the building parts that make up the exterior may seem organic, but they're actually a bit affected, considering their location. For example, a closer look at the hefty column of tree trunk used to frame the entry to a Bowman house would reveal that neither is it structurally essential nor is it cut from a tree that could be found on that land. As with all romantic architects, the possibility of creating a delightful or humorous composition is sometimes the only reason Bowman needs to add one more bracket or beam to a façade.

On the interiors, ordinary house-building becomes extraordinary, because nothing is taken for granted.

Waste not, want not is the best way to describe Bowman's interior wall construction. What homeowner could not use another set of shelves for books, trophies, or family photos? Wherever possible, Bowman exposes the wood members that hold up interior walls, and he assembles them into a beautiful matrix of cubbyholes. Rooms are positioned up and down, here and there, as if they had built on uneven ground so that there is a natural flow throughout the house. Bowman uses inventive construction techniques to make fantastic interior spaces that the child in everyone would love. The families that inhabit them enjoy enchanting, cave-like corners in seemingly impossible places. Some Bowman houses have fully suspended platforms large enough for an entire master bedroom, even if the home is tiny.

All three are master architects. One purpose of this book is to demystify what makes them astonishing, so that their work can be admired more widely. My other purpose is to present modern homes that embody feelings, expression, and a love of nature. It is important to see these architects together, for it is through comparison that their common threads—their approach to the landscape and construction technologies, and their use of memorable forms—are revealed. This is the only way to recognize the picturesque techniques and principles that are the foundations of their practice, since they take many different final forms. That says a great deal about the potential of the picturesque. What I have tried to illustrate here is an approach that allows the individual homeowner and the architect to invent a completely unique residence without abandoning all those things that are sentimental, poetic, and familiar.

TOP The bold shape terminating a hallway and the elimination of unnecessary details at a nineteenth-century Shaker village in South Union, Kentucky, are almost modern in feeling.

BOTTOM The shape of the ceiling gives the master bedroom its character in this house of my design in Dutchess County, New York. Shapes are sometimes most powerful when stripped to their essence.

PART I
MODERN TECHNIQUES

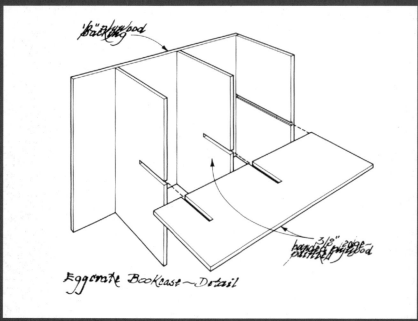

Hugh Newell Jacobsen's Egg crate bookshelf.

People generally think of modern houses as having little trim or ornamentation, lots of glass, and flat roofs whose proportions are exaggerated—sometimes extremely long, sometimes dramatically cantilevered, and sometimes exactly the opposite: hardly visible at all. The image of the modern house took root some seven decades ago, and although the avant-garde gave up the prohibition on sloped roofs more than half a century ago, the sense of what is "modern" has remained fairly consistent.

In contrast to modern houses, traditional houses are generally thought to have old-fashioned windows—either double-hung or casement—with multiple panes of glass. Instead of flat roofs, traditional houses tend to have gable or hip roofs, and on the interior and exterior, they often display classically-inspired trim and ornamentation. Many architects once expected that modern houses, which appeared on the scene around the dawn of the twentieth century, would eventually replace all the older styles of dwellings everywhere. And yet the revolution stalled. Modern design made an enormous impact on offices and other workplaces, but it never achieved its goal for the domestic side of life. In the eastern portions of the United States, you need only visit a few new residential developments to see that the traditional styles have remained firmly established. In the Northeast, where I live, the center-hall colonial remains the single most popular form of house, generations after the passing of Wright, Le Corbusier, and other heralds of new ways of designing.

When so much of life has changed, when four or five generations have passed since Americans traded in horse and buggy for a vehicle with a gasoline-powered engine, why hasn't the appearance of the modern house ever really caught on? One reason may be that many people find modern houses cold and alienating. A house needs to resonate with people's emotions if it is to be considered a home. Some people, it's true, get a thrill from the rigor and the other attributes of modern homes. They love their shapes, which can be unusual (they don't have to resemble a cube). Admirers of modernism tend to like industrial materials, and are happy at the absence of details that have already been used millions of times. These same attributes strike other people as harsh or downright intimidating. Novel or unfamiliar forms drive many people away, especially when the house's colors, textures, and finishes are also severe or out of the ordinary.

In my view, the techniques and details of good modern design can achieve tremendous beauty. When a modern approach is rendered in warm materials and is shaped into a romantic composition, the results can be more endearing than is the case with many traditional houses. A modern simple column made of exposed cedar can, in a woodland setting, be far more pleasing than a porch post designed in a classical style or one that's been ornately painted to mimic those of historical houses.

The architects whose work is presented here—Peter Bohlin, Hugh Newell Jacobsen, and Obie Bowman—all build modern houses employing modern details and modern techniques. The houses they create may not look nearly as familiar as a center-hall colonial, but they are not cold or alienating. These houses are built of wood and other materials found in their regions. They appear so natural in their settings that it is often difficult to tell when the houses were constructed, even though each is clearly modern in style. These houses have interiors that may be warm or cool, but are always sensual. The living quarters of these houses are filled with natural light, refreshed by cool breezes, and blessed with fantastic views, thanks to the modern use of glass and the decision to use open floor plans. Their unusual shapes and dramatic roofs give these houses a unique character, one not available in a traditional dwelling. Consequently, they can express something about their homeowners or their location and the inspiration that brought them into being. A traditional house can be tweaked, with a change of layout, to suit the site or the occupant's lifestyle better, but it cannot be as imaginative, as wholly creative, as a modern house can be. These architects whose work I examine in the pages that follow create houses that have the comfort and warmth of traditional houses, but that deliver the sometimes breathtaking advantages of modern design.

Modern houses are built with exciting shapes, with interiors that are not ornate, and with ample areas of glass. They wield dramatic assemblies of volumes; in some instances the volumes themselves are not out of the ordinary, but they are always bold. Detailing of the interiors spans the entire spectrum from highly traditional moldings to little or no trim, yet all of these houses emphasize a restrained palette; they are purposely limited in color and texture. They are built with a modern appreciation for the artistic qualities inherent in construction—the structure, joinery, and detailing not only hold the building together; they serve as compelling decoration. The ample expanses of glass typically are organized into a grid pattern of window panes, a signal of their modernity. The way the glass is arranged, proportioned, and divided gives the interior its comfortable atmosphere and enhances the view of the outdoors, at the same time preventing the occupants from feeling overly exposed. Properly handled, modern techniques are not cold or alienating; they are bold, artistic, and sensual.

BOLD SHAPES

Modern houses shy away from traditional embellishments. Instead they rely on bold shapes to make an impression. The more dramatic the shape, the more expressive the design is likely to be. A greatly simplified version of a traditional house shape can, for instance, make a strong visual impact and yet convey associations with the past. In New England, one possibility would be to use the shape of a colonial-era saltbox, without any of its historical trappings. Even when stripped of old-fashioned details such as small, multiple-pane windows, the saltbox form powerfully communicates a sense of "home" and affinity with its region.

The design of a modern home may be composed of a number of minimalist shapes. The bold shapes could be drawn from residential architecture, but they don't have to be. They might be fantastic, almost surreal; in some instances, they appear to defy gravity or to be unrelated to the house's construction system. Bold shapes can make for a house with a unique relationship to the landscape. Dramatic shapes have the ability to float over their adjacencies; a house with a large roof and giant overhangs can appear to hover above the first floor and the landscape. A crisply shaped house may seem to sit lightly on the land, looking as if it could easily be removed.

Boldness may also be achieved by giving a traditionally detailed house a remarkably steep roof or an oddly shaped footprint. You might enlarge the scale of traditional details, for instance employing an oversized chimney or extra-large dormers. The strong shapes used in a home's exterior and interior are often symbols that can relate to the house's setting. A towering chimney may remind you of nearby trees or of the history of the region. Dormers may have shapes that are exaggerated versions of a local tradition. They may also reflect the predilections of the owners for a particular profile, such as that of a barn. The modern romantic will balance the composition to consist of just a few strong shapes so that the house is expressive without being too confusing or showy.

LEFT A modern house in Adirondack style practically jumps out toward those who walk up the stone entry path. Despite being covered in rustic materials, the shape is bold—and made still more emphatic by being supported on a pair of stout logs. It has a gutsy quality.

ABOVE The paring back of overhangs almost to the point of disappearance allows the irregular wall profiles to stand out. The boldness is accentuated by the angles and asymmetry and by limiting the exterior to one material.

LEFT Asymmetry abounds in this simple building. The door is off-center, with a window on one side but not on the other. The fieldstone wall at the left is not closely balanced by the sloping wood wall at the right. The roof flaunts its irregular composition. All these elements create a bold appearance for what might otherwise have been a humble, old-fashioned-looking shed.

ABOVE Deep overhangs and thin posts with butt-jointed glass at the corner make the roof shape bold.

BELOW A busy composition of jutting triangles, a big, straight-up chimney, and a prominent bay window makes a forceful impression. All the visual commotion engenders assertiveness.

RIGHT A strong, angular profile gains further intensity from the rugged log supports.

ABOVE An incomplete stone wall looking like a ruin, and a trellis above part of it, make eye-catching forms in the landscape for a bold impression. Other shapes, such as the garage, can remain undemonstrative when elements in the foreground seize the viewer's attention.

ABOVE Breaking a big house into a series of simple, repeating volumes produces a powerful and confident composition. The repeating trellis form magnifies the house's potency.

FOLLOWING PAGES The shape of each pavilion is reduced to the fewest and sharpest strokes and then repeated for stunning effect.

RESTRAINED PALETTE

Simplicity is a trademark of the good modern home. One way of achieving simplicity on the interior and exterior—while achieving expressiveness in the home's details, textures, and compositions—is to limit variation in the house's materials and colors. The restrained palette ties everything together and generates simplicity. For example, let's suppose you were eager to surround yourself with natural materials and you wanted a certain eclectic touch, yet you also wanted your home to be modern. One way to achieve this on the interior would be with smooth wide-plank floors, walls made of knotty beadboard, ceilings edged with carved crown moldings, and giant grid-pattern windows made of yellow pine. Such an interior would abound with wood, and it would be eclectic in its aesthetics—crown moldings are more formal than either the plank floors or the knotty walls—yet the uniformity of materials and color would give the house a distinctly contemporary character. A house may have many features and details, but if they are monochromatic, it will appear as minimalist and modern as a house whose features and details are few.

Limiting the palette on the exterior accomplishes a number of objectives for the modern romantic. Without the distraction of multiple colors or finishes, the simple and powerful shapes of the house stand out. You can indulge in mixing contrasting kinds of details if you keep them all within a narrow spectrum of color and materials.

Traditional details might be chosen to express the heritage of the region or the personal taste of the homeowner and blended with modern details into a pleasing, consistent ensemble.

A home is often more harmonious with the landscape when the exterior adheres to a limited palette. The colors can be carefully chosen to match those of the natural features of the property. Materials can be selected according to what could be found in the surrounding landscape, such as local woods and stones. A colorful, showy house might compete with the views, the countryside, or neighboring buildings, whereas a good design that sticks to a simple palette would tend to fit in.

LEFT The use of just one material, redwood, unites a rambling, informal composition and creates an air of simplicity. Because each piece of wood has a subtly different color and grain and weathers unpredictably, the restrained palette is not dull—it's alluring.

ABOVE Humble, unshowy materials—cedar shakes covering the walls and roofs—let the crisply composed volumes of the house read strongly and make a modern impression.

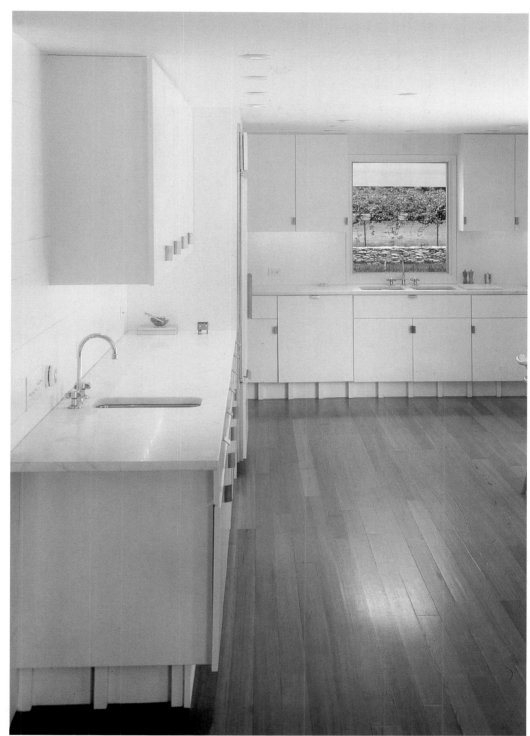

LEFT With the wall, the door, and the muntins of the large window painted white and with the redwood deck stained a soft gray, all that's needed to give this exterior some punch is a touch of another color—green trim around a tiny window.

ABOVE The floor of recycled antique heart pine and the other surfaces of white create a restrained aesthetic, so the splash of color of the view through the window stands out, the way a beautiful painting on a monochromatic wall would grab attention.

ABOVE A humble material such as plywood can achieve a rich, elegant look when one material is used throughout. Keeping everything a single color or tone makes a small space more aggressive and, conversely, makes a large space cozier. The smallest bit of light, in the gable window, becomes impressive when the palette is limited.

LEFT By using a single material—wood—in a single color, this interior's glowing floor, walls, and ceiling manifest a modern flair. The orange-toned wood looks as if it generates more warmth than the fireplace does. Because the old-fashioned barn doors conform to the monochromatic color palette, they do not undercut the modern effect.

ABOVE The impact of natural light is intensified when the interior is restricted to wood in a narrow color range. With few colors to compete for attention, the light becomes soft and soothing—refuting the notion that modern design is necessarily cold.

ABOVE Light has a powerful effect when the surfaces it strikes are limited to white painted wood and natural strip flooring. The light becomes intense and sensual.

ABOVE Two tones, wood and white, dramatize this room's contemporary character when sunlight pours in. The restrained use of color allows furnishings to take the foreground.

ABOVE Keeping the walls white calls attention to the texture of the brick and mortar and focuses attention on the jaggedness of the profile. In the sunlight, the walls become a perfect canvas for the trees' shadows, creating a chiaroscuro effect.

GRIDS OF GLASS

Owners of modern homes delight in patterns. They admire designs that repeat again and again, such as plaids, stripes, and grids. The patterns can produce a variety of effects, from calming to unnerving. One popular modern pattern is the grid, in which each unit is a rectangle of identical size. This pattern is supremely restful, so easy on the eyes that it can practically disappear even while it generates a calming sense of order and control.

Modern windows are often divided into grids of glass through the use of muntins. Long ago muntins were a necessity, since the only sizes in which glass could be produced at reasonable cost were tiny. For decades, however, manufacturers have been able to produce glass in large dimensions, so muntins have to a great extent become something chosen for aesthetic reasons rather than an essential element of window construction. In fact, it's now less costly to make a single pane measuring, say, four feet square than to assemble many small panes into a window of that size. Modern architects frequently divide windows into patterns composed of smaller pieces of glass because they want to layer a pattern over the openings in the wall. The muntins may be nothing but a decorative appliqué.

Window grid patterns do triple duty. First, they add to a house's walls a proportioning feature that can be a pleasing, recurring motif. The house may have windows that are all in similar sizes and shapes, or it may have a varied assortment of windows, but all divided into panes of one size. Another alternative is for the house to have a few unusually sized or shaped windows, serving as eye-catching features. Second, the grid patterns add a design feature to

a room's interior. A large window meticulously divided into a grid of glass can be a focal point, just as a stained glass window would be. Finally, a grid of glass can enhance enjoyment of a landscape by splitting the view into a series of framed scenes. And when no one piece of glass could be built large enough to take in the whole view, a. grid of windows can capture the panorama.

ABOVE Grids of glass above give an almost monumental character to this house. The effect is created with assemblies of inexpensive glass doors.

LEFT A division of a large window into pieces of equal size is a signature of modernists. It frames a series of small scenes within the overall view and creates an orderly pattern.

LEFT The industrial sash brings the outside in. The glass, with its lacy pattern, becomes a surface to be savored, not just a void. The glass grid steps down toward a sunken seating area in the corner.

TOP RIGHT Equal division of window panes is associated with modernists, but in some instances it could just as easily be part of a traditional house.

BOTTOM RIGHT Thin muntins tend to give a window a modern feeling.

PRECEDING PAGES A continuous grid spreads across several otherwise ordinary windows, generating a compelling modern quality for the interior of a rustic house. An expansive grid can create an aesthetic that is both relaxing and controlled.

ABOVE Rustic posts and beams standing free of the walls dramatize and decorate the interior beautifully. The structure adds visual interest while it adds a layer to the interior.

EXPOSED STRUCTURE

The classic modern home is known for its lack of decorative touches: the absence of crown moldings at the ceiling, the elimination of elaborate trim around windows or doors, and so on. This is not to say that those who design modern houses are uninterested in making things pleasing to the eye. Far from it. They recognize that the most beautiful aspect of the modern home may be the structure itself. The structure, when well designed and well executed, can be breathtaking. The interior is often designed so that the construction is exposed rather than concealed within the layers of the walls or hidden above the ceilings. Exposed structural parts such as beams and posts do not have decorative classical details added to them or carved into them, but they are detailed to be attractive to look at and to add character to the entire interior.

The modernist is often obsessed with the way the house is put together. After all, the origins of modernism lie in new construction technologies such as reinforced concrete, steel beams, and structural glass. The classic modernist is known for employing a universal approach to construction, whether the construction of a house or the construction of an airplane hanger.

Modernists, however, have become less dogmatic over the years, and there is now considerable diversity in the approaches they take. The architects whose work is presented here share a passion for beautiful and innovative construction, but they employ much humbler materials, such as wood and masonry. For most of their houses, they use methods of building that any skillful carpenter can execute. The results, though far less industrial than those associated with orthodox modernists, are no less impressive and are more suitable to the creation of comfortable homes in the countryside. Visible structure may be as rough-hewn as logs or it may be smooth and regular. It may consist almost entirely of wood, or it may use metal or other materials. The exposed structure not only adds drama to the interiors, but also helps establish the character of the home or express the nature of the surrounding landscape.

ABOVE A more restrained pattern of large and smaller beams gives a fascinating look to a pitched ceiling.

ABOVE   A lacy trellis creates a focal point and, because of its shadow pattern, adds interest to the wall below it. The slotted structure of the wall also generates aesthetic interest on its own.

RIGHT   Conspicuous brackets and posts make a modern house more flamboyant. The exposed metal straps exude a tough decorative quality.

ABOVE A wooden girder with its reinforcing diagonal bracing is a strong focal point in this interior, where the structure is used for dramatic purposes.

LEFT A straightforward layering of beams and joists gives modern character to an open interior. Even the built-ins reflect attention to how things are put together and contribute to the house's personality.

ABOVE The contrast between the log wall and the industrial truss system animates this interior. Timber posts and laminated wood beams complete the structural ensemble. Structure can be poetic.

ABOVE Oversized structural members in a very small house crown the upper part of the interior. Modern industrial materials and structure can energize a house.

ABOVE Traditional stick-frame construction, exposed in a renovation, creates a screen between two spaces.

RIGHT PHOTOS An unusual structure adds a layer to the interior, with the glass extending beyond it. The bold shape makes a spectacular focal point visible from both inside and out.

PERFECT LIGHT

Good homes only get better when they take advantage of light. Sunlight is often employed to create an image of perfection. A minimalist, modern interior lined with beautiful materials depends on the sun to achieve its purposes. Without the distractions of decorations, the light of the sun, ever changing through the day and across the year, brings to life magnificent and animated interiors. The power of the sun has a similarly magical effect on simple modern exteriors composed of strong volumes and clear profiles. Owners are known to walk outside at dawn or at sunset to observe the perfect light's effect on their homes, just as other people would head out to view the ocean.

Windows in a modern house can be strategically positioned, sized, and detailed to bring the right effect to bear on the exterior and interior and to manipulate the flow of sunlight into specific patterns and environments. Unlike traditional houses, modern dwellings allow almost complete flexibility in the size, type, and placement of windows. The ability to create the sense of perfect light does not depend on expensive materials or complicated construction. The progression of light across a yellow pine floor, as modulated by the pattern of the windows, can be enough to make a simple room look and feel wonderful. Rooms that are positioned and fenestrated to radiate with the dappled light of a woodland setting will be no less wonderful whether they are lined in birds-eye maple or plywood.

The evening sky, with the light of the stars and the moon, can make homes even more enchanting than they are in the daytime. Artificial lighting can make every home a jack-o-lantern. Striking compositions of windows that please or challenge the eye during the day can present a fascinating display of planes of light at night. The architecture of a romantic home is often defined by a few dramatic details that are blended into the overall composition of the house by limiting the colors and materials. With perfect lighting, these details can be highlighted at night, allowing them to stand out as theatrical sculpture.

ABOVE In a modern home, even ordinary lights on the interior serve to emphasize the composition of glass openings. They can create a jack-o'-lantern effect.

LEFT Because illumination is so important in modern design, interesting ways of providing artificial lighting are esteemed. These decorative lamps are made of white pine dowels and translucent glass.

ABOVE Artificial lighting causes a roof, already large, to appear even more prominent at night. Redwood surfaces and corrugated metal are invigorated by the light.

RIGHT Transparent, well-lit modern structures stand out against a dark sky, often becoming more impressive than they are during the day. The spaces in between can be delineated by the light.

ABOVE A pool of light bathes the living room, making a sensual oasis as dusk arrives. Transparency in modern design transforms interiors into showplaces.

LEFT A layer of light, whether natural or artificial, can make an interior space cozy, as exemplified by warm light from the side at the top of a landing.

PRECEDING PAGES A row of clerestory windows becomes extraordinary at night, when the wood seems to emit a yellow-orange glow. Effective lighting maximizes the effect of exposed structure.

PART II
ROMANTIC TECHNIQUES

Peter Bohlin's Tunipus Compound at Goosewing Farm.

The quintessential romantic house is a cottage nestled in the woods or overlooking the sea. It might have a steeply pitched roof embellished with dormers, and it may feature old-fashioned adornments such as window shutters and porch posts—perhaps even a weather vane. Most of us are drawn to a house of this sort because it appeals to something deeply rooted in our emotional makeup: it arouses feelings that we associate with a good home. Romantic homes can be sentimental; they do not shy away from allegorical details. But they are not defined by those elements either. The defining feature of the romantic house is its ability to stir people's senses.

How to create a romantic house should not be the big mystery that it currently is for most people. In this section, I lay out a series of techniques you can use to design a house that's expressive and emotionally satisfying Ideally, you would build the house in a setting that intimately relates to nature, since natural scenery—particularly a view of a valley, mountains, shoreline, or luxuriant vegetation—always stirs an emotional response. The house ought to seem at ease in its setting, which means it might lie partly concealed in earth and vegetation or might have only its roof and chimneys visible above the tree-tops as you approach. It might be positioned so that a visitor glimpses the house first from one direction, then from another; the path toward the entrance can tantalize the visitor with a succession of skillfully circumscribed and choreographed views. "The approach is like the roll of the drums," says Hugh Newell Jacobsen. In a conventional urban or suburban location, the possibilities are more constrained, yet even a small site often has the potential to incorporate a garden, an area left to grow wild, water, rocks, or some other feature that will summon up associations with nature and help create an enticing and not entirely predictable entry sequence. Wherever it's placed, the house should be imbued with sentiment.

Irregularity and surprise are important parts of the picturesque approach to design. The irregularity may take the form of unusual proportions—an exceptionally tall roof, or thinner-than-normal windows, or porch posts larger than structural necessity would dictate. The surprise may also come when a cozy entrance leads to a soaring interior, which in turn contains a diminutive alcove or an unusual staircase. There should be things that spark curiosity. Most houses oversell the big door and the grand façade—features that are not essential for stirring a response.

If a pair of shutters flanking a window will make the homeowners or their guests feel joyful or contented, the designer may well add them, regardless of whether they're necessary to protect against the weather. Materials, both inside and out, should be seductive; they should have a sensual quality, causing people to want to touch them or gaze at them. Natural materials such as wood and stone are especially effective at achieving this effect.

You can bring portions of nature indoors, either literally or figuratively. A house with a log wall is one example; the logs link the home to trees outdoors; at the same time, they give the house a hardy, rough-hewn personality. One advantage of materials from nature is that they open up the potential for many varieties of expression. Natural materials left rough and primitive generate one sensation; natural materials that have been smoothed and refined create quite another. Stonework might consist of chunky, odd-shaped boulders if a rustic atmosphere is the objective. They might be rocks that fit precisely together if you want a sense of order and sophistication. The feeling differs, but in each case, the inhabitants of the home will revel in a connection to nature.

Just as the approach to the house should be carefully orchestrated, movement through the interior should be manipulated to stir emotions or establish a mood. In many houses today, you open the front door and all the living areas are exposed to your gaze. Poof, the effect is finished. It's much better if you orchestrate a sequence of experiences. In a house by Obie Bowman, you may find yourself enveloped in a forest of gigantic log columns, then led into a cave-like space, and then—bang!—comes a stunning view, maybe a view of the ocean, perhaps just a graceful old oak tree. A good home doesn't have just one kind of space; it has a variety of them, organized so that movement from one to another is full of interest and incident.

Jacobsen, Bowman, and Peter Bohlin, the architects of the homes featured here, have collaborated with their clients to create unique designs that are both modern and romantic. Their homes have few, if any, traditional details, yet they convey as much sentiment as houses composed of more old-fashioned elements. The power of these homes to bring forth a broad range of emotions—from delight to curiosity to tranquility—comes from romantic techniques such as a carefully planned approach, thoughtful manipulation of how you move through the interior, and a reward at the end of the journey.

The romantic modernist will often tap into the seductive quality of natural materials such as wood and stone and may also find expressive character in raw construction materials such as metal and untreated concrete. Unlike the picturesque movement of the nineteenth century, which used decorative details and motifs to represent nature, the modernist uses the real thing, either by capturing expansive views of the natural surroundings or by incorporating boulders, tree trunks, and other specimens of nature into the house's construction. Bohlin's own home in northern Pennsylvania rests on a large stone. "It would have been so easy to move the house or cut the stone," Bohlin says. "But the house is so much better, so much more interesting, by accommodating the stone."

Romantic techniques can make people sigh with delight or cry with surprise. They go a long way toward creating the good home.

BLENDING
INTO THE LANDSCAPE

One of the finest compliments a new house can be given is that it appears to have always been there—that it looks "natural" in its setting. Looking as if it belongs is universally understood as a good thing. Yet most houses do not give that impression at all. Conventional building practices more often go in the opposite direction. Many builders go to great lengths to keep natural elements away from houses; they tame unruly landscapes, they raise the floors high off the damp ground, and they do their best to block the effects of hot sun, cold air, snow, and rain.

Romantic designers, by contrast, want to provide all the comforts of shelter while bringing the homeowners as close to nature as possible. Some romantic homes take the notion of being in touch with the landscape literally. Obie Bowman has designed grass-covered roofs, or tucked the house into the side of a cliff, or squeezed the house in between stands of trees. The aim of many romantic designers is to achieve a seamless connection between house and ground. In this approach, mature existing plantings are preserved wherever possible; they become focal points visible from the interior and, in some instances, elements around which the exterior is composed. A romantic architect may engineer the house to allow stones or trees to remain in their original locations, with the house built around them.

Another way of fitting into the landscape involves choosing colors and materials that blend the house into its setting. The design may not be subservient to the shape of the land, yet coloration and texture tie the house to the site. Still another technique involves positioning and laying out the house and designing its silhouette in relation to the existing or intended landscape but letting the house read as a clearly independent object. Hugh Newell Jacobsen's all-white houses often do this. When seen from the landscape, they clearly belong even though they stand out. There is no one single way to blend a home into the landscape; a variety of techniques can achieve that result.

ABOVE A house adopts the self-effacing technique of merging into the landscape, letting a covering of vegetation grow up over the roof. This low-profile tactic creates a distant view that pays respect to nature.

LEFT Peter Bohlin believes in "accommodation," leaving the rock in place and building directly on it. In a poetic manner, a column that rests on the rock is splintered into several supports, highlighting the massiveness of the rock.

LEFT There is a house in this scene, but it is totally subservient to the meadow, practically disappearing into the earth. Anyone who approaches gets a sense of nature, of grasses waving in the breeze beneath an expansive sky.

ABOVE The landscape here is jagged, and the house mimics its jaggedness. The house has eye-catching shapes, but does not insist upon being the main attraction. It is half tucked into the vegetation.

LEFT A house uses rustic features, including a randomly laid stone base, to harmonize with the landscape. The log column in the center of the façade echoes and pays tribute to the surrounding trees.

ABOVE A house on a cliff hunkers down below the broad-spreading canopy of a tree. The landscape, not the house, is the focus of attention.

RIGHT With foresight, a house can be planned around existing trees, as in this elegant example by Hugh Newell Jacobsen. The deck cantilevers over the roots, allowing the craggy tree to become a beautiful counterpoint to the house's crisp forms.

ABOVE An example of how a house can defer to the land. Obie Bowman has mastered the art of letting the earth extend right up onto the roof of the house.

LEFT The broad, soothing porch has a tropical quality, suggesting a man-made equivalent of the shady area beneath a tree. In this way, a house can reflect the climate of its landscape.

ABOVE A house keeps a low profile so that it does not compete with the terrain. From a distance, the house appears to nestle into its natural setting.

ABOVE   A gravel path and a wall made of concrete work well with the rocks and grasses of the abutting landscape. When the tone of the landscape is echoed in the building and paving materials, a house blends with its surroundings.

ABOVE A corridor in the landscape frames a view of a house straight on. If the landscape had a regimented allée of trees, the view would be extremely formal. This view is softer because the vegetation has been left in more or less natural condition.

LEFT A house looks perfectly relaxed in its setting, sitting low beneath the trees, framed by segments of a stone wall.

CONTROLLING THE APPROACH

A romantic architect makes a house expressive not only by designing the house according to lay of the land but also by controlling the way you move through the land to get to the house and by controlling what you see of the house as you do. Like designers of picturesque gardens, romantic architects prefer to keep the house from revealing itself in its entirety until a magic moment arrives; the buildup is almost like a striptease. A small part of the home is revealed from a distance; then a bend in the driveway may reveal a different, slightly larger part; and finally at another turn, the visitor sees the house in its entirety—dead on or at an angle, depending upon the results desired.

The seemingly pretentious movie scenes that feature an architect crouching down, wandering this way and that, and climbing a tree, trying to get the "feel of the land," are less exaggerated than you might think. A romantic architect wants to gain an intimate knowledge of the contours and qualities of the site and to control the approach. Ultimately, the approach is often laid out according to one of two models. The formal model calls for the entire house to be seen head on; in this method, the entrance is obvious. The picturesque method calls for the house to be seen from an angle or to be partly obscured. The entrance has to be searched for or is revealed subtly.

The picturesque approach is cinematic, and it enhances the qualities of the home and the property as a whole. It calls for imagining all the different experiences that a person could have on the way in. This might include giving the visitor a view of a garden gate with no house in sight, followed by a path that seems to lead to just a corner of the house, and then a spot where a big tree blocks the path, forcing a turn that reveals the front porch. Such experiences can be orchestrated on surprisingly small parcels of land leading to the most modest of homes.

ABOVE The approach leads toward and then curves past a modest part of what is actually a sizable house. Passing through the opening in the stone wall makes approaching the home an event. Later comes a view of the water.

LEFT A movie-like sequence shows how an elaborate house at first seems an almost incidental part of the landscape, then becomes a strong object when viewed head on, and becomes still more impressive when experienced from the smooth court.

ABOVE   This perspective emphasizes the primacy of the natural landscape and the house's position as a kind of refuge within it.

LEFT   The angled approach from a distance creates an air of mystery. Where is the front door? With relief, the visitor discovers the wooden footbridge, which leads directly to the entrance.

LEFT Once the visitor to the Peter Bohlin house shown on Page 84 comes inside, he finds the interior continuing the corridor-like effect that began on the footbridge. The entry path offers a controlled but welcoming environment, emotionally warm because of the wood glowing in the sun.

RIGHT The approach to a Hugh Newell Jacobsen house reveals a series of pavilions, each of equal prominence, sitting in a serene landscape of tall trees.

BOTTOM RIGHT As the visitor comes closer, the path leads into a slot between two pavilions. The narrow slot envelops the visitor and provides a sense of arrival in a reclusive but dignified little space.

MANIPULATING CIRCULATION

The feeling you get from a house is heavily influenced by how you move through its interior. In a romantic home, the aim is to foster patterns of circulation that will help make the house emotionally engaging. The layout of rooms, passages, and openings should force people to move through the house in a particular way.

Hugh Newell Jacobsen sometimes divides a house into a series of buildings, each containing one or two rooms, and connects them with a series of glass breezeways, just to make sure that it's as exciting to move through the house as to settle down within it. Peter Bohlin creates a home consisting mainly of one big room with an open floor plan; within it he then creates an area that from certain angles looks like a long corridor extending from one end of the house to the other. This forces movement to follow a carefully orchestrated path that draws attention to views of the landscape and emphasize special spots on the interior.

People are often attracted to sunken living rooms. What makes a sunken living room appealing is that you need to climb down into them. Although changes of levels throughout a design may not be practical for every homeowner, where they appear, they add delight. Obie Bowman designed a California house in which the visitor is immediately forced to turn the corner, climb a few steps, and turn again before discovering the living room. The twists and turns not only make the house fun to explore; they make the living room, with its breathtaking view of the Pacific, seem even more special upon arrival.

Staircases can be designed to give interesting overlooks, at times splashed with sunlight. Landings should be seen as opportunities, because with a bit of enlargement or embellishment or outlooks, they become distinctive places within a home. When paths intersect at a landing, the designer can make it a pivot point possessing great character. Moving through the house should be an experience, one that offers a variety of sensations. Architecture comes to life through movement that attracts your interest and engages your feelings.

LEFT An open staircase offering views in all directions makes every movement through this house stimulating. The play of sunlight on circulation spaces and adjoining areas enhances the experience.

ABOVE An angle toward the end of a passage, with a space around the corner, makes movement more interesting than a walk down a straight corridor.

LEFT Going up and down stairs and continuing around a corner can make a part of the house feel special. As this interior shows, the circulation does not have to be arranged in the most direct path possible. A more involved route may add to the pleasure. Openings off the circulation route modulate the views and make progression through the interior more enjoyable.

ABOVE *Having to find your way around the big column at the base of the stairs makes every trip through the interior more of an event. The column gives people an interesting object to touch or rest against, engaging another sense. The open railing lets anyone at the right watch people move through—another source of visual interest.*

LEFT *A corridor gains interest by passing inviting spaces on one side and then another. The protruding logs become a marker on journeys through the house. Glass at the end of the passage provides light and a view, pulling the person forward. The designer needs to consider whether the glassy end-wall will produce undesirable glare.*

ABOVE A small landing makes an interesting intersection where people turn, continue straight on, or occasionally rest. In this house, people experience various spaces as they go, and enjoy a view of the outdoors.

LEFT The spiraling motion of the stairs adds another dimension to movement. At the same time, it directs a person's gaze outdoors. The climb inside the house is coordinated with the hill outside.

WORSHIPING
NATURE FROM WITHIN

It might seem a stretch to say that a house or its occupants "worship" nature. But the romantic temperament takes enormous delight in earth, sky, sunlight, water, trees, flowers, and plant life of all kinds. A good home ought to express heartfelt feelings about nature, not only through the way the house looks on the outside, where it meets the land, but through the character of its interior as well.

One of the ways this is accomplished is by offering eye-catching connections between the interior living spaces and whatever lies beyond. A romantic responds to the idea of floor-to-ceiling glass or a sitting area extending outward toward the trees or an assemblage of corner windows that dissolves any sense of confinement. While sitting inside, you can be visually immersed in a rainstorm, snowfall, or bright, sunny weather. The goal, in at least part of the home, is to blur the distinction between indoors and out. The connection to nature intensifies when materials on the inside, such as a stone floor, continue beyond the house's walls. At the same time, it's a good idea to have some spaces that feel sheltered from the outdoors. Contrasts of feeling—openness and exposure in some areas, coziness and retreat in others—help make a home romantic.

Natural materials may be employed on the interior. Wood and stone communicate a sense of reverence toward nature. So does any natural material that's used as part of the house's structure, such as wood posts, beams, and brackets. Really mesmerizing emotional effects come, however, when nature in all its roughness is brought inside—as a wall built of logs, as a rock face, or some other element not wholly smoothed, tamed, and made mild.

In some instances, the structure may evoke the character of nature. Beams and brackets that support a roof may echo the lines of tree limbs. A column may suggest a tree trunk. Analogies such as these can enrich a home's emotional repertoire. Where possible, existing natural features, such as a stone outcropping, might be retained and incorporated into the house. In a romantic home, nature is close by.

ABOVE   *The bathtub is placed next to a sheltered window that lets the bather relax with a colorful view of flowers and plants.*

LEFT   *A tree trunk that was found on the property makes a magnificent feature in a house by Obie Bowman.*

ABOVE The use of materials from nature ties this house to its woodland setting. Details from the interior continue outside so that the house feels like an integral part of its environment. Note the roof brackets, which echo the shapes of the porch roof.

ABOVE The angled wall of glass pushes the interior out, blurring the demarcation between indoors and outdoors. The house seeks to become part of nature.

LEFT The floor flows ever so smoothly from the living room out to the patio and landscape, giving the view of nature primacy. A generous corner of glass is an effective tool for breaking up a house's solidity and making nature the center of attention.

TOP RIGHT Chunks of driftwood logs bring nature indoors. Complementing this, a round window funnels a view through a cliff and toward the water below.

BOTTOM RIGHT A large stone detail frames a fireplace. The stone, which is part of a shower, makes showering feel like an activity in the great outdoors.

LEFT The walls are so clear and translucent that the house melds with nature. Living here has a dramatic quality, a sense of hanging over the landscape.

TOP RIGHT A greenhouse-type roof makes the treetops feel like part of the interior.

BOTTOM RIGHT A view through a thin-mullioned window to a pond. The interior requires fewer decorative effects when the focal point is the landscape.

SEDUCTIVE MATERIALS

In the nineteenth century, when the world was in the midst of the industrial revolution, romantic designers countered the machine age by designing homes and furniture that highlighted beautiful materials such as handcrafted wood, handmade ceramic tiles, and blown glass. Those old-fashioned arts and crafts not only incorporated motifs relating to nature; they employed rich natural materials in their construction. Modern romantics are similarly fascinated by seductive natural materials such as wood and stone. However, today's architects don't stop there; they also admire the picturesque qualities of materials such as concrete, metal, and masonry.

Natural materials appear at their best when put together in the most practical way given their makeup. Granite stones can be nicely chiseled into blocks, so they look best when assembled like an old stone wall. Traditional methods are not the only way, however. A good modern architect may use materials in a more innovative fashion, to tell a story or make an impression. The architect knows that wood for siding can be cut and carved in a hundred ways that will show off its rich texture and soft profiles while still standing up to the elements. A modern romantic may take common building components such as bricks and stack them to form elegant patterns—without the custom shapes or details that a nineteenth-century romantic would have used. Concrete can be poured into rough

wood forms with a raised grain to produce wonderful, soft-looking wood patterns.

Two or three contrasting materials can become even more seductive when combined in a simple composition. Set against a glossy all-white interior, a slate floor appears richer than the most extravagant marble. A touch of elegant stainless steel or nickel on its details will make a staircase of inexpensive concrete seem exotic. A material that may not seem in keeping with a home's other materials may be introduced as a way of creating a focal point. A fireplace assembled from industrial materials will have a bigger impact than one made of stone when added to a cottage in which stone and wood abound. What matters is that the materials exude feeling.

ABOVE Stones possess sensual appeal because of their color, their irregular shapes, and the patterns they form both with one another and with sunlight or artificial light. Big rough-hewn blocks over window openings make a wall look as if it's stood there for many years.

LEFT An industrial material—Corten steel—has what architects would call "a dialogue" with the cedar of the seat. The two materials together emphasize the character of each other. Many materials can be allowed to show their age.

ABOVE Concrete takes its character from the forms, usually made of wood, that surround it when it's poured. The texture and pattern of the wood forms can be chosen to generate many different effects. The thin-profile roof sits on industrial metal brackets.

LEFT This is what's called "novelty" siding. The pattern, chosen for its beauty, produces soft profiles that are very appealing. Here the siding is painted white, against which the contrasting green window trim pops out.

ABOVE   The ends of a few charcoal-colored clinker bricks protrude from a wall of smooth red bricks. The tones of the two kinds of brick contrast nicely, and the two- and three-dimensional patterns give the wall visual punch.

LEFT   The roughness of poured concrete against the smoothness of a polished-steel handrail makes a striking combination. Both materials become more seductive through juxtaposition.

ABOVE Redwood on a deck, on board and batten siding, and on a sliding barn door gives a natural and warm quality to a courtyard. With such appealing materials on so many conspicuous surfaces, asphalt composition shingles suffice on the roof.

ABOVE *The concrete in this home exudes a natural feeling because it's been poured to take the pattern of wood forms. The interior stirs an emotional response because of its concrete and wood.*

PART III
EVOKING THE FAMILIAR

Obie Bowman's sketch of Pin Sur Mer.

All the houses collected in this book are modern. They are filled with natural light and views, thanks to large expanses of glass. They are built in a modern mode that favors simplicity and minimalism, even if they make extensive use of traditional residential materials such as wood, brick, and stone. All of this, in my opinion, is good. So the question is, how do modern houses become modern romantic homes—dwellings that will tap into people's feelings and engage their emotions? One of the most effective methods I've found involves having the new house evoke buildings and places that people are familiar with. If a new house brings to mind buildings or places that people already know and are fond of, the owners—and visitors, too—will feel attachment.

Peter Bohlin, Hugh Newell Jacobsen, and Obie Bowman, as modern architects, do not copy buildings from the past. They create houses that are unique, in most instances strikingly so. Yet the houses these three produce connect with people's sentiments, in large part because the designs summon up associations with historical or vernacular houses; they suggest something of the character of traditional houses or rural buildings, usually in the same region where the new houses are being built.

The shape of the house is the most basic tool for evoking the familiar. Countless "no-frills" center-hall colonials and saltboxes come into being every year because their shapes suggest classic all-American homes, satisfying people's understandings of what a house should be. In many cases, it's worth noting that although profiles of historical house styles are discernible in the new tract houses, the detailing that accompanied those shapes two or three centuries ago is absent, so although the new construction reads as a traditional home, it seems anemic—a weak and bland copy of the original.

Modern romantics approach the design process in a different way. They aren't interested in simply imitating houses from the past. They may use familiar silhouettes from older buildings, and may even group several such silhouettes together, but they give the houses fresh touches too, because they don't want the results to be boring. Modern romantics want to create something new and imaginative, no matter how much of the familiar may also be perceptible in the composition. Houses by Bohlin, Jacobsen, and Bowman recall shapes that people are accustomed to, but, equally important, they exhibit the passion, ingenuity, or humor that separates distinctive homes from cookie-cutter designs.

Each traditional shape possesses its own character. A skillful designer recognizes that character and works with it. Colonials and saltboxes have strong, simple profiles that are well suited to conveying a prim and proper expression; they may suggest a historic place or an established family. A cabin in the woods or a ranch-style house is a more relaxed source of imagery; it befits a more casual or rustic expression. Historically, houses often mimicked the shapes of religious buildings, as when the physical attributes of Gothic churches and Greek temples were applied to residences, for a theatrical kind of expression. The architects featured here often borrow and reinterpret the shapes of American agricultural buildings such as barns, grain silos, storage sheds, and lean-tos. You might question whether such shapes are appropriate for the homes of human beings, but I find them stimulating and usually successful. The advantage of such shapes is that they are familiar and yet they widen the variety of expression; they make possible a composition that's completely modern without being too disconcerting. Shapes can be derived from all sorts of sources.

Modern houses run the risk of feeling alien or weird if they utterly disregard convention. Consequently, in addition to playing with familiar shapes, the romantic modern house often uses elements such as columns, chimneys, windows, and siding patterns in ways that are familiar yet imaginative. The columns or chimneys may be greatly simplified, or they may have a rustic character, depending on the feeling the house is meant to express. Decorators use salvaged pieces of old houses, such as Victorian posts or brackets, as ornamental elements and artistic focal points in contemporary interiors. If the entire interior were Victorian in style, these elements would lose their impact. Employed strategically, however, they have a strong positive effect. For some of his rustic houses, Bowman has used sliding barn doors and timber columns that look like they were retrieved from construction sites. Jacobsen's refined designs in the eastern U.S. and Canada repeat one traditional compositional element over and over again on his stark white buildings as if it were a wallpaper pattern. Bohlin blends traditional elements such as old-fashioned window and door trim with extraordinarily modern windows and doors in his woodland homes. These contrasts help generate a romantic feeling.

The romantic house challenges you; it makes you stop and wonder. In this kind of house, the pure forms and abstract patterns of modernism are layered with picturesque touches. Often there is something intentionally odd about how the familiar or sentimental elements are used in the design. The chimney may be way too large for the house. The columns may be too few or too many to be holding up the porch roof. The windows may be off center. These peculiarities are there to hold your attention, to interest you each time you look at the house. They may also be there to express the uniqueness of the homeowner, the designer, or the site.

A pair of giant columns on a small house may express a certain haughtiness, while a series of curiously small windows on a large house may express humility. Symmetrical arrangements of odd renditions of familiar elements can make them less off-putting, yet still challenging. The same is true of seemingly haphazard arrangements of traditional elements. The goal of the modern romantic is to infuse character and delight into a home without abandoning the elegance, simplicity, and up-to-date spirit of modern design. Evoking and playing with the familiar make this combination possible.

GAFFNEY RESIDENCE

What the eye first gravitates toward in the Gaffney house is its tallest and most traditional element: a three-story façade of red cedar with semi-transparent stain that sits beneath a steep roof. The gable roof and the set of four-pane windows underneath it give this house in Romansville, Pennsylvania, a comforting sense of the familiar. But familiar in what way? Devoid of ornament, almost shorn of overhangs, does the Gaffney home recall a farmhouse, or does it bring to mind a barn? This house seems familiar, yet at the same time enigmatic.

Soon your eye moves toward a strikingly modern and transparent corner that dissolves the division between indoors and out. After taking in that two-story expanse of horizontally proportioned glass, you notice that the house doesn't stop there. To the right appears an irregular extension consisting not so much of solid walls as of big, revealing sheets of glass. These last forms have a free, organic feeling, more irregular than the three-story portion that first caught your eye.

Only after you've taken in all of this do you notice that far to the left is another appendage—a glass-topped area that butts up against a stone wall. Peter Bohlin's Gaffney residence is a remarkable compound of contrasting building forms and varied window shapes. Gables are not symmetrical. Windows do not quite line up. The plan is lightly skewed to accommodate views into the Chester County landscape. Bohlin believes that such subtle juxtapositions and distortions create a dreamlike ambience, stimulating the client's memories of experiences abroad and of childhood on a Midwestern farm.

LEFT *The impression of the Gaffney house changes markedly from day to night. When the sun is up, the solid, cedar-clad walls stand out most. At night, illumination in the interior turns the house into a glowing, transparent object.*

ABOVE *The owner and the house, which from a distance first appears almost gaunt, like some simple rural structures.*

ABOVE The sloped, gabled ceiling in the master bedroom exudes a spacious, modern feeling. The cross-ties give the space some pattern and scale and make it feel a bit more secure.

RIGHT Each window is proportioned differently, pragmatically maximizing the views of the landscape. Stone walls create bounded outdoor spaces and create landscape impressions that evoke experiences the client had abroad. There's just enough visible wall above the windows to generate a partial sense of enclosure. Near the small-pane windows, a lower ceiling makes the space feel cozier.

ABOVE A one-story stone wall and the presence of only a few, small windows in the taller, unadorned wall of wood make the western exposure reminiscent of a barn. But it is a barn with a modern glass roof overhanging the fieldstone wall.

ABOVE The living and dining spaces are bounded by a stone wall that's appealing to look at and to touch; it's a remnant of a barn that stood here. After entering the living-dining area, a visitor finds a view to the countryside opening up through a glass corner at the opposite end of the house.

KAHN RESIDENCE

The Kahn house in Lima, Ohio at first glance harks back to nineteenth-century Gothic Revival. The board-and-batten siding and the pointed central gable seem as quaintly all-American as Grant Wood's famous "American Gothic" painting. The symmetrical chimneys, the curious oval windows, and the high porch with pairs of skinny posts evoke the past.

Yet this is no literal reproduction. Despite latticework, divided-light windows, and other elements associated with a distant time, this is a modern house—abstracted, pared down, and painted pure white rather than the earth colors that A.J. Downing advocated 150 years ago. The continual back-and-forth between strikingly modern and old-and-familiar is what makes Hugh Newell Jacobsen's design so intriguing.

The view toward the entrance makes the house look as if it's not terribly large. But when the long and elaborate side of the house comes into view, a different character emerges. Suddenly the house looms larger, becoming more of a mansion than a Gothic cottage. With its two-story bays, still more chimneys, and its powerful symmetry, the Kahn house recalls the grand houses of the past—though the impression is given an unusual modern twist by the squaring off of all the chimneys and by the placement of a skylight at the roof's peak.

The contrast between modern touches and an older character pervades the interior as well. Transparency and openness dominate some views, but other perspectives feel more traditional. This is a house of contradictions, provoking varied sensations—in one moment closed in, and in the next instant wide open and filled with light.

LEFT   The narrow end of the Kahn house makes a friendly and generally old-fashioned impression, but the squared-off chimneys give the home a contemporary flair.

ABOVE   The undulating side is much grander. The narrow, elongated windows recall those from Victorian times. The whiteness and the simplicity of the chimney shapes, however, mark this as a contemporary home.

ABOVE The long, divided-light windows and folding wooden shutters evoke the feeling of a house from more than a century ago.

ABOVE When the shutters are folded together, the interior becomes airy, open, and bright, strongly modern in its atmosphere.

FOLLOWING PAGES At night the house becomes transparent and seems strikingly modern despite the procession of bays and divided-light windows.

WINDHOVER

indhover, at Sea Ranch in northern California, calls to mind the temple form that goes all the way back to ancient Greece. Versions of this house's great pediment, raised upon two massive columns, have appeared in thousands of banks, courthouses, churches, and (most relevant for residential design) grand houses—those that have graced American estates, plantations, and boulevards for well over 200 years.

What's most curious about Obie Bowman's design is that the strongly classical overtones are contradicted by the spareness, indeed primitiveness, of the materials and surfaces. Though the pediment above the entrance has a shape handed down through centuries of human culture, the house has a rustic air. The "columns" are tree trunks with only the bark removed. You can even tell where the branches used to grow. They give the house a muscular feeling, which is intensified by the columns' lack of either capitals or bases. The tone of the steps ties the building to the earth.

The bold temple shape helps a modest-size (1,570-square-foot) dwelling hold its own in a neighborhood of larger homes. Once you look past the columns, it becomes evident that this is, nonetheless, a thoroughly contemporary house. Glass abounds. Double-height glass beneath the pediment consists of inexpensive, stacked sliding glass doors. The interior is open and full of angles, with white gypsumboard walls contrasting against extensive areas of wood, so that although there is a hearty connection with nature, the overall esthetic is bracingly modern.

LEFT A very simple pediment of clear redwood bevel siding rests on two massive tree trunk-columns, with no capitals or bases to add decorum. The steps serve as pedestals for the columns, but their brown tone ties the house to the earth.

ABOVE The loft-like interior has a brawny feeling, with large exposed timbers running both orthogonally and at unusual angles. The overall effect is dynamic and contemporary.

LEFT The bedroom loft, with its asymmetrical skylight, features layer upon layer of wood components, which weave a pattern and generate a sense of depth. The effect resembles the intertwining of the branches in a forest.

ABOVE The fireplace is made of concrete formed from pieces of the bevel siding. The corrugated pipe makes the composition more dramatic, and it contrasts against the wall's 2-by-10 wood framing structure of Douglas fir, dyed red.

FOLLOWING PAGES Logs that drifted up from the water have been placed on bases of industrial metal, making an evocative, cave-like entrance. The lighting accentuates the entrance's depth. The modern cutaway created by expanses of glass treats the bedroom like a display case.

OLD MISSION COTTAGE

What could be more enticing than a cottage in the woods? This modest house on the western shore of Old Mission Peninsula in Lake Michigan's Grand Traverse Bay reinterprets the aged cabins the owners were fond of visiting during summer holidays. It reflects a simple and relaxed life of long, warm summer days, of quiet, cool evenings with a fire, and of sailing, reading, and baking fruit pies. The quintessential cottage shape, with a simple, steeply pitched roof sheltering the living space below, brings forth feelings that almost everyone has for cottages sitting amid the trees, their protruding roofs offering protection from the elements.

By enlarging the roof beyond usual cottage dimensions, Peter Bohlin magnifies the cottage image and at the same time makes way for a much more expansive interior. The series of doors on the front evokes something familiar—perhaps a camp dining hall where people gather for meals and conversation. The posts and beams are rugged, yet artful. The flaring, exaggerated brackets attached to the posts seem fanciful and light-hearted—a fitting emotion for a getaway place.

Approach closer and what next commands attention is the capacious steps—broader than the house itself. They extend outward in three directions in a welcoming gesture. Some might see them as a kind of waterfall in wood. At the same time, they bring to mind gregarious spots like amphitheaters and the steps of college buildings, where people relax while looking outward. The components of this house seem familiar, yet the ensemble is fresh and pleasurably surprising.

LEFT   The big gable roof evokes a sense of shelter and domesticity. Gables can be used in imaginative ways and in radically differing sizes.

ABOVE   The cottage reaches out with cascading wooden steps. Their tone distinguishes them from the ground. They seem to elevate the house, making it grander, while also harmonizing with the galvanized metal roof.

LEFT *The exposed rafters bring to mind rustic camp buildings and the welcome sense of shelter they provide. The walls accentuate the feeling that this is a cabin in the woods.*

ABOVE *Repetitive patterning, including the suggestion of grid upon grid through the far window, takes this house beyond the realm of elementary shelter. Note the gnarled tree trunk at the entrance to the room; here architecture worships nature..*

LEFT Dimness can produces a sense of refuge, while translucent wall panels create a bright area in an otherwise wooden, cabin-like interior. A pitched ceiling with exposed rafters seems a perfect counterpoint to the imaginatively designed concrete fireplace.

ABOVE Unlike the steps of many houses, which require people to enter and leave by a narrow, set route, these steps are generous in the freedom they offer. These steps also adapt to the undulations of the land.

PART IV
CREATING A SENSE OF PLACE

Peter Bohlin's sketch of Point House.

Romantic architects like to refer to the effects they create as "senses." A "sense of breeze" might be the experience of sitting in a room that makes you feel as if you're on a breezy tropical island, when in fact you're in a northern city. A "sense of height" is the feeling that a room or space or building is very tall, far beyond its actual dimensions. One of the perceptions that designers especially prize is a "sense of place"—the feeling that you're somewhere special, whether it's on top of a mountain or in a town square. A romantic house can create a sense of place even if the property is not large, does not possess a panoramic view, and does not comprise an entire compound.

The predominant way of creating a "sense" is by intensifying people's perceptions. You can create a sense of height by positioning short objects next to the walls that you want to appear tall. You can create a sense of distance by placing certain features, such as a pond, in the landscape's middle ground, thus making the background stretch farther away. To generate a sense of place, you can select the most auspicious of the locations available and then intensify the experience of getting to, and moving through, the house and the property.

A landscape architect might intensify the approach to the house by creating an *allée*—a walk or path between two rows of formally-planted trees or shrubs that are at least twice as high as the walk or path is wide. An architect tends to intensify the approach by carefully choosing where to position the house on the land and by giving the house a shape and a layout that will make it feel like a special spot. Usually there are just one or two best routes for reaching a property, whether by foot or by vehicle. Down a street or road, over a hill, or around a bend, the house comes into view. The romantic architect shapes the house and composes its windows and doors with a keen awareness of which part of the house will be seen first, second, and so on. The house may be designed to give one impression from far off and a different impression up close. For example, from a distance, visitors may not be able to discern where the front door is, but once they come near, they discover a conspicuous and welcoming entryway. This technique makes a person feel confused or curious while searching for the door, thus intensifying the delight of finding the inviting entrance once it becomes visible.

If this sounds a bit manipulative, keep in mind that romantic houses are intended to be sensual, making impressions on the occupants and visitors. A walk through a romantic house can be a journey that includes vistas, surprises, and illusions, such as short distances that seem much longer than they are, and tall spaces that seem even taller. Photography, even video photography, can rarely capture the full effect of moving in and about a romantic house, for it is an emotional experience, forged by visual, tactile, and kinetic sensations as you advance.

Once you have entered the house, the romantic architect continues massaging and managing your feelings. In Hugh Newell Jacobsen's design for the Palmedo residence, the front door is in a part of the house that stands forward from the house's main body, connected only by a glass-walled, short corridor. Its separation from the rest of the dwelling requires the guest to enter, then go through a transparent passage—in effect, to go outside again—and then walk forward to get into the main component of the house, which of course is the real destination. It's a shocking experience that intensifies the experience of arriving. The jolt is not unlike the frustrations that often accompany travel to special places. Since the journey through the front door is only for visitors (the occupants usually enter from another direction) there is no reason the entry sequence can't be a little contorted to convey a "sense of place." Obie Bowman is fond of making both the visitors and the owners climb up or down or find their way around corners after they enter a house, for no other reason than to intensify their experiences. At Peter Bohlin's own home in northeastern Pennsylvania, the sense of place is accentuated by striking views or unusual ways of presenting them, such as a kitchen window with Alice-in-Wonderland proportions, magically suspended in air.

These three modern romantics share an admiration for a modern flow of spaces, one room open to next, but arranged in just the right way so as to control your every step. The genteel homes of Jacobsen, the beatnik homes of Bowman, and the traditional-to-progressive homes of Bohlin are all built from plans that set rooms apart at angles, or that separate them with small rooms or spaces. Views or focal points pull you one way or another so that moving through a house conjures up the feeling of walking through the woods or ascending a mountain.

Jacobsen designs into each house a stunning experience that is akin to suddenly commanding a panoramic view from a mountaintop. His goal is for the visitor and the homeowner to be bowled over upon discovering it. Not everyone can do that. But a modern romantic strives to lay out a house so that a person moves through it according to a plan that leads them to a spot offering an impressive view of the countryside, a breathtaking fireplace, or just a beautiful tree. That special spot, sometimes referred to as the point of arrival, is an essential ingredient in a "sense of place."

TUNIPUS COMPOUND

A good home can use any of a number of techniques to create a sense of place. At Tunipus Compound in Little Compton, Rhode Island, the landscape—70 acres, much of it meadow, surrounded by water on three sides—sets the scene. Parts of the landscape are defined by elements such as a low concrete wall, which forms a semi-enclosed courtyard linking the guesthouse (a renovated and expanded nineteenth-century cottage) to the garage. Often people realize that a place is special when they are obliged to take a journey through its constituent parts. Here visitors must walk through an open courtyard to reach the house's entrance. Even a journey of a few dozen feet registers on people's consciousness.

Peter Bohlin designed Tunipus Compound with broad, uninterrupted slopes of roof that allow the upper portions of the walls to retreat into shadow. Mild colors, undemanding shapes, familiar materials such as cedar shingles, and a semi-protected courtyard all contribute to the sense that this home is a refuge.

What causes Tunipus Compound to rise above the ordinary is the surprises that unfold upon venturing into the house. The interior, though sheltering, is unmistakably modern. The ceiling was peeled away to reveal the joists. The wall practically evaporates around a prominent fireplace framed by views into the landscape beyond. Once a warren of small rooms, the interior now is open and spacious. The crispness of the interior leaves no doubt that this is, after all, a distinctive setting.

LEFT The simple slope of a cedar shake roof and the quiet exterior walls, seen across a meadow, make Tunipus Compound a relaxing place.

ABOVE A low concrete wall encloses a courtyard, which establishes a connection between the guesthouse in the distance and the garage in the foreground. A boardwalk from the garage to the house's front door makes approaching the entrance an event.

LEFT A glossy floor of Douglas fir and a ceiling dramatically opened up to expose recycled Douglas fir stock trusses with finish boards on the bottom chord make the interior distinctive. The ceiling opening starts narrow near the private bedroom area and widens as it approaches a fireplace in the most public part of the house.

ABOVE The progression through the house leads toward a dramatically designed fireplace, which commands attention, yet also allows people to see past it to the sunlight and trees. The slatted Douglas fir screen around the fireplace is a recurrent motif, an example of the appeal of patterns.

LEFT A blocky chimney of zinc-coated copper, butted up against the symmetrical windows, contrasts strikingly with traditional shingles. The finish material at the lower portion of the exterior wall is cedar clapboard with a painted finish.

ABOVE Water is a restful feature, one that usually makes a property feel larger. Because the concrete wall is weathered and eroded, the setting feels relaxed rather than pristine and formal.

PREVIOUS PAGES The ceiling offers a mesmerizing pattern of beams and diagonal bracing. An ordinary, box-like interior has become dynamic. In every direction a person looks, the views are interesting enough to generate a sense of place.

SOHN RESIDENCE

One way to create a sense of place is to break a large house into a series of smaller components. Architects often liken the resulting components to a "village," since the overall effect resembles a cluster of individual dwellings. Hugh Newell Jacobsen, a master of such assemblages, used that technique in the playful Sohn residence overlooking Lake Simcoe in southern Ontario.

Whereas a conventional house has a front, a back, and two sides, the Sohn residence is variegated, containing a series of spaces between segments, making the dwelling less of a monolith and more of a "place." The 1½-story components have a diminutive scale, like that of cabins in a camp. Because each unit shares the same spare esthetic and has the same pristine white on its exterior, they read as a unified group; small variations such as differing window sizes and non-uniform chimney locations force the eye to go back and forth, measuring and comparing. This engagement with the viewer is part of the fascination of a picturesque house.

The Sohn house is traditional in its use of shapes such as gabled roofs and prominent chimneys and in its choice of divided-light windows, but it is undeniably modern in its abstraction and in its severe detailing. Ornament is almost entirely absent. Chimneys are clad in tongue-and-groove clear cedar—very smooth. On the interior, the variety of spaces reinforces the sense of place even while the pervasive whiteness, the openness to the landscape, and the abundance of sunlight mark this as a truly modern design.

LEFT   The relationships among the series of components are simple, yet intriguing, drawing a person in. The entrance walkway threads into one of the passages between similar, but not identical, white gabled segments. The wood exteriors and wood chimneys unify the parts.

ABOVE   Lining up the components so that they define the edge of a shared landscape generates a village-like effect. The complex, linked by glass-walled passages, runs parallel with a treed embankment overlooking Lake Simcoe.

LEFT A spare, white interior space feels expansive because it extends all the way to the underside of the roof.

ABOVE The rhythmic window panes capture a succession of views for a picturesque effect. The lower panes frame a series of tree trunks, while upper panes capture views of their crowns. The window in the gable makes the sky yet another focal point.

ABOVE The interior becomes alluring at dusk when selective lighting makes the niches in the walls stand out. Suddenly the room achieves a feeling of depth. The design allows a simple, almost severe room to become visually rich and complex.

RIGHT The mantel evokes some of the feeling of a traditional hearth, but because it's abstracted—pared down into a simple horizontal with no moldings for support—the entire wall reads as the fireplace-surround. The modern wall-to-wall grid of bookcases bestows additional character on the room.

LEFT *The consistent whiteness belies the complexity with which some of the parts come together. The skylight-cupola above the factory-coated white aluminum roof introduces a picturesque element to the composition.*

ABOVE *The movement of sun and shadow modulates the feeling of the exterior throughout the day, and gives the simple forms interesting nuance.*

TIN ROOF

This house at Sea Ranch on the northern California coast generates a sense of place by offering the inhabitants a series of small outdoor areas sheltered from strong winds off the Pacific. The reddish concrete pavers make visitors feel that they've arrived even before they reach the entrance. The irregular path of approach—the grid of paving stones is incomplete and interspersed with vegetation—accentuates the feeling that getting to this house is something of an adventure. The guestroom door is positioned in an alcove, making a small space off a larger space—a very effective way to create a succession of experiences.

The contrast between the sheltering exterior, much of it clad in wood or in corrugated metal that conjures up associations with agricultural structures, and the interior, dramatically open and abounding in natural light, arouses a strong reaction—relief and delight—when people finally enter. The materials are familiar, but used in ways that provoke surprise, as with the corrugated metal pipe that serves as fireplace flues. A visitor's emotions oscillate between being soothed by the glowing expanses of wood and made curious by the complex ceilings and walls.

An intriguing combination of coziness and spaciousness keeps sensations alert in this interior by Obie Bowman. The biggest effects are reserved for after a person is safely ensconced indoors. Outdoors, the house doesn't try to upstage its formidable surroundings of rocky cliff, rough meadow, and weather-beaten conifers.

LEFT The high, sheltering walls and the disarmingly irregular arrangement of plants and decorative objects, such as a salvaged cupola, make the terrace at Tin Roof a place with character. Visitors climb concrete steps (illuminated at night by a light in the dark circle in the photo's extreme right) to approach a barn-like guestroom door.

ABOVE The house defers to the rugged, windblown landscape in this view from a distance. The integrity of the surroundings is preserved.

ABOVE Corrugated pipe and distinctive plant life make an exotic entryway.

RIGHT Reductive materials such as the wood of the floor, ceiling, and walls and the stone of the fireplace are calculated to stimulate an emotional response from the inhabitants. At the same time, the corrugated steel culvert pipes rising from the fireplace provoke curiosity; they intriguingly shroud a double flue.

LEFT Wood everywhere—on the floors, the walls, and the ceiling, not to mention the exposed beam—suffuses the bathroom with a welcoming warmth.

ABOVE The slope of the roof gives the bedroom loft an appealing snugness even though the room itself is open and receives a good deal of light.

ABOVE A house that seemed to recede into the background during the day becomes extroverted and bright at night. The framing of a vista makes the house feel in charge of its setting.

RIGHT In daytime, the house remains a quiet part of its surroundings.

TOM & KARIN'S PLACE

In a wooded setting at California's Sea Ranch, Tom & Karin's place is a small and simple main volume, but with steeply sloped appendages attached to its gray-shingled sides. These "add-ons," projecting outward and visually anchoring the building to its site, generate the sense that this is a compound of buildings. As you go around the house, the varying scale of the parts and the way they come together create unique places in an otherwise homogeneous woodland.

The way the deck with a built-in bench seat and a hot tub extends into the shaded area beneath the trees accentuates the feeling that the house is entirely at home in the landscape. The add-ons suggest familiar shapes, such as agricultural sheds. The western red cedar siding on the angular extensions further intensifies the house's sense of being at one with the surroundings.

Despite the familiar elements, Obie Bowman has created a strongly modern house, especially inside, where a cantilevered master bedroom—cliff-like and suffused with light from above—overlooks a double-height living room. Because the house has so much glass, the interior cliff is clearly visible from outside, strengthening the house's sense of belonging to the site. Grids of wood organize storage spaces on the walls. Details such as the lamp above the dining table, projecting from the bottom chord of a Douglas fir truss, introduce interesting surprises. This is a house that is loose and unpredictable, exuding almost a funhouse quality.

LEFT The appendages stop short of the roofline and the corners, thus staying visually subordinate to the house's main volume. This reduces the apparent size of the house, which has 1,595 square feet of heated space and 509 square feet of unheated space.

ABOVE Trees come close to the house, accentuating its woodsy character.

LEFT Areas of brightness lead the eye through the ground floor and up to the peak, generating curiosity and wonder. A visitor looks upward and wonders what it's like in the platform area suspended above the main floor.

ABOVE The combination of views is rich: in the foreground, the warmly gridded wall, beyond, the bedroom platform, and above, a shaft of light.

LEFT Indoors and outdoors are intimately connected. This view shows the cantilevered bedroom inside and reflections of trees on the glass.

ABOVE A spa in the woods, open to the foliage on two sides and protected by a low wooden enclosure on the other two sides.

GREENE RESIDENCE

The Greene residence on the Chesapeake Bay's eastern shore in Maryland divides a substantial house into a series of components, each with its own volume. Whereas the Sohn house in Canada is a group of gable-roofed pavilions all in a neat row, the Green house shows that pavilions can be arranged at a series of angles, thus making the collection appear even more strikingly village-like. The five pavilions gather together as if huddling against the elements; the result is a welcoming courtyard.

Pyramidal roofs give each component a conspicuous stillness, an eye-stopping stability. The walls are redolent of the restrained houses that people built in the area in the eighteenth century, their windows no larger than absolutely necessary. An atmosphere of purity and decorum pervades the setting. Hugh Newell Jacobsen turns the absence of substantial trees—ordinarily a disadvantage—into an asset, by spreading a pavement of pea gravel around the base of the structures, thereby tying everything together.

As you explore the house, the calm initial impression is contradicted by surprises such as a sheer corner of butt-jointed tinted glass giving the occupants unconstrained views of the water. The glassy cutaway corner and other details are as restrained, in modern style, as the white clapboard aesthetic is in traditional style. Upon entering, a visitor discovers that this house is anything but traditional. Soaring interiors with divided-light windows far higher than in an old house creates the drama for which the best modern romantics are famous.

LEFT The segments of this spread-out residence meet the ground with spare, quietly contrasting forms of steps. The pea gravel all around unifies the composition.

ABOVE The five-part house exudes restraint, perhaps even humility. The house might have overpowered the landscape if the five volumes had been a single, large structure.

ABOVE From the fireplace up, the interior soars three stories, with light from a dormer window catching the eye. Behind the free-standing fireplace wall is an open hall.

ABOVE The angled end of the fireplace wall, paneled to give it human scale and a bit of traditional feeling, leads the eye to the glassy corner of the room and out toward the bay.

ABOVE The owner's second-floor painting studio, cozy and almost colonial in scale, offers respite from the expansiveness of the main floor.

ABOVE Space flows around the main staircase in a part of the interior that exudes a loft-like feeling. The effect has a sculptural quality.

LEFT Traditional white clapboard suddenly gives way to sheet glass where the bay demands to be seen. The simplicity of traditional design meets the simplicity of the modern.

ABOVE The landscape slips past the house and at the same time tucks into the folds of the house, for inside/outside continuity. It's as if a person could step through the glass and onto the patio.

PART V

COLLABORATING WITH THE LAND

Altering the land and the landscape is easy. Machinery can cut down a small forest or rearrange a hillside in a few days. Soft, marshy ground or solid rock that's seemingly unsuitable for building upon is a slight nuisance the homeowners can surmount if they're willing to take on some additional expense. Dry yellow deserts can be turned into lush landscapes of green almost overnight with installation of mature plantings and simple irrigation systems. So why would modern architects be at all anxious about the land and landscape when they are free of its constraints?

The answer is that the existing landscape usually possesses character—perhaps undervalued—and is more an asset than a detriment if you adopt a romantic approach to design. Romantic architects see raw, undisturbed land as a potential collaborator in the creation of unique houses, dwellings that gain much of their distinctiveness from the limitations and idiosyncrasies of the site.

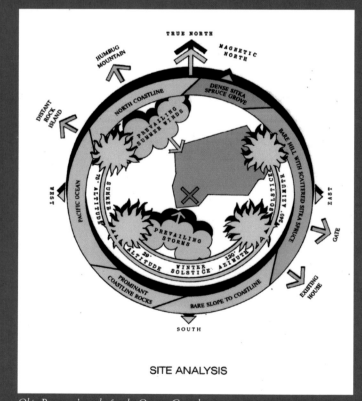

Obie Bowman's study for the Oregon Coast house.

Every piece of land offers the sun, the night sky, breezes, contours, scents, sounds, colors, and the presence or absence of water. Even a property in the middle of the densest city has some relationship to nature. The ideal partner for romantic solutions is, of course, untouched countryside. A rocky cliff covered in moss or barnacles, a rolling meadow of wild grasses and flowers, a thicket of birches—these are landscapes with valuable attributes that can too easily be destroyed as the land is developed. This is why Obie Bowman frets with his "'personal paradox': a desire to build and a need to work with rather than against the natural landscape." Hugh Newell Jacobsen and Peter Bohlin operate slightly differently; they have no fear of manipulating the natural landscape, but like skilled and experienced gardeners, they begin with an intense reverence for the land's qualities.

"Worshiping nature" would not be too strong a description of the designs of these three. You can sense it in how their houses are photographed. Often the landscape, more than the architecture, seems to be the central feature. "It's the landscape that drives it," says Jacobsen, discussing the inspiration for his seemingly formal, almost rigid designs. "The secret is the garden." Bohlin's designs emerge from geometric layouts based on rules of proportion—annotated with picturesque observations about the rising and setting sun or the vistas of the landscape.

The elements—sun, wind, and weather—are fundamental aspects of nature to be respected and capitalized upon. Any home can benefit from a sunny spot for breakfast or a shady spot for spending warm summer afternoons. Interiors are designed as theaters for the play of natural light. Bowman begins every project with a personal and expressive diagram that charts each element onto a map of the property so that he can best tap into each of their potential effects. Patterns of windows become displays of sun shadows on hardwood floors. Shadowy spaces lead to intensely lit ones, like scenes in a suspense movie. Combined with other elements such as cool or fragrant breezes, the sun and shade animate the house, making one room or another a favorite spot at certain times and in certain seasons.

When skillfully harnessed, the sun exerts a powerful impact on a home's appearance. One of the most surprising features of Jacobsen's seemingly stoic houses is how expressive they become with the changing of the day. The slick surfaces of white painted wood or bricks put on new faces from morning through evening as the sun casts ever-changing shadows on the houses' sculptured exteriors, set against the surrounding landscape. In some instances, the ground itself is a focal point of romantic design. Bowman solves his paradox of building and yet preserving the landscape by wrapping his homes with the land, bringing the ground up and over the roof. Jacobsen creates great carpets of gravel to set his homes upon, recalling the good, solid surface of farmyards. For Bohlin an outcropping of rock is as glorious as any vista, and rather than frame it as a view, he pays tribute to it by "accommodating it." He does not remove it; he builds around it.

A steeply sloped site costs less to acquire than a more level one, and for the romantic architect this is a lucky break. Never satisfied with homes that permit only one expression, a romantic appreciates the ability of a steep hill to put a house low to the ground on one side and high in the air on another. This applies to interiors as well. Embracing the multiplicity of the land's traits rather than forcing uniformity onto it brings forth a more dynamic and romantic solution. Allowing vegetation, whether it's a single large tree or a marshy lowland, to be preserved or "accommodated," as Bohlin would say, is in tune with romantic ideal and also with the capabilities of modern construction. Traditional methods of building disturbed the earth, often eliminating all the plantings not only within the house's footprint but also within twenty or thirty feet of its perimeter. By using reinforced concrete, laminated wood posts and beams, steel cables, or other structural components, the modern romantic can allow large parts of their homes to float above the ground. That is one of the methods of collaborating with the land.

Whether to merge into the land, float above it, establish a man-made base, or make a strong contrast against the terrain is a choice that must be worked out for each romantic house. No matter which route is selected, in one way or another the romantic designer collaborates with the land.

OREGON COAST HOUSE

his house overlooking the Pacific Ocean at Gold Beach, Oregon, is designed with the landscape and protection from 100-mile-per-hour winds uppermost in mind. Obie Bowman organized the house so that the occupants can enjoy generous and varied views—of the ocean, of the mammoth rocks known as sea stacks that jut from the water, and of the green, often foggy coastline. The walls facing the ocean are largely transparent. But rather than being composed of sheets of glass as large as possible—the stance that modernists have often taken—most of the glazing is divided into segments that are four feet wide at most and in some instances only a couple of feet high.

The pervasive grid of wood imposes structure on the views, encouraging the eye to focus on particular parts of the scene. One window—or set of windows—frames a slice of land, another a view of sky, still another a mix of sea and sky. Besides isolating the different views, the wooden grid, with its natural tones, makes the interior feel warmer and more sheltering.

Massive log buttresses give the impression of holding the house in place, adding a dramatic sense of structural tension that makes the home feel as strong and rugged as the terrain it commands. Visually, the timbers step the house down into the landscape, grounding it. The interior contains a mix of large open spaces and intimate spots, such as a built-in window seat, where a person can nestle in and feel secure, right next to the big view.

LEFT Like the prow of a ship, the triangular end of the house looks out to sea. Window framing of various sizes, and log posts beyond, subdivide the large vista into a series of more focused views.

ABOVE Posts and diagonal buttresses of Port Orford cedar logs form a kind of exoskeleton, dramatizing the house and making it feel amply braced against the winds.

LEFT The large opening from the living room to a bedroom helps the 1,860-square-foot house feel more spacious.

ABOVE Giant timbers on the outside make the setting seem wilder by suggesting that without such structural heroics, the house might not stand.

LEFT A structural grid wall, minimalist but warm, creates a pattern for the living room while at the same time providing useful display space.

ABOVE The bedroom can be closed off with steel-framed sliding doors. Their cement fiberboard panels have been stained with chemicals and garden-variety fertilizers for a beautiful mottled look.

ABOVE *The cantilevered prow sits high enough to command views of the sea stacks and the ocean. The division of the windows into a series of panes makes the sitting area feel less vulnerable.*

ABOVE Sometimes a house must be designed boldly if it's to live up to the character of its setting. Port Orford cedar log buttresses step the house at Gold Beach into the landscape.

PALMEDO RESIDENCE

Though you may not realize it instantly, given the tabletop-flat terrain, the Palmedo house on Long Island, New York, works extremely closely with its landscape. The visitor's approach is straight on, with the squared-off forecourt providing a meticulously ordered base for a perfectly symmetrical house. House and setting resonate with each other. What later bowls the visitor over is that Jacobsen has devised a sequence of movement and views that links people to the landscape at every turn.

The gabled entry portion exists principally to bring you inside and then send you through a glass-walled outdoor passage—reinforcing the primacy of the setting—before you reach the main body of the house.

Once you're fully inside, you find the walls and the generous windows arranged to bring the outdoors in. Dissolving the barriers between inside and out has long been a goal of modern architects. In the living room, a corner of butt-jointed glass running all the way down to the floor makes the walls practically melt away. The landscape—first the manmade portion, where the corner of a patio echoes the corner of the room, and then the larger natural portion, where a view of Long Island Sound is framed by trees—shows itself to be the star of the show.

By not giving away the full view at the start, by withholding parts of it, adding dynamic twists, placing some windows where they capture only sky, and then letting certain views explode outward, this home intensifies the feeling of involvement in the landscape.

LEFT The large, butt-jointed corner window allows the view to come powerfully inside the living room.

ABOVE Seen head-on from the entry drive, the house makes a strong, yet playful impression, thanks to its symmetrical but almost cartoon-like series of parts, each echoing its neighbor.

LEFT Cutting out the corner makes the view more important than the house itself. This technique makes sense in a scenic location.

ABOVE Crisp corners with virtually no overhangs make a dramatic modern volume.

ABOVE The contrast between floor-to-ceiling wall and floor-to-ceiling glass invariably draws a person to the light.

ABOVE The dining area enjoys unobstructed views through glass doors on one side and a patterned exposure to sunlight in the opposite direction.

ABOVE The entry section stands in front of the main body of the house. The sheer glass of the connecting passage contrasts strikingly against the multitude of tiny panes in the other windows.

ABOVE At night the glow from the windows makes a beckoning impression. A staircase in the front section connects to the house's second floor.

GEORGIAN BAY RETREAT

The Georgian Bay retreat in Cedar Ridge, Ontario, fits into the trees. The aim is to intrude as little as possible on the landscape, inserting the house so that nature appears to have the upper hand. Bohlin deferred to the landscape by refraining from cutting the white birches and other trees that conceal the house in summer and that form a substantial border of trunks and branches through the winter, when the house's long, white front wall merges with the snow.

"Wall" is not quite the right word. It's really a plane, floating free, as advocated by modernists such as Mies van der Rohe. This technique reduces a wall's apparent weight and decreases the building's boxiness. Reducing the walls to planes helps to break down their bulk and let the sur-

rounding woods dominate the atmosphere all the more. Likewise, the simple slope of roof diminishes the magnitude of the roof and lets it float ever so lightly in its natural setting.

The house does arouse feelings, however. When people enter, they discover the dramatic use of exposed wood trusses, handsome details such as lines of black bolts holding the timbers together, and the beautiful effect of natural light coming in from above. The warm sheen of the wood that's employed throughout—in the post-and-beam construction, in the underside of the roof, in the flooring, and in the cabinetry—reinforces the woodland character and offers a sense of refuge from the cold. Exposed wood and articulated structure seem at home amid the trees.

LEFT The lean-to roof and the floating wall plane take the massiveness out of the house and allow nature to appear untrampled-upon.

ABOVE The long white wall blends in with the snow and the birches, making the house a comfortable part of the woodland.

FOLLOWING PAGES Exposed wood structure make the house feel well-suited to its surroundings.

ABOVE Exposed structure, which might be considered a modernist characteristic, doubles as a beautiful, intricate pattern in the central corridor.

ABOVE An alcove-like family room along the corridor feels like an extension of a log cabin.

LEFT The house seems reclusive with its wall nestled behind the vegetation and with the roof practically disappearing into shadow.

ABOVE An orange glow, like that of embers in a campfire, evokes the feeling of being outdoors.

MCKINNEY HOUSE

nother way to collaborate with the landscape is to insert the house tightly into the trees and use modernist construction techniques that avoid touching the ground. In the McKinney house on Figure Eight Island on the Outer Banks of North Carolina, Jacobsen runs a wood deck out from the perimeter of the house, letting people perch among the sunlight and vegetation and allowing the house to feel like it's delicately floating rather than subjugating the land. Decks are inherently lighter in feeling than traditional porches, which rest with a fully expressed heaviness of gravity upon the soil.

Where a mature tree stands, the deck has been built around it, amplifying the feeling that the house is subordinate to the landscape. In instances like this, the tree becomes more than just a tree; it becomes a kind of living sculpture, its form celebrated by being allowed to rise through the flooring. It provokes the realization that the house, which is usually conceived as permanent, has been rendered subservient to nature, whose individual specimens are temporary but whose realm outlasts the dwellings of human beings.

Abstraction and glass, both used to great effect by modernists over the decades, can reduce a house's imprint on its setting. In the McKinney home, gable endwalls are completely opened up, diminishing the building's mass and also making the outdoors a more powerful visual and emotional influence on the inhabitants. In a house like this, people live outdoors much of the time, and even when they retreat to the interior, they still feel connected to the great outdoors.

LEFT The cantilevering of the deck avoids damaging the root systems and lets people feel as if they're living in the trees.

ABOVE The roofscape, with its chimneys, makes a strong impression, but because the house is engulfed in foliage, the landscape loses none of its lush appeal.

ABOVE Pavilions with a bridge-like connection make minimal intrusions on the landscape. The multiple-pane window is a familiar object to be lingered over, whereas the gable containing a triangle of glass is emphatically a bold shape.

ABOVE The eye insists on going out when there are no dividers in the glass to stop it and when the deck outside seems a natural continuation of the floor.

LEFT Building the deck around mature trees ties the house to the landscape, greatly enhancing the enjoyment that people get from it.

RIGHT The walls and ceiling look so light that they might fly away in a breeze. Abstracted forms seem relatively weightless.

BELOW RIGHT A tree very close to the house casts a dappled light, softening the solid surfaces.

PINS SUR MER

Frank Lloyd Wright believed the proper place to build a house was on the brow of a hill rather than on top of it. That way, the occupants would enjoy the view, but landscape would remain dominant and undefiled. In one sense, Pins Sur Mer by Obie Bowman, overlooking Schooner Gulch and Bowling Ball Beach in Point Arena, California, violates Wright's dictum; it stands on the top of a bluff above the Pacific. But Pins Sur Mer—roughly, Pines on the Sea—actually is more reticent than some of Wright's designs. Though the house occupies a fantastic setting, the dwelling hunkers down in a stand of evergreens, almost completely hidden. This is a masterly example of the art of staying in the background and letting nature remain the primary attraction.

The house form itself epitomizes calm. The gently pitched hip roof makes a soothing presence, deferring to the drama of nature's spectacle. The roof hovers above walls of glass, which maximize the occupants' enjoyment of the stunning locale.

Despite the avoidance of flamboyance on the exterior, the house is full of feeling within. Great tree trunks rise in its center. Their heft and their placement at the core of the home make it seem that the house is not just in harmony with nature; it worships nature. The tree columns are so majestic, they could be the supports of a rustic temple. The extensive use of Douglas fir amplifies the sense that the house is both of nature and in nature.

LEFT Pins Sur Mer sits inconspicuously amid a grove of Bishop pines, preserving the drama of the oceanside bluff.

ABOVE The house appears serene beneath its gradually pitched roof. The residents enjoy great views through large expanses of glass.

LEFT Rustic detailing, including a barn-like door and floors of recycled antique oak, imbue the interior with character.

ABOVE The owners requested covered sitting porches on at lest two sides. Because the porches ran the risk of making the interior too dark, the decision was made to arrange the rooms around a high, skylit entry, surrounded by interior openings allowing light to pour into every room.

ABOVE In contrast to the impression created by the low, horizontal lines of the roof, the interior is tall and generous in feeling.

RIGHT Four huge pine columns give a temple-like quality to what is the cultural center of the home—an area containing an upright piano and a wall of bookshelves complete with a rolling ladder.

FOREST HOUSE

The traditional method of building in difficult terrain involves putting in a solid—and massive-looking—foundation, usually of stone or concrete. That technique usually proves effective from an engineering point of view, but in some locations, it draws too much attention to the house and to the imposing, sometimes bulky-looking base. What's often better is to use a lighter style of construction that will allow the house to appear as if it's floating above its site.

Peter Bohlin did that with the Forest house, which sits astride a hillside in West Cornwall, Connecticut. Concrete piers support the house while conveying the illusion of near-weightlessness. The industrial sash of the windows—simple and straightforward—brings the feeling of the woods inside. The way the exterior walls are pared away for a huge window assembly gives the house an abstract character that reduces the house's visual intrusion on nature. The intense color of the sash and of a few other elements provides just enough liveliness to make it clear that the house is not trying entirely to camouflage its presence. The drooping lamp over the beginning of the entrance ramp brings to mind a bright flower blooming in the moistness of the forest.

The house touches the land as little as possible, a good tactic when the objective is to make the landscape seem unconquered and pristine. The ultra-simple decks and entrance walkway, and the reduction of the roof to nothing more than a sharp angle, complete the remarkable effect.

LEFT Outdoors merges with indoors, thanks to the huge expanses of industrial sash, which wrap the corner.

ABOVE The house appears to float above the landscape on its concrete piers. The window area is such a continuous expanse of industrial sash that the house appears barely enclosed.

LEFT The floor and the window area step down together, creating good places for sitting and looking out.

ABOVE A wood bridge avoids stepping on the landscape.

LEFT The green-stained house has such a simple shape that it doesn't compete with its setting for attention. The boardwalk-like decks reach into the woods.

ABOVE From a bedroom, the forest floor looks strangely beautiful, like the bottom of an ocean. Cantilevering and large areas of glass let the residents enjoy the setting to the utmost.

PLATES

GAFFNEY RESIDENCE

PAGE 114

THIRD-FLOOR PLAN

K. Den/Bedroom

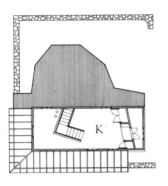

SECOND-FLOOR PLAN

H. Bedroom

I. Balcony

J. Closet

SOUTH ELEVATION

FIRST-FLOOR PLAN

A. Courtyard

B. Entry

C. Kitchen

D. Dining Room

E. Living Room

F. Closet

G. Utility

NORTH ELEVATION

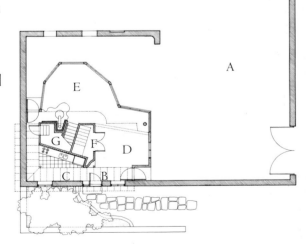

KAHN RESIDENCE

PAGE 120

SITE PLAN

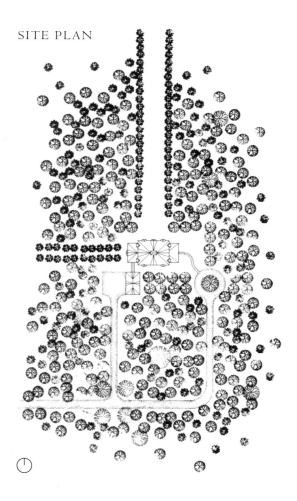

SECOND-FLOOR PLAN

K. Bedroom

L. Bath

M. Master Bedroom

N. Master Bath

FIRST-FLOOR PLAN

A. Entry

B. Porch

C. Dining Room

D. Kitchen

E. Powder Room

F. Play Room

G. Living Room

H. Music Room

I. 2nd Living Room

J. Upper Hall

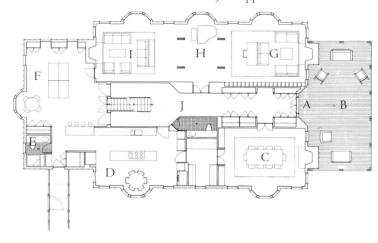

WINDHOVER

PAGE 126

SECOND-FLOOR PLAN

I. Loft

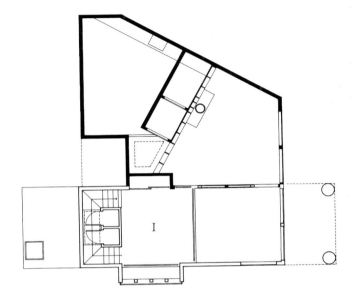

SITE PLAN

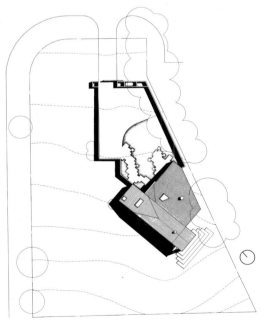

FIRST-FLOOR PLAN

A. Living Room
B. Bath
C. Bedroom
D. Bunkroom
E. Bath
F. Kitchen
G. Dining
H. Deck

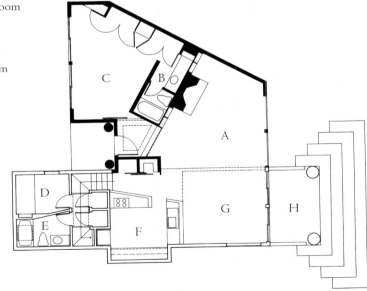

OLD MISSION COTTAGE

PAGE 132

SECOND-FLOOR PLAN

H. Master Bedroom

I. Walk-in Closet

J. Bedroom

K. Open to Below

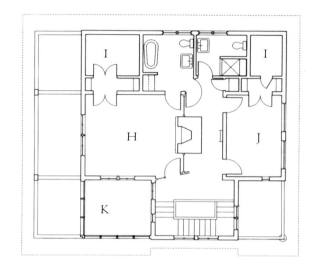

FIRST-FLOOR PLAN

A. Porch

B. Entry

C. Kitchen

D. Pantry

E. Dining Room

F. Living Room

G. Screened Porch

WEST ELEVATION

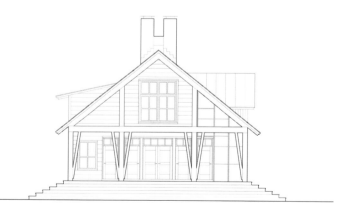

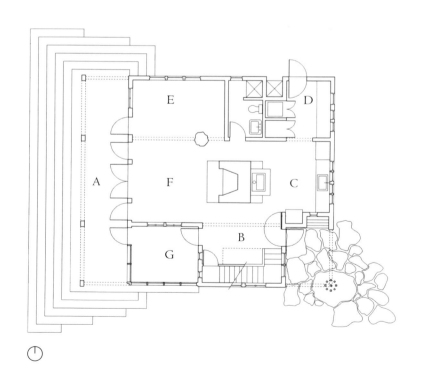

TUNIPUS COMPOUND

PAGE 140

FLOOR PLAN

A. Front Deck
B. Entry
C. Living
D. Dining
E. Kitchen

F. Back Deck
G. Study
H. Bedroom
I. Master Bedroom
J. Outdoor Shower

K. Wetland
L. Office
M. Workshop
N. Storage Room
O. Driveway

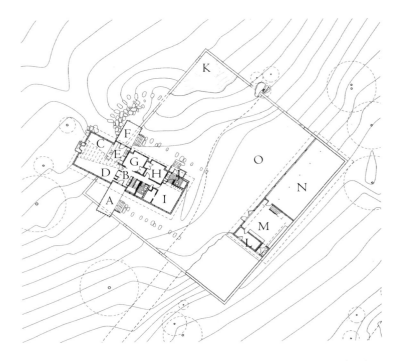

SITE PLAN

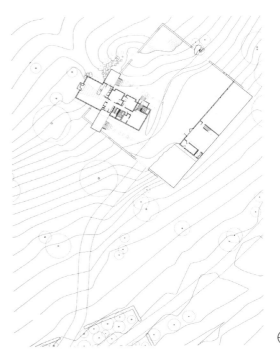

SOHN RESIDENCE

PAGE 148

FLOOR PLAN

A. Garage
B. Laundry
C. Dining Room
D. Living Room

E. Porch
F. Master Bedroom
G. Bedroom
H. Playroom

I. Guest Bedroom
J. Library
K. Kitchen

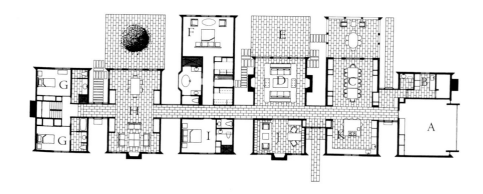

SITE PLAN

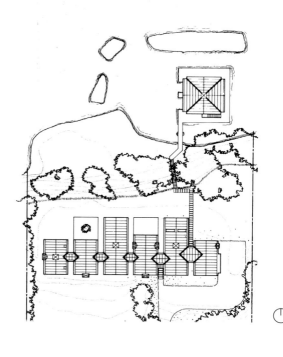

ELEVATION

TIN ROOF

PAGE 156

SITE PLAN

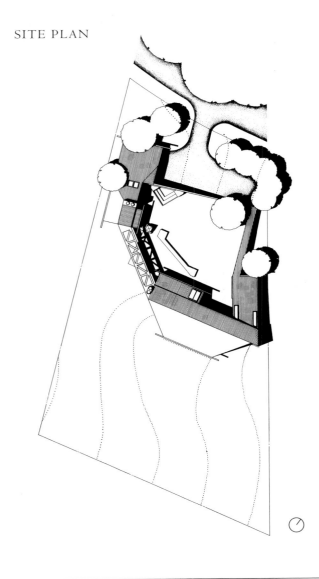

FLOOR PLAN

A. Courtyard

B. Garage

C. Stacked Guest Rooms

D. Living/Dining

E. Kitchen

F. Bedroom

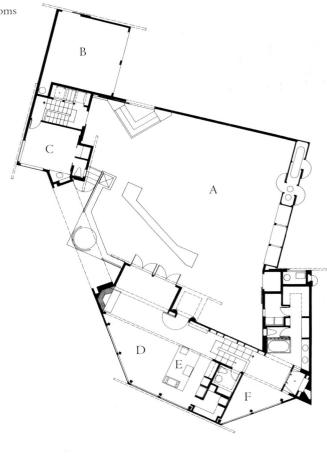

TOM & KARIN'S PLACE

PAGE 164

SITE PLAN

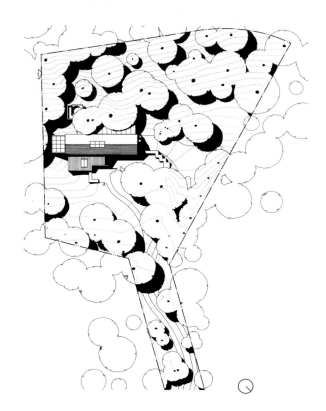

UPPER-LEVEL PLAN

K. Guest Bedroom

L. Loft

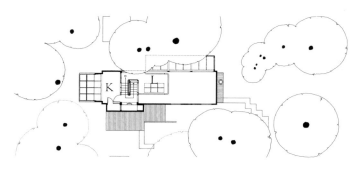

MAIN-LEVEL PLAN

B. Entry

C. Living Room

D. Terrace

E. Kitchen

F. Guest Bath

G. Dining

H. Master Bath

I. Master Bedroom

J. Spa

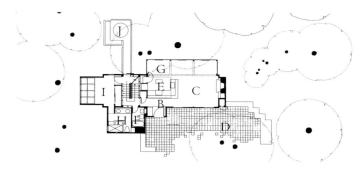

LOWER-LEVEL PLAN

A. Tandem Garage

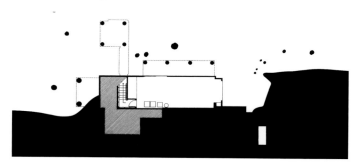

GREENE RESIDENCE

PAGE 170

SECOND-FLOOR PLAN

L. His Studio
M. Her Studio
N. Open
O. Guest Suite
P. Bath
Q. Staff Quarters

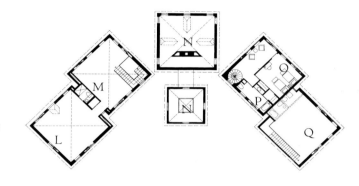

FIRST-FLOOR PLAN

A. Entry
B. Powder Room
C. Bath
D. Master Bedroom
E. Library
F. Living Room
G. Dining Room
H. Kitchen
I. Garage
J. Pool
K. Cabana

SITE PLAN

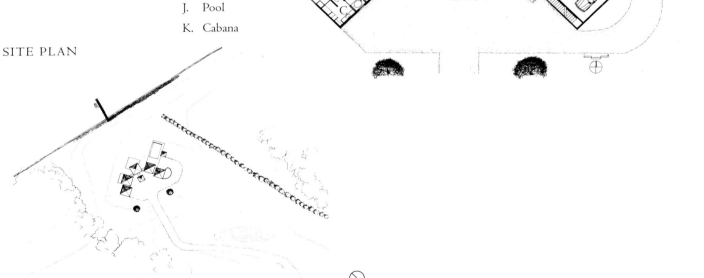

OREGON COAST HOUSE

PAGE 180

FLOOR PLAN

A. Entry

B. Living

C. Bedroom

D. Kitchen

E. Bedroom

F. Deck

G. Garage

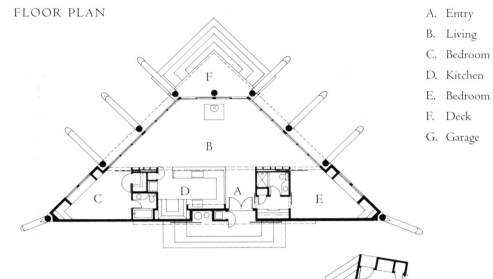

SITE PLAN

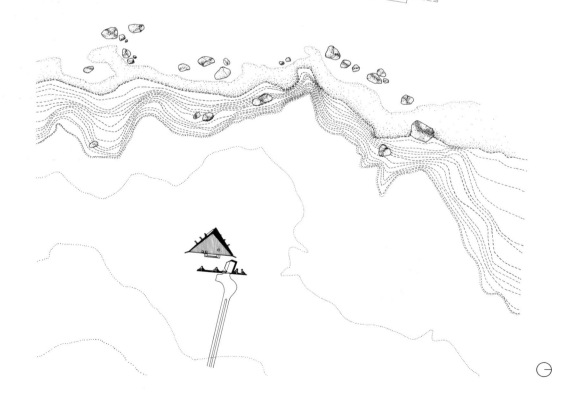

PALMEDO RESIDENCE

PAGE 188

SECOND-FLOOR PLAN

M. Roof

N. Open Boy

O. Bath

P. Study

Q. Open

R. Stair

S. Link Entry

T. Open Above

FIRST-FLOOR PLAN

A. Entry

B. Stair

C. Open Above

D. Master Bedroom

E. Master Bath

F. Powder Room

G. Gallery

H. Library

I. Living Room

J. Entrance Hall

K. Dining Room

L. Kitchen

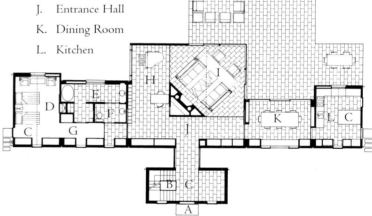

SITE PLAN

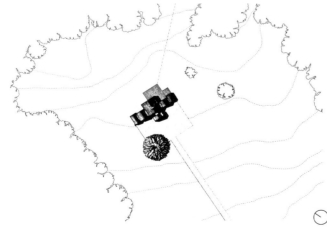

GEORGIAN BAY RETREAT

PAGE 196

SECOND-FLOOR PLAN

I. Bedroom

J. Recreation

K. Sauna

L. Utility

FIRST-FLOOR PLAN

A. Deck

B. Entry

C. Kitchen

D. Dining Room

E. Living Room

F. Fireplace Sitting

G. Master Bedroom

H. Bedroom

NORTH ELEVATION

SOUTH ELEVATION

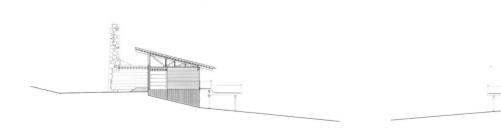

MCKINNEY HOUSE

PAGE 204

FIRST-FLOOR PLAN

I. Dressing Room

J. Master Bedroom

K. Living Room

L. Bath

M. Kitchen

N. Dining Room

O. Breakfast Room

P. Studio

Q. Guest Bedroom

R. Guest Bath

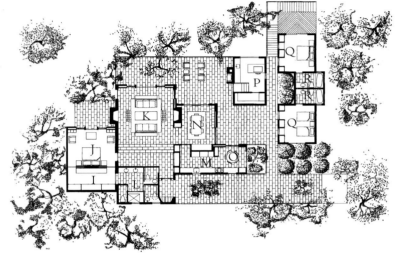

LOWER-LEVEL PLAN

A. Covered Area

B. Cabana

C. Sauna

D. Mechanical

E. Play Room

F. Guest Bedroom

G. Guest Bath

H. Open

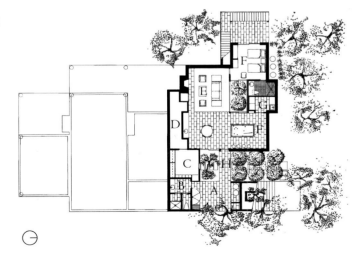

PINS SUR MER

PAGE 210

FLOOR PLAN

A. Entry

B. Entry Porch

C. Bath

D. Bedroom

E. Kitchen

F. Dining

G. Living

H. Bath

I. Bedroom

J. Porch

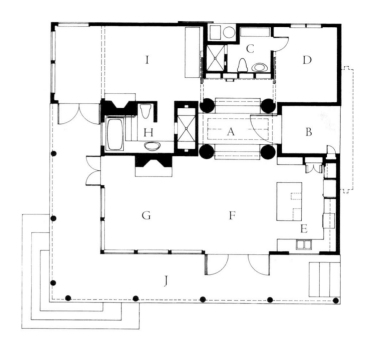

SITE PLAN

FOREST HOUSE

PAGE 216

SECOND-FLOOR PLAN

F. Open to Below

G. Study

H. Storage

I. Guest Bedroom

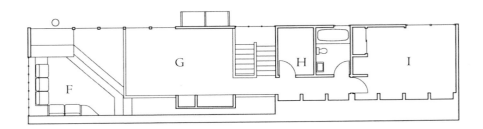

FIRST-FLOOR PLAN

A. Living Room

B. Dining Room

C. Kitchen

D. Utility

E. Master Bedroom

CREDITS